Advances in Dielectrics

Series editor

Friedrich Kremer, Leipzig, Germany

Aims and Scope

Broadband Dielectric Spectroscopy (BDS) has developed tremendously in the last decade. For dielectric measurements it is now state of the art to cover typically 8–10 decades in frequency and to carry out the experiments in a wide temperature and pressure range. In this way a wealth of fundamental studies in molecular physics became possible, e.g. the scaling of relaxation processes, the interplay between rotational and translational diffusion, charge transport in disordered systems, and molecular dynamics in the geometrical confinement of different dimensionality—to name but a few. BDS has also proven to be an indispensable tool in modern material science; it plays e.g. an essential role in the characterization of Liquid Crystals or Ionic Liquids and the design of low-loss dielectric materials.

It is the aim of "Advances in Dielectrics" to reflect this rapid progress with a series of monographs devoted to specialized topics.

Target Group

Solid state physicists, molecular physicists, material scientists, ferroelectric scientists, soft matter scientists, polymer scientists, electronic and electrical engineers.

More information about this series at http://www.springer.com/series/8283

Ranko Richert
Editor

Nonlinear Dielectric Spectroscopy

Springer

Editor
Ranko Richert
School of Molecular Sciences
Arizona State University
Tempe, AZ
USA

ISSN 2190-930X ISSN 2190-9318 (electronic)
Advances in Dielectrics
ISBN 978-3-030-08484-4 ISBN 978-3-319-77574-6 (eBook)
https://doi.org/10.1007/978-3-319-77574-6

Printed on acid-free paper

This Springer imprint is published by the registered company Springer International Publishing AG part of Springer Nature
The registered company address is: Gewerbestrasse 11, 6330 Cham, Switzerland

Preface

Dielectric spectroscopy has a long history of characterizing the magnitude and time or frequency dependence of the polarization that results from an external electric field. Technical developments have facilitated access to a broad range of timescales and frequencies, covering at least the range from 1 ps to 1 year in terms of observed equilibrium relaxation times. This broadband property together with the high resolution and measurement automation have turned dielectric relaxation measurements into a standard tool for characterizing the dynamics of a wide range of materials by measuring the permittivity, ε. In a typical experiment, polarization $P = \chi\varepsilon_0 E$ is proportional to the magnitude of the applied field E, implying that the susceptibility $\chi = \varepsilon - 1$ remains independent of the amplitude of the electric field. In fact, many experimental reports do not specify the field amplitude, because it is considered irrelevant for the results. At sufficiently high electric fields, however, the dielectric behavior will depend on the field magnitude in this nonlinear regime.

The term "nonlinear dielectric effect" refers to any signature of deviations from the linear correlation between polarization P and external electric field E. The interest in studying such nonlinear features goes back to P. Debye and his book on *Polar Molecules* published in 1929. At the time, only dielectric saturation was a known nonlinear effect, observed as a reduction in the amplitude of permittivity. About 10 years later, the chemical effect was recognized by Piekara, which amounted to an increase in amplitudes. Subsequently, it has been discovered that also time constants can be affected by high fields, leading to accelerated or frustrated dynamics, depending on the type of field used, alternating versus static. The slowing down of dielectric relaxation by static electric fields in simple liquids was not discovered until 2014. In recent years, tremendous advances have been made regarding both the high-resolution measurements of nonlinear dielectric effects and their understanding in terms of theoretical and modeling approaches.

The aim of this book is to introduce the ideas and concepts of *Nonlinear Dielectric Spectroscopy*, outline its history, and provide insight into the present state of the art of the experimental technology and understanding of nonlinear dielectric effects. Emphasis will be on what can be learned from nonlinear experiments that could not be derived from the linear counterparts. It will become clear that nonlinear dielectric spectroscopy can be used as a tool to measure structural recovery or physical aging, as well as potential connections between dynamics and thermodynamic variables such as enthalpy and entropy. Supercooled liquids in their viscous regime are ideal candidates for investigating nonlinear effects, because they are particularly sensitive to changes in temperature, and are thus expected to be sensitive to changes in the electric field. Other interesting materials to be covered are plastic crystals and complex liquids near criticality. It is also to be pointed out that, compared with other techniques such as mechanical shear experiments, the nonlinear regime of dielectric spectroscopy is special in the sense that the energies involved always remain small compared with thermal energies.

Theoretical approaches to nonlinear effects are particularly complicated because the tools available for the linear regime no longer apply. As a result, there is no single generally accepted theory regarding nonlinear dielectric responses of real liquids. Various approaches to nonlinear dielectric features have been reported, and the different aspects will be communicated in the first three chapters. The remaining chapters focus more on the experimental aspects, involving different experimental techniques and a range of materials such as liquids, supercooled liquids, plastic crystals, electrolytes, ionic liquids, and polymeric materials. The reader will notice that the contributions will offer different or even conflicting views on how to interpret the results observed with nonlinear dielectric spectroscopy. This feature reflects the present state of research activities, indicating that this field still bears numerous unresolved questions that warrant further research on nonlinear dielectric spectroscopy for years to come.

The editor is grateful to all the contributors to this volume for a smooth and effective collaboration on this joint project. Support from the staff of Springer and from the Series Editor, F. Kremer, is also gratefully acknowledged.

Tempe, USA
January 2018

Ranko Richert

Contents

Nonlinear Dielectric Response of Polar Liquids

Dmitry V. Matyushov

Abstract The linear dielectric constant of a polar molecular material is mostly the function of the molecular dipole moment and of the binary correlations between the dipoles. The dielectric response becomes nonlinear for a sufficiently strong electric field gaining a dielectric decrement proportional, in the lowest order, to the squared field magnitude. The alteration of the dielectric response with the electric field is governed by a combination of binary and three- and four-particle dipolar correlations and thus provides new structural information absent in the linear response. Similar higher order correlations between the molecular dipoles enter the temperature derivative of the linear dielectric constant. Mean-field models, often applied to construct theories of linear dielectric response, fail to account for these multi-particle correlations and do not provide an adequate description of the nonlinear dielectric effect. Perturbation theories of polar liquids offer a potential resolution. They have shown promise in describing the elevation of the glass transition temperature by an external electric field. The application of such models reveals a fundamental distinction in polarization of low-temperature glass formers close to the glass transition and high-temperature, low-viscous liquids. The dielectric response of the former is close to the prescription of Maxwell's electrostatics where surface charge is created at any dielectric interface. On the contrary, rotations of interfacial dipoles are allowed in high-temperature liquids, and they effectively average the surface charge out to zero. Models capturing this essential physics will be required for the theoretical description of the nonlinear dielectric effect in these two types of polar materials.

1 Introduction

This chapter discusses theoretical approaches to nonlinear response of polar materials to the externally applied electric field. The domain of linear theories is limited by the assumption of a linear scaling of the macroscopic dipole moment M with the applied

D. V. Matyushov (✉)
Department of Physics and School of Molecular Sciences,
Arizona State University, PO Box 871504, Tempe, AZ 85287, USA
e-mail: dmitrym@asu.edu

R. Richert (ed.), *Nonlinear Dielectric Spectroscopy*, Advances in Dielectrics
https://doi.org/10.1007/978-3-319-77574-6_1

field. The response to an external field depends on the geometry of the sample through the polarization of the sample's surface. On the contrary, experimental evidence suggests that the response to the Maxwell field is local and independent of the sample shape. Thc dielectric susceptibility, connecting M to the Maxwell field, and the dielectric constant ε defined from the susceptibility are both material properties.

The dependence of the dipole induced in a macroscopic sample on its shape is a direct consequence of the long-range, $\propto r^{-3}$, scaling of the interaction energy between the dipoles. The same distance dependence enters the pair distribution functions [23] describing intermolecular correlations. This long-range scaling, therefore, has to be eliminated in the theories of local material properties. In the linear regime, this is achieved by the Kirkwood–Onsager equation [4], which allows the cancelation of the long-ranged correlations through a linear combination of the longitudinal and transverse components of the polar response [35]. This result directly follows from liquid-state theories operating in terms of pair distribution functions [24]. We follow here a somewhat different approach and, for the sake of pedagogy, arrive at the Kirkwood–Onsager result from the general electrostatic considerations applied to a slab sample of the dielectric. This derivation is contrasted with the spherical geometry of the sample commonly used [4, 17] following Kirkwood's original work [27]. Before discussing the issues pertinent to the linear dielectric response, we start with exact relations for the thermodynamics of polarized dielectrics, which are not limited by the linear response approximation. Those can be found in the standard textbooks on dielectrics [25, 31], as well as in more specialized monographs [4, 17]. We, however, combine the thermodynamic results with statistical fluctuation relations usually not provided in the standard sources.

The Kirkwood–Onsager equation for the linear dielectric constant connects it to the variance of the sample's macroscopic dipole moment. This variance involves binary correlations between the molecular dipoles quantified through the Kirkwood correlation factor g_K, which is connected to a specific angular projection of the pair correlation function. However, already the temperature derivative of the dielectric constant brings the demand on the theory to a new level since it requires orientational correlations of the order higher than binary and cannot be described on the same level of theory as the dielectric constant itself [47]. Orientational correlations of even higher order, between three or four distinct dipoles in the liquid, are required for the description of the first nonlinear correction to the dielectric constant [18, 44], which makes the dielectric response depend on the electric field. Both binary density and binary orientational correlations, along with the higher order orientational correlations, enter the observable decrement of the dielectric constant, which scales quadratically with the Maxwell field. Moreover, the binary and higher order correlations strongly compensate each other in the final result for the dielectric function decrement. No simple approximation, such as a mean-field model, can therefore be applied to this problem.

The overall change of the dielectric constant with the Maxwell field can generally be represented as a product of the Binder parameter [30] for the dipole moment measured along the field and the number of particles N in the sample. The Binder parameter is designed to gauge the deviation of the global statistics of a chosen extensive property from the Gaussian statistics expected for a macroscopic material far from points of global instability (phase transitions and criticality). From this perspective, the decrement of the dielectric constant with increasing electric field is a fundamental parameter giving access to non-Gaussian fluctuations of the macroscopic dipole moment not observable in the thermodynamic limit $N \to \infty$ [44]. More specifically, it gives access to the first expansion term of the Binder parameter in N^{-1}. This parameter provides, at the expense of significantly increased complexity, a glimpse of the material properties not accessible in linear response.

A significant assumption adopted in the derivation of the nonlinear dielectric response presented below and silently adopted in the literature is that, being a function of the external field, the dielectric function remains a local material property, i.e., a property independent of the sample shape. Experiment suggests that this is a reasonable assumption, but the current state of the theory does not allow a direct calculation of highest order many-particle correlations involved. It is therefore important to present a theoretical treatment, even incomplete, that consistently leads to local nonlinear response functions. Perturbation expansions for the thermodynamics of polar liquids in the form of Padé-truncated perturbation series [22, 62] allow such a derivation. We show below how to apply this theoretical approach to derive the linear dielectric response. This formalism provides a new solution for the local field acting on a molecule in the polar liquid. The typically applied approximation, derived from solving the dielectric boundary-value problem, for the so-called cavity field [4] is generally inconsistent with atomistic computer simulations and the new analytical theory provides a better agreement with the numerical results. A fundamental issue appearing in the analysis of the experimental data is a significant distinction between the local dielectric response exhibited by high-temperature and low-temperature liquids. It appears that Lorentz's concept of a virtual interface, producing no surface charges, is more reliable at high temperatures, while slower relaxing liquids, close to the glass transition, fall into the domain of solid-like interfaces envisioned by the Maxwell view of dielectric polarization [48].

The distinction between the Lorentz's and Maxwell's views of interfacial polarization comes in direct focus in an attempt to understand the elevation of the glass transition temperature by an external field. This theory is based on the use of the Padé-truncated perturbation theories of polar liquids [22] to formulate the configurational entropy of a polar glass former [45]. The application of the electric field lowers the configurational entropy thus shifting the glass transition temperature upward. The amount of the shift is, however, strongly affected by whether the Lorentz or Maxwell view of the local field in the dielectric is adopted. The Maxwell result turns out to be more consistent with experimental evidence. This outcome strongly suggests that the boundary conditions applied in the dielectric response problem significantly depend on the ability of the interfacial polar molecules to average out the surface charge on the observation window of the experiment. A supercooled liquid near the glass transition suppresses the orientational motions, thus leading to solid-like Maxwell boundary conditions. We discuss this problem in more detail at the end of this chapter.

2 Thermodynamic and Statistical Relations

Dielectric experiments performed with plane capacitors report the free energy stored inside the capacitor and are usually presented in the form of capacitance, from which the dielectric constant is derived. The electrostatic free energy is quadratic in the Maxwell field $E = \Delta\phi/d$ determined experimentally as the ratio of the electric potential drop on the capacitors plates $\Delta\phi$ and the distance d between them (Fig. 1). When the capacitor volume is V, the equation for the electrostatic Helmholtz free energy becomes [31] (in Gaussian units)

$$F_E = \frac{V}{8\pi}\varepsilon_E E^2. \tag{1}$$

Equation (1) is formally exact since it represents the electrostatic free energy, generally nonlinear in the Maxwell field, in the form of the capacitance

$$C = \varepsilon_E V/(4\pi d^2) \tag{2}$$

including an unknown function $\varepsilon_E = \varepsilon_E(E)$.

Very few exact relations can be established without resorting to the linear response when $\varepsilon_E = \varepsilon$ is the dielectric constant of the material. Alternatively, for relatively weak fields relevant for most experimental conditions (up to ~300 kV/cm [56]), $\varepsilon_E = \varepsilon + \Delta\varepsilon_E$ can be expanded in the powers of E. The first nonvanishing term in the series is quadratic in E for isotropic materials. The dielectric constant decrement $\Delta\varepsilon_E$ is therefore linear in E^2

$$\Delta\varepsilon_E = -aE^2. \tag{3}$$

The proportionality constant a is known as the Piekara coefficient [7]. It is usually anticipated to be positive and, in such cases, is associated with dielectric saturation, when, loosely speaking, dipoles diminish their ability to respond in high fields. This mechanism is usually discussed in the framework of the mean-field Langevin model. However, negative values have been observed as well [7, 38], and those are generally related to multiparticle dipolar correlations in a polar liquid [18] as we discuss in more detail below.

Before these approximations are introduced, one can start with exact thermodynamic relations. The alteration of the Helmhotz free energy of the bulk material performed at constant volume (subscript "V") is given by the following relation [31]:

$$\delta F_V = -S\delta T + \frac{V}{4\pi}E\delta D. \tag{4}$$

Here, $D = E + 4\pi\langle M\rangle_E/V$ is the dielectric displacement and $\langle M\rangle_E$ is the dipole moment induced in the material in the presence of the external field. We denote M as the dipole moment of a macroscopic sample along the direction of the field, which we align along the z-axis of the laboratory frame (Fig. 1). The z-projection

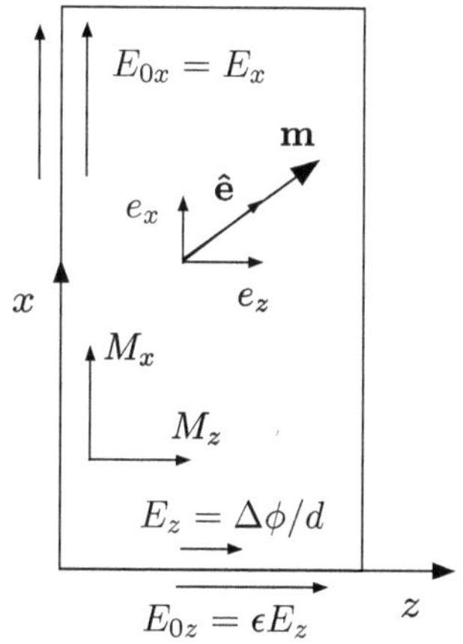

Fig. 1 Cartoon of the slab geometry of the dielectric sample with the projections of the dipole moment M_z and M_x along the corresponding axes. The drop of the electrostatic potential $\Delta\phi$ at the capacitor's plates creates the Maxwell field $E = E_z = \Delta\phi/d$ inside the dielectric. The boundary conditions for the slab preserve the electric field across the slab plane, $E_{0x} = E_x$, while the Maxwell field is reduced by ϵ relative to the external field in the z-direction perpendicular to the slab. Also shown is the unit vector $\hat{\mathbf{e}}$ along the molecular dipole $\mathbf{m}$ and its corresponding projections on the axes

will be explicitly indicated where its omission can lead to confusion with other Cartesian components. For the rest of our discussion, we consider the plane capacitor geometry commonly used in the experimental setup. In this geometry, the dielectric displacement is equal to the field E_0 of the external charges on the capacitor's plates. If the charge density (charge per unit area) is σ_0, one obtains from the standard electrostatic arguments [31] $D = E_0 = 4\pi\sigma_0$. This connection implies that varying D and E_0 is achieved by charging the plates.

Equation (4) leads to the connection between the external field and the Maxwell field through the relation

$$E = (4\pi/V)\,(\partial F/\partial E_0)_{V,T} = E_0 - 4\pi\langle M\rangle_E/V. \tag{5}$$

One additionally obtains the Maxwell relation between the variation of the entropy with the external field and the temperature derivative of the induced dipole moment $\langle M\rangle_E$

$$\left(\frac{\partial S}{\partial E_0}\right)_{V,T} = \left(\frac{\partial\langle M\rangle_E}{\partial T}\right)_{V,E_0}. \tag{6}$$

The dipole moment $\langle M\rangle_E$ induced in the material by the external field is the target of statistical theories of dielectrics [17]. It is generally calculated by recognizing that the external perturbation produced by the field of external charges is $-\mathbf{M}\cdot\mathbf{E}_0 = -ME_0$ [31]. The induced dipole is then calculated as the statistical average

$$\langle M \rangle_E = [Q(E_0)]^{-1} \int M e^{-\beta H_0 + \beta M E_0} d\Gamma, \tag{7}$$

where H_0 is the Hamiltonian of the unperturbed dielectric and

$$Q(E_0) = \int e^{-\beta H_0 + \beta M E_0} d\Gamma \tag{8}$$

is the partition function of the polar material in the external field. The phase space element $d\Gamma$ involves all degrees of freedom of the macroscopic sample over which integration is performed and $\beta = (k_B T)^{-1}$ is the inverse temperature. From this definition, one can immediately convert the Maxwell relation in Eq. (6) to the fluctuation relation involving the correlation of the dipole moment with the system Hamiltonian

$$T \left(\frac{\partial S}{\partial E_0} \right)_{V,T} = \langle \delta M \beta \delta H \rangle_E, \tag{9}$$

where $H = H_0 - M E_0$, $\delta M = M - \langle M \rangle_E$, and $\delta H = H - \langle H \rangle_E$.

The use of a linear relation between the Maxwell and external fields, $E = \varepsilon E_0$, significantly simplifies the thermodynamics of dielectrics. The free energy of the capacitor becomes

$$F_E = \frac{V}{8\pi} E D. \tag{10}$$

This free energy is a sum of the component describing the electric field in vacuum (the first summand) and the free energy of polarizing the dielectric (the second summand)

$$F_E = \frac{V}{8\pi} E_0^2 - \tfrac{1}{2} E_0 \langle M \rangle_E. \tag{11}$$

One can next calculate the entropy of electrostatic polarization, which is expressed through the derivative of the dielectric constant with temperature taken at the constant volume [17, 31]

$$T S_E = F_E \left(\frac{\partial \ln \varepsilon}{\partial \ln T} \right)_V. \tag{12}$$

The constant volume temperature derivative can be related to the more often available temperature derivative at constant pressure through the thermodynamic relation [47]

$$\left(\frac{\partial \varepsilon}{\partial T} \right)_V = \left(\frac{\partial \varepsilon}{\partial T} \right)_P + \frac{\alpha_P}{\beta_T} \left(\frac{\partial \varepsilon}{\partial P} \right)_T. \tag{13}$$

Here, α_P is the isobaric expansivity and β_T is the isothermal compressibility. The results of calculating this correction from experimental data [37] are listed in Table 1 for a number of common polar liquids. The derivatives at constant volume and pres-

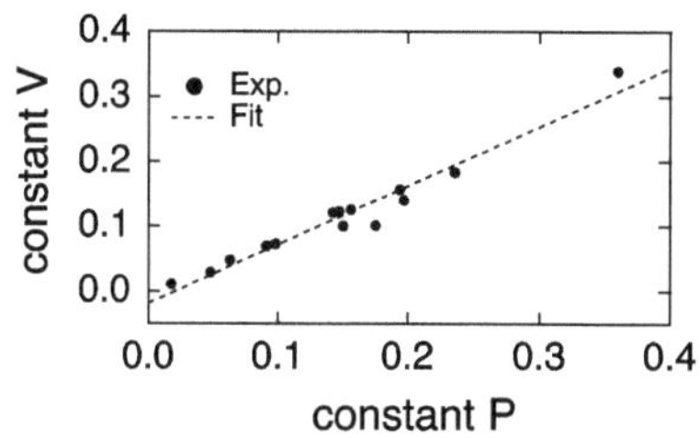

Fig. 2 $-(\partial \ln \varepsilon / \partial \ln T)_V$ versus $-(\partial \ln \varepsilon / \partial \ln T)_P$ shown for experimental data in Table 1 (points, formamide is omitted). The dashed line refers to a linear fit through the points with the equation $-0.019 + 0.908x$

sure follow an approximately linear relation shown in Fig. 2 for the polar liquids listed in Table 1.

The temperature derivative of the dielectric constant can be connected to the correlation between the dipole moment and energy (Hamiltonian) fluctuations by using Eq. (9). Since the right-hand side in Eq. (12) is taken from linear response, the same approximation needs to be applied to Eq. (9). By keeping only terms quadratic in the electric field, one arrives at a fluctuation relation for the linear response entropy

$$TS_E = \frac{\beta E_0^2}{2} \langle M^2 (\beta \delta H_0 - 1) \rangle . \tag{14}$$

Here, we use angular brackets without the subscript to designate a statistical ensemble average in the absence of the external field and have assumed $\langle M \rangle = 0$, which is true for isotropic materials without spontaneous polarization (ferroelectricity). Also note that since entropy is extensive, one has to require that both correlators, $\langle M^2 \delta H_0 \rangle$ and $\langle M^2 \rangle$, scale with the number of particles as $\propto N$.

The external field E_0 is weak compared to local fields in a polar material in all practical cases. A perturbation expansion of the statistical averages in terms of the external perturbation $H' = -ME_0$ often applies, as in fact done in deriving Eq. (14). The resulting expansion is in the powers of E_0. On the contrary, an expansion in the powers of the Maxwell field is required for the local material properties, such as ε_E and in the corresponding definition of the Piekara coefficient in Eq. (3). The difficulty of connecting the response to the external field E_0, following from statistical theories, to the response to the Maxwell field required by local properties and measured experimentally is shared by all theories of dielectrics [12, 17]. Arriving at the material dielectric constant from fluctuation relations requires connecting E to E_0. This connection depends on the chosen geometry of the sample [17] due to charges produced at its surface by the polarizing external field. Given the final result must be independent of the sample shape, one commonly resorts to calculating $\langle M \rangle_E$ for a specific shape for which the connection between E and E_0 is particularly simple. Typical derivations are performed either for a spherical dielectric sample or for a spherical region inside a continuous dielectric [27]. A generalization to an arbitrary

Table 1 List of liquid properties [37] used in the analysis

Liquid	ε	ε_∞	m^a	$(\partial\varepsilon/\partial T)_P{}^b$	$(\partial\varepsilon/\partial T)_V{}^b$	$(\partial\varepsilon/\partial P)_T{}^c$	$\alpha_P{}^d$	$\beta_T{}^e$
Formamide	109.5	2.091	3.37	−1.653	−1.560	0.0493	0.75	0.399
Methanol	35.87	1.760	2.87	−0.156	−0.126	0.0319	1.19	1.248
Ethanol	24.55	1.848	1.66	−0.147	−0.123	0.0255	1.09	1.153
n-propanol	20.33	1.915	3.09	−0.142	−0.121	0.0197	1.09	1.025
t-butanol	12.47	1.919	1.66	−0.175	−0.101	0.058	1.26	0.989
Water	78.46	1.776	1.834	−0.360	−0.339	0.037	0.26	0.457
Propylene carbonate	64.92	2.014	4.94	−0.236	−0.184	0.0325	0.95	0.590
Ethylene glycol	40.7	2.047	2.31	−0.194	−0.157	0.0233	0.62	0.392
Nitromethane	32.7	1.902	3.56	−0.197	−0.140	0.0392	1.14	0.790
Acetone	20.7	1.839	2.69	−0.098	−0.073	0.023	1.45	1.324
Acetonitrile	35.94	1.798	3.92	−0.150	−0.100	0.0385	1.38	1.070
Benzonitrile	25.2	2.326	4.18	−0.091	−0.070	0.0159	0.83	0.621
Pyridine	12.91	2.271	2.37	−0.063	−0.048	0.0107	1.02	0.715
1,1-dichloroethane	10.0	1.997	1.82	−0.048	−0.029	0.0163	1.33	1.148
Chloroform	4.9	2.079	1.15	−0.018	−0.011	0.0055	1.29	1.033

^{a}in D, ^{b}in K^{-1}, ^{c}in MPa^{-1}, ^{d}in 10^{-3} K^{-1}, ^{e}in GPa^{-1}

shape cut from a liquid volume can be found in the paper by Høye and Stell [24]. Here, we follow a somewhat alternative route by deriving all equations for the linear and nonlinear dielectric response for a slab geometry of the sample. We show below how this approach can be applied to deriving the linear Kirkwood–Onsager equation for the dielectric constant ε, but first start with some general relations which can be obtained solely from the assumption that the nonlinear dielectric polarization is small compared to the linear one.

One can start with an empirical relation representing the dipole moment induced in the sample as a series expansion in odd powers of the Maxwell field [7, 56]

$$\frac{\langle M \rangle_E}{V} = \chi_1 E + \chi_3 E^3, \tag{15}$$

where χ_1 is the linear dielectric susceptibility $\varepsilon = 1 + 4\pi\chi_1$. This relation can be substituted into the thermodynamic link between the field of external charges E_0 and the Maxwell field E in Eq. (5), which leads to the relation between E_0 and E

$$E_0 = \varepsilon E + 4\pi\chi_3 E^3. \tag{16}$$

From this formula, one obtains for the variation of the external field

$$\delta D = \delta E_0 = \varepsilon\delta E + 12\pi\chi_3 E^2 \delta E. \tag{17}$$

The resulting connection between δE_0 and δE can be used to integrate the free energy in Eq. (4) to arrive at Eq. (1) in which the nonlinear dielectric function becomes

$$\varepsilon_E = \varepsilon + 6\pi\chi_3 E^2. \tag{18}$$

One therefore obtains for the Piekara coefficient in Eq. (3)

$$a = -6\pi\chi_3. \tag{19}$$

3 Linear Response

The connection between the linear dielectric constant and the variance of the sample dipole moment is provided by the Kirkwood–Onsager equation [4, 17]. We adopt here the slab geometry of the sample (Fig. 1) and assume that the electric field can be aligned either along the z-axis, as is done in the standard dielectric setup [4], or along the x-axis, which would typically correspond to experiments with absorption of light propagating perpendicular to the slab plane [46]. The first-order perturbation expansion in Eqs. (7) and (8) yields

$$\langle M_\alpha \rangle_E = \beta \langle M_\alpha^2 \rangle E_{0\alpha}, \tag{20}$$

where $\alpha = x, z$ specifies Cartesian projections. For the slab geometry, the connection between E_0 and E depends on whether the field is perpendicular or parallel to the slab plane [25]: $E_{0z} = \varepsilon E_z$ and $E_{0x} = E_x$ (Fig. 1). Since the response to the Maxwell field, $4\pi \langle M_\alpha \rangle_E = V(\varepsilon - 1)E_\alpha$, is invariant in respect to the projection taken for an isotropic dielectric, one can write the dielectric constant in terms of the variance of the total dipole moment of the sample $\langle \mathbf{M}^2 \rangle = \langle M_z^2 \rangle + 2\langle M_x^2 \rangle$. This procedure leads to the Kirkwood–Onsager equation

$$\frac{(\varepsilon - 1)(2\varepsilon + 1)}{9\varepsilon} = \frac{4\pi}{9V}\beta \langle \mathbf{M}^2 \rangle. \tag{21}$$

The variance of the sample dipole in the absence of the external field is independent of the shape of a macroscopic sample and the Kirkwood–Onsager equation can be used for any macroscopic material [24]. A useful connection between $\langle \mathbf{M}^2 \rangle$ and variances of the corresponding projections follows from this derivation

$$\langle \mathbf{M}^2 \rangle = (2\varepsilon + 1)\langle M_z^2 \rangle = \frac{2\varepsilon + 1}{\varepsilon}\langle M_x^2 \rangle. \tag{22}$$

The dipole moment $\mathbf{M}$ in Kirkwood–Onsager equation is the total dipole of the sample, including all permanent and induced molecular dipoles. We will discuss the separation into two components below and first start off by neglecting the molecular polarizability and the corresponding induced dipoles. The dipole moment $\mathbf{M}$ then becomes the sum of N molecular permanent dipoles $\mathbf{m}_j$: $\mathbf{M} = \sum_{j=1}^{N} \mathbf{m}_j$. The left-hand side of Eq. (21) becomes

$$(4\pi/9V)\beta \langle \mathbf{M}^2 \rangle = y g_K, \tag{23}$$

where

$$y = (4\pi/9)\beta m^2 \rho, \tag{24}$$

$\rho = N/V$ is the number density, and m is the magnitude of the molecular dipole (see Table 1 for the typical values). The parameter y plays the role of the effective density of permanent dipoles in the liquid, while g_K in Eq. (23) is the Kirkwood correlation factor [27]

$$g_K = N^{-1} \sum_{i,j} \langle \hat{\mathbf{e}}_i \cdot \hat{\mathbf{e}}_j \rangle. \tag{25}$$

It is given in terms of the unit vectors $\hat{\mathbf{e}}_j$ along the direction of the molecular dipole $\mathbf{m}_j$ (Fig. 1) and thus defines the average cosine of the angles between all pairs of dipoles in the liquid. We discuss an alternative definition of the Kirkwood factor below in connection to the nonlinear dielectric effect.

The average over the orientations of the dipoles can be expressed in terms of the pair correlation function [22, 23] of the liquid $h(1, 2) = h(r_{12}, \boldsymbol{\omega}_1, \boldsymbol{\omega}_2)$, which depends on the distance r_{12} between two dipoles and their orientations $\boldsymbol{\omega}_1$ and $\boldsymbol{\omega}_2$. The scalar product $\Delta(1,2) = \hat{\mathbf{e}}_1 \cdot \hat{\mathbf{e}}_2$, when averaged over the orientations of the dipoles $\boldsymbol{\omega}_1$ and $\boldsymbol{\omega}_2$, creates the projection $h^{\Delta}(r)$ [22, 23] of the pair correlation function on the rotational invariant $\Delta(1,2)$

$$h^{\Delta}(r) = 3\langle \Delta(1,2) h(1,2)\rangle_{\omega_1,\omega_2}. \tag{26}$$

The Kirkwood factor is usually expressed as the $k = 0$ value of the Fourier transform of the Δ-projected correlation function $\tilde{h}^{\Delta}(k = 0)$

$$g_K = 1 + (\rho/3)\tilde{h}^{\Delta}(k = 0). \tag{27}$$

The correlation function $h^{\Delta}(r)$ is short-ranged, which implies that it decays to zero faster than r^{-3} of the dipole–dipole interactions in the liquid [63]. In fact, another projection of the pair correlation function on a rotational invariant, the projection on $D(1,2) = 3(\hat{\mathbf{e}}_1 \cdot \hat{\mathbf{r}}_{12})(\hat{\mathbf{r}}_{12} \cdot \hat{\mathbf{e}}_2) - (\hat{\mathbf{e}}_1 \cdot \hat{\mathbf{e}}_2)$, decays as r^{-3} at long distances, $h^D(r) \to (\varepsilon - 1)^2/(4\pi\varepsilon\rho y r^3)$ [63], where y is given by Eq. (24) and $\hat{\mathbf{r}}_{12} = \mathbf{r}_{12}/r_{12}$. Linear combinations of the Fourier-transformed Δ- and D-projections enter the longitudinal (L) and transverse (T) structure factors of the polarization fluctuations in the liquid [53], which depend on the wavevector k

$$\begin{aligned} S^L(k) &= 1 + (\rho/3)\left[\tilde{h}^{\Delta}(k) + 2\tilde{h}^{D}(k)\right], \\ S^T(k) &= 1 + (\rho/3)\left[\tilde{h}^{\Delta}(k) - \tilde{h}^{D}(k)\right]. \end{aligned} \tag{28}$$

Both $S^L(k)$ and $S^T(k)$ are long-ranged, but the trace of the structure factor

$$g_K(k) = 1 + (\rho/3)\tilde{h}^{\Delta}(k) = \tfrac{1}{3}\left[S^L(k) + 2S^T(k)\right] \tag{29}$$

eliminates the long-range dipolar correlations present in $\tilde{h}^D(k)$ and leads to Eq. (27) at $k \to 0$. Examples of $S^{L,T}(k)$, as well as the k-dependent Kirkwood factor $g_K(k)$, are shown in Fig. 3 for SPC/E force-field model of water [60].

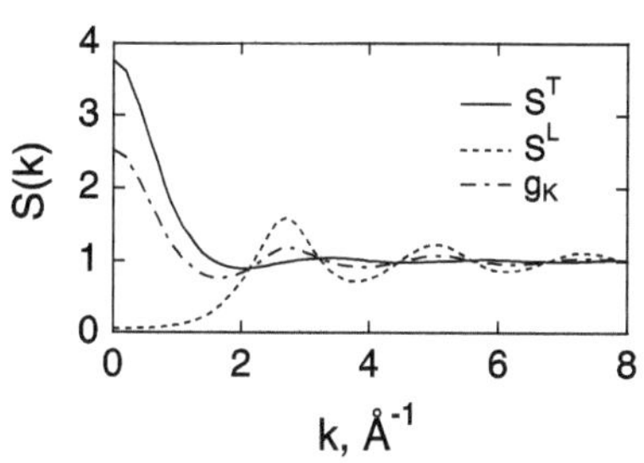

Fig. 3 Longitudinal (L) and transverse (T) structure factors and $g_K(k)$ from Eq. (29) for SPC/E water at $T = 300$ K from molecular dynamics simulations [60]

The physical meaning of the Kirkwood derivation [27] and of the corresponding Onsager–Kirkwood relation [Eq. (21)] is, therefore, in eliminating the long-range dipole–dipole correlations, scaling as r^{-3} and producing the dependence of the result on the sample shape [24]. The linear combination of the longitudinal (z-axis) and transverse (x-axis) responses reduces the problem to the local correlations represented by $h^{\Delta}(r)$. The locality of the linear dielectric constant ε is, therefore, directly linked to the locality of $h^{\Delta}(r)$ or of the corresponding direct correlation function $c^{\Delta}(r)$ [54]. This simple notion raises an objection to a recently proposed interpretation of second-order light scattering [10] from electrolytes in terms of long-ranged $h^{\Delta}(r)$ [8]. If a long-ranged component with $h^{\Delta}(r) \propto r^{-3}$, or slower asymptote, would indeed exist, the dielectric constant of an electrolyte would cease to be a material property, for which no evidence exists either experimentally or theoretically [6]. The derivation of the Kirkwood–Onsager equation outlined here also makes it clear that the linear combination of the longitudinal and transverse projections, leading to Eqs. (27) and (29), is specific to linear response and cannot be extended to the definition of the nonlinear dielectric function ε_E. It is therefore not justified to use the Kirkwood–Onsager equation to connect ε_E to $\langle \mathbf{M}^2 \rangle_E$ [29] in the general nonlinear scenario.

The use of linear response and the relations between the variance of the dipole moment and the dielectric constant provide a fluctuation relation for the derivative of the dielectric constant with temperature. Combining Eqs. (12) and (14), one arrives at the following fluctuation relation for the logarithmic derivative of the dielectric constant with respect to temperature:

$$\left(\frac{\partial \ln \varepsilon}{\partial \ln T} \right)_V = 1 + \varepsilon \left(\frac{4\pi\beta^2}{V} \langle M^2 \delta H_0 \rangle - 1 \right), \tag{30}$$

where H_0 is the total energy (Hamiltonian) of the unperturbed liquid, see Eq. (7).

Equation (30) clearly shows that the temperature derivative of the dielectric constant reports on dipolar correlations of the order higher than the binary correlations affecting the linear dielectric constant through the Kirkwood factor. The nonlinear dielectric effects, which we consider next, are also strongly influenced by the higher order correlations and can potentially deliver structural and dynamic information not available from the linear dielectric constant [34, 56]. Before we proceed to that next topic, we first present the result for the temperature derivative of the dielectric constant connecting it to numerical simulations of polar liquids.

Since spherical cutoff is commonly employed in atomistic simulations of liquids [2] one needs fluctuation relations in terms of the spherically symmetric vector dipole moment. For that purpose, Eqs. (12) and (14) can be rewritten in the form

$$\left(\frac{\partial \ln \varepsilon}{\partial \ln T} \right)_V = \frac{3\varepsilon}{2\varepsilon^2 + 1} M_T, \tag{31}$$

where

$$M_T = \frac{4\pi\beta}{3V}\langle \mathbf{M}^2(\beta\delta H_0 - 1)\rangle. \tag{32}$$

The dimensionless parameter M_T in this equation can be calculated from experimental temperature derivatives of the dielectric constant listed in Table 1 for molecular liquids [37]. The results are shown in Fig. 4, where M_T follows an approximately linear scaling with the dielectric constant, $M_T \simeq -0.75\varepsilon$, (dashed line in Fig. 4). This scaling suggests $\varepsilon(T) \propto T^{-9/8}$, which is quite close to the commonly used [67] empirical approximation $\varepsilon(T) \propto T^{-1}$. The latter empirical result is in fact in perfect agreement with M_T values obtained by excluding H-bonding liquids listed in Table 1 (polar liquids shown by closed points in Fig. 4), which yields $M_T \simeq -0.64\varepsilon$. These empirical observations suggest a fluctuation relation which should approximately hold for high-temperature polar liquids

$$\frac{2\pi\beta}{\varepsilon V}\langle \mathbf{M}^2(\beta\delta H_0 - 1)\rangle \simeq -1. \tag{33}$$

What is currently not known is whether deviations from this relation can be used to signal an "abnormal" behavior of a liquid. An obvious target for using such fluctuation relations in numerical simulations is to detect intermittent structural fluctuations, such as formation of hydrogen-bonded rings in alcohols [61], when the Kirkwood correlation factor gains a significant temperature dependence.

From the thermodynamic perspective, the empirical relation $\varepsilon(T) \propto T^{-1}$ implies that the entire free energy of electrostatic polarization is applied to the polarization entropy

$$F_E = -TS_E \tag{34}$$

and the internal energy of electric polarization vanishes, $U_E = 0$. There seems to be no fundamental reason why this result should hold in an extended range of temperatures. This implies that $\varepsilon(T) \propto T^{-1}$ has to be violated, as indeed was found in recent experiments with propylene carbonate derivatives [66]. It is also violated in simula-

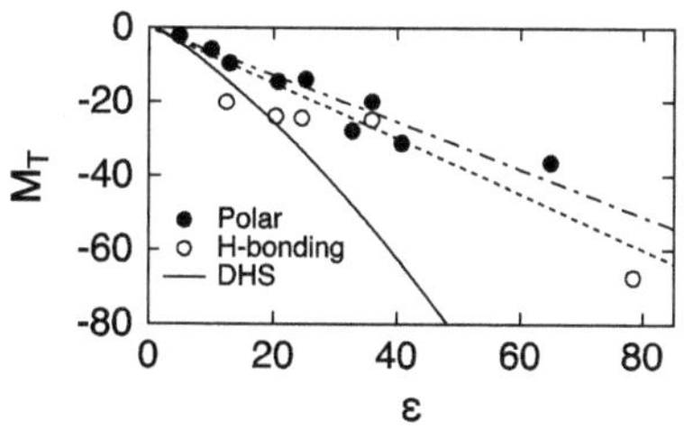

Fig. 4 M_T calculated from Eqs. (31) and (32) for polar liquids (closed points) and alcohols and water (open points, formamide omitted) at $T = 298$ K. The experimental data are taken from Ref. [37] and the dashed line, $M_T = -0.75\varepsilon$, is the fit to all data. The dash-dotted line, $M_T = -0.64\varepsilon$, is the fit to the polar liquids only. The solid line refers to the fluid of dipolar hard spheres (DHS) [40]

tions of the model fluid of dipolar hard spheres [40], as is shown by the solid line in Fig. 4. The origin of the empirical temperature scaling is the β-factor in the right-hand side of the Kirkwood–Onsager equation [Eq. (21)], which results in $\varepsilon(T) \propto T^{-1}$ at $\varepsilon \gg 1$ if $\langle \mathbf{M}^2 \rangle$ or g_K are temperature-independent. This, however, is not the case when dielectric data are collected over an extended range of temperatures. Figure 5 shows $\langle \mathbf{M}^2 \rangle(T) \simeq m^2 N g_K(T)$ for glycerol [47].

The temperature dependence of $\langle \mathbf{M}^2 \rangle$ leads not only to the violation of the $\varepsilon(T) \propto T^{-1}$ empirical rule, but also violates a more fundamental requirement typically imposed on macroscopic variables and known as the Nyquist [51], or fluctuation–dissipation [28], theorem. It prescribes that the variance of a macroscopic extensive variable A is proportional to the number of particles in the system N and temperature T: $\langle (\delta A)^2 \rangle \propto NT$, $\delta A = A - \langle A \rangle$. The linear scaling of $\langle \mathbf{M}^2 \rangle$ with N is usually observed and is very essential to our arguments below regarding the non-linear dielectric effect. On the contrary, the temperature scaling of the macroscopic dipole moment clearly violates the Nyquist theorem: it decreases with increasing T instead of the anticipated linear increase. In contrast to the dipole moment, thermal fluctuations of the liquid density behave in accord with the Nyquist theorem (Fig. 5). The $k = 0$ value of the density–density structure factor

$$S(0) = \langle (\delta N)^2 \rangle_\Omega / \langle N \rangle_\Omega = \beta^{-1} \rho \beta_T \tag{35}$$

describes fluctuations $\delta N = N - \langle N \rangle_\Omega$ of the number of particles N in a given volume Ω relative to the average value $\langle N \rangle_\Omega$. For a sufficiently large Ω and a macroscopic N, one expects $S(0) \propto T$ in a qualitative agreement with observations [9, 11, 14].

The microscopic origin of $\langle \mathbf{M}^2 \rangle(T)$ shown in Fig. 5 and shared by many polar liquids [66] has never been clarified. More theoretical studies of the temperature effect on the dielectric constant and the Kirkwood factor are required, but they all share the same difficulty of accounting for multiparticle orientational correlations entering the fluctuation relations in Eqs. (31) and (32). While the Kirkwood factor reflects only binary correlations, its temperature dependence requires understanding orientational correlations of higher order.

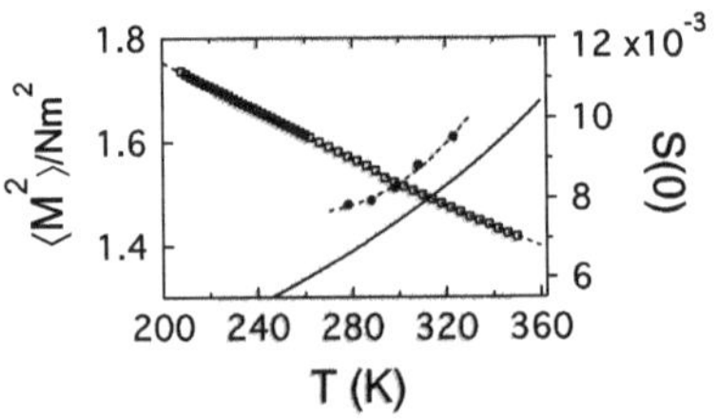

Fig. 5 $\langle \mathbf{M}^2 \rangle / (N m^2)$ for glycerol versus temperature ($m = 4.6$ D) [47]. The solid line, referring to the right axis, shows $S(0)$ calculated from the adiabatic bulk modulus [9, 11]. The filled points refer to direct measurements of β_T [14]. The dashed lines are polynomial fits through the experimental points

4 Molecular Polarizability

As mentioned above, the variance $\langle \mathbf{M}^2 \rangle$ in the Kirkwood–Onsager equation (21) represents the sum of all permanent and induced, due to electronic and vibrational polarizability, molecular dipoles in a macroscopic sample. The separation of the two components is achieved by introducing the effective polarity parameter y_e into the modified Kirkwood–Onsager equation

$$\frac{(\varepsilon - 1)(2\varepsilon + 1)}{9\varepsilon} = y_e g_K . \tag{36}$$

The polarity parameter y_e needs special care for polarizable molecular liquids.

The problem of incorporating the molecular polarizability as a separate component of the dielectric response, distinct from the permanent dipoles, occupied already early studies by Onsager [52] and Fröhlich [17]. The most widely used relation for the dielectric constant is due to Fröhlich who considered the model of permanent dipoles of the liquid immersed in the polarizable continuum characterized by the high-frequency dielectric constant ε_∞ (Table 1). The Kirkwood–Fröhlich equation then reads

$$\frac{(\varepsilon - \varepsilon_\infty)(2\varepsilon + \varepsilon_\infty)}{\varepsilon(\varepsilon_\infty + 2)^2} = y g_K . \tag{37}$$

This equation achieves a specific form for the dipole moment in the liquid state m', which is typically enhanced compared to the gas-phase dipole moment m due to molecular polarizability α [4]. The polarizability itself is enhanced from the gas-phase value α to the condensed-phase value α'.

The prescription for obtaining m' in Fröhlich's approach is just one of many possible formulations of mean-field theories for screening the permanent charges by induced molecular dipoles [24]. Microscopic mean-field theories of polarizable liquids [63, 65] allow far better estimates of m' and α' than Fröhlich's ansatz. Different routes to these parameters can be summarized in terms of a single effective polarity parameter y_e in Eq. (36), which becomes [4, 63]

$$y_e = (4\pi/9)\beta\rho(m')^2 + (4\pi/3)\rho\alpha' . \tag{38}$$

A significant advantage of this formulation is that it preserves the left-hand side of the Kirkwood–Onsager equation [Eqs. (21) and (36)], shifting the focus to the calculation or experimental measurement of the effective condensed-phase dipole moment m' and polarizability α'. Such calculations of y_e applying the Wertheim theory [65] of polarizable liquids are listed in Table 2. These results are then used in Eq. (36) to calculate g_K from experimental ε (Table 2). Since molecular polarizability directly enters the calculations, the sensitivity to the choice of ε_∞, typically complicating estimates of g_K from the Kirkwood–Fröhlich equation [37], is avoided when Eq. (36) is used instead.

Table 2 Nonlinear response coefficients b_P [subscript "P" refers to constant pressure in Eq. (49)] calculated from experimental Piekara coefficients [37]

Liquid	y_e	g_K	b_P	b_V^{MF}	$b_V^{\mathrm{MF,a}}$	$b_V^{(2)}$
Methanol	2.80	2.54	0.276	0.018	0.742	−0.54
Ethanol	1.93	2.74	0.269	0.024	0.675	−0.53
Propanol	1.51	2.92	0.324	0.028	0.661	−0.52
t-butanol	1.21	1.90	0.287	0.051	0.508	−0.54
Water	6.65[b]	2.61	0.068	0.008	0.915	−0.52
1,1-dichloroethane	1.35	1.55	0.228	0.054	0.450	−0.44
Chlorophorm	1.53	0.60	0.149	0.104	0.293	4.53

[a]With the cavity field susceptibility determined from the perturbation expansions according to Eq. (72). [b]$y = 6.22$ for SPC/E force-field water used to produce the structure factors in Fig. 3

5 Non-Gaussian Fluctuations of the Macroscopic Dipole

Before proceeding to the derivation of the Piekara coefficient, let's first lake a look at the general question of what such a parameter can tell us about the statistics of dipolar fluctuations in a macroscopic material. The starting point here is the partition function of a dielectric material in Eq. (8). One can consider the dipole moment M as a stochastic variable and define the cumulant generating function [21]

$$e^{g(E_0)} = \langle e^{\beta M E_0} \rangle = \frac{Q(E_0)}{Q(0)}. \tag{39}$$

The function $g(E_0)$ is given by the cumulant series in the field E_0 with the coefficients defined in terms of cumulants K_n of the stochastic variable M

$$g(E_0) = \sum_{n=2}^{\infty} \frac{(\beta E_0)^n}{n!} K_n. \tag{40}$$

The sum starts with $n = 2$ because $\langle M \rangle = 0$ and the second cumulant K_2 is equal to the dipole moment variance

$$K_2 = \langle M^2 \rangle. \tag{41}$$

The fourth cumulant $K_4 = \langle M^4 \rangle - 3\langle M^2 \rangle^2$ can be written as $K_4 = -3\langle M^2 \rangle^2 U_N$ with

$$U_N = 1 - \langle M^4 \rangle / (3\langle M^2 \rangle^2). \tag{42}$$

The parameter U_N quantifies the deviation of the statistics of M from the Gaussian statistics; $U_N = 0$ when M is a Gaussian stochastic variable. The subscript "N" specifies that certain scaling of this parameter with the number of molecules N is anticipated, in analogy to a similar parameter considered in the theory of critical phenomena [30]. As mentioned above, the variance of the dipole moment of the bulk material scales linearly with N, $\langle M^2\rangle \propto N$. This scaling implies that the statistics of the macroscopic dipole moment M of a thermodynamically stable phase, characterized by a single free energy minimum, is Gaussian ($U_N \to 0$) in the thermodynamic limit $N \to \infty$. Only close to the critical point of transition to spontaneous polarization (such as ferroelectricity) can the system visit a finite number of alternative free energy minima. Transitions between these alternative configurations make the statistics of the macroscopic dipole non-Gaussian, characterized by a nonzero U_N [30].

From Eqs. (39) and (40), the average dipole moment induced by the external field is

$$\langle M\rangle_E = \sum_{n=1}^{\infty} \frac{(\beta E_0)^n}{n!} K_{n+1}. \tag{43}$$

By truncating the series after the second expansion term, one obtains for the uniform polarization of the sample

$$\frac{\langle M\rangle_E}{V} = \beta E_0 \frac{\langle M^2\rangle}{V} - (\beta E_0)^3 \frac{\langle M^2\rangle^2}{2V^2\rho} B_V, \tag{44}$$

where

$$B_V = N U_N \tag{45}$$

and the subscript "V" specifies the constant volume conditions.

Since $U_N \to 0$ at $N \to \infty$, it can be represented by a series in powers of N^{-1}. Therefore, in contrast to U_N itself, B_V in Eq. (45) is finite at $N \to \infty$ and gives access to the $O(N^{-1})$ expansion term of U_N in powers of N^{-1}. We conclude that the cubic expansion term, connecting the induced dipole with E_0^3, characterizes deviations from the Gaussian statistics of the dipole moment in finite-size samples. Such deviations cannot be observed by direct measurements of the dipole moment performed on macroscopic samples, but are accessed through the measurements of the third-order susceptibility connected to the Piekara coefficient [Eq. (19)]. We next show that such third-order susceptibility fundamentally reflects correlations between many distinct dipoles in the liquid and cannot be reduced to the binary correlations sufficient for linear response.

6 Nonlinear Dielectric Response

The derivation of the nonlinear dielectric response in terms of the external field E_0 [Eq. (44)] can be recast in terms of the Maxwell field E. This is achieved by using the connection between $\langle M_z^2\rangle$ and the total dipole variance $\langle \mathbf{M}^2\rangle$ from Eq. (22) and the connection between the external and Maxwell field given by expansion (16). By keeping only the terms up to $\propto E^3$, the substitution of Eqs. (16), (22), (23), and (33) into (44) leads to the equation for χ_3

$$\chi_3 = \frac{9y_e g_K}{2\varepsilon + 1}\chi_3 - \frac{\beta\varepsilon^3}{2\rho}\left(\frac{9y_e g_K}{4\pi(2\varepsilon + 1)}\right)^2 B_V. \tag{46}$$

By applying the Kirkwood–Onsager equation [Eq. (36)], one arrives at

$$\chi_3 = -\frac{\beta\varepsilon^2\Delta\varepsilon^2}{2(4\pi)^2\rho}B_V, \tag{47}$$

where $\Delta\varepsilon = \varepsilon - 1$.

The final result for the dielectric constant decrement can be written in terms of the reduced free energy density of the electric field in the capacitor

$$f_E = (\beta\varepsilon/8\pi\rho)E^2 \tag{48}$$

It is easy to see [Eq. (10)] that f_E is the free energy of the electrostatic field per molecule of the liquid divided by $k_B T$. In terms of this natural energy scale, the dielectric increment becomes [see Eq. (19)]

$$\Delta\varepsilon_{E,V} = -\Delta\varepsilon^2 f_E b_V, \tag{49}$$

where we have defined a new dimensionless nonlinear response parameter

$$b_V = \tfrac{3}{2}\varepsilon B_V. \tag{50}$$

The subscript "V" in $\Delta\varepsilon_{E,V}$ specifies the dielectric increment at constant volume of the liquid, which does not incorporate potential electrostriction when measurements are done at constant pressure [16, 36]. Most data are collected at $P =$ Const, but the correction from $V =$ Const to $P =$ Const is very minor [44].

The reason for scaling B_V with ε in Eq. (50) is the observation that b_V is nearly constant for many molecular liquids, in contrast to the Piekara coefficient. Table 2 lists b_P ($P =$ Const) values calculated from reported Piekara coefficients [37] according to Eq. (49). If the relative invariance of $b_{P,V}$ among molecular liquids (with water being a notable exception) and/or thermodynamic conditions is confirmed, the formulation in terms of the reduced cumulant b_V provides a significant insight since it anticipates a simple scaling for the Piekara coefficient

$$a \propto \beta\varepsilon\Delta\varepsilon^2 V_m, \tag{51}$$

where V_m is the molar volume.

The simultaneous constancy of b_V and $(\partial \ln \varepsilon/\partial \ln T)_V$ suggests another empirical relation

$$\Delta\varepsilon_{E,V} \simeq 0.3\Delta\varepsilon^2 S_E/(k_B N), \tag{52}$$

where S_E is the polarization entropy in Eq. (12) (linear response). There might be no fundamental reason for this correlation since the fluctuation relations for the nonlinear dielectric decrement and polarization entropy are clearly distinct. It is also limited to high-temperature liquids listed in Table 2, which have supplied the proportionality coefficient in this equation.

The parameter B_V in Eq. (50) is the reduced cumulant quantifying the non-Gaussian character of dipole moment fluctuations in the dielectric [Eqs. (42) and (45)]. As we show below, it involves complex many-particle correlations between the dipoles in the liquid. Before we turn to a fully microscopic derivation, it is useful to derive an expression for b_V in the framework of a mean-field theory of dielectrics.

One constructs a mean-field theory by reducing the instantaneous polarization of the liquid created by the external field and thus depending on liquid's configuration by an average local field produced by all dipoles oriented by the field [19, 64]. The prescription introduced by Onsager [52] for this local field is in terms of the cavity field experienced by a dipole placed inside a physical dielectric cavity carved from the dielectric. Such a cavity field $E_c = \chi_c E_0$ is the product of the cavity field susceptibility χ_c and the external field E_0 [40]. If, following Onsager [52], one assigns the cavity to a single molecule, the dielectric models suggest the relation [25]

$$\chi_c = 3/(2\varepsilon + \varepsilon_\infty). \tag{53}$$

Here, the high-frequency dielectric constant ε_∞ appears from the assumption that the entire liquid outside a given molecule carries the dielectric constant ε, while only the electronic polarizability, characterized by ε_∞, exists inside the molecule. The solution of the dielectric boundary volume problem can be formulated in terms of the dielectric drop $\varepsilon/\varepsilon_\infty$ at the dividing dielectric surface [31], which leads to Eq. (53).

Based on the assumption of independent liquid dipoles experiencing the local field E_c, one calculates the average dipole of the liquid in the form of the Langevin function $L(x) = \coth(x) - 1/x$

$$\langle M \rangle_E = NmL\left(\beta m \chi_c E_0\right). \tag{54}$$

A clear advantage of this formulation is that it allows both linear and nonlinear dielectric response for an induced dipole. The Langevin model also allows an extension from the static response to rotational dynamics of the mean-field dipole described as Brownian rotational diffusion [13]. Solutions of such equations give access to nonlinear frequency-dependent susceptibilities. What is absent from these dynamic theories is the dynamics of the cavity susceptibility, the static limit of which is given by Eq. (53). It accounts for the collective effect of all dipoles surrounding a given target dipole and polarized by the external field. While modeling χ_c is still a theoretical challenge, as we discuss below, it is not the main difficulty of the mean-field models. The main deficiency comes from the assumption of an ideal gas of dipoles representing a dense polar liquid. Nevertheless, one readily arrives at the nonlinear dielectric effect by expanding Eq. (54) in powers of E_0. The lowest order dielectric increment is given then by Eq. (49) with the mean-field formula for the parameter b_V

$$b_V^{\mathrm{MF}} = \frac{2}{5}\chi_c. \tag{55}$$

From the perspective of mean-field theories, the problem of calculating the non-linear dielectric response is reduced to the problem of the local cavity field, which we consider in more detail below. If the prescription given by Eq. (55) is adopted, $b_V^{\mathrm{MF}} \simeq 3/(5\varepsilon)$ at $\varepsilon \gg 1$. This result comes in direct contradiction with reported instances [7, 38] of negative b_V, which are missed altogether by the mean-field theories. The values of b_V^{MF} obtained by combining Eqs. (53) and (55) are also too low compared to observations (Table 2). The use of the cavity susceptibility derived below to replace the continuum expression (53) somewhat improves the agreement with observations (Table 2). However, from the fundamental perspective, the Piekara coefficient is the result of compensatory influence of binary and higher order dipolar correlations completely missed by the mean-field theories, as we discuss next.

In contrast to the Kirkwood–Onsager equation for the linear dielectric constant, the nonlinear dielectric response expressed in terms of the reduced fourth-order cumulant [Eq. (44)] does not clearly separate the long-range correlations, carrying the potential dependence on the sample shape, from the short-range correlations. One has to prove that the resulting parameter B_V involves short-range correlations only. This is achieved by splitting the fourth-order cumulant of the sample dipole moment into a sequence of terms of increasing correlation order

$$\begin{aligned} m^{-4}\langle M^4\rangle = \frac{N}{5} + \sum_{i\neq j}\left[3\langle e_{iz}^2 e_{jz}^2\rangle + 4\langle e_{iz} e_{jz}^3\rangle\right] \\ +6\sum_{i\neq j\neq k}\langle e_{iz}e_{jz}e_{kz}^2\rangle + \sum_{i\neq j\neq k\neq m}\langle e_{iz}e_{jz}e_{kz}e_{mz}\rangle. \end{aligned} \tag{56}$$

Here, e_{iz} is the z-projection of the unit vector of the molecular dipole. The two terms in the brackets in the second summand can be expressed in terms of the established binary correlations in the liquid, while the last two terms represent triple- and fourth-order correlations between distinct dipoles in the liquid, which have to be short-ranged. For the terms in the second summand in Eq. (56), one derives

$$\begin{aligned} 3\sum_{i\neq j}\langle e_{iz}^2 e_{jz}^2\rangle &= N^2/3 + (N/3)\rho\tilde{h}^0(0), \\ 4\sum_{i\neq j}\langle e_{iz} e_{jz}^3\rangle &= (4N/5)(S^L(0)-1). \end{aligned} \tag{57}$$

In this equation, $h^0(r) = g^0(r) - 1$ is the angular isotropic pair correlation function of the liquid and $\tilde{h}^0(k)$ is its Fourier transform. Further, $S^L(0)$ is the $k=0$ value of the longitudinal structure factor given by Eq. (28).

One can repeat the same derivation for the x-projection of the dipole moment, which yields an equation identical to Eq. (56) upon the substitution $e_{iz} \to e_{ix}$. Correspondingly, Eq. (57) converts to

$$\begin{aligned} 3\sum_{i\neq j}\langle e_{ix}^2 e_{jx}^2\rangle &= N^2/3 + (N/3)\rho\tilde{h}^0(0), \\ 4\sum_{i\neq j}\langle e_{ix} e_{jx}^3\rangle &= (4N/5)(S^T(0)-1). \end{aligned} \tag{58}$$

The inequality between the z- and x-projections for $\langle e_{i\alpha}e_{j\alpha}^3\rangle$, $\alpha = x, z$ in Eqs. (57) and (58) was clarified by Fulton [20].

The cumulant expansion in terms of the external field E_0 in Eq. (44) can now be written separately for $\langle M_z\rangle_E$ and $\langle M_x\rangle_E$ by assuming that the same external field E_0 is applied along both directions. Combining them as $\langle M_z\rangle_E + 2\langle M_x\rangle_E$ leads to the total dipole moment of the sample $\mathbf{M}$ in the linear expansion term in Eq. (44) and to the elimination of the long-range component of the dipolar correlation function h^D from the linear combination of the longitudinal and transverse responses of the polar liquid in Eqs. (57) and (58). The result is the analogue of Eq. (44)

$$\langle M_z\rangle_E + 2\langle M_x\rangle_E = \beta E_0\langle \mathbf{M}^2\rangle + (\beta E_0)^3(K_4/2), \tag{59}$$

where

$$K_4 = \frac{2m^4N}{15}\left[H^{(2)} + H^{(3,4)}\right]. \tag{60}$$

In Eq. (60), the binary correlations are collected into the $k = 0$ density–density structure factor $S(0)$ given by Eq. (33) and the Kirkwood factor according to the relation

$$H^{(2)} = 6(g_K - 1) + \tfrac{5}{2}S(0) - 1. \tag{61}$$

Further, the third- and fourth-order correlations [last two summands in Eq. (56)], which are currently challenging to compute [44], are collected into the component $H^{(3,4)}$ in Eq. (60). Importantly, K_4 carries the expected $\propto N$ scaling, implying that both $H^{(2)}$ and $H^{(3,4)}$ are intensive parameters.

Combining Eq. (59) with the expansion of the dipole moment in terms of the powers of the Maxwell field in Eq. (15) and accounting for the connection between the Maxwell field and the external field for the slab sample, one arrives at the following expression for the nonlinear expansion coefficient:

$$\chi_3 = \frac{3\varepsilon^3}{2\varepsilon^3+1}\frac{\beta^3K_4}{6V} \simeq \frac{\beta^3K_4}{4V}. \tag{62}$$

This equation can be used directly to estimate the Piekara coefficient in Eq. (19). Alternatively, one can use Eq. (62) to obtain the dimensionless reduced cumulant b_V in Eq. (50), which becomes

$$b_V = -\frac{1}{10}\left(\frac{2\varepsilon+1}{g_K\varepsilon}\right)^2\left[H^{(2)} + H^{(3,4)}\right]. \tag{63}$$

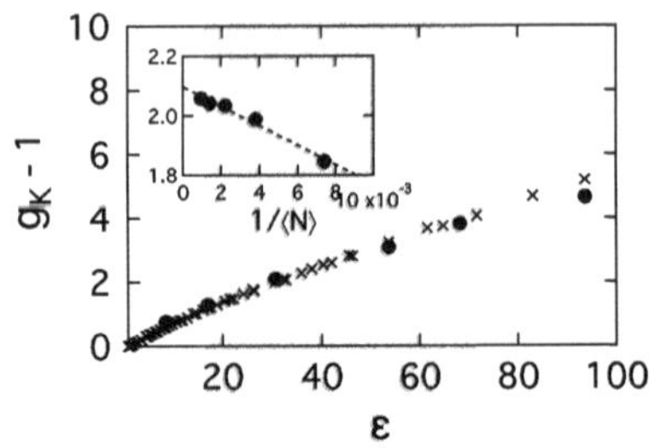

Fig. 6 g_K calculated from Eq. (64) (circles) and from the dielectric constant according to the Kirkwood–Onsager equation [Eq. (21)] by using the reported simulation data [40]. The inset shows $g_K - 1$ at $\varepsilon_s = 30.6$ versus the reciprocal number of particles $1/\langle N \rangle$ within a spherical cutoff equal to the half of the length of the simulation cell. The dashed line is the linear regression through the simulation points. Its intercept for each liquid polarity is reported by a filled circle in the main panel. All points refer to Monte Carlo simulations of fluids of dipolar hard spheres with varying dipole moment [46]

The estimates of b_V based on experimental Piekara coefficients suggest that both the binary correlations incorporated into $H^{(2)}$ and the higher order correlations entering $H^{(3,4)}$ are equally important [44] (Table 2). Specifically, $H^{(2)}$ are usually positive and the overall positive values of b_V observed for many polar liquids arise from the compensating effect of the negative $H^{(3,4)}$ component.

Equations (57) and (58) provide a definition of the Kirkwood factor alternative to commonly used in terms of the average cosines between the liquid dipoles [Eq. (25)]

$$g_K = \frac{5}{3N} \sum_{\alpha} \sum_{i,j} \langle e_{i\alpha} e_{j\alpha}^3 \rangle. \tag{64}$$

Here, $\alpha = x, y, z$ specifies three Cartesian components of the unit vector of the molecular dipole (Fig. 1). Calculating the Kirkwood factor from finite-size numerical simulations turned out to be a nontrivial computational problem [33, 49]. It is not clear at the moment if Eq. (64) provides a superior route for computations. However, when sufficient sampling is achievable, the standard route [49] to g_K through Eqs. (21) and (23) is consistent with Eq. (64). Figure 4 illustrates this point for a fluid of dipolar hard spheres [46].

7 Perturbation Theories of Polar Liquids

Thermodynamic functions of polar liquids are obviously affected by interactions of molecular multipoles. One successful approach to calculate the free energy of multipolar interactions is through Padé-truncated perturbation series introduced by Stell and coworkers [22, 62]. The idea is to expand the free energy of a macroscopic sample in the anisotropic interaction energy H_a while adopting the isotropic distribution functions obtained with the isotropic part on intermolecular interactions as reference. The free energy of the liquid

$$F = F_0 + \Delta F = F_0 - F_2 + F_3 + \cdots \tag{65}$$

becomes a sum of the reference isotropic part F_0 and the perturbation expansion for the polar part collected into ΔF. The expansion terms can be directly calculated, for instance: $F_2 = (\beta/2)\langle H_a^2 \rangle$ and $F_3 = (\beta^2/6)\langle H_a^3 \rangle$. The terms beyond F_3 involve high-order dipolar correlations, for which no computational or formal approaches have been developed. The perturbation expansion has to be truncated, and the Padé truncation [22, 62] offers the following form

$$\Delta F = -\frac{F_2}{1 + F_3/F_2}. \tag{66}$$

This approximation is exact for the first two expansion terms and generates a sign-alternating infinite series, as expected from the general properties of the infinite perturbation series for dipolar liquids [62].

This procedure, which performs exceptionally well for polar liquids [22], can be extended to a dipolar fluid placed in the external electric field [45]. Since the external field is typically weak, the solution of the problem is achieved by applying the mean-field approximation. It replaces the instantaneous field of all dipoles in the liquid polarized by the external field with a local cavity field $\mathbf{E}_c$ acting on each dipole

$$H_a(E_0) = H_a - \sum_j \mathbf{m}_j \cdot \mathbf{E}_c. \tag{67}$$

The new definition of the anisotropic interaction $H_a(E_0)$ can be used in the perturbation expansion to replace ΔF with the polarization free energy ΔF_E in the presence of the field. It turns out that from two terms, F_2 and F_3, used in the Padé form only F_2 is affected by the field. The cumulant $\langle H_a^2 \rangle$ is replaced with

$$\langle H_a(E_0)^2 \rangle = \langle H_a^2 \rangle + (m^2/3)N\chi_c^2 E_0^2. \tag{68}$$

Consequently, ΔF_E is given by Eq. (66) in which one replaces F_2 with $F_2(E_0) = (\beta/2)\langle H_a(E_0)^2 \rangle$. This formulation of the theory, despite the use of the cavity field susceptibility, is a significant step forward compared to the Langevin equation [Eq. (54)]. It explicitly takes into account the orientational correlations between the dipoles in the liquid, while they are totally neglected in the one-particle Langevin formula.

The inspection of Eq. (66) shows that its combination with Eq. (68) leads to the polarization free energy nonlinear in the external field E_0. Both linear and nonlinear relations for the dipole induced in the sample by the external field can be established based on this formalism. We present here the linear results, followed the theory's application to the problem of the field effect on the glass transition.

The dipole moment induced in the sample by the external field E_0 is obtained as the derivative of the polarization free energy [see Eq. (5)]

$$\langle M \rangle_E = -\left(\frac{\partial \Delta F_E}{\partial E_0}\right)_{V,T}. \tag{69}$$

One obtains for a given component of the dipole moment in the linear response approximation

$$\frac{4\pi}{V}\langle M_z \rangle_E = 3y\chi_c^2 E_{0z}\frac{1+2z}{(1+z)^2}, \tag{70}$$

where

$$z = \frac{F_3}{F_2} = \frac{\beta}{3}\frac{\langle H_a^3 \rangle}{\langle H_a^2 \rangle}. \tag{71}$$

Since the free energy has been established for an isotropic liquid, one has to repeat the derivative for each Cartesian component and then follow the procedure outlined above for the slab sample to establish the connection of the perturbation expansion to the dipole moment variance $\langle \mathbf{M}^2 \rangle$. The result of this procedure is a new equation for the cavity susceptibility

$$\chi_c = \sqrt{g_K}\frac{1+z}{\sqrt{1+2z}}. \tag{72}$$

The terms F_2 and F_3 in the perturbation expansion for the free energy are well established for simple model liquids with known reference distribution functions describing isotropic interactions in the liquid [22]. For a fluid of dipolar hard spheres,

$$z = \frac{y}{4\pi}\frac{I_{\mathrm{TD}}(\rho^*)}{I_6(\rho^*)}. \tag{73}$$

Here, $I_n = 4\pi \int_0^\infty g_0(r)(dr/r^{n-2})$ is the two-particle perturbation integral calculated based on the pair distribution function $g_0(r)$ of the reference system. Correspondingly, $I_{TD}(\rho^*)$ is the three-particle perturbation integral involving dipolar interactions between three separate molecular dipoles. Both functions are tabulated as polynomials of the reduced density $\rho^* = \rho\sigma^3$, where σ is the diameter of the hard sphere [32]

$$\begin{aligned} I_{\mathrm{TD}}(x) &= 16.4493 + 19.8096x + 7.4085x^2 - 1.0792x^3 - 0.9901x^4 - 1.0249x^5, \\ I_6(x) &= 4.1888 + 2.8287x + 0.8331x^2 + 0.0317x^3 + 0.0858x^4 - 0.0846x^5. \end{aligned} \tag{74}$$

The Kirkwood factor $g_K(y)$ is often known from numerical simulations and can be used to test Eq. (72). Figure 7 compares the results of independent simulations of the cavity field in dipolar fluids [40] to the analytical formula. The Maxwell solution given by Eq. (53) agrees with simulations only at small ε (dashed line in Fig. 7). On the other hand, Eq. (72), does not capture the initial drop of the cavity field, but

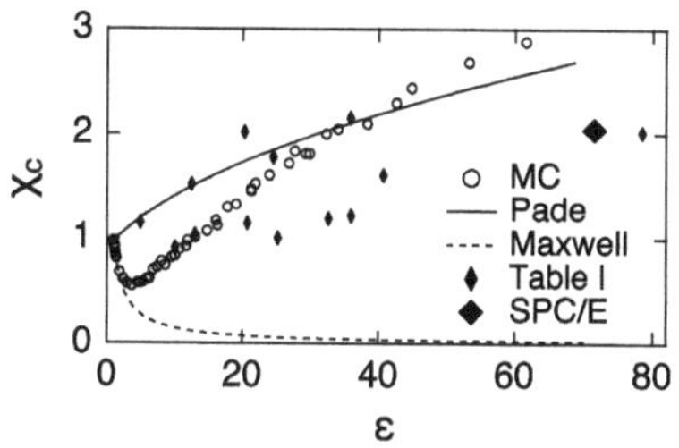

Fig. 7 Cavity field at molecules of a homogeneous fluid of dipolar hard spheres calculated from Monte Carlo simulations (MC) [40], from the Maxwell solution for a cavity carved from a dielectric (Maxwell, Eq. (53) with $\varepsilon_\infty = 1$), and from Padé-based perturbation expansion leading to Eq. (72) (Pade). The small filled diamonds refer to the calculations done according to Eq. (72) for liquids listed in Table 1. The large diamond is the cavity field calculated from molecular dynamics simulations of the Lennard–Jones sphere with the size equal to the size of the water molecule immersed in SPC/E force-field water ($\varepsilon = 71.5$) [39]. The results for molecular liquids and SPC/E water are reported at $T = 298$ K

provides an overall better description of the shape of $\chi_c(\varepsilon)$. The distinction between the perturbation and Maxwell results is most significant at $\varepsilon \gg 1$ when their ratio scales as $\varepsilon^{3/2}$. The analytical calculations are also extended to the liquids listed in Table 1 (filled diamonds in Fig. 7). Empirical Kirkwood factors listed in Ref. [37] and the hard-sphere diameters from Ref. [59] were used in this rather crude calculation, which does not include higher order molecular multipoles [22] in addition to a number of other approximations. Nevertheless, the calculations yield the cavity field for water essentially coinciding with that produced by molecular dynamics simulations for a water-like Lennard–Jones sphere placed in SPC/E water [39] [a large diamond in Fig. 6 at the same level as the small diamond referring to Eq. (72)]. The water calculation does not coincide with the SPC/E result because the dielectric constant of the latter is shifted to $\varepsilon = 71.5$ at $T = 298$ K. Overall, these data, even though currently limited, indicate that continuum prescription for the cavity field susceptibility, given by Eq. (53), has a very limited range of applicability (small ε) and is bound to fail for most polar molecular liquids.

There is a fundamental reason for the failure of continuum estimates of the cavity field in molecular liquids. The standard model, going back to Maxwell [48], assumes that electric field leads to surface charge at any dividing surface. Maxwell thought of a dielectric in terms of two mutually neutralizing fluids carrying positive and negative charge. Within this model, the external field shifts one fluid relative to the other, thus creating positive and negative lobes of the surface charge at any closed surface within the dielectric (Fig. 8a). The current view of a polarized dielectric is in terms of molecular dipoles aligned by the field and creating the surface charge through the corresponding ends of the dipoles exposed to the dielectric surface [25] (Fig. 8b). While this latter view is probably correct for solid or strongly viscous materials (see below), it can hardly provide the correct physical picture for high-temperature polar liquids. The conceptual difficulty here is that the external fields are weak and aligning energies supplied by them cannot compete with thermal agitation. The issue

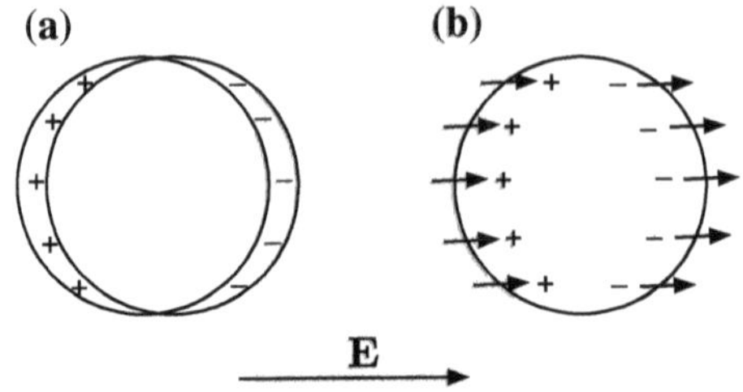

Fig. 8 Schematic representation of the origin of the surface charge at dielectric interfaces in the Maxwell model of deformable positively and negatively charged fluids (**a**) and in the model of oriented surface dipoles (**b**)

of timescales is hidden here, as is often the case with seemingly static problems [15]. If the surface dipoles can rotate, through thermal agitation, on the experimental timescale, the surface alignment averages out to zero and no surface charge is produced. A molecular cavity with surface charge then effectively turns into the virtual cavity considered by Lorentz [4], which does not carry surface charge. This likely does not happen for low-temperature liquids close to the glass transition since the relaxation time is very close to the observation time and some residual surface charge must be preserved (see below).

The dielectric Lorentz cavity is a macroscopic construct that considers a large volume of polarized liquid separated from the rest of the polarized liquid without producing a real physical interface (virtual cavity) [4]. The cavity field susceptibility in such virtual cavity is

$$\chi_c = \frac{\varepsilon + 2\varepsilon_\infty}{3\varepsilon}. \tag{75}$$

Comparing this equation to the Maxwell result in Eq. (53), one can see that the main qualitative distinction between two results is that the Maxwell cavity field strongly screens the field of external charges, $\chi_c \simeq 3/(2\varepsilon)$, while the Lorentz cavity susceptibility reduces to a constant $\chi_c \to 1/3$ at $\varepsilon \gg 1$. This latter limit is indeed reached in simulations of large solutes which do not significantly perturb the structure of the liquid and thus mimic the Lorentz cavity [39, 40] (Fig. 9). A significant point here is that the liquid molecules in the surface layer are not restrained in their molecular motions and average the surface charge out to zero on the observation time.

The notion that molecular interfaces of liquids at sufficiently high temperatures do not carry surface charge has direct impact on observations where polarization of the interface is probed by experiment. One such observable property is the absorption of radiation by solutions. The radiation in the THz domain of frequencies is fast enough to allow dynamic freezing of the dipole moment of a large solute, but provides a sufficient observation window for the water molecules to relax. Figure 10 illustrates the distinction between the Maxwell and Lorentz cavity susceptibilities used to calculate absorption of THz radiation by aqueous solutions of lysozyme with changing

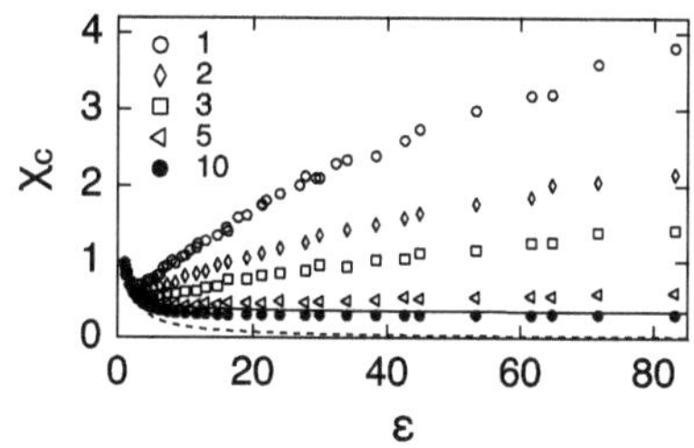

Fig. 9 $\chi_c(\varepsilon)$ for hard-sphere cavities of varying size in liquids of dipolar hard spheres with changing dipole moment (points) [40]. The legend lists the ratio of the cavity diameter to that of the liquid. Lines refer to the Lorentz [solid, Eq. (75)] and Maxwell [dashed, Eq. (53)] cavity susceptibilities

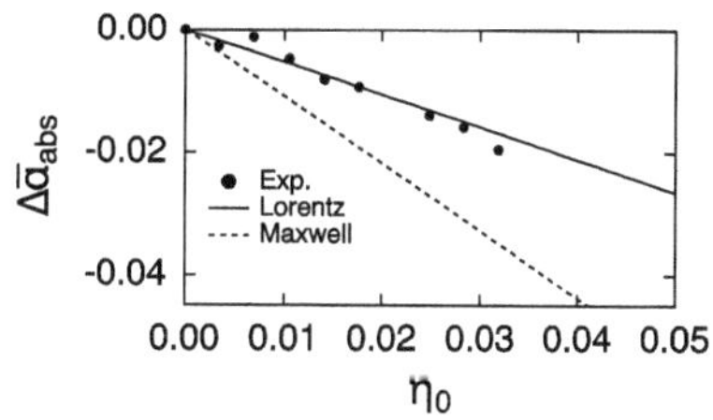

Fig. 10 $\Delta\bar{\alpha}_{\mathrm{abs}} = \alpha_{\mathrm{abs}}/\alpha_{\mathrm{w}} - 1$ calculated from the absorption coefficient of the solution, α_{abs}, and water, α_{w}, versus the volume fraction η_0 of lysozyme in solution. The use of the Lorentz cavity susceptibility [Eq. (75), solid line] provides a better description of the experimental results (points [50]) than the Maxwell cavity susceptibility [Eq. (53), dashed line]

concentration [50]. Even though one might expect that the protein–water interface is too complex to allow any simple model, it appears that waters in the hydration shells are sufficiently disordered to produce an overall Lorentz cavity field when the electric field of radiation is applied [41]. The Lorentz susceptibility then provides a satisfactory account of the change in the absorption coefficient against the volume fraction of the protein in solution (Fig. 10). This outcome can be rationalized for large solutes creating cavities approaching the macroscopic limit envisioned by the Lorentz virtual cavity construction. For local fields acting on individual molecules inside the bulk liquid, the Lorentz result is hardly applicable, as is indeed seen from simulations and calculations shown in Figs. 7 and 9. Microscopic perturbation theories give superior description in this case. The current formulation, however, does not anticipate nearly frozen orientational dynamics of low-temperature liquids near the point of glass transition as we discuss next.

8 Effect of the Electric Field on Glass Transition

Electric field elevates the temperature of glass transition T_g, and the change $\Delta T_g \propto E^2$ scales quadratically with the applied field. Glass transition is commonly viewed as a dynamical phenomenon requiring crossing of the main relaxation process of the glass former with the experimental observation time [3]. Therefore, the alteration of the glass transition temperature is measured from the effect of the electric field on the relaxation time of the main relaxation process [56]. However, an alternative perspective on the glass transition views the dynamic slowing down as merely a manifestation of thermodynamic changes in the system related to shrinking of the configuration space available to the system with lowering temperature [1]. This perspective places the configurational entropy S_c of the material in the forefront as the main property to consider when addressing the approach to the glass transition [42, 57]. The temperature of the laboratory glass transition also turns out to be close to the thermodynamic Kauzmann temperature T_K at which configurational entropy vanishes. If the thermodynamic view of glass transition is adopted, all effects of thermodynamic and external conditions on glass transition are reduced to the corresponding effects on the configurational entropy. When applied to the effect of the electric field, one approximates ΔT_g by the corresponding shift of the Kauzmann temperature: $\Delta T_K \simeq \Delta T_g$ (Fig. 11).

In the canonical Gibbs ensemble, the configurational entropy is the logarithm of the density of states evaluated at the average energy of the system $\bar{E}$

$$S_c = k_{\mathrm{B}} \ln \left[\Omega(\bar{E}) \right]. \tag{76}$$

The density of states $\Omega(E)$ in turn enters the canonical partition function in the form of the Laplace transformation from the variable of energy E to the variable of inverse temperature β

$$e^{-\beta F(\beta)} = \int_0^{\infty} \Omega(E) e^{-\beta E} dE. \tag{77}$$

If the functional form $F(\beta)$ is known, this information can be used to calculate $\Omega(E)$ by inverse Laplace transform. This opportunity presents itself for the Padé-truncated perturbation expansion for polar liquids [45] discussed above. The free energy in Eq. (66) can be rewritten as

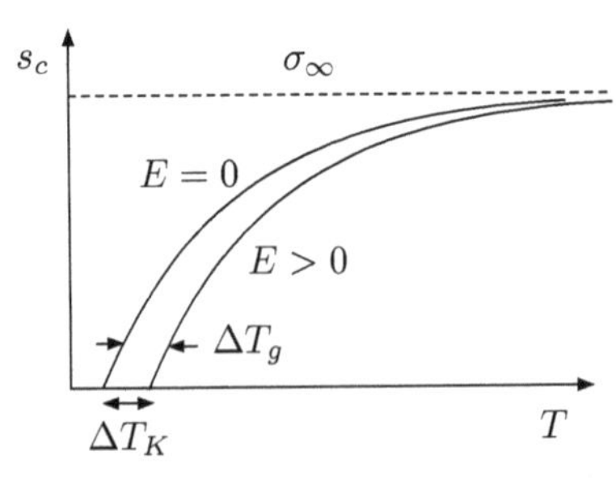

Fig. 11 Schematic representation of the effect of the electric field on the temperature dependence of the configurational entropy and the depression of the glass transition caused by the shift of the Kauzmann temperature

$$\Delta F \propto -N\frac{z^2}{1+z}, \tag{78}$$

where z, as defined by Eqs. (71) and (73), is proportional to β with the proportionality coefficient determined by perturbation theories. This functional form allows an exact inverse Laplace transform and the calculation of the density of states related to orientational degrees of freedom of the dipoles in the liquid. The corresponding configurational entropy becomes [45]

$$s_c = S_c/(k_\mathrm{B}N) = \sigma_\infty + s_p, \tag{79}$$

where

$$s_p = -\frac{s_0}{(1+T/T')^2} \tag{80}$$

is the entropy of the dipolar interactions in the liquid

$$s_p = -\frac{1}{k_\mathrm{B}N}\left(\frac{\partial \Delta F}{\partial T}\right)_V. \tag{81}$$

In Eqs. (79) and (80), $T/T' = 1/z$ and σ_∞ defines the configurational entropy per molecular dipole at $T \to \infty$ [57] (Fig. 11). Further, the dimensionless entropy s_0 of disordering the dipoles when changing the temperature from $T = 0$ to $T \to \infty$ is constructed from the second-order and third-order cumulants of the anisotropic intermolecular interaction H_a

$$s_0 = (9/2N)\langle H_a^2\rangle^3/\langle H_a^3\rangle^2. \tag{82}$$

As is seen from Eq. (79), the appearance of the Kauzmann temperature $T_K > 0$ is related to the requirement $\sigma_\infty < s_0$.

Since the high-temperature plateau σ_∞ is not affected by the field, the change in the configurational entropy induced by the electric field is given by the corresponding change in the entropy of the polar liquid [Eq. (79)]

$$\Delta s_{cE} = \Delta s_{pE}, \tag{83}$$

where the subscript "E" specifies the effect of the electric field on the corresponding quantities. The change in the entropy of dipolar interactions Δs_{pE} can be approximated by $S_E/(k_\mathrm{B}N)$ in Eq. (12) [26].

The configurational entropy is decreased in the presence of the field (Fig. 11), with the resulting upward shift in the temperature at which s_c becomes zero, i.e., the Kauzmann temperature. The dependence on the electric field enters the configurational entropy through the parameter z in Eq. (71), which depends on E_0^2 through $\langle H_a(E_0)^2\rangle$ in Eq. (68). From this connection, the relative change of T_K is obtained as [45]

$$\sigma_\infty \frac{\Delta T_K}{T_K} = \frac{\beta_K^2}{6\sqrt{2}} \frac{(mE_c)^2}{1+z}, \tag{84}$$

where $\beta_K = (k_B T_K)^{-1}$. By assuming $\Delta T_K / T_K \simeq \Delta T_g / T_g$ and employing the Kirkwood–Onsager equation [Eq. (33)], this relation can be brought to the form

$$\sigma_\infty \frac{\Delta T_g}{T_g} = \frac{(\varepsilon_g - 1)(2\varepsilon_g + 1)}{3\sqrt{2} g_K} \chi_c^2 \frac{f_E}{1+z}, \tag{85}$$

where $\varepsilon_g = \varepsilon(T_g)$ and f_E is the electrostatic free energy per molecule given by Eq. (48) and reduced with $k_B T_g$. Note that the nonlinear dielectric effect discussed above in terms of the Langevin model [Eqs. (54) and (55)] and the elevation of T_g are both affected by χ_c^2. This is also the case for nonlinear spectroscopic techniques [5]. These observables are, therefore, sensitive to the models applied to describe χ_c.

The theoretical prediction for the shift of the glass transition temperature is strongly dependent on the model used for the cavity susceptibility χ_c. If one assumes the high-temperature model for this function and employs Eq. (72), the result is

$$\sigma_\infty \frac{\Delta T_g}{T_g} \simeq f_E \frac{\sqrt{2}\varepsilon_g^2}{3} \frac{1+z}{1+2z} \propto \beta \varepsilon_g^3 V_m. \tag{86}$$

If, on the contrary, the Maxwell model with $\chi_c \simeq 3/(2\varepsilon_g)$ [Eq. (53)] is adopted, one obtains

$$\sigma_\infty(\Delta T_g / T_g) \simeq (3/2\sqrt{2} g_K) f_E/(1+z) \propto \beta(\varepsilon_g / g_K) V_m. \tag{87}$$

The cubic scaling of ΔT_g with the dielectric constant ε_g in Eq. (86) is not supported by the presently available data for the elevation of T_g by the external field [58]. For instance, for 2-methyltetrahydrofuran glass former with $T_g = 97.5$ K and $\varepsilon_g \simeq 16.8$ one obtains $(\sigma_\infty / f_E)(\Delta T_g / T_g)(1+z) \simeq 1.0$ [58], where $\sigma_\infty = 9.9$ and $z = T'/T_K = 0.81$ [45]. This result is more consistent with the Maxwell limit in Eq. (87) than with the cavity field from perturbation theories used in Eq. (86). This comparison suggests that the Maxwell model for the cavity susceptibility is a better representation of polarization of liquids close to the glass transition. The comparison of high-temperature and low-temperature limits for the cavity field also suggests that $\chi_c(T)$ must show a substantial drop on approach to T_g. We are not aware of any experiments reporting $\chi_c(T)$ over an extended range of temperatures.

From a more fundamental perspective, a significant impact of the model adopted for the cavity field on the calculated ΔT_g is an indication that different theoretical descriptions of interfacial polarization might be required for high- and low-temperature liquids. The distinction between the Maxwell and Lorentz results for the cavity field is a consequence of different types of boundary conditions imposed on the solution of the Poisson equation in the dielectric boundary-value problem [25, 43]. The high-temperature liquids average out the orientations of the dipoles in the interface, thus mostly eliminating the surface charge. This physics leads to a significant reduction of the screening imposed by a polar liquid on the field of the external charges. In contrast, thermal motions are nearly frozen dynamically near T_g and the surface charge at the dielectric dividing surface is preserved. The same outcome can be manifested at interfaces where slowing of the rotational thermal motions of the interfacial dipoles is achieved by intermolecular interactions between the liquid and the substrate. The classical problem of ion solvation is a physically relevant example. Similar considerations might apply to the frequency-dependent cavity field susceptibility, which might show a dynamical crossover at frequencies corresponding to the rate of collective relaxation of the interface. The dynamic response of the cavity field has not yet received attention in the literature either theoretically or experimentally. The general issues of the role of collective dynamics and structure of interfaces in the observable dielectric properties still pose significant challenges to our understanding of the dielectric response and remain open questions requiring further studies.

9 Conclusions

Nonlinear response of polar materials to an external electric field provides information not available in the linear regime. The main fundamental distinction of the nonlinear dielectric susceptibility is the access to nontrivial high-order orientational correlations of molecular dipoles. Microscopic understanding of the nonlinear response is mostly an uncharted territory. Experimental challenges of resolving a relatively weak nonlinear response and separating it from complications arising from heating and other potential artifacts are still significant. While substantial progress of experimental techniques has been achieved in recent years [34, 55, 56], theory still largely lags behind. This comes in a stark contrast to many recent advances of computational techniques in understanding microscopic correlations underlying observables interrogated by experiment [23]. The reason for this lack of progress is in astounding difficulties encountered in computing many-particle molecular correlations in condensed-phase materials. Analytical approximations for multi-body correlations are nearly nonexistent and direct numerical simulations typically fail to converge the corresponding correlation functions. This is a new frontier for simulations of polar materials currently driven by advances in experiment reviewed in this volume.

Despite significant theoretical difficulties, some emergent opportunities have been outlined here. In particular, temperature derivatives of linear response functions provide information about triple and fourth-order correlations. While these observables already give access to nonbinary dipolar correlations, they might be potentially easier to compute compared to the nonlinear dielectric function. The development of formal theories is still required, in particular in the form of the connection between observables produced in linear and nonlinear response. The cavity susceptibility, which strongly affects the mean-field nonlinear response and field-induced elevation of the glass transition temperature, is one such property of interest. It enters many spectroscopic observables and can potentially be accessed independently to test formal theories. We have shown here that this function crosses from the high-temperature

Lorentz limit to the low-temperature Maxwell limit when the relaxation time of the liquid dipoles approaches the observation window and dipolar screening in polarized interfaces becomes more prominent.

Acknowledgements This research was supported by the National Science Foundation (CHE-1800243). The author is grateful to Ranko Richert for many fruitful discussions.

References

1. G. Adam, J.H. Gibbs, On the temperature dependence of cooperative relaxation properties in glass-forming liquids. J. Chem. Phys. **43**, 139 (1965)
2. M.P. Allen, D.J. Tildesley, *Computer Simulation of Liquids* (Clarendon Press, Oxford, 1996)
3. C.A. Angell, Formation of glasses from liquids and biopolymers. Science **267**, 1924–1935 (1995)
4. C.J.F. Böttcher, *Theory of Electric Polarization*, vol. 1 (Elsevier, Amsterdam, 1973)
5. D.M. Burland, R.D. Miller, C.A. Walsh, Second-order nonlinearity in poled-polymer systems. Chem. Rev. **94**, 31–75 (1994)
6. D.Y.C. Chan, D.J. Mitchell, B.W. Ninham, A model of solvent structure around ions. J. Chem. Phys. **70**(6), 2946–2957 (1979)
7. A. Chełkowski, *Dielectric Physics* (Elsevier Scientific Pub. Co., Amsterdam, 1980)
8. Y. Chen, H.I. Okur, N. Gomopoulos, C. Macias-Romero, P.S. Cremer, P.B. Petersen, G. Tocci, D.M. Wilkins, C. Liang, M. Ceriotti, S. Roke, Electrolytes induce long-range orientational order and free energy changes in the H-bond network of bulk water. Sci. Adv. **2**(4), e1501891 (2016)
9. T. Christensen, N.B. Olsen, Determination of the frequency-dependent bulk modulus of glycerol using a piezoelectric spherical-shell. Phys. Rev. B **49**(21), 15396–15399 (1994)
10. K. Clays, A. Persoons, Hyper-Rayleigh scattering in solution. Phys. Rev. Lett. **66**, 2980–2983 (1991)
11. L. Comez, D. Fioretto, F. Scarponi, G. Monaco, Density fluctuations in the intermediate glass-former glycerol: a Brillouin light scattering study. J. Chem. Phys. **119**(12), 6032 (2003)
12. P. Debye, Dielektrische Sättigung und Behinderung der freien Rotation in Flüssigheiten. Z. Phys. Chem. **36**, 193–194 (1935)
13. J.L. Déjardin, Y.P. Kalmykov, Nonlinear dielectric relaxation of polar molecules in a strong ac electric field: steady state response. Phys. Rev. E **61**(2), 1211–1217 (2000).
14. G.I. Egorov, D.M. Makarov, Volumetric properties of binary liquid-phase mixture of (water+glycerol) at temperatures of (278.15 to 323.15) K and pressures of (0.1 to 100) MPa. J. Chem. Thermodyn. **79**, 135–158 (2014)
15. R.P. Feynman, *Statistical Mechanics* (Westview Press, Boulder, CO, 1998)
16. H.S. Frank, Thermodynamics of a fluid substance in the electrostatic field. J. Chem. Phys. **23**(11), 2023–2032 (1955)
17. H. Fröhlich, *Theory of Dielectrics* (Oxford University Press, Oxford, 1958)
18. R.L. Fulton, On the theory of nonlinear dielectrics. J. Chem. Phys. **78**(11), 6865–6876 (1983)
19. R.L. Fulton, The nonlinear dielectric behavior of water: comparisons of various approaches to the nonlinear dielectric increment. J. Chem. Phys. **130**(20), 204503 (2009)
20. R.L. Fulton, Linear and nonlinear dielectric theory for a slab: the connections between the phenomenological coefficients and the susceptibilities. J. Chem. Phys. **145**(8), 084105 (2016)
21. C.W. Gardiner, *Handbook of Stochastic Methods* (Springer, Berlin, 1997)
22. C.G. Gray, K.E. Gubbins, *Theory of Molecular Liquids* (Clarendon Press, Oxford, 1984)
23. J.P. Hansen, I.R. McDonald, *Theory of Simple Liquids*, 4th edn. (Academic Press, Amsterdam, 2013)

24. J.S. Høye, G. Stell, Statistical mechanics of polar systems. II. J. Chem. Phys. **64**(5), 1952–1966 (1976)
25. J.D. Jackson, *Classical Electrodynamics* (Wiley, New York, 1999)
26. G.P. Johari, Effects of electric field on the entropy, viscosity, relaxation time, and glass-formation. J. Chem. Phys. **138**(15), 154503 (2013)
27. J.G. Kirkwood, The dielectric polarization of polar liquids. J. Chem. Phys. **7**(10), 911–919 (1939)
28. R. Kubo, The fluctuation-dissipation theorem. Rep. Prog. Phys. **29**, 255–284 (1966)
29. P.G. Kusalik, Computer simulation study of a highly polar fluid under the influence of static electric fields. Mol. Phys. **81**(1), 199–216 (1994)
30. D.P. Landau, K. Binder, *Monte Carlo simulations in statistical physics* (Cambridge University Press, Cambridge, 2000)
31. L.D. Landau, E.M. Lifshitz, *Electrodynamics of Continuous Media* (Pergamon, Oxford, 1984)
32. B. Larsen, J.C. Rasaiah, G. Stell, Thermodynamic perturbation theory for multipolar and ionic liquids. Mol. Phys. **33**, 987 (1977)
33. P. Linse, G. Karlström, Dipolar order in molecular fluids: II. Molecular influence. J. Stat. Phys. **145**(2), 418–440 (2011)
34. P. Lunkenheimer, M. Michl, T. Bauer, A. Loidl, Investigation of nonlinear effects in glassy matter using dielectric methods. Eur. Phys. J. Special Topics **226**(14), 3157–3183 (2017)
35. P. Madden, D. Kivelson, A consistent molecular treatment of dielectric phenomena. Adv. Chem. Phys. **56**, 467–566 (1984)
36. Y. Marcus, Electrostriction in electrolyte solutions. Chem. Rev. **111**(4), 2761–2783 (2011)
37. Y. Marcus, *Ions in Solution and their Solvation* (Wiley, New Jersey, 2015)
38. Y. Marcus, G. Hefter, On the pressure and electric field dependencies of the relative permittivity of liquids. J. Sol. Chem. **28**(5), 575–592 (1999)
39. D.R. Martin, A.D. Friesen, D.V. Matyushov, Electric field inside a "Rossky cavity" in uniformly polarized water. J. Chem. Phys. **135**, 084514 (2011)
40. D.R. Martin, D.V. Matyushov, Microscopic fields in liquid dielectrics. J. Chem. Phys. **129**, 174508 (2008)
41. D.R. Martin, D.V. Matyushov, Terahertz absorption of lysozyme in solution. J. Chem. Phys. **146**, 084502 (2017)
42. L.M. Martinez, C.A. Angell, A thermodynamic connection to the fragility of glass-forming liquids. Nature **410**, 663 (2001)
43. D.V. Matyushov, Electrostatics of liquid interfaces. J. Chem. Phys. **140**, 224506 (2014)
44. D.V. Matyushov, Nonlinear dielectric response of polar liquids. J. Chem. Phys. **142**, 244502 (2015)
45. D.V. Matyushov, Configurational entropy of polar glass formers and the effect of electric field on glass transition. J. Chem. Phys. **145**, 034504 (2016)
46. D.V. Matyushov, Response to "Comment on 'Nonlinear dielectric response of polar liquids"' [J. Chem. Phys. **144**, 087101 (2016)]. J. Chem. Phys. **144**, 087102 (2016)
47. D.V. Matyushov, R. Richert, Communication: temperature derivative of the dielectric constant gives access to multipoint correlations in polar liquids. J. Chem. Phys. **144**, 041102 (2016)
48. J.C. Maxwell, *A Treatise on Electricity and Magnetism*, vol. 1. Dover Publications, New York (1954, sec. 63)
49. M. Neumann, Computer simulation and the dielectric constant at fimite wavelength. Mol. Phys. **57**, 97 (1986)
50. F. Novelli, S. Ostovar Pour, J. Tollerud, A. Roozbeh, D.R.T. Appadoo, E.W. Blanch, J.A. Davis, Time-domain THz spectroscopy reveals coupled protein-hydration dielectric response in solutions of native and fibrils of human lysozyme. J. Phys. Chem. B **121**, 4810–4816 (2017)
51. H. Nyquist, Thermal agitation of electric charge in conductors. Phys. Rev. **32**(1), 110–113 (1928)
52. L. Onsager, Electric moments of molecules in liquids. J. Am. Chem. Soc. **58**, 1486–1493 (1936)
53. F.O. Raineri, H. Resat, H.L. Friedman, Static longitudinal dielectric function of model molecular fluids. J. Chem. Phys. **96**, 3068 (1992)

54. J.D. Ramshaw, Exsistence of the dielectric constant in regid-dipole fluids: the direct correlation function. J. Chem. Phys. **57**, 2684 (1972)
55. R. Richert, Supercooled liquids and glasses by dielectric relaxation spectroscopy. Adv. Chem. Phys. **156**, 101–195 (2015)
56. R. Richert, Nonlinear dielectric effects in liquids: a guided tour. J. Phys.: Cond. Matter **29**(36), 363001–363025 (2017)
57. R. Richert, A.C. Angell, Dynamics of glass-forming liquids. V. On the link between molecular dynamics and configuration entropy. J. Chem. Phys. **108**, 9016 (1998)
58. S. Samanta, R. Richert, Electrorheological source of nonlinear dielectric effects in molecular glass-forming liquids. J. Phys. Chem. B **120**(31), 7737–7744 (2016)
59. R. Schmid, D.V. Matyushov, Entropy of attractive forces and molecular nonsphericity in real liquids: a measure of structural ordering. J. Phys. Chem. **99**, 2393 (1995)
60. S. Seyedi, D.R. Martin, D.V. Matyushov (unpublished)
61. L.P. Singh, R. Richert, Watching hydrogen-bonded structures in an alcohol convert from rings to chains. Phys. Rev. Lett. **109**(16), 167802 (2012)
62. G. Stell, Fluids with long-range forces: toward a simple analytic theory, in *Statistical Mechanics. Part A: Equilibrium Techniques*, ed. by B.J. Berne (Plenum, New York, 1977)
63. G. Stell, G.N. Patey, J.S. Høye, Dielectric constants of fluid models: statistical mechanical theory and its quantitative implementation. Adv. Chem. Phys. **48**, 183–328 (1981)
64. J.H. van Vleck, On the role of dipole-dipole coupling in dielectric media. J. Chem. Phys. **5**(7), 556–568 (1937)
65. M.S. Wertheim, Theory of polar fluids: V. Thermodynamics and thermodynamic perturbation theory. Mol. Phys. **37**(1), 83–94 (1979)
66. A.R. Young-Gonzales, K. Adrjanowicz, M. Paluch, R. Richert, Nonlinear dielectric features of highly polar glass formers: derivatives of propylene carbonate. J. Chem. Phys. **147**(22), 224501–224511 (2017)
67. C. Zhang, G. Galli, Dipolar correlations in liquid water. J. Chem. Phys. **141**(8), 084504–084506 (2014)

Nonlinear Dielectric Relaxation in AC and DC Electric Fields

P. M. Déjardin, W. T. Coffey, F. Ladieu and Yu. P. Kalmykov

Abstract Current theories of the nonlinear static and dynamic dielectric susceptibilities of polar fluids subjected to strong AC and DC electric fields are reviewed with specific emphasis on those extending Debye's theory of linear dielectric relaxation of an assembly of polar molecules. The inclusion of intermolecular interactions in this theory as well as nonlinear dielectric relaxation in the presence of time-dependent fields is discussed. In particular, we emphasize the role played by intermolecular interactions in the determination of the macroscopic dielectric properties of a polar fluid via microscopic calculations, in both the linear and nonlinear responses.

1 Introduction

A well-founded microscopic theory of the electric polarization (both static and dynamic) of polar fluids is essential to understanding many dielectric and electro-optical relaxation phenomena and as such was initiated by Debye [1]. He first calculated the static susceptibility of an assembly of noninteracting rigid dipoles obtaining a result which is essentially a replica of Langevin's theory of paramagnetism and so is called the Langevin–Debye theory. He then extended the calculation to include the linear dielectric susceptibility of noninteracting polar molecules subjected to a weak AC electric field, which unlike the static situation poses a nonequilibrium problem. In order to accomplish this, he treated the effects of the heat bath surrounding a dipole via the rotational diffusion model. This is based on a generalization of Einstein's 1905 [2] theory of the translational Brownian motion [3] to rotation on the unit sphere and to include the effects of a weak AC field applied along an axis chosen

P. M. Déjardin (✉) · Y. P. Kalmykov
Laboratoire de Mathématiques et de Physique (LAMPS, EA4217), Université de Perpignan via Domitia, Perpignan 66860, France
e-mail: dejardip@univ-perp.fr

W. T. Coffey
Department of Electronic and Electrical Engineering, Trinity College, Dublin 2, Ireland

F. Ladieu
SPEC, CEA, CNRS, Université Paris-Saclay, Bât. 772, Gif-sur-Yvette 91191, France

R. Richert (ed.), *Nonlinear Dielectric Spectroscopy*, Advances in Dielectrics
https://doi.org/10.1007/978-3-319-77574-6_2

as the Z-axis. Thus, in the dynamical Debye theory as with that of Einstein, inertial effects are negligible and the rotation of the molecule is described by a random walk over small angular orientations. Later, his original calculation was generalized (using perturbation theory) to cover nonlinear phenomena in polar dielectrics subjected to strong AC and indeed other external fields [4–6]. As specific examples, we cite both the dynamic Kerr and nonlinear dielectric effects [4–8]. In particular, in nonlinear dielectric relaxation, depending on the particular form of the stimulus chosen, additional terms in the fundamental, third, etc., harmonic appear in the polarization response [5–7], which have been confirmed by experiment [4, 9–12]. Furthermore, the dynamical Debye theory has also been extended (by exact numerical solution using matrix continued fractions) to include nonlinear effects in *arbitrarily large* external fields [13–16]. Nevertheless, assemblies of *noninteracting* dipoles are still assumed implying that the Debye theory and its extensions may not be used for *dense* dipolar systems, where intermolecular interactions are significant. Although the treatment of the latter is much more involved, several methods are still available. For example, the dielectric relaxation of polar nematic liquid crystals may be regarded as the rotational Brownian motion in the Maier–Saupe uniaxial anisotropy potential [17, 18], leading to an Arrhenius-like escape over a barrier process due to the shuttling action of the rotational Brownian motion giving rise to reversal of a dipole occurring in all the dynamical responses. Such a mechanism was first identified by Kramers [19] in the context of the translational Brownian motion, and was recognized by Debye in the context of *normal* dispersion and absorption in solids. The method comprises the *static mean field* approach. However, such a treatment, although of restricted applicability because it ignores local order effects, is easily visualized and permits quantitative evaluation of dielectric parameters. Thus, it qualitatively demonstrates the effect of intermolecular interaction on these, an effect which must be included for the purpose of comparison with experimental data [20]. Yet another advantage is that it also yields the nonlinear response of assemblies of noninteracting uniaxial single-domain ferromagnetic particles [21]. Unfortunately, the static mean field method still ignores *dynamical effects* due to intermolecular interactions.

In contrast, the *dynamical mean field* method reveals dynamical effects due to intermolecular interactions manifesting themselves *at the nonlinear response level only* [22]. These novel predictions are interesting as they are *qualitatively* similar to observations of *supercooled* polar liquid nonlinear dielectric response measurements, namely non-monotonic behavior of the nonlinear response moduli, without corresponding modification of the linear response at low frequencies.

A succinct account of nonlinear dielectric effects in liquids has recently been given by Richert [23], who emphasized the growing importance of such measurements in so far as they can characterize many polar fluids in various states. Here, we review nonlinear dielectric response calculations based on further developments of the Debye theory which are accomplished by generalizing it to include both strong electric fields and intermolecular interactions. The chapter is organized as follows: in Sects. 2 and 3, we review the methods used in [7] for the nonlinear dynamic dielectric susceptibilities of a gas of noninteracting dipoles subjected to strong DC and AC electric fields. Next two mean field models of interaction are described in Sects. 4 and 5, while Sect. 6 is devoted to internal field corrections. These are the only corrections needed as a dielectric liquid always occupies the entire empty space between the electrodes of the measuring device. The depolarizing field effect is first discussed. Next, Sect. 7 shows how to include both static and dynamical intermolecular correlations.

2 Nonlinear Dielectric Response of Noninteracting Polar Molecules to a Strong AC Electric Field

We consider the *nonlinear* AC (alternating current) stationary response of an assembly of noninteracting polar molecules (electric dipoles) undergoing rotational Brownian motion due to the heat bath and also acted upon by a strong external AC field **E**(t). Moreover, we suppose without loss of generality that **E** is directed along the Z-axis of the laboratory coordinate system so that axial symmetry is preserved. This treatment is a simple extension of the work of Debye and is essentially due to Coffey and Paranjape [7]. The starting point of the theory is the rotational diffusion equation (the Smoluchowski equation, a particular form of the Fokker–Planck equation) due to Debye for the surface distribution function of the dipole orientations on the unit sphere when embedded in a heat bath, viz.,

$$2\tau_D \frac{\partial W}{\partial t} = \frac{1}{\sin\vartheta} \frac{\partial}{\partial\vartheta} \left[\sin\vartheta \left(\frac{\partial W}{\partial\vartheta} + \beta W \frac{\partial V}{\partial\vartheta} \right) \right]. \tag{1}$$

In Eq. (1) $T_D = \varsigma \,/\, (2kT)$ is the rotational diffusion time also called the Debye relaxation time, and expresses the given fluctuation–dissipation relation which exists between the magnitude of the Brownian *Schwankung* of the angle ϑ and the temperature T and friction constant ς, i.e., $\overline{\vartheta^2}/\,(4\delta t) = kT/\varsigma$. Here $\beta = (kT)^{-1}$, k is Boltzmann's constant, ϑ is the angle a (tagged) dipole moment makes with the externally applied uniform electric field **E** (t), W (ϑ, t) is the surface probability density of orientations of a dipole, and V (ϑ, t) is the potential of axially symmetric applied external torques. Here, it is simply that of the interaction of a dipole with the electric field, namely

$$V\,(\vartheta, t) = -\mu E\,(t) \cos\vartheta, \tag{2}$$

where $\boldsymbol{\mu}$ is the dipole moment of a molecule, and E (t) is the amplitude of the electric field. The polarization in the field direction is then

$$P\,(t) = \rho_0 \mu \int_0^{\pi} \cos\vartheta\; W\,(\vartheta, t) \sin\vartheta d\vartheta. \tag{3}$$

Equation (3) is evaluated via the statistical moment method, which consists in expanding $W(\vartheta, t)$ as a series of Legendre polynomials $P_n(\cos\vartheta)$ [24], i.e.

$$W(\vartheta, t) = \sum_{n=0}^{\infty} \left(n + \frac{1}{2}\right) f_n(t)\, P_n(\cos\vartheta), \tag{4}$$

and then using the recurrence and orthogonality properties of the P_n, thereby yielding an infinite hierarchy of differential-recurrence relations for the statistical moments $f_n(t)$, viz.,

$$\frac{2\tau_D}{n(n+1)} \dot{f}_n(t) + f_n(t) = \frac{\xi(t)}{2n+1} \left[f_{n-1}(t) - f_{n+1}(t)\right], \; n > 0 \tag{5}$$

Here $f_0 = 1$, $\xi(t) = \beta\mu E(t)$ and Eq. (3) by orthogonality can be rewritten using Eq. (4) as

$$P(t) = \rho_0 \mu f_1(t). \tag{6}$$

Now, even for strong electric field intensities, $\xi(t) < 1$, thus the hierarchy of Eq. (5) may be solved by iterating a perturbation series, yielding

$$f_n(t) = f_n^{(0)} + \sum_{k=1}^{\infty} f_n^{(k)}(t), \tag{7}$$

where the superscript (k) indicates the desired order in the field strength, yielding the perturbed equations

$$n(n+1)\, f_n^{(0)} = 0, \tag{8}$$

$$\frac{2\tau_D}{n(n+1)} \dot{f}_n^{(k)}(t) + f_n^{(k)}(t) = \frac{\xi(t)}{2n+1} \left[f_{n-1}^{(k-1)}(t) - f_{n+1}^{(k-1)}(t)\right], \tag{9}$$

The hierarchy of recurrence Eq. (9) is solved subjected to the initial condition $f_n^{(k)}(-\infty) = 0$ since we are interested in the steady-state regime only. Now, the solution of Eq. (8) is obvious since it is a simple algebraic equation. Thus $f_n^{(0)} = 0$ for all $n \neq 0$. Moreover, $f_0^{(0)} = f_0 = 1$ and $f_0^{(k)} = 0,\ k > 0$. Hence, the linear response of the polarization is given explicitly by $f_1^{(1)}(t)$, that is

$$f_1^{(1)}(t) = \frac{1}{3\tau_D} \int_{-\infty}^{t} \mathrm{e}^{-\frac{t-t_1}{\tau_D}} \xi(t_1)\, \mathrm{d}t_1. \tag{10}$$

Since $f_n^{(1)}(-\infty) = 0$, we have

$$f_n^{(1)}(t) = 0, \; n \neq 1 \tag{11}$$

Likewise, we have the quadratic response functions $f_n^{(2)}(t)$, viz.,

$$f_n^{(2)}(t) = 0,\ n \neq 2, \tag{12}$$

and

$$f_2^{(2)}(t) = \frac{1}{5\tau_D^2} \int_{-\infty}^{t} \int_{-\infty}^{t_2} \mathrm{e}^{-\frac{3(t-t_2)}{\tau_D}} \xi(t_2)\, \mathrm{e}^{-\frac{t_2-t_1}{\tau_D}} \xi(t_1)\, \mathrm{d}t_1 \mathrm{d}t_2. \tag{13}$$

Moreover, via Eq. (9) for $n = 1$ and $k = 3$, we have the cubic polarization dynamical response, viz.,

$$f_1^{(3)}(t) = -\frac{1}{15\tau_D^3} \int_{-\infty}^{t} \int_{-\infty}^{t_3} \int_{-\infty}^{t_2} \mathrm{e}^{-\frac{t-t_3}{\tau_D}} \xi(t_3)\, \mathrm{e}^{-\frac{3(t_3-t_2)}{\tau_D}} \xi(t_2)\, \mathrm{e}^{-\frac{t_2-t_1}{\tau_D}} \xi(t_1)\, \mathrm{d}t_1 \mathrm{d}t_2 \mathrm{d}t_3 \tag{14}$$

Now, specializing to the pure AC field $E(t) = E_0 \cos \omega t$ so that $\xi(t) = \xi_0 \cos \omega t$ with $\xi_0 = \beta \mu E_0$, the polarization Eq. (6) can be written as

$$P(t) = P^{(1)}(t) + P^{(3)}(t),$$

where $P^{(1)}(t)$ is the linear polarization response given by

$$P^{(1)}(t) = \frac{\rho_0 \mu \xi_0}{3\left(1 + \omega^2 \tau_D^2\right)} \left(\cos \omega t + \omega \tau_D \sin \omega t\right), \tag{15}$$

which is the result of Debye, while the cubic polarization $P^{(3)}(t)$ is

$$P^{(3)}(t) = \frac{\rho_0 \mu \xi_0^3}{60\left(1 + \omega^2 \tau_D^2\right)\left(9 + 4\omega^2 \tau_D^2\right)} \left\{ \frac{\left(13\omega^2 \tau_D^2 - 27\right) \cos \omega t - 2\omega \tau_D \left(21 + \omega^2 \tau_D^2\right) \sin \omega t}{3\left(1 + \omega^2 \tau_D^2\right)} \right.$$
$$\left. + \frac{\left(17\omega^2 \tau_D^2 - 3\right) \cos 3\omega t + 2\omega \tau_D \left(3\omega^2 \tau_D^2 - 7\right) \sin 3\omega t}{\left(1 + 9\omega^2 \tau_D^2\right)} \right\}. \tag{16}$$

This equation represents one of the most important results of the nonlinear Debye theory, demonstrating that for strong AC field amplitudes, the linear response (15) is corrected by the first term in the right-hand side of Eq. (16), while the second term predicts the existence of a third harmonic in the polarization response. This result was confirmed experimentally 20 years after its publication [12]. When $\omega = 0$, $P(t) = P_s$ is time-independent and is given by the two first terms in the Taylor expansion of the Langevin function, viz.,

$$P_s = \rho_0 \mu \left(\frac{\xi_0}{3} - \frac{\xi_0^3}{45} \right), \tag{17}$$

The treatment described here must be further refined if the electric field comprises both a DC part and an AC part, since then the static conditions are no longer given by Eq. (8). This is the subject of the next section.

3 Nonlinear Dielectric Response in Superimposed AC and DC Electric Fields

The basic equation is still the rotational Smoluchowski Eq. (1) as given by Debye, with polarization given by Eq. (3) or (6), save that

$$V(\vartheta, t) = -\mu\,(E_S + E_0 \cos\omega t)\cos\vartheta, \tag{18}$$

where E_S is the static electric field amplitude, also supposed uniform and applied in the direction of the AC field. The differential-recurrence relations (5) now become

$$\frac{2\tau_D}{n(n+1)}\dot{f}_n(t) + f_n(t) = \frac{\xi_S + \xi(t)}{2n+1}\left[f_{n-1}(t) - f_{n+1}(t)\right],\ n > 0. \tag{19}$$

Where $\xi_S = \beta\mu E_S$.

We further assume that all transients due to the (sudden) application of the DC field E_S have disappeared, so that the assembly of dipoles has reached equilibrium in the absence of the AC field. Furthermore, starting from this configuration which represents the stationary state of the system before $E_0 \cos\omega t$ is applied, we then consider the *new stationary state* (i.e., all transient effects due to the application of $E_0 \cos\omega t$ have also disappeared) obtained in the presence of *both* fields. Hence, we can also use the perturbation expansion, Eq. (7) yielding the perturbation equations

$$n(n+1)\,f_n^{(0)} = \frac{n(n+1)\,\xi_S}{2n+1}\left[f_{n-1}^{(0)} - f_{n+1}^{(0)}\right], \tag{20}$$

$$\begin{aligned}\tfrac{2\tau_D}{n(n+1)}\tfrac{\mathrm{d}}{\mathrm{d}t} f_n^{(k)}(t) + f_n^{(k)}(t) &= \tfrac{\xi_S}{2n+1}\left[f_{n-1}^{(k)}(t) - f_{n+1}^{(k)}(t)\right] \\ &+ \tfrac{\xi(t)}{2n+1}\left[f_{n-1}^{(k-1)}(t) - f_{n+1}^{(k-1)}(t)\right].\end{aligned} \tag{21}$$

Equation (20) can be solved using continued fractions, allowing one to express the static moments as ratios of modified Bessel functions [3]. However, we avoid this here because we can use the condition $\xi_S < 1$, yielding a perturbation expansion of all the $f_n^{(k)}(t)$ in terms of the powers of the DC field strength. Thus, we write, in an obvious notation

$$f_n^{(0)} = \sum_{q=0}^{\infty} f_n^{(0,q)}, \tag{22}$$

$$f_n^{(k)}(t) = \sum_{q=0}^{\infty} f_n^{(k,q)}(t). \tag{23}$$

Thus, the perturbation Eqs. (20) and (21) become $\left(f_0^{(0,0)} = f_0^{(0)} = f_0 = 1\right)$,

$$n(n+1) f_n^{(0,q)} = \frac{n(n+1)\xi_S}{2n+1}\left[f_{n-1}^{(0,q-1)} - f_{n+1}^{(0,q-1)}\right], \tag{24}$$

$$\begin{aligned} \tfrac{2\tau_D}{n(n+1)} \dot{f}_n^{(k,q)}(t) + f_n^{(k,q)}(t) &= \tfrac{\xi_S}{2n+1}\left[f_{n-1}^{(k,q-1)}(t) - f_{n+1}^{(k,q-1)}(t)\right] \\ &+ \tfrac{\xi(t)}{2n+1}\left[f_{n-1}^{(k-1,q)}(t) - f_{n+1}^{(k-1,q)}(t)\right], \quad k > 0. \end{aligned} \tag{25}$$

The polarization response to third order in the field strength is

$$P(t) = P^{(1)}(t) + P^{(2)}(t) + P^{(3)}(t), \tag{26}$$

where

$$P^{(1)}(t) = \rho_0\mu\left(f_1^{(1,0)}(t) + f_1^{(0,1)}\right) \tag{27}$$

is the linear polarization response, while

$$P^{(2)}(t) = \rho_0\mu\left(f_1^{(2,0)}(t) + f_1^{(1,1)}(t) + f_1^{(0,2)}\right), \tag{28}$$

is the quadratic polarization response (expected to vanish), and finally

$$P^{(3)}(t) = \rho_0\mu\left(f_1^{(3,0)}(t) + f_1^{(2,1)}(t) + f_1^{(1,2)}(t) + f_1^{(0,3)}\right) \tag{29}$$

is the cubic polarization response. The nonlinear polarization (26) is explicitly determined by solving Eq. (24) up to $q = 3$. For $q = 0$, we have $n(n+1) f_n^{(0,0)} = 0$ implying $f_n^{(0,0)} = 0,\ n \neq 0$. For $q = 1$, Eq. (24) become

$$n(n+1) f_n^{(0,1)} = \frac{n(n+1)\xi_S}{2n+1}\left[f_{n-1}^{(0,0)} - f_{n+1}^{(0,0)}\right], \tag{30}$$

with nonvanishing solution

$$f_1^{(0,1)} = \frac{\xi_S}{3}. \tag{31}$$

which is the first term in the Taylor expansion of the Langevin function. For $q = 2$, we have

$$n(n+1) f_n^{(0,2)} = \frac{n(n+1)\xi_S}{2n+1}\left[f_{n-1}^{(0,1)} - f_{n+1}^{(0,1)}\right], \tag{32}$$

with solution

$$f_2^{(0,2)} = \frac{\xi_S^2}{15}. \tag{33}$$

which is the first term in the Taylor expansion of $\langle P_2 \rangle_0$. Finally, for $q = 3$, we have

$$n(n+1) f_n^{(0,3)} = \frac{n(n+1)\xi_S}{2n+1}\left[f_{n-1}^{(0,2)} - f_{n+1}^{(0,2)}\right], \tag{34}$$

so that

$$f_1^{(0,3)} = -\frac{\xi_S^3}{45}, \; f_3^{(0,3)} = \frac{\xi_S^3}{105}, \tag{35}$$

as expected.

Next, we evaluate $f_1^{(1,0)}(t)$, the linear response to the AC field. Obviously, we see that $f_n^{(1,0)}(t) = 0$ save for $n = 1$, thus Eq. (25) becomes

$$\tau_D \dot{f}_1^{(1,0)}(t) + f_1^{(1,0)}(t) = \frac{\xi(t)}{3}, \tag{36}$$

with steady-state solution the Debye response, viz.,

$$f_1^{(1,0)}(t) = \frac{\xi_0}{3\left(1+\omega^2\tau_D^2\right)}\left(\cos\omega t + \omega\tau_D \sin\omega t\right). \tag{37}$$

Thus the linear polarization is

$$P^{(1)}(t) = \frac{\rho_0 \mu}{3}\left[\xi_S + \frac{\xi_0}{\left(1+\omega^2\tau_D^2\right)}\left(\cos\omega t + \omega\tau_D \sin\omega t\right)\right]. \tag{38}$$

This result is physically acceptable, since in the linear response approximation, the steady-state DC and AC responses simply *superimpose*.

We now calculate the quadratic polarization response (28). Clearly, $f_1^{(0,2)} = 0$ by our earlier arguments, while the two remaining functions in Eq. (28) must satisfy the differential equations

$$\tau_D \dot{f}_1^{(1,1)}(t) + f_1^{(1,1)}(t) = 0,$$
$$\tau_D \dot{f}_1^{(2,0)}(t) + f_1^{(2,0)}(t) = 0,$$

with steady-state solutions $f_1^{(1,1)}(t) = 0$ and $f_1^{(2,0)}(t) = 0$. Therefore, as expected,

$$P^{(2)}(t) = 0. \tag{39}$$

Finally, we evaluate $P^{(3)}(t)$ as given by Eq. (29). We already have $f_1^{(0,3)}$ since it is given by Eq. (35). The three remaining functions satisfy

$$\tau_D \dot{f}_1^{(3,0)}(t) + f_1^{(3,0)}(t) = -\frac{\xi(t)}{3} f_2^{(2,0)}(t), \tag{40}$$

$$\tau_D \dot{f}_1^{(2,1)}(t) + f_1^{(2,1)}(t) = -\frac{\xi_S}{3} f_2^{(2,0)}(t) - \frac{\xi(t)}{3} f_2^{(1,1)}(t), \tag{41}$$

$$\tau_D \dot{f}_1^{(1,2)}(t) + f_1^{(1,2)}(t) = -\frac{\xi_S}{3} f_2^{(1,1)}(t) - \frac{\xi(t)}{3} f_2^{(0,2)}. \tag{42}$$

The determination of $f_1^{(3,0)}(t)$, $f_1^{(2,1)}(t)$ and $f_1^{(1,2)}(t)$ requires knowledge of $f_2^{(2,0)}(t)$ and $f_2^{(1,1)}(t)$ which satisfy

$$\tau_D \dot{f}_2^{(2,0)}(t) + 3 f_2^{(2,0)}(t) = \frac{3\xi(t)}{5} f_1^{(1,0)}(t), \tag{43}$$

$$\tau_D \dot{f}_2^{(1,1)}(t) + 3 f_2^{(1,1)}(t) = \frac{3\xi_S}{5} f_1^{(1,0)}(t) + \frac{3\xi(t)}{5} f_1^{(0,1)}. \tag{44}$$

We infer that the DC field does not affect the 3ω component of the nonlinear polarization in the cubic response approximation, due to Eq. (37) and because $f_1^{(3,0)}(t)$ is the sole term in the nonlinear polarization containing 3ω terms. Equations (40)–(44) then yield

$$f_1^{(3,0)}(t) = \frac{\xi_0^3}{60\left(1+\omega^2\tau_D^2\right)\left(9+4\omega^2\tau_D^2\right)} \left\{ \frac{\left(13\omega^2\tau_D^2 - 27\right)\cos\omega t - 2\omega\tau_D\left(21+\omega^2\tau_D^2\right)\sin\omega t}{3\left(1+\omega^2\tau_D^2\right)} \right.$$
$$\left. + \frac{\left(17\omega^2\tau_D^2 - 3\right)\cos 3\omega t + 2\omega\tau_D\left(3\omega^2\tau_D^2 - 7\right)\sin 3\omega t}{\left(1+9\omega^2\tau_D^2\right)} \right\}, \tag{45}$$

i.e., the original Coffey–Paranjape result, and along with this the additional terms

$$f_1^{(2,1)}(t) = -\frac{\xi_0^2\xi_S}{90} \frac{\left(27+7\omega^2\tau_D^2\right)}{\left(1+\omega^2\tau_D^2\right)\left(9+\omega^2\tau_D^2\right)}$$
$$+ \frac{\xi_0^2\xi_S}{30} \frac{\left(8\omega^6\tau_D^6 + 62\omega^4\tau_D^4 + 153\omega^2\tau_D^2 - 81\right)\cos 2\omega t}{\left(1+\omega^2\tau_D^2\right)\left(9+\omega^2\tau_D^2\right)\left(1+4\omega^2\tau_D^2\right)\left(9+4\omega^2\tau_D^2\right)}$$
$$- \frac{2\xi_0^2\xi_S}{15} \frac{\omega\tau_D\left(4\omega^4\tau_D^4 + 22\omega^2\tau_D^2 + 63\right)\sin 2\omega t}{\left(1+\omega^2\tau_D^2\right)\left(9+\omega^2\tau_D^2\right)\left(1+4\omega^2\tau_D^2\right)\left(9+4\omega^2\tau_D^2\right)}, \tag{46}$$

$$f_1^{(1,2)}(t) = -\frac{\xi_0\xi_S^2}{45} \frac{\left(27+\omega^2\tau_D^2 - 2\omega^4\tau_D^4\right)\cos\omega t + \omega\tau_D\left(42+19\omega^2\tau_D^2+\omega^4\tau_D^4\right)\sin\omega t}{\left(1+\omega^2\tau_D^2\right)^2\left(9+\omega^2\tau_D^2\right)}. \tag{47}$$

Furthermore, for $\omega = 0$, we have

$$P^{(3)}(t) = -\frac{\rho_0\mu\left(\xi_0+\xi_S\right)^3}{45}. \tag{48}$$

This is simply the second term of the Taylor expansion for the Langevin function with two superimposed DC fields.

Now, Eqs. (45)–(47) should be commented upon. First, on inspection of Eq. (46), application of a DC field in addition to an AC one causes a 2ω harmonic term to appear on the resulting DC response. This nonlinear frequency-dependent effect is completely different, however, from that due to the dynamic Kerr effect because there, an AC field alone is required to create a frequency-dependent DC term (the square law nonlinearity rectifies the applied field). Nevertheless, the *qualitative* frequency behavior is the same for both phenomena. Second, on inspection of Eq. (45), an extra term oscillating at the fundamental is superimposed on the pure AC response. Third, these formulas pertain with obvious changes in notation to magnetic relaxation of blocked ferrofluids [3]. Finally, we see that the DC field does not affect the third harmonic term at all *at this level of approximation*. However, if the *pentic* response is considered, the fundamental and the third harmonic will also be affected. In contrast to the original Coffey–Paranjape formulas, these results have been obtained only recently [25].

4 Account of Interactions via a Mean Field Potential

We shall now treat intermolecular interactions via a mean field static potential. The basic idea has been alluded to by Fröhlich [26], and relies on the *Ansatz* that intermolecular interactions may be represented by a (mean field) symmetric double-well potential. Thus, a way of including them in the dynamical Eq. (1) is to choose a potential exhibiting two wells in a cycle of the motion. Then, guided by the work of Maier and Saupe [17] at equilibrium and Martin, Meier and Saupe [17] for time-dependent situations in nematic liquid crystals, we merely rewrite $V(\vartheta, t)$ as

$$V(\vartheta, t) = K \sin^2 \vartheta - \mu (E_S + E(t)) \cos \vartheta, \tag{49}$$

where K represents an intermolecular interaction strength. We remark that disparate physical problems can be modeled using Eqs. (1) and (49), e.g., the nonlinear relaxation of (noninteracting) magnetic nanoparticles, with application to magnetic hyperthermia and information storage, or equally well the dielectric relaxation of polar nematic liquid crystals. Numerical and analytical calculations have been undertaken recently [27], which we now summarize.

The electric polarization is still given by Eq. (3), however, the differential-recurrence relations become [3]

$$\begin{aligned} \frac{2\tau_D}{n(n+1)} \dot{f}_n(t) + \left[1 - \frac{2\sigma}{(2n-1)(2n+3)}\right] f_n(t) &= \frac{\xi_S + \xi(t)}{2n+1} \left[f_{n-1}(t) - f_{n+1}(t)\right] \\ &+ 2\sigma \left[\frac{(n-1)}{(2n-1)(2n+1)} f_{n-2}(t) - \frac{(n+2)}{(2n+1)(2n+3)} f_{n+2}(t)\right], \end{aligned} \tag{50}$$

where $\sigma = \beta K$. Again specializing to a pure AC field, we seek the solution of Eq. (50) as a perturbation series in the AC field amplitude (cf. Eq. 7), so yielding the perturbed equations

$$\begin{aligned}\frac{2\tau_D}{n(n+1)}\dot{f}_n^{(k)}(t) + f_n^{(k)}(t) &= \frac{\xi_S}{2n+1}\left[f_{n-1}^{(k)}(t) - f_{n+1}^{(k)}(t)\right]\\ &+\frac{\xi(t)}{2n+1}\left[f_{n-1}^{(k-1)}(t) - f_{n+1}^{(k-1)}(t)\right]\\ +2\sigma\left[\frac{n-1}{(2n-1)(2n+1)}f_{n-2}^{(k)}(t)\right. &- \left.\frac{n+2}{(2n+1)(2n+3)}f_{n+2}^{(k)}(t)\right],\quad k>0,\end{aligned} \tag{51}$$

with the stationary values

$$f_n^{(0)} = Z^{-1}\int_0^{\pi} P_n(\cos\vartheta)\, e^{-\sigma\sin^2\vartheta+\xi_S\cos\vartheta}\sin\vartheta d\vartheta, \tag{52}$$

and the partition function

$$Z = \int_0^{\pi} \mathrm{e}^{-\sigma\sin^2\vartheta+\xi_S\cos\vartheta}\sin\vartheta d\vartheta. \tag{53}$$

However, it is no longer possible to solve the hierarchy of Eq. (51) by simple straightforward iteration, because of the mathematical complexity caused by the coupling between the seven kinds of terms involved. Nevertheless, as demonstrated in [27], we may formally solve these equations by writing them in matrix form. Thus, we introduce the column vectors in which the n dependence is subsumed,

$$\mathbf{c}^{(0)} = \begin{pmatrix} f_1^{(0)}\\ f_2^{(0)}\\ \vdots\\ f_n^{(0)}\\ \vdots\end{pmatrix},\ \mathbf{c}^{(k)}(t) = \begin{pmatrix} f_1^{(k)}(t)\\ f_2^{(k)}(t)\\ \vdots\\ f_n^{(k)}(t)\\ \vdots\end{pmatrix},\ k>0. \tag{54}$$

Hence, Eq. (51) become the forced matrix differential equations

$$\tau_D\dot{\mathbf{c}}^{(1)}(t) + \mathbf{A}\mathbf{c}^{(1)}(t) = \xi(t)\,\mathbf{c}_1, \tag{55}$$

$$\tau_D\dot{\mathbf{c}}^{(k)}(t) + \mathbf{A}\mathbf{c}^{(k)}(t) = \xi(t)\,\mathbf{B}\mathbf{c}^{(k-1)}(t), \tag{56}$$

where

$$\mathbf{c}_1 = \frac{1}{3}\begin{pmatrix} 1 \\ 0 \\ \vdots \\ 0 \\ \vdots \end{pmatrix} + \mathbf{B}\mathbf{c}^{(0)}. \tag{57}$$

The matrix elements of the time-independent system matrix **A** and driving force matrix **B** in Eqs. (55)–(57) are thus given by

$$(\mathbf{A})_{nm} = \frac{n\,(n+1)}{2}\left(-\frac{2\sigma\,(n-1)}{(2n-1)\,(2n+1)}\delta_{nm+2} + \left[1 - \frac{2\sigma}{(2n-1)\,(2n+3)}\right]\delta_{nm} \right.$$
$$\left. + \frac{\xi_S}{2n+1}\left(\delta_{nm-1} - \delta_{nm+1}\right) + \frac{2\sigma\,(n+2)}{(2n+1)\,(2n+3)}\delta_{nm-2}\right), \tag{58}$$

$$(\mathbf{B})_{nm} = \frac{n\,(n+1)}{2\,(2n+1)}\left(\delta_{nm+1} - \delta_{nm-1}\right) \tag{59}$$

(δ_{nm} is Kronecker's delta). The solution of the matrix Eqs. (55) and (56) is obtained by quadratures. We have, as in the scalar case, with $\mathbf{c}^{(k)}\,(-\infty) = \mathbf{0},\;\; k > 0$,

$$\mathbf{c}^{(1)}\,(t) = \frac{1}{\tau_D}\int_{-\infty}^{t} \xi\left(t'\right) \mathrm{e}^{-\mathbf{A}\frac{t-t'}{\tau_D}}\,\mathbf{c}_1 \mathrm{d}t', \tag{60}$$

while

$$\mathbf{c}^{(k)}\,(t) = \frac{1}{\tau_D}\int_{-\infty}^{t} \xi\left(t'\right) \mathrm{e}^{-\mathbf{A}\frac{t-t'}{\tau_D}}\,\mathbf{B}\mathbf{c}^{(k-1)}\left(t'\right) \mathrm{d}t',\; k > 1. \tag{61}$$

Next, iterating Eq. (61) twice yields vector-valued time-ordered integral representations of the vector quadratic and cubic responses analogous to the scalar case. We have

$$\mathbf{c}^{(2)}\,(t) = \frac{1}{\tau_D^2}\int_{-\infty}^{t}\int_{-\infty}^{t'} \xi\left(t'\right)\xi\left(t''\right) \mathrm{e}^{-\mathbf{A}\frac{t-t'}{\tau_D}}\,\mathbf{B}\mathrm{e}^{-\mathbf{A}\frac{t'-t''}{\tau_D}}\,\mathbf{c}_1 \mathrm{d}t''\mathrm{d}t', \tag{62}$$

$$\mathbf{c}^{(3)}\,(t) = \frac{1}{\tau_D^3}\int_{-\infty}^{t}\int_{-\infty}^{t'}\int_{-\infty}^{t''} \xi\left(t'\right)\xi\left(t''\right)\xi\left(t'''\right) \mathrm{e}^{-\mathbf{A}\frac{t-t'}{\tau_D}}\,\mathbf{B}\mathrm{e}^{-\mathbf{A}\frac{t'-t''}{\tau_D}}\,\mathbf{B}\mathrm{e}^{-\mathbf{A}\frac{t''-t'''}{\tau_D}}\,\mathbf{c}_1 \mathrm{d}t'''\mathrm{d}t''\mathrm{d}t' \tag{63}$$

Furthermore, for $\xi(t) = \xi_0 \cos \omega t$, Eqs. (61), (62), and (63) can be written in forms suitable for numerical computational purposes [27]. We have from Eqs. (61)–(63):

$$\mathbf{c}^{(1)}(t) = \xi_0 \mathrm{Re}\left[\boldsymbol{\varphi}_1^{(1)}(\omega)\, e^{i\omega t}\right], \tag{64}$$

$$\mathbf{c}^{(2)}(t) = \frac{\xi_0^2}{2} \mathrm{Re}\left[\boldsymbol{\varphi}_0^{(2)}(\omega) + \boldsymbol{\Phi}_2^{(2)}(2\omega)\, \boldsymbol{\varphi}_0^{(2)}(\omega)\, e^{2i\omega t}\right], \tag{65}$$

$$\begin{aligned}\mathbf{c}^{(3)}(t) = \frac{\xi_0^3}{4} \mathrm{Re}\Big\{ &\Big(2\mathrm{Re}\left[\boldsymbol{\Phi}_1^{(3)}(\omega)\right] \boldsymbol{\varphi}_0^{(2)}(\omega) + \boldsymbol{\Phi}_1^{(3)}(\omega)\, \boldsymbol{\Phi}_2^{(2)}(2\omega)\, \boldsymbol{\varphi}_0^{(2)}(\omega)\Big) e^{i\omega t} \\ &+ \boldsymbol{\Phi}_1^{(3)}(3\omega)\, \boldsymbol{\Phi}_2^{(2)}(2\omega)\, \boldsymbol{\varphi}_0^{(2)}(\omega)\, e^{3i\omega t}\Big\},\end{aligned} \tag{66}$$

where

$$\boldsymbol{\varphi}_1^{(1)}(\omega) = \mathbf{G}(\omega)\, \mathbf{c}_1, \; \boldsymbol{\varphi}_0^{(2)}(\omega) = \mathbf{A}^{-1}\mathbf{B}\, \boldsymbol{\varphi}_1^{(1)}(\omega), \tag{67}$$

$$\boldsymbol{\Phi}_2^{(2)}(\omega) = \mathbf{G}(\omega)\, \mathbf{A}, \; \boldsymbol{\Phi}_1^{(3)}(\omega) = \mathbf{G}(\omega)\, \mathbf{B}, \tag{68}$$

$$\mathbf{G}(\omega) = (\mathbf{A} + i\omega\tau_D \mathbf{I})^{-1}, \tag{69}$$

and $\mathbf{I}$ is the identity matrix. In writing Eqs. (64)–(69), we have supposed that the transition matrix $\exp(\mathbf{A}t)$ satisfies the condition

$$\lim_{t \to -\infty} \mathrm{e}^{\mathbf{A}t} = \mathbf{0}, \tag{70}$$

because *all the eigenvalues of the system matrix* $\mathbf{A}$ *are real and positive due to the properties of the Smoluchowski operator* [28]. Furthermore, the vectors $\boldsymbol{\varphi}_1^{(1)}(\omega)$ and $\boldsymbol{\varphi}_0^{(2)}(\omega)$ in Eq. (67) can also be written as linear and second-order nonlinear generalized normalized susceptibilities $X_{n1}(\omega)$ and $X_{n0}^{(2)}(\omega)$, viz.,

$$\boldsymbol{\varphi}_1^{(1)}(\omega) = \begin{pmatrix} \chi_{11} X_{11}(\omega) \\ \chi_{21} X_{21}(\omega) \\ \chi_{31} X_{31}(\omega) \\ \vdots \end{pmatrix}, \; \boldsymbol{\varphi}_0^{(2)}(\omega) = \begin{pmatrix} \chi_{12} X_{10}^{(2)}(\omega) \\ \chi_{22} X_{20}^{(2)}(\omega) \\ \chi_{32} X_{30}^{(2)}(\omega) \\ \vdots \end{pmatrix} \tag{71}$$

with $X_{n1}(0) = 1$ and $X_{n0}^{(2)}(0) = 1$, while $\chi_{n1} = \left[\boldsymbol{\varphi}_1^{(1)}(0)\right]_n$ and $\chi_{n2} = \left[\boldsymbol{\Phi}_1^{(3)}(0)\, \boldsymbol{\varphi}_0^{(2)}(0)\right]_n$ are the corresponding static susceptibilities.

Although the foregoing matrix solutions facilitate numerical evaluation of nonlinear responses, they do not permit a qualitative understanding of the relaxation dynamics. These can be qualitatively understood however via the so-called two-mode approximation, originating in the *large* separation of the timescales of the fast intra-well and slow over-barrier (or inter-well) relaxation modes in the asymmetric double-well mean field potential (i.e., Eq. 49) with $E(t) = 0$). Here, we simply write down these two-mode approximations for the first and second-order responses and deduce from them the cubic one. Details can be found in [27].

It has become well established in the last two decades [3, 29] that the linear AC response to of dipolar systems undergoing (overdamped) rotational Brownian motion in a field of force essentially comprises two processes, namely

(a) A slow Arrhenius over-barrier relaxation process, with the same timescale for all linear response functions, which is represented here by the slowest decaying eigenvalue of the transition matrix $\exp(-\mathbf{A}t) = L^{-1}\left[(s\mathbf{I}+\mathbf{A})^{-1}\right]$, where L^{-1} denotes the inverse Laplace transform,
(b) A fast intra-well relaxation process which is not thermally activated and is near degenerate, with a characteristic timescale depending on the order of the linear response function considered.

Thus, we write for the general matrix elements of the linear response

$$f_n^{(1)}(t) = \xi_0 \chi_{n1} \mathrm{Re}\left[X_{n1}(\omega)\, \mathrm{e}^{i\omega t}\right], \tag{72}$$

where the scalar representations of χ_{n1} and $X_{n1}(\omega)$ are [3]

$$\chi_{n1} = \langle P_n P_1 \rangle_0 - \langle P_n \rangle_0 \langle P_1 \rangle_0, \tag{73}$$

$$X_{n1}(\omega) = \frac{\Delta_{n1}}{1 + i\omega/\lambda_1} + \frac{1 - \Delta_{n1}}{1 + i\omega\tau_W^{(n1)}}. \tag{74}$$

Here, Δ_{n1} is the weight of the thermally activated process specific to the function $f_n^{(1)}(t)$, and $\tau_W^{(n1)}$ is the timescale of the short time near degenerate intra-well processes also specific to the function $f_n^{(1)}(t)$. These parameters are defined by [3]

$$\Delta_{n1} = \frac{\tau_{n1}/\tau_{\mathrm{eff}}^{(n1)} - 1}{\lambda_1 \tau_{n1} - 2 + \left(\lambda_1 \tau_{\mathrm{eff}}^{(n1)}\right)^{-1}}, \tag{75}$$

$$\tau_W^{(n1)} = \tau_{\mathrm{eff}}^{(n1)} \frac{\lambda_1 \tau_{n1} - 1}{\lambda_1 \tau_{\mathrm{eff}}^{(n1)} - 1}, \tag{76}$$

where in terms of low and high frequency limits of the generalized linear susceptibility

$$\tau_{n1} = \lim_{\omega \to 0} (\omega \chi_{n1})^{-1} \, \mathrm{Im}\left(\left[\varphi_1^{(1)}(\omega)\right]_n\right), \tag{77}$$

$$\tau_{\mathrm{eff}}^{(n1)} = \lim_{\omega \to \infty} \frac{\chi_{n1}}{\omega} \mathrm{Im}\left(\left[\varphi_1^{(1)}(\omega)\right]_n\right)^{-1} \tag{78}$$

The quadratic response functions $f_n^{(2)}(t)$ can also be represented in similar scalar fashion using the two-mode approximation. For example, we cite $f_1^{(2)}(t)$ for the polarization of the quadratic nonlinear response that is given by

$$f_1^{(2)}(t) = \frac{\xi_0^2}{2}\chi_{12}\mathrm{Re}\left[X_{10}^{(2)}(\omega) + X_{12}(2\omega)\, X_{10}^{(2)}(\omega)\, \mathrm{e}^{2i\omega t}\right], \tag{79}$$

where

$$X_{10}^{(2)}(\omega) = \frac{\Delta_{10}}{1+i\omega/\lambda_1} + \frac{1-\Delta_{10}}{1+i\omega\tau_W^{(10)}}, \tag{80}$$

$$X_{10}^{\prime(2)}(\omega) = \frac{\Delta_{10}^{\prime}}{1+i\omega/\lambda_1} + \frac{1-\Delta_{10}^{\prime}}{1+i\omega\tau_W^{(10)}} \tag{81}$$

$$X_{12}(\omega) = \frac{\Delta_{12}}{1+i\omega\tau_{12}} + \frac{1-\Delta_{12}}{1+i\omega\tau_W^{(12)}} \tag{82}$$

The parameters Δ_{10} and $\tau_W^{(10)}$ can be evaluated via

$$\Delta_{10} = \frac{\tau_{10}/\tau_{10}^{\mathrm{eff}} - 1}{\lambda_1\tau_{10} - 2 + \left(\lambda_1\tau_{10}^{\mathrm{eff}}\right)^{-1}}, \tag{83}$$

$$\tau_W^{(10)} = \tau_{\mathrm{eff}}^{(10)}\frac{\lambda_1\tau_{10} - 1}{\lambda_1\tau_{\mathrm{eff}}^{(10)} - 1}, \tag{84}$$

where the characteristic times τ_{10} and $\tau_{\mathrm{eff}}^{(10)}$ are determined by

$$\tau_{10} = \lim_{\omega\to 0}(\omega\chi_{12})^{-1}\,\mathrm{Im}\left(\left[\varphi_0^{(2)}(\omega)\right]_1\right), \tag{85}$$

$$\tau_{\mathrm{eff}}^{(10)} = \lim_{\omega\to\infty}\frac{\chi_{12}}{\omega}\mathrm{Im}\left(\left[\varphi_0^{(2)}(\omega)\right]_1\right)^{-1} \tag{86}$$

However, unlike Eqs. (75) and (76), analytic equations for the parameters $\Delta_{10}^{\prime}$, $\tau_W^{\prime(10)}$, Δ_{12}, τ_{12} and $\tau_W^{(12)}$ are unknown. Therefore, in Eqs. (81) and (82), they are treated as adjustable. In this way, the cubic polarization response $f_1^{(3)}(t)$ can be rewritten as

$$f_1^{(3)}(t) = \frac{\xi_0^3}{4}\chi_{13}\mathrm{Re}\left\{\left(2\mathrm{Re}\left[X_{13}(\omega)\right]X_{10}^{\prime(2)}(\omega) + X_{13}(\omega)\,X_{12}(\omega)\,X_{10}^{\prime(2)}(\omega)\right)\mathrm{e}^{i\omega t} + X_{13}(3\omega)\,X_{12}(2\omega)\,X_{10}^{\prime(2)}(\omega)\,\mathrm{e}^{3i\omega t}\right\} \tag{87}$$

with $X_{10}^{\prime(2)}(\omega)$ given by Eq. (81) and

$$X_{13}(\omega) = \frac{\Delta_{13}}{1+i\omega/\lambda_1} + \frac{1-\Delta_{13}}{1+i\omega\tau_W^{(13)}}, \tag{88}$$

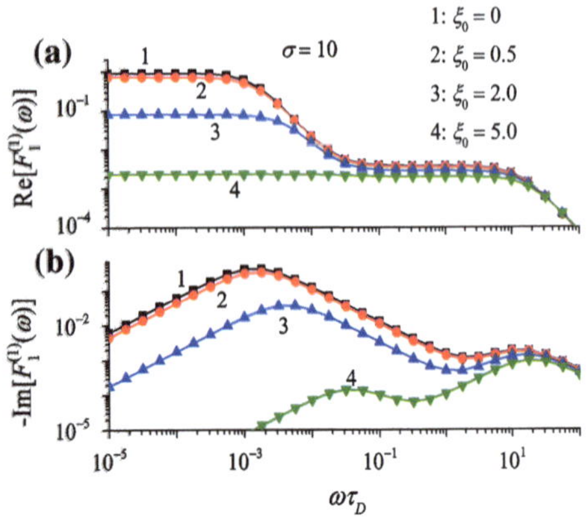

Fig. 1 (Color on line) Real (**a**) and imaginary (**b**) parts of the linear susceptibility $F_1^{(1)}(\omega) = \chi_{11} X_{11}(\omega)$ versus the normalized frequency $\omega\tau_D$ for various DC field amplitudes ξ_0 with anisotropy parameter $\sigma = 10$. Solid lines: the matrix solution. Symbols: the two-mode approximation

where Δ_{13} and $\tau_W^{(13)}$ are again adjustable parameters.

As indicated by Figs. 1 and 2, the two-mode approximation formulas yield excellent agreement with the exact numerical solution obtained via various matrix methods. The introduction of a distribution of relaxation times in the above calculations is discussed in [27]. The treatment as outlined may be used to any order in perturbation theory in the field strength. Thus, it may directly be applied both to nonlinear dielectric relaxation of polar nematic liquid crystals and to nonlinear magnetic relaxation of noninteracting single-domain ferromagnetic particles, and indeed to all polar systems where the interaction field is static. However, if this field is replaced by a *mean* field accounting for the *dynamics* of the dipole, then pronounced new features appear which are revealed *in the nonlinear response only*. We now review these.

5 Dynamical Mean Field Effects in the Nonlinear Dielectric Response

Here, dynamical effects due to interactions are accounted for in first approximation, as inspired by Berne [30]. He, by solving the Poisson equation with natural boundary conditions, demonstrated that the collective tumbling of an assembly of interacting dipoles was described by a nonlinear Fokker–Planck equation, where the orientational pair distribution function is systematically unity. Consequently, the potential has dynamical features related to the time-dependent orientational probability density. The Fokker–Planck (Smoluchowski) Eq. (1) is formally unchanged, however V is now replaced by

$$V(\vartheta, t) = U_s(\vartheta, t) + V_{\text{int}}(\vartheta, t), \tag{89}$$

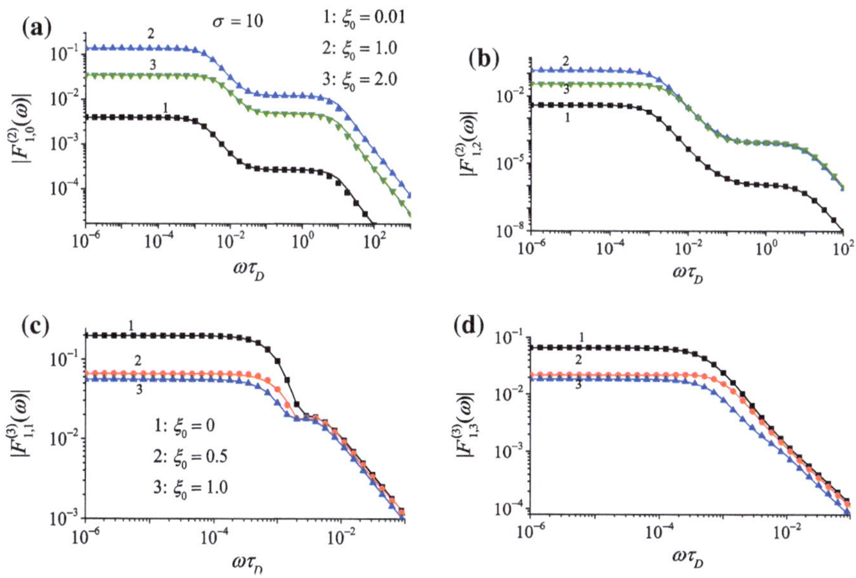

Fig. 2 (Color on line) Modulus of **a** the DC component of the nonlinear dielectric response $\left|F_{1,0}^{(2)}(\omega)\right|$, **b** the second harmonic component of the nonlinear dielectric response $\left|F_{1,2}^{(2)}(\omega)\right|$, **c** the fundamental component of that response $\left|F_{1,1}^{(3)}(\omega)\right|$, and **d** the third harmonic component $\left|F_{1,3}^{(3)}(\omega)\right|$ versus $\omega\tau_D$ for various DC field amplitudes ξ_0 with $\sigma = 10$. Solid lines: matrix solution. Symbols: two-mode approximation using the fitting parameters as described in the text

where U_s in Eq. (89) as usual contains the orientational terms due to the externally applied fields, while in the dipolar approximation [22], the interaction field is represented by

$$V_{\text{int}}(\vartheta, t) = \frac{4\pi\rho_0\mu^2}{3}\cos\vartheta f_1(t). \tag{90}$$

Under these conditions, we may write

$$\beta V(\vartheta, t) = -\xi(t)\cos\vartheta + \lambda\cos\vartheta f_1(t), \tag{91}$$

where

$$\lambda = 4\pi\beta\rho_0\mu^2/3 \tag{92}$$

is 4π times the linear Langevin susceptibility of an ideal gas of dipoles. Thus, using Eqs. (1) and (91), we have the intrinsically *nonlinear* differential-recurrence relations as opposed to the linear result Eq. (5)

$$\frac{2\tau_D}{n(n+1)}\dot{f}_n(t) + f_n(t) = \frac{1}{2n+1}\left[\xi(t) - \lambda f_1(t)\right]\left[f_{n-1}(t) - f_{n+1}(t)\right]. \tag{93}$$

The perturbation expansion (7) to cubic order in the field strength then yields the scheme

$$f_n^{(0)} = \frac{\lambda f_1^{(0)}}{2n+1}\left[f_{n+1}^{(0)} - f_{n-1}^{(0)}\right], \tag{94}$$

$$\frac{2\tau_D \dot{f}_n^{(1)}(t)}{n(n+1)} + f_n^{(1)}(t) = \frac{\xi(t) - \lambda f_1^{(1)}(t)}{2n+1}\left[f_{n-1}^{(0)} - f_{n+1}^{(0)}\right] + \frac{\lambda f_1^{(0)}}{2n+1}\left[f_{n+1}^{(1)}(t) - f_{n-1}^{(1)}(t)\right], \tag{95}$$

$$\begin{aligned}\frac{2\tau_D \dot{f}_n^{(2)}(t)}{n(n+1)} + f_n^{(2)}(t) &= \frac{\xi(t) - \lambda f_1^{(1)}(t)}{2n+1}\left[f_{n-1}^{(1)}(t) - f_{n+1}^{(1)}(t)\right] + \frac{\lambda f_1^{(2)}(t)}{2n+1}\left[f_{n+1}^{(0)} - f_{n-1}^{(0)}\right] \\ &+ \frac{\lambda f_1^{(0)}}{2n+1}\left[f_{n+1}^{(2)}(t) - f_{n-1}^{(2)}(t)\right],\end{aligned} \tag{96}$$

and

$$\begin{aligned}\frac{2\tau_D \dot{f}_n^{(3)}(t)}{n(n+1)} + f_n^{(3)}(t) &= \frac{\xi(t) - \lambda f_1^{(1)}(t)}{2n+1}\left[f_{n-1}^{(2)}(t) - f_{n+1}^{(2)}(t)\right] + \frac{\lambda f_1^{(3)}(t)}{2n+1}\left[f_{n+1}^{(0)} - f_{n-1}^{(0)}\right] \\ &+ \frac{\lambda f_1^{(2)}(t)}{2n+1}\left[f_{n+1}^{(1)}(t) - f_{n-1}^{(1)}(t)\right] + \frac{\lambda f_1^{(0)}}{2n+1}\left[f_{n+1}^{(3)}(t) - f_{n-1}^{(3)}(t)\right].\end{aligned} \tag{97}$$

We desire $f_1^{(1)}(t)$ and $f_1^{(3)}(t)$. Thus [22] we have, specializing to a pure AC field

$$f_1^{(1)}(t) = \xi_0\left[\alpha_1^{\prime(1)}(\omega)\cos\omega t + \alpha_1^{\prime\prime(1)}(\omega)\sin\omega t\right] \tag{98}$$

with

$$\alpha_1^{\prime(1)}(\omega) = \frac{1}{(3+\lambda)(1+\omega^2\tau_1^2)}, \tag{99}$$

$$\alpha_1^{\prime\prime(1)}(\omega) = \frac{\omega\tau_1}{(3+\lambda)(1+\omega^2\tau_1^2)}, \tag{100}$$

and

$$\tau_1 = \frac{3\tau_D}{3+\lambda}. \tag{101}$$

Clearly, the linear response to the AC stimulus in the dynamical mean field picture is still essentially of Debye type. In contrast, however, the nonlinear response $f_1^{(3)}(t)$ is now given by

$$f_1^{(3)}(t) = \xi_0^3\left[\alpha_3^{\prime(1)}(\omega)\cos\omega t + \alpha_3^{\prime\prime(1)}(\omega)\sin\omega t + \alpha_3^{\prime(3)}(\omega)\cos 3\omega t + \alpha_3^{\prime\prime(3)}(\omega)\sin 3\omega t\right], \tag{102}$$

where

$$\alpha_3^{\prime(1)}(\omega) = -\frac{3}{20}\,\frac{(3+\lambda)^2(4\lambda-39)\,\omega^4\tau_1^4 + (378+522\lambda+51\lambda^2)\omega^2\tau_1^2 + 729}{(3+\lambda)^4\,[81+4(3+\lambda)^2\,\omega^2\tau_1^2](1+\omega^2\tau_1^2)^3}, \tag{103}$$

$$\alpha_3^{\prime\prime(1)}(\omega) = \frac{\omega\tau_1\left\{81(2\lambda-21) + 3(3+\lambda)\left[(\lambda(\lambda-45)-198) - (3+\lambda)^2\omega^2\tau_1^2\right]\omega^2\tau_1^2\right\}}{10(3+\lambda)^4\left(81+4(3+\lambda)^4\,\omega^2\tau_1^2\right)\left(1+\omega^2\tau_1^2\right)^3}, \tag{104}$$

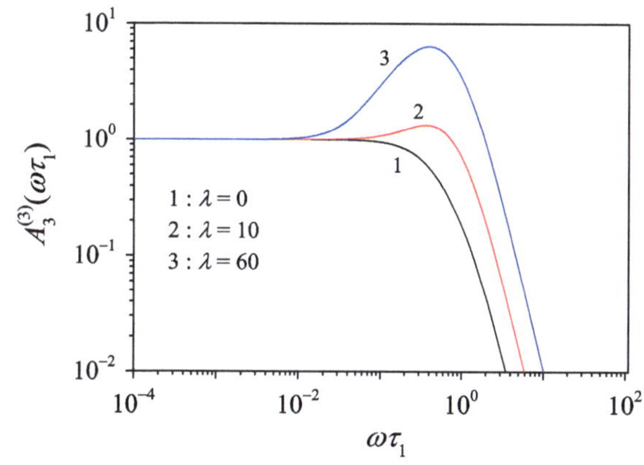

Fig. 3 Normalized modulus of the normalized nonlinear response 3ω component $A_3^{(3)}(\omega\tau_1) = \frac{1}{\alpha_3'^{(3)}(0)}\sqrt{\left(\alpha_3'^{(3)}(\omega\tau_1)\right)^2 + \left(\alpha_3''^{(3)}(\omega\tau_1)\right)^2}$ as a function of $\omega\tau_1$ for various λ. Note that $\alpha_3'^{(3)}(0) \propto (3+\lambda)^{-4}$, i.e., decreases upon cooling. The value $\lambda = 0$ is the Coffey–Paranjape result

$$\alpha_3'^{(3)}(\omega) = \frac{9\left[(3+\lambda)^2(51+20\lambda)\omega^6\tau_1^6 + 3(279 - 4\lambda(\lambda(6+\lambda) - 21))\omega^4\tau_1^4 - 3(\lambda(78+\lambda) - 99)\omega^2\tau_1^2 - 81\right]}{20(3+\lambda)^4\left[81 + 4(3+\lambda)^2\omega^2\tau_1^2\right]\left[1+\omega^2\tau_1^2\right]^3\left[1+9\omega^2\tau_1^2\right]} \tag{105}$$

$$\alpha_3''^{(3)}(\omega) = \frac{9\omega\tau_1\left\{9(2\lambda - 21) + \left(\lambda^3 - 243\lambda - 297\right)\omega^2\tau_1^2 - 3(3+\lambda)\left[(3+\lambda(19+4\lambda)) - (3+\lambda)^2\omega^2\tau_1^2\right]\omega^4\tau_1^4\right\}}{10(3+\lambda)^4\left[81 + 4(3+\lambda)^2\omega^2\tau_1^2\right]\left[1+\omega^2\tau_1^2\right]^3\left[1+9\omega^2\tau_1^2\right]} \tag{106}$$

For $\lambda = 0$, these formulas become the usual nonlinear response of noninteracting dipoles to alternating electric fields. However, cf. Figures 3 and 4, they strongly deviate from the *known results for large interactions*, thereby revealing pronounced dynamical effects due to intermolecular interactions, which must be investigated via nonlinear response measurements. In particular, the humped-back shape of the nonlinear response moduli found at large λ reveals the non-monotonic behavior of that response for interacting molecules.

This finding is in marked contrast to that of the previous section, where the modulus of the nonlinear response is *monotonic* for all interaction strengths. Now, recalling that the mean field approximation is a poor representation of long range intermolecular interactions, shorter interaction ranges could be modeled using the model under discussion by superimposing a $P_2(\cos\vartheta)\, f_2(t)$ term in the interaction potential (90). This Ansatz then leads to a Martin–Maier–Saupe-type model as pertains to dielectric relaxation of polar nematic liquid crystals. In particular, as the amplitude of the P_2 term is increased, the humped-back shape disappears, implying that Eqs. (103)–(106) can represent at best the "trivial" contribution to Ladieu's toy model of nonlinear dielectric relaxation of supercooled liquids [31]. This conclusion may be drawn because according to the present theory, on decreasing the temperature, the humped-back behavior of the nonlinear response spectrum vanishes, while experimental data on glycerol exhibit the *opposite* behavior [22, 32]. Finally, quantitative comparison

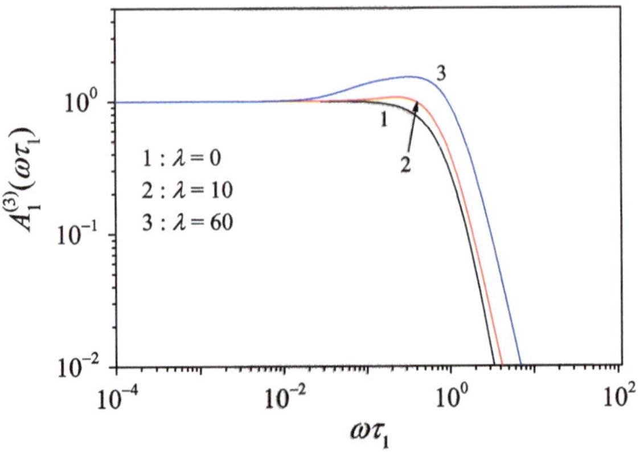

Fig. 4 Normalized modulus of the ω nonlinear polarizability component $A_3^{(1)}(\omega\tau_1) = \frac{1}{\alpha_3^{\prime(1)}(0)}\sqrt{\left(\alpha_3^{\prime(1)}(\omega\tau_1)\right)^2 + \left(\alpha_3^{\prime\prime(1)}(\omega\tau_1)\right)^2}$ as a function of $\omega\tau_1$ for various λ. Note that $\alpha_3^{\prime(1)}(0) \propto (3+\lambda)^{-4}$, i.e., decreases upon cooling. $\lambda = 0$ is the Coffey–Paranjape result

of Eqs. (103)–(106) with experiment implies that we are using them for $-3 < \lambda \leq 3$, meaning that they can now only pertain to *low densities*. This behavior agrees with that expected from the mean field (or random phase) approximation [33], although that occasionally predicts a large density effect. Consequently, the only feasible way to treat dynamical effects of the intermolecular interactions in the above calculation is to abandon the mean field approximation entirely so that intermolecular orientational correlations can be fully included. This task is much more involved and is beyond the scope of our review.

6 Depolarizing Field and Internal Field

First, we recall various electrostatic concepts, e.g., the depolarizing and internal fields, as they may be important in explaining experimental data. We start with the depolarizing field. As much as possible, we loosely follow Brown's excellent presentation of the subject [34].

(a) The depolarizing field.

Consider a capacitor polarized due to a constant voltage imposed between its electrodes in vacuo. Consequently, an electric field $\mathbf{E}_{vac}$ exists between the electrodes. However, insertion of a dielectric will cause a decrease in the voltage between the electrodes (with respect to the field measured in vacuo). The origin of the decrease is that polarization charges (of opposite signs) appear at the surfaces of the dielectric which interface with the electrodes in order to ensure global electro-neutrality of the overall structure (capacitor + dielectric). Consequently, one says that the dielectric is polarized, and the electric field inside the dielectric $\mathbf{E}_\mathrm{d}$ has an opposite direction to that of $\mathbf{E}_\mathrm{vac}$. Hence, the overall electric field between the plates is decreased, thereby explaining the voltage decrease due to the insertion of the dielectric, hence the name "depolarizing field" for $\mathbf{E}_\mathrm{d}$ [35]. Standard electrostatics shows that the depolarizing field $\mathbf{E}_\mathrm{d}$ is systematically proportional to the polarization vector of the substance, and points in an opposite direction to that of the polarization vector of the dielectric. Moreover, $\mathbf{E}_\mathrm{d}$ depends on the sample shape. Thus

$$\mathbf{E}_\mathrm{d} = -\overleftrightarrow{D}_p \cdot \mathbf{P}, \tag{107}$$

where $\overleftrightarrow{D}_p$ is the depolarization tensor. The total field *inside* the dielectric (called the Maxwell field) is then $\mathbf{E} = \mathbf{E}_\mathrm{vac} + \mathbf{E}_\mathrm{d}$, and the various electric susceptibilities as well as the linear and nonlinear permittivities in electromagnetic theory are defined *with respect to this field*. For example, in the linear case in the AC regime, we have

$$\mathbf{P}^{(1)}(\omega) = \overleftrightarrow{\boldsymbol{\chi}}^{(1)}(\omega) \cdot \mathbf{E}(\omega), \tag{108}$$

where $\overleftrightarrow{\boldsymbol{\chi}}^{(1)}(\omega)$ is the linear susceptibility Cartesian tensor. For a macroscopic spherical isotropic sample, this tensor becomes a scalar so that the linear polarization can be linked to the vacuum field $\mathbf{E}_\mathrm{vac}$, e.g., for a pure AC field in vacuo, we have

$$\mathbf{P}^{(1)}(\omega) = \frac{3}{4\pi}\left(\frac{4\pi\chi^{(1)}(\omega)}{3 + 4\pi\chi^{(1)}(\omega)}\right) \cdot \mathbf{E}_\mathrm{vac}. \tag{109}$$

Since it is believed that the linear susceptibility (and therefore the linear complex permittivity) of a sample is an intensive quantity, determining it under the assumption of a specific shape is relatively unimportant since calculations for two different shapes lead to the same expression [22, 36]. In particular, the expression for the linear complex permittivity is the same both for an infinite thin dielectric sheet and for a sphere, yielding [22, 36]

$$\varepsilon(\omega) - 1 = 4\pi\chi^{(1)}(\omega) = \frac{\lambda}{1 + i\omega\tau_D}, \tag{110}$$

where λ is given by Eq. (92). This equation yields in particular the Langevin–Debye equation for an assembly of purely polar molecules $\varepsilon(0) - 1 = \lambda$ in the static regime. Now, the nonlinear susceptibilities may depend on the sample shape; however, if a liquid completely fills the vacuum between the electrodes, such corrections are unnecessary, because then *the applied field coincides with the Maxwell field*. However, for strongly polar liquids, the Langevin–Debye equation $\varepsilon(0) - 1 = \lambda$ noticeably disagrees with experiment, if the actual value of the dipole moment is used. This is also true for any trivial modification of this equation. An explanation of this discrepancy was given by Lorentz. He conjectured that a typical molecule in a *dense* system does not experience the applied field as a consequence of the discrete nature of matter, instead it experiences that field plus the sum of the electric fields due to all the other molecules, thus automatically leading to the concept of *internal or local field*.

(b) *The internal field*.

We have seen that this field was introduced into the literature by Lorentz. Furthermore, it is a convenient tool in the *classical microscopic theory of dielectrics* [36] because at the molecular level, matter can no longer be regarded as continuous. Although, strictly speaking, the calculation of this field should be quantum-mechanical, nevertheless the translational and rotational motion of the molecules does not depart markedly from classical behavior. Thus, we will continue with the classical treatment, and again following Brown's discussion [34], excluding the motion of atoms or ions *within* the molecules.

By definition, the internal field $\mathbf{E}_\ell$ is the field at the position of a specific (tagged) molecule *due to all charges* except those attached to that molecule. Now, although writing a general expression for the polarization (i.e., a *macroscopic quantity*) in *microscopic* terms is relatively straightforward (i.e., the statistical average of the vector sum of all molecular dipoles times the number of molecules per unit volume or concentration), it is not at all simple to relate the internal field $\mathbf{E}_\ell$ to the Maxwell field $\mathbf{E}$, or even to its average value [34]. Only for solutions of polar molecules in nonpolar solvents is this distinction unimportant and then only in the simplest cases can such a relation be established. These have been considered in detail by Lorentz and later by Onsager.

(i) *Outline of the Lorentz method for* $\mathbf{E}_\ell$ *for polar dielectrics*

We proceed as follows [34]: we construct a macroscopic sphere of radius R (i.e., *large* with respect to intermolecular distances, but *small* with respect to the overall macroscopic size of our sample), with center taken as the location of a typical molecule where the local field is calculated. The local field can then be divided into two parts: that due to matter *outside* the sphere $\mathbf{E}_{\text{out}}$ and that due to matter *inside* it $\mathbf{E}_{\text{in}}$. Under quite general conditions, we have from electrostatics

$$\mathbf{E}_{\text{out}} = \mathbf{E} + \frac{4\pi\mathbf{P}}{3}. \tag{111}$$

The computation of $\mathbf{E}_{\text{in}}$ is more difficult, as it must account in some way for the spatial arrangement of the molecules near the (tagged) one at which $\mathbf{E}_\ell$ is calculated. Lorentz showed, assuming that the molecules are arranged at the sites of a simple cubic lattice, that

$$\mathbf{E}_{\text{in}} = \mathbf{0}. \tag{112}$$

Hence, if the molecules near our (tagged) one are also situated at the sites of such a lattice, Lorentz finds that

$$\mathbf{E}_\ell = \mathbf{E}_{\text{in}} + \mathbf{E}_{\text{out}} = \mathbf{E} + \frac{4\pi\mathbf{P}}{3}. \tag{113}$$

If the arrangement differs from the simple cubic one, then we have

$$\mathbf{E}_\ell = \mathbf{E} + \frac{4\pi\mathbf{P}}{3} + \mathbf{E}_{\text{in}}, \tag{114}$$

where $\mathbf{E}_{\text{in}}$ is unknown. For a dielectric consisting of purely polar molecules, Eq. (113) leads to the Debye–Lorentz equation of state, viz.

$$\frac{\varepsilon - 1}{\varepsilon + 2} = \frac{\lambda}{3}, \tag{115}$$

where $\varepsilon = \varepsilon(0)$. Since $\lambda \propto T^{-1}$, Eq. (115) predicts a transition from an *unpolarized* to a *spontaneous polarized* state at a finite temperature. For water in particular, this temperature coincides with room temperature, therefore giving rise to a fundamental criterion whereby Eqs. (113) and (115) must be rejected for polar dielectric liquids with large dielectric constants. Finally, Eq. (114) is not used in practice because of the extreme difficulty in evaluating $\mathbf{E}_{\text{in}}$.

(ii) *Outline of Onsager's method of calculating* $\mathbf{E}_\ell$ *for polar dielectrics.*

As already alluded to above, the Lorentz method of calculating ε for polar substances must be rejected. Now, Onsager [37] remarked (full details are available in [39]) that the effect of long range dipole–dipole interactions is not accounted for properly in Lorentz's computation of $\mathbf{E}_\ell$. In effect, he modified Lorentz's method to include the effects of the *surroundings of the tagged molecule of permanent dipole moment* $\boldsymbol{\mu}$ *on the local field at this molecule*. In order to calculate ε, he used a model originally proposed by Bell [38] for a spherical dipolar molecule *embedded in a dielectric*. This model is a rigid point dipole situated at the center of a *macroscopic* empty spherical cavity of radius a in a dielectric continuum with permittivity equal to the bulk permittivity ε. The radius of the cavity is determined from the close-packing condition

$$\frac{4\pi\rho_0 a^3}{3} = 1,$$

so that the volume of the cavity is that available to each molecule. Now the dipole $\boldsymbol{\mu}$ itself creates a dipolar field that *polarizes the surroundings*. The resulting polarization of the surroundings in turn *induces a uniform field in the cavity* which is called *the reaction field* $\mathbf{R}$. For a spherical cavity, the uniform field $\mathbf{R}$ has the same direction as the dipole moment in the cavity (if the cavity is not spherical, this is not so), thus, for purely polar molecules, Onsager can write $\mathbf{R} = f\boldsymbol{\mu}$. Furthermore, if a uniform electric field $\mathbf{E}$ is imposed on the dielectric by external sources, standard electrostatics [35] shows that the field in the *empty* cavity (i.e., with no dipole in it) is not equal to $\mathbf{E}$. This field is called the cavity field $\mathbf{G}$ and for a spherical cavity, is collinear with

$\mathbf{E}$, so that we have $\mathbf{G} = g\mathbf{E}$. Then Onsager writes the *overall* field in the *spherical* cavity in the presence of $\mathbf{E}$ and the tagged dipole $\boldsymbol{\mu}$ *due to the surroundings* as

$$\mathbf{E}_\ell = \mathbf{G} + \mathbf{R} = g\mathbf{E} + f\boldsymbol{\mu}, \tag{116}$$

The coefficients g and f can be calculated via electrostatics [35]. For a spherical cavity of radius a in an *infinite* dielectric, we have (details in [39, 47])

$$g = \frac{3\varepsilon}{2\varepsilon + 1}, \quad f = \frac{2(\varepsilon - 1)}{(2\varepsilon + 1)a^3}. \tag{117}$$

Now, at equilibrium, the reaction field gives rise to zero torque on the tagged dipole because the term $-\boldsymbol{\mu} \cdot \mathbf{R}$ contributes only a constant to its orientational potential energy. Then, by *equating* the macroscopic polarization from electrostatics with that obtained via statistical mechanics, Onsager finds (in the linear regime)

$$(\varepsilon - 1)\mathbf{E} = \lambda g\mathbf{E}. \tag{118}$$

Thus, by using Eq. (117), we have Onsager's equation describing dielectrics consisting of pure polar molecules, viz.,

$$\frac{(\varepsilon - 1)(2\varepsilon + 1)}{3\varepsilon} = \lambda. \tag{119}$$

For polar and *isotropically* polarizable molecules, Eq. (119) is only slightly modified. Using $\varepsilon_\infty = n^2$ (where n is the refractive index of the medium) and the Lorenz–Lorentz equation [39], Onsager finds

$$\frac{(\varepsilon - \varepsilon_\infty)(2\varepsilon + \varepsilon_\infty)}{\varepsilon(\varepsilon_\infty + 2)^2} = \frac{\lambda}{3}. \tag{120}$$

Equation (120) may be used to determine the permittivity of assemblies of pure polar molecules. However, for water at 25 °C, Eq. (120) yields $\varepsilon \approx 30$, while the experimental value is 78.5. Nevertheless, Onsager's method has the great advantage of removing the unphysical ferroelectric Curie point predicted by the Lorentz–Debye formula (115). The lack of full agreement with the experimental value of ε suggests that Eq. (120) should be improved.

(iii) *The Kirkwood–Fröhlich formula for the relative permittivity*

Onsager's Eq. (120) was generalized by Kirkwood [40] to a cavity containing a very large number of interacting molecules and he obtained in so doing a much more acceptable value for the relative permittivity of water. However, Fröhlich [39] presented a more systematic derivation valid for all assemblies of polar molecules, which may be summarized as follows. We regard our entire macroscopic specimen as a very large sphere of radius b placed in a uniform field and select from it a smaller yet still macroscopic sphere of radius a, such as $a << b$. The inner sphere is treated on a *microscopic* basis (i.e., via *statistical mechanics*), while the large surrounding shell is treated as a *continuous* dielectric medium, i.e., on a macroscopic basis. The entire system (inner sphere + surroundings) is assumed to obey the laws of classical statistical mechanics. Regarding the inner sphere, the total dipole moment of an ensemble of N charges is

$$\mathbf{M}\left(\mathbf{X}\right)=\sum_{i=1}^{N}\mathrm{e}_i\mathbf{r}_i, \tag{121}$$

where $\mathbf{r}_i$ is the displacement of charge number i and e_i is its charge. Now, an atom or molecule contains several elementary charges (s in total) and $\mathbf{X}$ is the ensemble of the associated displacements $\mathbf{r}_i$. Following Fröhlich, we term an atom or molecule of the inner sphere *a cell* and label such a cell j, and assume that each cell makes *the same contribution to the polarization* in the direction of the applied field $\mathbf{E}$ inside the shell. The dipole moment of the cell j is, applying Eq. (121) to the cell j

$$\mathbf{m}\left(\mathbf{x}_j\right)=\sum_{k=1}^{s}\mathrm{e}_{jk}\mathbf{r}_{jk} \tag{122}$$

with obvious notations. Then, the total dipole moment of the inner sphere comprising N cells is

$$\mathbf{M}\left(\mathbf{X}\right)=\sum_{j=1}^{N}\mathbf{m}\left(\mathbf{x}_j\right)=\sum_{j=1}^{N}\sum_{k=1}^{s}e_{jk}\mathbf{r}_{jk} \tag{123}$$

Now the mean total dipole moment of the inner sphere in the direction of $\mathbf{E}$ is

$$\langle\mathbf{M}\cdot\mathbf{e}\rangle=\frac{1}{Z}\int_0^{\pi}\int_N\left(\mathbf{M}\left(\mathbf{X}\right)\cdot\mathbf{e}\right)\mathrm{e}^{-\beta U(\mathbf{X},\mathbf{E})}\sin\vartheta\, d\mathbf{X}d\vartheta, \tag{124}$$

where $U\left(\mathbf{X},\mathbf{E}\right)$ is the potential energy of the system of molecules inside the inner sphere in the presence of $\mathbf{E}$, ϑ is the angle between $\mathbf{M}$ and $\mathbf{E}$, $\mathbf{e}$ is a unit vector in the direction of $\mathbf{E}$, and Z is the partition function defined by

$$Z=\int_0^{\pi}\int_N \mathrm{e}^{-\beta U(\mathbf{X},\mathbf{E})}\sin\vartheta\,\mathrm{d}\mathbf{X}\mathrm{d}\vartheta.$$

Hence, in the linear approximation in $\mathbf{E}$, we have (details in [39])

$$\langle\mathbf{M}\cdot\mathbf{e}\rangle=\frac{3\varepsilon}{2\varepsilon+1}\frac{\beta}{3}\left\langle M^2\right\rangle_0, \tag{125}$$

where $\langle M^2\rangle_0$ *is the mean square value of the dipole moment of the inner sphere in the absence of the electric field*. By equating the (macroscopic) polarization obtained via electrostatics with that given by Eq. (125), we then have

$$\frac{(\varepsilon-1)(2\varepsilon+1)}{3\varepsilon}=\frac{4\pi\beta}{3\upsilon}\langle M^2\rangle_0, \tag{126}$$

where υ is the volume of the dielectric. The last equation is a perfectly general result, and is *the equation of state for linear dielectrics*. Put succinctly, it yields ε in terms of the mean square fluctuations of the dipole moment of a *macroscopic* spherical specimen of the dielectric *embedded in a large volume of the same dielectric*. These fluctuations in the dipole moment are the *total fluctuations* from *all causes*, because in the dielectric several mechanisms of polarization may be operative [39].

Equation (126) may further be specialized to identify a specific mechanism, namely the contribution of the displacement (or distortional) polarization to the total dipole moment of a molecule, which is the sum of the permanent and the induced dipole moment. This postulate assumes that the contribution of this mechanism to the permittivity may be treated by separating the overall polarization into a systematic term essentially due to the permanent dipoles of the molecules and a term due to the elastic displacement of all charges. For simplicity, the latter mechanism is then treated on a continuous basis, assuming that *for this mechanism* only the inner sphere is filled with material having static relative permittivity ε_∞. The sum of the cavity and reaction fields yields

$$\mathbf{E}_\ell=\frac{3\varepsilon\mathbf{E}}{2\varepsilon+\varepsilon_\infty}+\frac{2(\varepsilon-\varepsilon_\infty)}{a^3(2\varepsilon+\varepsilon_\infty)}\mathbf{M}, \tag{127}$$

plus the field of the dipole **M**. Since both the dipole and the reaction fields contribute only a constant to the orientational potential energy, Eq. (126) becomes (details again in [39])

$$\frac{(\varepsilon-\varepsilon_\infty)(2\varepsilon+\varepsilon_\infty)}{3\varepsilon}=\frac{4\pi\beta\langle M^2\rangle_0}{3\upsilon}. \tag{128}$$

Now, the evaluation of the static permittivity from Eq. (128) requires the calculation of $\langle M^2\rangle_0$. On using Eq. (123) and confining ourselves to terms linear in **E**, we have

$$\langle M^2\rangle_0=\sum_{j=1}^{N}Z_0^{-1}\int_N \mathbf{m}(\mathbf{x}_j)\cdot\mathbf{M}(\mathbf{X})\,\mathrm{e}^{-\beta U(\mathbf{X},\mathbf{0})}\mathrm{d}\mathbf{X}, \tag{129}$$

where

$$Z_0=\int_N \mathrm{e}^{-\beta U(\mathbf{X},\mathbf{0})}\mathrm{d}\mathbf{X}$$

is the partition function in the absence of the external field. On introducing the notation

$$\mathrm{d}\mathbf{X}_j = \mathrm{d}\mathbf{x}_1 \ldots \mathrm{d}\mathbf{x}_{j-1}\mathrm{d}\mathbf{x}_{j+1} \ldots \mathrm{d}\mathbf{x}_N$$

so that

$$\mathrm{d}\mathbf{X} = \mathrm{d}\mathbf{X}_j \mathrm{d}\mathbf{x}_j,$$

we can rewrite Eq. (129) with some algebra as

$$\left\langle M^2 \right\rangle_0 = \sum_{j=1}^{N} \int_j \mathrm{d}\mathbf{x}_j \mathbf{m}\left(\mathbf{x}_j\right) \cdot \frac{\int_{N-1} \mathbf{M}\left(\mathbf{X}\right) \mathrm{e}^{-\beta U(\mathbf{X},\mathbf{0})} \mathrm{d}\mathbf{X}_j}{\int_{N-1} \mathrm{e}^{-\beta U(\mathbf{X},\mathbf{0})} \mathrm{d}\mathbf{X}_j} \frac{\int_{N-1} \mathrm{e}^{-\beta U(\mathbf{X},\mathbf{0})} \mathrm{d}\mathbf{X}_j}{\int_{N} \mathrm{e}^{-\beta U(\mathbf{X},\mathbf{0})} \mathrm{d}\mathbf{X}}.$$

Next, by introducing the probability $p\left(\mathbf{x}_j\right)$ of finding the jth cell with the set of displacements $\mathbf{x}_j$

$$p\left(\mathbf{x}_j\right) = \frac{\int_{N-1} \mathrm{e}^{-\beta U(\mathbf{X},\mathbf{0})} \mathrm{d}\mathbf{X}_j}{\int_{N} \mathrm{e}^{-\beta U(\mathbf{X},\mathbf{0})} \mathrm{d}\mathbf{X}}$$

and $\mathbf{m}^*\left(\mathbf{x}_j\right)$ the mean moment of the sphere given that its jth cell has a set of *fixed* displacements $\mathbf{x}_j$ so that

$$\mathbf{m}^*\left(\mathbf{x}_j\right) = \frac{\int_{N-1} \mathbf{M}\left(\mathbf{X}\right) \mathrm{e}^{-\beta U(\mathbf{X},\mathbf{0})} \mathrm{d}\mathbf{X}_j}{\int_{N-1} \mathrm{e}^{-\beta U(\mathbf{X},\mathbf{0})} \mathrm{d}\mathbf{X}_j},$$

one may express the mean square dipole fluctuations $\left\langle M^2 \right\rangle_0$ as *a sum of statistical averages* over the *jth cell only*. In other words, we have

$$\left\langle M^2 \right\rangle_0 = \sum_{j=1}^{N} \int_j \mathrm{d}\mathbf{x}_j \mathbf{m}\left(\mathbf{x}_j\right) \cdot \mathbf{m}^*\left(\mathbf{x}_j\right) p\left(\mathbf{x}_j\right) \mathrm{d}\mathbf{x}_j. \tag{130}$$

Next, it may be shown from electrostatics [41] that the dipole moment induced in a sphere by a dipole residing in a cavity in that sphere is independent of the size of the latter. This result is extremely important because ε is an intensive quantity (therefore independent of the size and shape of the dielectric, and the calculations are easiest for spherical shapes). This result is true even if the cavity is not concentric with the surrounding spherical shell, so that the precise location of the cavity in the dielectric is unimportant provided *it is taken as spherical*. Next, let $\mathbf{m}_s^*$ denote the dipole moment of a sphere surrounding the jth cell. Thus, if $\mathbf{m}_s^*$ can be obtained by treating the jth cell *as a point dipole in a spherical cavity surrounded by a continuous dielectric*, then

$$\mathbf{m}_s^* = \mathbf{m}^*,$$

where by definition $\mathbf{m}^*$ is the dipole moment of the *entire sphere*. Thus, we have

$$\mathbf{m}^* = \mathbf{m},$$

so that Eq. (128) reduces to Onsager's Eq. (120). Therefore, *we must assume that* $\mathbf{m}^* \neq \mathbf{m}$, since Onsager's equation *does not predict* ε *quantitatively*, meaning that it is impossible to treat the jth cell as a point dipole *surrounded by a continuous dielectric*. In other words, $\mathbf{m}^* \neq \mathbf{m}$ if and only if

- The shape of the jth cell differs from that of a sphere, a hypothesis that we do not make, otherwise the electrostatic part of the calculation becomes very difficult,
- The region surrounding the jth cell *cannot be treated on a macroscopic basis*, a hypothesis that we will maintain in our calculation of ε because then we can handle the surroundings of the jth cell by the methods of (classical) statistical mechanics.

Furthermore, an important consequence of all the electrostatic considerations made above is that $\mathbf{m}^*$ is independent of the position of the jth cell as long as this cell is so far removed from the bounding surface of the dielectric so that it *allows its interaction with the outside to be treated on a macroscopic basis*. Of course, for an infinite dielectric, this last condition is always true. Bearing in mind all the above hypotheses, we have

$$\left\langle M^2 \right\rangle_0 = N \left\langle \mathbf{m} \cdot \mathbf{m}^* \right\rangle_0,$$

since each cell contributes equally to the polarization. Consequently, Eq. (128) becomes

$$\frac{(\varepsilon - \varepsilon_\infty)(2\varepsilon + \varepsilon_\infty)}{3\varepsilon} = \frac{4\pi\beta\rho_0}{3} \left\langle \mathbf{m} \cdot \mathbf{m}^* \right\rangle_0. \tag{131}$$

In this equation, $\rho_0 = N/v$ is the *number of cells* per unit volume of the dielectric, and $\mathbf{m}$ and $\mathbf{m}^*$ now refer to nonelectronic displacements. Having derived Eq. (131), we can obtain the so-called Kirkwood–Fröhlich equation by first choosing the cell j in such a way that it contains only one dipolar molecule of dipole moment $\boldsymbol{\mu}$, meaning that the orientations of the dipoles are the only variables. We now define

$$\mathbf{m}^* = \boldsymbol{\mu}^*, \tag{132}$$

where $\boldsymbol{\mu}^*$ is the average dipole moment of the sphere when the tagged dipole $\boldsymbol{\mu}$ is held in a *fixed* orientation. Now, in a liquid, in the absence of an applied field, all dipolar directions are equivalent therefore

$$\langle \mathbf{m} \cdot \mathbf{m}^* \rangle_0 = \langle \boldsymbol{\mu} \cdot \boldsymbol{\mu}^* \rangle_0$$

and we must also have

$$\langle \boldsymbol{\mu} \cdot \boldsymbol{\mu}^* \rangle_0 = \boldsymbol{\mu} \cdot \boldsymbol{\mu}^*. \tag{133}$$

Finally, if the interactions with *nearest* neighbors only are considered, then $\boldsymbol{\mu}^*$ is the sum of the moment $\boldsymbol{\mu}$ of the tagged dipole held in a *fixed orientation* relative to its neighbors and the *average of the sum of the moments of its nearest neighbors*. Hence, if z represents the average number of nearest neighbors, we have

$$\boldsymbol{\mu} \cdot \boldsymbol{\mu}^* = \mu^2 \left(1 + z \langle \cos \gamma \rangle_{Av}\right) = \mu^2 g_K,$$

where γ is the angle between neighboring dipoles, and g_K is called the Kirkwood correlation factor, so that Eq. (128) becomes

$$\frac{(\varepsilon - \varepsilon_\infty)(2\varepsilon + \varepsilon_\infty)}{3\varepsilon} = \frac{4\pi \rho_0 \mu^2 g_K}{3kT}. \tag{134}$$

Furthermore, the value of the dipole moment to be used in Eq. (134) is the dipole moment of a molecule embedded in a medium of dielectric constant ε_∞. This moment is related to the vacuum moment by the equation [39]

$$\mu = \frac{\varepsilon_\infty + 2}{3} \mu_g, \tag{135}$$

where μ_g is the dipole moment of the molecule in vacuo. Hence, by combining Eqs. (134) and (135), we finally have the Kirkwood–Fröhlich equation, viz.,

$$\frac{(\varepsilon - \varepsilon_\infty)(2\varepsilon + \varepsilon_\infty)}{\varepsilon (\varepsilon_\infty + 2)^2} = \frac{\lambda g_K}{3}. \tag{136}$$

By accounting for nearest neighbor contributions as described above, Kirkwood obtained $\varepsilon = 67$ for water at 25 °C, a far more acceptable value. By including both nearest and next-nearest neighbors in the evaluation of g_K, Oster and Kirkwood [42] found $\varepsilon = 78.5$, in excellent agreement with experiment. Now, we describe the generalization of the Kirkwood–Fröhlich equation to the frequency-dependent (complex) permittivity in the linear approximation.

(iv) The dynamical equation for the linear complex permittivity

Any theory of the linear complex permittivity $\varepsilon(\omega)$ of polar fluids must include the effect of the local field at the dynamical level. This calculation is much more involved than its static counterpart, because the dynamics of the internal field are generally unknown [39] and in addition are a function of $\varepsilon(\omega)$, i.e., the property one is trying to calculate. However, we may proceed in a general sense by establishing a relation between the *time-dependent* dipole moment of the dielectric and the complex permittivity. The dipole moment induced in a dielectric body induced by a very small time-dependent external electric field is, at any time (assuming that both quasi-electrostatics and linear response obtain)

$$\mathbf{M}(t) = \int_0^t \mathbf{E}(t-x)\,\frac{\mathrm{d}a}{\mathrm{d}x}(x)\,\mathrm{d}x, \tag{137}$$

where $a(t)$ is the step response of the body and $\mathbf{E}(t) = \mathbf{0}$ for $t < 0$. Following Scaife [41], we introduce the *aftereffect function* $b(t)$ defined by

$$b(t) = \begin{cases} a(\infty) - a(t) & (t > 0) \\ 0 & (t < 0) \end{cases}, \tag{138}$$

so that the polarizability $\alpha(\omega)$ of the body is given by

$$\alpha(\omega) = \int_0^\infty \dot{a}(t)\,\mathrm{e}^{-i\omega t}\mathrm{d}t = -\int_0^\infty \dot{b}(t)\,\mathrm{e}^{-i\omega t}\mathrm{d}t. \tag{139}$$

We must now relate $a(t)$ (or $b(t)$) to the induced time-dependent dipole moment $\mathbf{M}(t)$. This is accomplished by via the fluctuation–dissipation theorem which we explain as follows. First, we remark that by applying the Kramers–Kronig relations to $\alpha(\omega) = \alpha'(\omega) - i\alpha''(\omega)$, viz.,

$$\alpha'(\omega) = \frac{2}{\pi}\int_0^\infty \frac{z\alpha''(z)}{z^2-\omega^2}\mathrm{d}z, \quad \alpha''(\omega) = -\frac{2}{\pi}\int_0^\infty \frac{\omega\alpha'(z)}{z^2-\omega^2}\mathrm{d}z,$$

we have, at zero frequency

$$\alpha'(0) = \frac{2}{\pi}\int_0^\infty \frac{\alpha''(\omega)}{\omega}\mathrm{d}\omega = \frac{\beta}{3}\left\langle M^2\right\rangle_0. \tag{140}$$

Now, denoting time averages by an overbar, we have by ergodicity

$$\left\langle M^2\right\rangle_0 = \overline{M^2} = \lim_{T'\to\infty}\frac{1}{T'}\int_{-T'/2}^{T'/2} \mathbf{M}(t)\cdot\mathbf{M}(t)\,\mathrm{d}t$$

However, by the Parseval–Plancherel theorem, we must also have

$$\left\langle M^2 \right\rangle_0 = \frac{1}{2\pi} \int\limits_{-\infty}^{\infty} \lim_{T' \to \infty} \frac{\left| \tilde{\mathbf{M}}(\omega) \cdot \tilde{\mathbf{M}}^*(-\omega) \right|}{T'} \mathrm{d}\omega = \frac{1}{2\pi} \int\limits_{-\infty}^{\infty} S_M(\omega)\, \mathrm{d}\omega, \tag{141}$$

where the star denotes the complex conjugate, the tilde denotes the Fourier transform of $\mathbf{M}(t)$, i.e.,

$$\tilde{\mathbf{M}}(\omega) = \int\limits_{-T'/2}^{T'/2} \mathbf{M}(t)\, \mathrm{e}^{-i\omega t} \mathrm{d}t,\ T' \to \infty,$$

and $S_M(\omega)$ is by definition the spectral density of the fluctuations of the dipole $\mathbf{M}$. Hence via Eqs. (140) and (141), we then have the fluctuation–dissipation theorem (FDT),

$$6\alpha''(\omega) = \beta\omega S_M(\omega). \tag{142}$$

Thus, we have related the *dissipative part* $\alpha''(\omega)$ *of the frequency-dependent complex polarizability* to the *spectral density of the spontaneous fluctuations in the dipole moment at equilibrium* of the dielectric body. In deriving the FDT, we have asserted that *macroscopic fluctuations decay according to macroscopic laws*.

Now, on introducing the autocorrelation function of the dipole $C_M(t)$ defined by

$$C_M(t) = \lim_{T' \to \infty} \frac{1}{T'} \int\limits_{-T'/2}^{T'/2} \mathbf{M}\left(t - t'\right) \cdot \mathbf{M}\left(t'\right) \mathrm{d}t' = \overline{\mathbf{M}(t - t') \cdot \mathbf{M}(t')}. \tag{143}$$

so that by the Wiener–Khintchine theorem [3], $C_M(t)$ and $S_M(\omega)$ are Fourier cosine transform pairs, hence recalling that $C_M(t)$ is *even* in time and $S_M(\omega)$ is *even* in frequency we have

$$C_M(t) = \frac{1}{\pi} \int\limits_0^{\infty} S_M(\omega) \cos(\omega t)\, \mathrm{d}\omega = \frac{6}{\pi\beta} \int\limits_0^{\infty} \frac{\alpha''(\omega)}{\omega} \cos(\omega t)\, \mathrm{d}\omega \tag{144}$$

In writing the foregoing equation, we have used the FDT Eq. (142). Thus, by Fourier inversion, Eq. (144) obviously yields $\alpha''(\omega)$ in terms of the Fourier cosine transform of the aftereffect function $b(t)$, viz.,

$$\alpha''(\omega) = \frac{\beta\omega}{3} \int\limits_0^{\infty} C_M(t) \cos(\omega t)\, \mathrm{d}t = \omega \int\limits_0^{\infty} b(t) \cos(\omega t)\, \mathrm{d}t$$

where we have used Eq. (139). Thus $b(t)$ and $\mathbf{M}(t)$ are related via

$$3b(t) = \beta C_M(t) \tag{145}$$

Now, by ergodicity and stationarity in time, we also have

$$C_M(t) = \langle \mathbf{M}(0) \cdot \mathbf{M}(t) \rangle_0 .$$

Thus, the polarizability may be expressed in terms of the total dipole moment fluctuations as

$$\alpha(\omega) = \frac{\beta}{3} \left\{ \left\langle M^2 \right\rangle_0 - i\omega \int_0^\infty \langle \mathbf{M}(0) \cdot \mathbf{M}(t) \rangle_0 \, \mathrm{e}^{-i\omega t} \mathrm{d}t \right\} . \tag{146}$$

This is the commonly known Kubo relation generalizing that of Fröhlich at zero frequency, viz., $\alpha'(0) = \beta \left\langle M^2 \right\rangle_0 /3$ to the frequency-dependent case, as obtained using the new method of Scaife [47]. Furthermore Scaife (see [47], Chap. 7, Eq. (7.217) has also shown by means of rather involved calculations that Eq. (128) may be generalized to the frequency-dependent case yielding

$$\frac{[\varepsilon(\omega) - \varepsilon_\infty][2\varepsilon^*(\omega) + \varepsilon_\infty][2\varepsilon(\omega) + 1]}{\varepsilon(\omega)[2\varepsilon^*(\omega) + 1](\varepsilon_\infty + 2)^2} = \frac{4\pi\alpha(\omega)}{3\upsilon}. \tag{147}$$

Here, we have separated the displacement polarization mechanism as before, and $\alpha(\omega)$ is the polarizability of a sphere in vacuo (i.e., calculated by means of classical statistical mechanics). Any further calculation necessitates a detailed investigation of the dynamics of the fluctuation phenomena at the microscopic level.

In this context, Nee and Zwanzig [43] included the fact that in the time-dependent situation the reaction field **R** lags behind the dipole. Thus, they obtained, for the dynamical Onsager model (i.e., ignoring dynamical correlations) the Fatuzzo–Mason equation [44], viz.,

$$\frac{[\varepsilon(\omega) - \varepsilon_\infty][2\varepsilon(\omega) + \varepsilon_\infty]}{3\lambda\varepsilon(\omega)} = \frac{1}{1 - i\omega\tau_D - \left(1 - \frac{\varepsilon_\infty}{\varepsilon}\right)\left(\frac{\varepsilon(\omega) - \varepsilon}{2\varepsilon(\omega) + \varepsilon_\infty}\right)}. \tag{148}$$

Hence, they were able to reproduce the experimental complex permittivity of glycerol at −60 °C. However, their derivation of Eq. (148) is open to objection. In effect, when a dipole rotates, it produces a time-dependent field outside the cavity, and energy is dissipated to the surroundings because of dielectric loss. Therefore, the dipole moment slows down: this is dielectric friction. This frictional effect is in addition to the local effects of van der Waals forces, which are usually represented by the frictional term in the Langevin equation [3]. The problem is then to determine how dielectric friction combines with the Stokes–Einstein friction of the rotational Brownian motion [39]. Therefore, the most rigorous treatment of the linear complex permittivity to date is that of Scaife, Eq. (147) [47].

(v) *The nonlinear static susceptibilities and the local field*

The range of validity of both the Onsager and the Kirkwood–Fröhlich equations has been described in Fröhlich's own words [39]: a molecule occupies a sphere of radius a, its polarizability is isotropic and no saturation effects can take place. In other words, provided the external field is small and static, the above treatment of the local field holds. If the field is increased so that the dielectric behaves nonlinearly, all the above treatment must be revisited. Here, we summarize the inclusion of internal field effects on nonlinear susceptibilities. Thus, we briefly review the main results already obtained for the cubic dielectric increment of polar fluids. In this context, Onsager mentions that his local field formula (116) is not suitable for estimating nonlinear effects due to strong electric fields. The reason for that is in the absence of free charges, the electric displacement vector **D** obeys the Maxwell equation:

$$\nabla \cdot \mathbf{D} = 0, \tag{149}$$

while the Maxwell field obeys the irrotational condition

$$\nabla \times \mathbf{E} = \mathbf{0} \tag{150}$$

so that $\mathbf{E} = -\nabla\Phi$, where Φ is the electrostatic potential. However, Eqs. (149) and (150) cannot be solved without a constitutive equation linking **D** and **E**, which for nonlinear dielectrics is

$$\mathbf{D} = \varepsilon_R \left(E^2\right) \mathbf{E}. \tag{151}$$

Furthermore, we are interested in the cubic dielectric increment only, thus we may expand the field-dependent permittivity $\varepsilon_R \left(E^2\right)$ as

$$\varepsilon_R \left(E^2\right) \approx \varepsilon \left(1 - \kappa E^2\right), \tag{152}$$

where κ is the relative cubic nonlinear dielectric increment and ε is the constant linear permittivity as before. Now, it is found [45] that κ is of the order of 10^{-18} m^2/V^2 and is usually positive (normal saturation), however, negative values have also been observed (anomalous saturation) [45–47]. By analogy with the Langevin theory of paramagnetism, the phenomenon described by Eq. (152) is called dielectric saturation. From Eqs. (149)–(152), we see that the electrostatic potential Φ no longer satisfies Laplace's equation, instead satisfying [45]

$$\left(1 - \kappa \left(\nabla\Phi\right)^2\right) \nabla^2\Phi = \kappa \left(\nabla\Phi \cdot \nabla\right) \left(\nabla\Phi\right)^2 \tag{153}$$

The details of the approximate perturbative solution of the nonlinear Eq. (153) are given elsewhere [45], and may be summarized as follows. Because of the smallness of κ, one may seek the solution of Eq. (153) by perturbation methods, i.e., we may write

$$\Phi = \Phi^{(0)} + \kappa\,\Phi^{(1)}. \tag{154}$$

Here, $\Phi^{(0)}$ is the solution of Laplace's equation while $\Phi^{(1)}$ is the perturbed part of the electrostatic potential because the dielectric is no longer linear. Including the effect of the polarizability of the molecules, we then find that the Kirkwood–Fröhlich equation is modified to (details in [45])

$$\begin{aligned}\varepsilon - \varepsilon_\infty = {} & \left(\frac{3\varepsilon}{2\varepsilon + \varepsilon_\infty}\right)\frac{4\pi\beta\left\langle M^2\right\rangle_0}{3\upsilon} + 2\kappa\varepsilon\left(\frac{2\varepsilon - 5\varepsilon_\infty}{(2\varepsilon+\varepsilon_\infty)^4}\right)\frac{(4\pi)^3\beta\left\langle M^4\right\rangle_0}{15\upsilon^3} \\ & + \frac{(4\pi)^2\kappa\varepsilon\,(\varepsilon_\infty - 1)\left\langle M^2\right\rangle_0}{(2\varepsilon+1)\,\upsilon^2}\left[\frac{2}{3}\frac{2\varepsilon - 5\varepsilon_\infty}{(2\varepsilon+\varepsilon_\infty)^3} - \frac{32\pi\beta\varepsilon}{5\,(2\varepsilon+\varepsilon_\infty)^4}\frac{\left\langle M^4\right\rangle_0}{\upsilon\left\langle M^2\right\rangle_0}\right],\end{aligned} \tag{155}$$

while the nonlinear dielectric increment is

$$\frac{\Delta\varepsilon}{E^2} = \kappa\varepsilon = -\frac{4\pi\beta^3}{90\upsilon Q}\left(\frac{2\varepsilon+1}{2\varepsilon+\varepsilon_\infty}\right)\left(\frac{3\varepsilon}{2\varepsilon+\varepsilon_\infty}\right)^3\left(3\left\langle M^4\right\rangle_0 - 5\left\langle M^2\right\rangle_0^2\right) \tag{156}$$

where $\Delta\varepsilon$ is the absolute nonlinear dielectric increment, and

$$\begin{aligned}Q = {} & 1 + \frac{1}{5}\frac{(\varepsilon_\infty - 1)\left(28\varepsilon^3 - 66\varepsilon^2\varepsilon_\infty - 60\varepsilon\varepsilon_\infty^2 - 37\varepsilon_\infty^3\right)}{(2\varepsilon+\varepsilon_\infty)^4} \\ & + \frac{4\pi\beta\left\langle M^2\right\rangle_0}{15\upsilon\,(2\varepsilon+\varepsilon_\infty)^5}\left\{\left(28\varepsilon^3 - 66\varepsilon^2\varepsilon_\infty - 60\varepsilon\varepsilon_\infty^2 - 37\varepsilon_\infty^3\right)(2\varepsilon+1) - 54\varepsilon\,(\varepsilon_\infty - 1)\left(4\varepsilon^2 + 2\varepsilon\varepsilon_\infty + 3\varepsilon_\infty^2\right)\right\} \\ & + \frac{(4\pi)^2\beta^2\left(9\left\langle M^4\right\rangle_0 - 5\left\langle M^2\right\rangle_0^2\right)}{75\upsilon^2\,(2\varepsilon+\varepsilon_\infty)^6}\left\{6\varepsilon^2(\varepsilon_\infty - 1)(2\varepsilon - 5\varepsilon_\infty) - \varepsilon\,(2\varepsilon+1)\left(4\varepsilon^2 - 8\varepsilon\varepsilon_\infty + 13\varepsilon_\infty^2\right)\right\} \\ & + 3\frac{(4\pi)^3\beta^3\varepsilon^2\left(3\left\langle M^6\right\rangle_0 - 5\left\langle M^2\right\rangle_0^2\left\langle M^4\right\rangle_0\right)}{25\upsilon^3\,(2\varepsilon+\varepsilon_\infty)^7}\left(4\varepsilon^2 - 18\varepsilon\varepsilon_\infty + 10\varepsilon - 5\varepsilon_\infty\right).\end{aligned} \tag{157}$$

Now, the extra terms in the modified Kirkwood–Fröhlich Eq. (155) can usually be ignored in practice, especially for liquids with large permittivity. In contrast, the expression for the nonlinear dielectric increment cannot be simplified. Here, it is impossible to proceed without a statistical model allowing the calculation of the statistical averages in Eqs. (156) and (157).

Other attempts to calculate $\Delta\varepsilon/E^2$ were made before that of Coffey and Scaife. Following an earlier remark of Van Vleck [48], a first calculation of the nonlinear dielectric increment was attempted by Thiébaut [49], assuming that intermolecular correlations are negligible. Orientational correlations were included in the formula derived with the assumptions of Van Vleck by Kielich [50] before Thiébaut's work, then later by Barriol and Greffe [51], and Böttcher [46], with the result

$$\frac{\Delta\varepsilon}{E^2} = \kappa\varepsilon = \frac{4\pi\rho_0\beta^3\mu^4}{45}\frac{\varepsilon^4(\varepsilon_\infty + 2)^4}{(2\varepsilon+\varepsilon_\infty)^2\left(2\varepsilon^2 + \varepsilon_\infty^2\right)}R_S, \tag{158}$$

where

$$R_S = \frac{3\left\langle M^4\right\rangle_0 - 5\left\langle M^2\right\rangle_0^2}{2\rho_0 \upsilon \mu^4} \tag{159}$$

is called the Piekara–Kielich correlation factor [46, 50], which also arises in the Coffey–Scaife formula. However, unlike the Kirkwood factor, R_S may be negative or positive, indicating that an anomalous saturation effect may on occasion dominate the normal saturation effect. Now, Van Vleck's derivation [48] assumes that

$$\varepsilon_R\left(E^2\right) \approx \varepsilon\left(1 - \kappa E^{(0)2}\right), \tag{160}$$

where $\mathbf{E}^{(0)} = -\nabla\Phi^{(0)}$ is the field existing in a *linear* dielectric. However, as already pointed out by Onsager and later by Brown [34], no logical grounds exist for making this assumption about a nonlinear dielectric and then replacing $\varepsilon_R\left(E^2\right)$ by $\varepsilon_R\left(E^{(0)2}\right)$ both in the cavity and reaction field factors (117), because this is equivalent to assuming that *the dielectric is linear in the overall sense*, despite the fact that one is attempting to calculate *a nonlinear property*. Hence, *one cannot merely assume that Eq.* (160) *holds for a nonlinear dielectric*. A more complete theoretical discussion is given in [45].

Now, although Eq. (160) is, stricto sensu, incorrect, comparison of Eq. (158) with experiment (with $R_s = 1$) shows that agreement may sometimes be achieved, especially in substances where the Kirkwood correlation factor has value 1 [45]. Marked differences between the Thiébaut [49] and Coffey–Scaife formulas and experimental data occur occasionally [45], nevertheless the disagreement arises for both formulas from the same sources of uncertainty, namely, either experimental errors or the lack of accounting for orientational correlations. Thus, Eq. (156) demonstrates clearly that the local field manifests itself quite differently for the relative permittivity and for the nonlinear cubic dielectric increment.

Regarding dynamical susceptibilities, all we may anticipate is that the single dielectric increment splits in two dynamical responses: one at the fundamental and one at the third harmonic.

7 A Perspective: The Dean-Kawasaki Approach

As alluded to previously, it is very difficult, if not impossible, to calculate the dynamics of the internal field exactly. Nevertheless, the various correlation effects may still be calculated by using a specific many-body method, comprising the Dean-Kawasaki method [52, 53]. The latter naturally extends Berne's approach [54] so far as the mean field approximation is relaxed. For simplicity, we consider an assembly of dipoles that are distributed at random however with positions fixed in space so that only the rotational degrees of freedom are relevant. The stochastic equation describing the collective tumbling of the dipoles is then [53]

$$2\tau_D \frac{\partial \rho}{\partial t}(\mathbf{u}, t) = \nabla_{\mathbf{u}} \cdot \left[\beta\rho(\mathbf{u}, t)\nabla_{\mathbf{u}} \frac{\delta F}{\delta\rho}(\mathbf{u}, t) + \boldsymbol{\gamma}(\mathbf{u}, t) \right], \tag{161}$$

where $\boldsymbol{\gamma}(\mathbf{u}, t)$ is a random Gaussian white noise vector field with statistical properties

$$\overline{\boldsymbol{\gamma}(\mathbf{u}, t)} = \mathbf{0}, \tag{162}$$

$$\overline{\gamma_i(\mathbf{u}, t)\gamma_j(\mathbf{u}', t)} = \tau_D^{-1}\overline{\rho(\mathbf{u}, t)}\delta_{ij}\delta(\mathbf{u} - \mathbf{u}')\delta(t - t'), \ (i, j) = x, y, z. \tag{163}$$

Here, the overbar denotes an average over the distribution of the realizations of the noise field $\boldsymbol{\gamma}$, δ_{ij} is Kronecker's delta while $\delta(\mathbf{u} - \mathbf{u}')$ and $\delta(t - t')$ are Dirac delta functions, ρ is defined by

$$\rho(\mathbf{u}, t) = \sum_{i=1}^{N} \delta(\mathbf{u} - \mathbf{u}_i(t)), \tag{164}$$

$\mathbf{u}_i(t)$ is the orientation of dipole i with dynamics governed by its individual rotational Langevin equation, $F = F[\rho]$ is a free energy functional and is also a (stochastic) functional of ρ (the compact notation $\delta F/\delta\rho$ holds for a functional derivative taken in the usual way for deterministic quantities). If only pair interactions are retained, the free energy functional F for a pure species may be restricted to just an entropic term, a field orientational term and (long range) pair intermolecular interactions, viz.,

$$F[\rho](t) = kT \int \rho(\mathbf{u}, t)\ln\rho(\mathbf{u}, t)\,d\mathbf{u} - \mu E(t)\int (\mathbf{u}\cdot\mathbf{e})\,\rho(\mathbf{u}, t)\,d\mathbf{u} + \frac{1}{2}\int\int \rho(\mathbf{u}, t)\,U_m(\mathbf{u}, \mathbf{u}')\,\rho(\mathbf{u}', t)\,d\mathbf{u}d\mathbf{u}', \tag{165}$$

where $U_m(\mathbf{u}, \mathbf{u}')$ is the interaction energy for a single pair of dipoles. By defining the orientational one-body and pair densities W and W_2 by the equations

$$\overline{\rho(\mathbf{u}, t)} = W(\mathbf{u}, t), \ \overline{\rho(\mathbf{u}, t)\rho(\mathbf{u}', t)} = W_2(\mathbf{u}, \mathbf{u}', t), \tag{166}$$

and averaging Eq. (161) over the distribution of the realizations of the noise field, we have the partial integrodifferential equation

$$2\tau_D \frac{\partial W}{\partial t}(\mathbf{u}, t) = \nabla_{\mathbf{u}} \cdot [\nabla_{\mathbf{u}} W(\mathbf{u}, t) + \beta W(\mathbf{u}, t)\nabla_{\mathbf{u}} V_i(\mathbf{u})] + \beta\nabla_{\mathbf{u}} \cdot \int \nabla_{\mathbf{u}} U_m(\mathbf{u}, \mathbf{u}')\,W_2(\mathbf{u}, \mathbf{u}', t)\,d\mathbf{u}', \tag{167}$$

where $V_i(\mathbf{u}, t) = -\mu E(t)(\mathbf{u} \cdot \mathbf{e})$ is the mean electrostatic orientational energy due to the molecular field (i.e., the orientational electrostatic energy as seen by a molecule). As written, Eq. (167) is just a rotational Fokker–Planck (Smoluchowski) equation forced by pair interactions. However, it may also be regarded as a nonlinear integrodifferential equation for W the orientational single-body density, because W_2 may be written in most general form as

$$W_2\left(\mathbf{u}, \mathbf{u}', t\right) = W(\mathbf{u}, t)\, W\left(\mathbf{u}', t\right) g\left(\mathbf{u}, \mathbf{u}', t\right),$$

where $g\left(\mathbf{u}, \mathbf{u}', t\right)$ is the dynamical orientational pair distribution function. In particular, Eq. (167) has been used to evaluate the temperature dependence of the dielectric constant of water and methanol, giving satisfactory agreement between the Kirkwood–Fröhlich formula and experimental data without any fitting parameter [55]. Thus, it appears that higher nonlinear correlation factors could also be computed with this method.

8 Conclusion

We have reviewed a number of methods for the calculation of the linear and nonlinear polarization responses to externally applied fields, both for noninteracting and interacting molecules. In this way, we have emphasized the role that may be played by dynamical interaction effects and the possible importance of the internal field effects in these nonlinear responses. In particular, we have given a simple method whereby thermally activated effects could be included in the theory. Moreover, we have also indicated how dynamical effects due to intermolecular interactions may alter the nonlinear polarizability spectra *without affecting the linear response*. The inclusion of internal field effects in these nonlinear spectra is absolutely nontrivial and is left for future investigation. Here, we have accounted for intermolecular interactions at the mean field level only, thereby effectively neglecting intermolecular interactions. However, we have also indicated how the collective tumbling of the dipolar system may be treated on the basis of the Dean-Kawasaki formalism, because this allows the inclusion of static and dynamic correlations at the molecular level. This formalism is essentially equivalent to the Bogolyubov–Born–Green–Kirkwood–Yvon formalism [33] treatment by diffusion processes, in which inertial effects are neglected completely. In effect, the short-range van der Waals forces are accounted for using a white noise approximation in the manner of Langevin [3], while the long range forces are treated explicitly. In particular, the Dean-Kawasaki formalism is able to reproduce the nonlinear integrodifferential equation obeyed by the equilibrium pair distribution function [32], in turn reducing to the Born–Green equation [33] when the Kirkwood superposition principle is used. Therefore, for the purpose of modeling long range interaction potentials, the various correlation factors occurring in Eqs. (136) and (156) can be computed. These tasks are left for future research.

Acknowledgements We are indebted to Profs F. van Wijland L.F. Cugliandolo for helpful conversations and for having introduced us to the Dean-Kawasaki formalism.

References

1. P. Debye, *Polar Molecules* (Chem. Catalog. Co., New York, 1929; Reprinted Dover, New York, 1954)
2. A. Einstein, *Investigations on the Theory of the Brownian Movement* (Methuen, London, 1926, reprinted by Dover, New York, 1954)
3. W.T. Coffey, Yu.P. Kalmykov, *The Langevin Equation*, 4th edn. (World Scientific, Singapore, 2017)
4. H. Benoit, Contribution à l'étude de l'effet Kerr présenté par les solutions de macromolécules rigides. Ann. Phys. **6**, 561 (1951)
5. H. Watanabe, A. Morita, Kerr effect relaxation in high electric fields. Adv. Chem. Phys. **56**, 255 (1984)
6. A. Morita, On nonlinear dielectric relaxation. J. Phys. D Appl. Phys. **11**, 1357 (1978)
7. W.T. Coffey, B.V. Paranjape, Dielectric and Kerr effect relaxation in alternating electric fields. Proc. Roy. Ir. Acad. A **78**, 17 (1978)
8. V. Rosato, G. Williams, Dynamic Kerr effect and dielectric relaxation of polarizable dipolar molecules, J. Chem. Soc. Faraday Trans. 2 **77**, 1767 (1981)
9. T. Furukawa, M. Tada, K. Nakajima, I. Seo, Nonlinear dielectric relaxation in a vinylidene cyanide/vinyl acetate copolymer, Jpn. J. Appl. Phys. **27**, 200 (1988); T. Furukawa and K. Matsumoto, Nonlinear dielectric relaxation spectra for polyvinyl acetate, *ibid.*, **31**, 840 (1992)
10. C. Crauste-Thibierge, C. Brun, F. Ladieu, D. l'Hôte, G. Biroli and J.-P. Bouchaud, Evidence of Growing Spatial Correlation at the Glass Transition from Nonlinear Response Experiments, Phys. Rev. Lett. **104**, 165703 (2010); C. Brun, F. Ladieu, D. l'Hôte, M. Tarzia, G. Biroli and J.-P. Bouchaud, Nonlinear dielectric susceptibilities : Accurate determination of the growing correlation volume in a supercooled liquid, Phys. Rev. B **84**, 104204 (2011)
11. H. Block, E.F. Hayes, Dielectric behavior of stiff polymers in solution when subjected to high voltage gradients. Trans. Faraday Soc. **66**, 2512 (1970)
12. K. De Smet, L. Hellemans, J.F. Rouleau, R. Corteau, T.K. Bose, Rotational relaxation of rigid dipolar molecules in nonlinear dielectric spectra, Phys. Rev. E **57**, 1384 (1998); P. Kędziora, J. Jadżyn, K. De Smet, L. Hellemans, Linear and nonlinear dipolar relaxation of 4,4′-n-hexylcyanobiphenyl, J. Mol. Liquids **80**, 19 (1999); J. Jadżyn, P. Kędziora, L. Hellemans, Phys. Lett. A **251**, 49 (1999); P. Kędziora, J. Jadżyn, K. De Smet, L. Hellemans, Nonlinear dielectric relaxation in non-interacting dipolar systems, Chem. Phys. Lett. **289**, 541 (1998)
13. Yu. L. Raikher, V.I. Stepanov and S.V. Burylov, Nonlinear orientation-optical effects in a dipolar particle suspension, Kolloid. Zh. **52**, 887 (1990) [Colloid. J. USSR, **52**, 768 (1990]; Low-frequency dynamics of the orientational birefringence in a suspension of dipolar particles, J. Coll. Interface Sci. **144**, 308 (1991)
14. J.L. Déjardin, Yu. P. Kalmykov, P.M. Déjardin, Birefringence and dielectric relaxation in strong electric fields, Adv. Chem. Phys. **117**, 275 (2001); J.L. Déjardin, Yu. P. Kalmykov, Nonlinear dielectric relaxation of polar molecules in a strong ac electric field: Steady state response, Phys. Rev. E **61**, 1211 (2000); Steady state response of the nonlinear dielectric relaxation and birefringence in strong superimposed ac and dc bias electric fields: Polar and polarizable molecules, J. Chem. Phys. **112**, 2916 (2000); Yu. P. Kalmykov, Matrix method of calculation of the Kerr effect transient and ac stationary responses of arbitrarily shaped macromolecules, J. Chem. Phys. **131**, 074107 (2009)

15. R.B. Jones, Transient and steady linear response of dielectric particles in a high bias field subject to a weak AC probe field, J. Phys. Cond. Matter, **14**, 7719 (2002); B.U. Felderhof, R.B. Jones, Nonlinear response of a dipolar system with rotational diffusion to an oscillating field, *ibid*. **15**, S1363 (2003); Mean field theory of the nonlinear response of an interacting dipolar system with rotational diffusion to an oscillating field, *ibid*. **15**, 4011 (2003)
16. R. Richert, Frequency dependence of dielectric saturation. Phys. Rev. E **88**, 062313 (2013)
17. G. Meier, A. Saupe, Dielectric relaxation in nematic liquid crystals, Mol. Cryst. **1**, 515 (1966); A.J. Martin, G. Meier, A. Saupe, Extended Debye theory for dielectric relaxation in nematic liquid crystals, Symp. Faraday Soc. **5**, 119 (1971); P.L. Nordio, G. Rigatti, and U. Segre, Dielectric relaxation theory in nematic liquids, Mol. Phys., 25, 129 (1973); B.A. Storonkin, Theory of dielectric relaxation in nematic liquid crystals, Kristallografiya **30**, 841 (1985) [Sov. Phys. Crystallogr. **30**, 489 (1985)]; A. Kozak, J.K. Moscicki, G. Williams, On dielectric relaxation in liquid crystals, Mol. Cryst. Liq. Cryst. **201**, 1 (1991)
18. W.T. Coffey, D.S.F. Crothers, Y.P. Kalmykov, J.T. Waldron, Exact solution for the extended Debye theory of dielectric relaxation of nematic liquid crystals, Physica A **213**, 551 (1995); Yu.P. Kalmykov, W.T. Coffey, Analytical solutions for rotational diffusion in the mean field potential: application to the theory of dielectric relaxation in nematic liquid crystals, Liquid Cryst. **25**, 329 (1998); W.T. Coffey, Yu.P. Kalmykov, Rotational diffusion and dielectric relaxation in nematic liquid crystals, Adv. Chem. Phys. **113**, 487 (2000)
19. H.A. Kramers, Brownian Motion in a field of force and the diffusion model of chemical reactions. Physica **7**, 284 (1940)
20. J. Jadzyn, G. Czechowski, R. Douali, and C. Legrand, On the molecular interpretation of the dielectric relaxation of nematic liquid crystals, Liquid Cryst. **26**, 1591 (1999); S. Urban, A. Würflinger, and B. Gestblom. On the derivation of the nematic order parameter from the dielectric relaxation time, Phys. Chem. Chem. Phys. **1**, 2787 (1999); S. Urban, B. Gestblom, W. Kuczyński, S. Pawlus, and A. Würflinger, Nematic order parameter as determined from dielectric relaxation data and other methods, Phys. Chem. Chem. Phys. **5**, 924 (2003); S. Urban, B. Gestblom, and S. Pawlus, Dielectric properties of 4-methoxy-4'-cyanobiphenyl (1 OCB) in the supercooled isotropic and nematic phases, Z. Naturforsch. A, **58**, 357 (2003); K. Merkel, A. Kocot, J.K. Vij, G.H. Mehl, and T. Meyer, Orientational order and dynamics of the dendritic liquid crystal organo-siloxane tetrapodes determined using dielectric spectroscopy, Phys. Rev. E, **73**, 051702 (2006)
21. P.M. Déjardin and Yu.P. Kalmykov, Relaxation of the magnetization in uniaxial single-domain ferromagnetic particles driven by a strong ac magnetic field, J. Appl. Phys. **106**, 123908 (2009); S.V. Titov, P.M. Déjardin, H. El Mrabti, and Yu. P. Kalmykov, Nonlinear magnetization relaxation of superparamagnetic nanoparticles in superimposed ac and dc magnetic bias fields, Phys. Rev. B **82**, 100413(R) (2010); H. El Mrabti, S.V. Titov, P.M. Déjardin, and Yu.P. Kalmykov, Nonlinear stationary ac response of the magnetization of uniaxial superparamagnetic nanoparticles, J. Appl. Phys. **110**, 023901 (2011); N. Wei, D. Byrne, W.T. Coffey, Yu.P. Kalmykov, and S.V. Titov, Nonlinear frequency-dependent effects in the dc magnetization of uniaxial magnetic nanoparticles in superimposed strong alternating current and direct current fields, J. Appl. Phys. **116**, 173903 (2014)
22. P.M. Déjardin, F. Ladieu, Nonlinear susceptibilities of interacting polar molecules in the self-consistent field approximation. J. Chem. Phys. **140**, 034506 (2014)
23. R. Richert, Nonlinear Dielectric effects in liquids: a guided tour. J. Phys: Cond. Mat. **29**, 363001 (2017)
24. R. Courant, D. Hilbert, *Methoden der Mathematischen Physik*, vol. 1 (Springer, Berlin, 1924)
25. W.T. Coffey, Yu.P. Kalmykov, N. Wei, Nonlinear normal and anomalous response of non-interacting electric and magnetic dipoles subjected to strong AC and DC bias fields. Nonlinear Dyn. **80**, 1861 (2014)
26. H. Fröhlich, *Theory of Dielectrics*, 2nd edn. (Oxford University Press, Oxford, 1958)
27. N. Wei, P.M. Déjardin, YuP Kalmykov, W.T. Coffey, External DC bias field effects in the nonlinear AC response of permanent dipoles in a mean field potential. Phys. Rev. E **93**, 042208 (2016)

28. H. Risken, *The Fokker-Planck Equation*, 2nd edn. (Springer, Berlin, 1989)
29. D.A. Garanin, Integral relaxation time of single-domain ferromagnetic particles. Phys. Rev. E **54**, 3250 (1996)
30. B.J. Berne, A self-consistent theory of rotational diffusion, J. Chem. Phys. **62**, 1154 (1975); M. Warchol and W.E. Vaughan, Forced rotational diffusion of linear molecules. Nonlinear aspects, J. Chem. Phys. **71**, 502 (1979)
31. F. Ladieu, D. L'Hôte, C. Brun, Nonlinear dielectric susceptibilities in supercooled liquids: a toy model. Phys. Rev. B **85**, 184207 (2012)
32. P.M. Déjardin and F. Ladieu, unpublished work (2014)
33. J.P. Hansen, I. McDonald, *Theory of Simple Liquids*, 3rd edn. (Elsevier, Amsterdam, 2006)
34. W.F. Brown, Dielectrics, in *Handbuch der Physik*, vol. 17, ed. by S. Flügge (Springer, Berlin, 1956), p. 1
35. J.A. Stratton, *Electromagnetic Theory* (McGraw-Hill, New York, 1941)
36. P. Madden, D. Kivelson, A consistent molecular treatment of dielectric phenomena. Adv. Chem. Phys. **56**, 497 (1984)
37. L. Onsager, Electric moments of molecules in Liquids. J. Am. Chem. Soc. **58**, 1486 (1936)
38. R.P. Bell, The electrostatic energy of dipole molecules in different media. Trans. Faraday Soc. **27**, 797 (1931)
39. M.W. Evans, G.J. Evans, W.T. Coffey, P. Grigolini, *Molecular dynamics and the theory of broad*, vol. Spectroscopy (Wiley, New York, 1982)
40. J.G. Kirkwood, The dielectric polarization of polar liquids. J. Chem. Phys. **7**, 911 (1939)
41. B.K.P. Scaife, *Complex Permittivity* (English Universities Press, London, 1971)
42. G. Oster, J.G. Kirkwood, The influence of hindered molecular rotation on the dielectric constants of water, alcohols, and other polar liquids. J. Chem. Phys. **11**, 175 (1943)
43. T.S. Nee, R. Zwanzig, Theory of dielectric relaxation in polar liquids. J. Chem. Phys. **58**, 6353 (1970)
44. E. Fatuzzo, P.R. Mason, A calculation of the complex dielectric constant of a polar liquid by the librating molecule method. Proc. Phys. Soc. **90**, 729 (1967)
45. W.T. Coffey, B.K.P. Scaife, On the theory of dielectric saturation of polar fluids, Proc. Roy. Irish. Acad. A **76**, 195 (1976); W.T. Coffey, B.K.P. Scaife, On the solution of some potential problems for a nonlinear dielectric, J. Electrostatics **1**, 193 (1975); W.T. Coffey, Ph.D. Thesis (The University of Dublin, 1975)
46. C.J.F. Böttcher, *Theory of Electric Polarization*, vol. I (Elsevier, Amsterdam, 1973)
47. B.K.P. Scaife, *Principles of Dielectrics*, 2nd edn. (Clarendon, Oxford, 1998)
48. J.H. Van Vleck, On the role of dipole-dipole coupling in dielectric media. J. Chem. Phys. **5**, 556 (1937)
49. J.M. Thiébaut, Thèse de 3e Cycle, (Nancy, 1968)
50. S. Kielich, Semi-macroscopic treatment of the theory of nonlinear phenomena in dielectric liquids subjected to strong electric and magnetic fields. Acta Phys. Pol. **17**, 239–255 (1958)
51. J. Barriol, J.L. Greffe, Relations between the Microscopic Model of Polarizable Dielectrics and the Macroscopic Model II: Dielectric Saturation and Fluctuation of the Dielectric Moment. J. Chim. Phys. **66**, 575 (1969)
52. K. Kawasaki, Stochastic Model of slow dynamics in supercooled liquids and dense colloidal suspensions, Physica A **208**, 35 (1994); D.S. Dean, Langevin equation for a system of interacting Langevin processes, J. Phys. A : Math. Gen. **29**, L613 (1996)
53. L.F. Cugliandolo, P.M. Déjardin, G.S. Lozano, F. van Wijland, Stochastic dynamics of collective modes for Brownian dipoles. Phys. Rev. E **91**, 032139 (2015)
54. B.J. Berne, A self-consistent theory of rotational diffusion. J. Chem. Phys. **62**, 1154 (1975)
55. P.M. Déjardin, Y. Cornaton, P. Ghesquière, C. Caliot, R. Brouzet, Calculation of linear and nonlinear orientational correlation factors of polar liquids from the rotational Dean-Kawasaki equation. J. Chem. Phys. **148**, 044504 (2018)

Stochastic Models of Higher Order Dielectric Responses

Gregor Diezemann

Abstract The nonlinear response for systems exhibiting Markovian stochastic dynamics is calculated using time-dependent perturbation theory for the Green's function, the conditional probability to find the system in a given configuration at a certain time given it was in another configuration at an earlier time. In general, the Green's function obeys a so-called master-equation for the balance of the gain and loss of probability in the various configurations of the system. Using various models for the reorientational motion of molecules it is found that the scaled modulus of the third-order response, X_3, shows a hump-like behavior for random rotational motion in some cases and it exhibits "trivial" behavior, a monotonuos decay from a finite zero-frequency value to a vanishing high-frequency limit, if the model of isotropic rotational diffusion is considered. For the time-honored model of dipole reorientations in an asymmetric double-well potential, it is found that X_3 exhibits a peak in a certain temperature range around a characteristic temperature at which the zero-frequency limit vanishes. The fifth-order modulus X_5 shows hump-like behavior in two distinct temperature regimes located at temperatures, where X_3 behaves trivially. For a trap model with a Gaussian density of states, a model that exhibits some features of glassy relaxation, both nonlinear response functions can exhibit either trivial or hump-like behavior depending on the specific choice for some model parameters. The height of the peak shows various temperature dependencies from increasing with temperature, decreasing or a temperature-independent behavior.

Keywords Nonlinear dielectric relaxation · Stochastic models · Molecular reorientations · Asymmetric double-well potential model · Trap model

G. Diezemann (✉)
Institut für Physikalische Chemie, Universität Mainz,
Duesbergweg 10-14, 55128 Mainz, Germany
e-mail: diezeman@uni-mainz.de

R. Richert (ed.), *Nonlinear Dielectric Spectroscopy*, Advances in Dielectrics
https://doi.org/10.1007/978-3-319-77574-6_3

1 Introduction

Broadband dielectric spectroscopy is a very powerful method to investigate the dynamical behavior of condensed matter systems over an extremely broad frequency range [1]. This is particularly important for systems like supercooled liquids and glasses where the dynamics spans timescales from very fast dynamics to extremely slow processes like the primary relaxation or the normal mode relaxation in polymers [2, 3]. In order to study the dynamically heterogeneous nature of glassy relaxation, however, one has to apply frequency-selective techniques [4–7]. Apart from probing the system at more than two times, as in higher dimensional NMR experiments [8–10], also the application of strong electric fields allows to monitor dynamical heterogeneities via the nonresonant hole-burning techniques [11, 12]. Similar techniques have also been used to monitor magnetic [13] and mechanical [14] hole-burning.

Motivated by a theoretical prediction of Bouchaud and Biroli [15], in the past a number of experimental studies of the nonlinear dielectric response have been conducted, for a very recent review see [16]. The idea was that the modulus of the third-order response function is related to the number of cooperatively rearranging particles N_{corr} participating in the primary relaxation process. As a function of frequency, the modulus exhibits a peak near the mean relaxation frequency and this was attributed to the existence of "glassy correlations" [17]. In this interpretation of the nonlinear response, the height of the peak is a measure of N_{corr} and experimental results have been analyzed along these lines [17, 18]. On the other hand, a monotonous decay of the modulus from a finite low-frequency value to zero at high frequencies was interpreted as being indicative of the lack of spatial correlations in the system and was assumed to indicate "trivial" behavior [18], i.e., the reorientational motion molecules lacking cooperativity.

Other studies of the nonlinear dielectric relaxation have been interpreted in terms of models that do not exhibit any spatial structure like the Box model, further phenomenological models [16, 19–21] and also the mode coupling theory of the glass transition [22]. Other models used to compute the cubic response of glasses devoid of spatial structure are a toy model based on the assumption of the existence of correlated particles [23] and a phenomenological model for the primary relaxation [24]. Additionally, the nonlinear response for molecules undergoing isotropic reorientational diffusion has been calculated and the resulting modulus does not exhibit a hump-like structure [25] but shows trivial behavior only. Quite recently, the fifth-order nonlinear susceptibilities of supercooled glycerol and propylene carbonate have been determined giving strong support for the interpretation in terms of cooperatively moving particles [26].

We have computed the third-order and the fifth-order response for a well-known model exhibiting glassy relaxation, the so-called trap model, and found that for certain values of the model parameters a hump is found in both cases [27, 28]. In addition for the model of dipoles reorienting in an asymmetric double-well potential humps are observed. These calculations were performed using time-dependent perturbation theory for the propagator of the underlying stochastic dynamics and will be reviewed in the present chapter. As mentioned above, the nonlinear response theory of dipole reorientations has been worked out for various models of rotational diffusion [25, 29–31] and has also been extended to include long-range dipolar interactions [32]. In most of these calculations, it is assumed that the stochastic reorientations of the molecules can be described in terms of a Fokker–Planck equation (FPE) [33, 34]. The perturbation theory presented in the present chapter generalizes these approaches to the treatment of the propagator of the more generally applicable master-equation (ME) [33, 35].

In general, the dielectric polarization is related to the expectation value of the electric dipole moment of the sample, see e.g., [1]. Here, we will not consider details of the theory of dielectrics, but we take the following simplified view. The response to an external electrical field $E(t)$ is determined by the change of the energy of the system due to the alignment of the dipole moment M in the field, $\propto (-M \cdot E(t))$. The dielectric response is determined by the time-dependent average $\langle M(t)\rangle$. Throughout the present chapter, we assume that M depends on a set of relevant coordinates $q(t)$, $M(t) = M(q(t))$. Although not necessary for the general discussion, the coordinates $q(t)$ are chosen to represent a Markov process and the stochastic dynamics is described by a ME or a FPE [33–35]. The coordinates $q(t)$ might represent one or more relevant angles defining the orientation of a molecule in a laboratory—fixed frame, in the most general case three Eulerian angles, or $q(t)$ can also represent some collective variables characterizing the relevant configurations of the system. With this, the expectation value can be written as

$$\langle M(t)\rangle = \int \mathrm{d}q M(q)p(q,t). \tag{1}$$

Here, $p(q, t)$ denotes the probability to find the value q at time t. Therefore, the response is determined by the time evolution of $p(q, t)$. General results do only exist in the linear response regime, where the fluctuation–dissipation theorem (FDT) holds, see e.g. Ref. [1]. The FDT relates the linear response $R(t, s)$ to the time derivative of the equilibrium correlation function of the variable conjugated to the applied field (the dipole moment), $C_M(t, s)$, which can be written for times $t \geq s$:

$$R(t,s) = \left.\frac{\delta\langle M(t)\rangle}{\delta E(s)}\right|_{E=0} = \beta\frac{\partial C_M(t,s)}{\partial s}. \tag{2}$$

Here, $\beta = (k_B T)^{-1}$ (k_B is the Boltzmann constant, in the following set to unity) and $C_M(t, s)$ is given by

$$C_M(t,s) = \langle M(t)M(s)\rangle = \int \mathrm{d}q \int \mathrm{d}q' M(q)M(q')G(q,t|q',s)p(q',s), \tag{3}$$

where $G(q, t|q', s)$, denotes the conditional probability (Green's function) to find the value q at time t, given it had the value q' at time s. Generally, the time-dependent probability can be computed from

$$p(q,t) = \int dq' G(q,t|q',s)p(q',s). \tag{4}$$

If one is interested in the generally nonlinear response of the system to a field applied at time s, one has to extend the calculation to the desired order in the field E and then uses

$$\langle M(t)\rangle = \int dq \int dq' M(q) G^{(E)}(q,t|q',s)p(q',s). \tag{5}$$

Note, that Eq. (4) cannot be used here, because the Green's function is to be calculated in the presence of the field and the a priori distribution $p(q',s)$ is the one before the field is switched on. Here, using an expansion of $G^{(E)}(q,t|q',s)$ linear in the field yields the linear response, and using the nth order gives the corresponding nonlinear order.

In the present chapter, we will discuss the high-order (third- and fifth-order) response for systems exhibiting a stochastic dynamics that can be assumed to be Markovian. In the following section, we will briefly recall some aspects of simple models of stochastic dynamics and then formulate the time-dependent perturbation theory that will be used to calculate the relevant response functions according to Eq. (5).

2 Markovian Stochastic Dynamics

As mentioned above, if the stochastic dynamics can be considered to be Markovian the Green's function $G(q,t|q',t')$, obeys the ME:

$$\dot{G}(q_k,t|q_l,t_0) = \int dq_n \mathcal{W}(q_k,q_n,t) G(q_n,t|q_l,t_0). \tag{6}$$

Here, the elements of the master-operator, $\mathcal{W}(q_n,q_k,t)$ are given by [33, 35]

$$\mathcal{W}(q_k,q_l,t) = W(q_k,q_l,t) - \delta(q_k - q_l)\int dq_n W(q_n,q_l,t), \tag{7}$$

where $W(q_k,q_l,t)$ is the transition rate from q_l to q_k at time t.

Equation (6) has to be solved with the initial condition $G(q_k,t_0|q_l,t_0) = \delta(q_k - q_l)$, where $\delta(x)$ denotes the Dirac delta function. The explicit form of the ME reads as

$$\dot{G}(q_k,t|q_l,t_0) = -\int dq_n W(q_n,q_k,t)G(q_k,t|q_l,t_0) + \int dq_n W(q_k,q_n,t)G(q_n,t|q_l,t_0).$$

Here, the first term describes the loss of probability in 'state' q_k due to all transitions out of this state with the effective escape rate $\int dq_n W(q_n, q_k, t)$ and the second terms gives account of the gain in probability in q_k.

The derivation of a FPE starting from the more general ME is quite straightforward and the result can be written in the form [33, 35]

$$\dot{G}(q_k, t|q_l, t_0) = \mathcal{L}(q_k, t)G(q_k, t|q_l, t_0), \tag{8}$$

where $\mathcal{L}(q_k, t)$ denotes the FP-operator. Generally, the FPE is used if one considers diffusion processes and the ME can be used to treat problems with larger jump lengths. For further information regarding the properties of the ME and the FPE, we refer to textbooks [33–36].

If the dynamics is stationary, all one-time quantities do not depend on time, such as the stationary distributions, $p^{\text{stat.}}(q, t) = p^{\text{eq}}(q)$, and two-time quantities only depend on the time difference, e.g., $C_M(t, s) = C_M(t - s)$. Furthermore, the same holds for the Green's function, $G(q, t|q', s) = G(q, t - s|q')$ and the long-time limit of $G(q, t|q')$ yields the equilibrium population, $\lim_{t\to\infty} G(q, t|q') = p^{\text{eq}}(q)$.

Molecular Reorientations: Rotational diffusion, rotational random jumps

In the past, a number of models for the reorientational motion of molecules in supercooled liquids have been considered and these models have been used in the interpretation of (linear) dielectric response functions and NMR relaxation times and spectra, see e.g. Refs. [37, 38]. Among these models are the well-known model of rotational diffusion, the model of random reorientational jumps, and also models treating rotational jumps of finite width.

Rotational diffusion: If one considers the isotropic rotational diffusion of molecules, one has the FPE:

$$\dot{G}(w, t|w_0) = D_{\text{R}} \frac{1}{\sin^2\theta} \left[\sin\theta \cdot \partial_\theta \left(\sin\theta \cdot \partial_\theta\right) + \partial_\varphi^2\right] G(w, t|w_0), \tag{9}$$

where $w = (\theta, \varphi)$ and D_{R} denotes the diffusion coefficient. The solution of this equation is well known and is given by the series

$$G(w, t|w_0) = \sum_{l=0}^{\infty} \sum_{m=-l}^{l} Y_{l,m}(w) Y_{l,m}^*(w_0) \mathrm{e}^{-D_{\text{R}} l(l+1)t} \tag{10}$$

with $Y_{l,m}(w)$ representing a spherical harmonics. From this expressions, rank-dependent time correlation functions can be computed and one finds that they decay exponentially according to $C_l(t) = \langle Y_{l,0}(w(t)) Y_{l,0}^*(w(0)) \rangle = C_l(0)\mathrm{e}^{-D_{\text{R}} l(l+1)t}$ [37]. If one compares the relaxation times $\tau_l = \int dt_0^\infty C_l(t)$ relevant for dielectric relaxation ($l = 1$) and NMR ($l = 2$), one finds $\tau_1/\tau_2 = 3$ at variance with most experimental results obtained for supercooled liquids, where more typically $\tau_1/\tau_2 \simeq 1$ is found [38].

Rotational random jumps: A l-independent result for the rotational correlation functions is found in the framework of the model of rotational random jumps. Here, it is assumed that any rotation completely decorrelates the orientational degrees of freedom and therefore the master-operator has the following form:

$$\mathcal{W}(w, w_0) = \Gamma \left[\frac{1}{4\pi} - \delta(w - w_0) \right] \tag{11}$$

This means that a transition out of an orientation w_0 takes place with a rate Γ and ends at any other orientation on the sphere with equal probability ($(4\pi)^{-1}$). The solution reads as

$$G(w, t|w_0) = \frac{1}{4\pi} + \mathrm{e}^{-\Gamma t} \left[\delta(w - w_0) - \frac{1}{4\pi} \right] \tag{12}$$

and in this case one finds for the rotational correlation functions the l-independent expression $C_l(t) = \mathrm{e}^{-\Gamma t}$.

More general models for the rotational motion incorporating rotational jumps have been considered, e.g., in Refs. [37, 39, 40] and references therein and in some cases also the heterogeneous nature of the rotational relaxation in supercooled liquids has been taken into account explicitly [38, 41–43].

The asymmetric double-well potential model

A quite different model that often has been used to model dielectric relaxation is the model of dipole reorientations in an asymmetric double-well potential (ADWP). The dipole can assume two orientations separated by a barrier, cf. Fig. 1. The dynamics of this system is then described as a process of diffusive barrier crossing and therefore one has to consider the FPE for the diffusion in the presence of a potential $V(q, t)$. This time-dependent potential usually contains a time-independent term that is given by the ADWP and the time-dependent coupling to the external fields. In this case, the FP-operator has the following generic form:

$$\mathcal{L}(q, t) = D\partial_q \mathrm{e}^{-\beta V(q,t)} \partial_q \mathrm{e}^{\beta V(q,t)} = \beta D \partial_q V'(q, t) + D\partial_q^2. \tag{13}$$

If one considers the double-well potential with deep wells as in Fig. 1, one can discriminate between two types of dynamical processes. The intra-well relaxation is fast and can be modeled by the well-known Ornstein–Uhlenbeck process [33]. The diffusive barrier crossing can be treated using Kramers theory and gives rise to transition rates that are of an Arrhenius form with the barrier as an apparent activation free energy. We will denote the two orientations corresponding to the minima of the ADWP by '1' and '2', characterized by polar angles $\theta_1 = \theta$ and $\theta_2 = (\theta + \pi)$ [12, 27, 28, 44]. The transition rates between these two states are given by $W_{12} = W\mathrm{e}^{-\beta\Delta/2}$ and $W_{21} = W\mathrm{e}^{+\beta\Delta/2}$. Here Δ denotes the asymmetry, cf. Fig. 1, and W is the hopping rate in the symmetric case, $W = W_0\mathrm{e}^{-\beta V}$, with the average activation energy V.

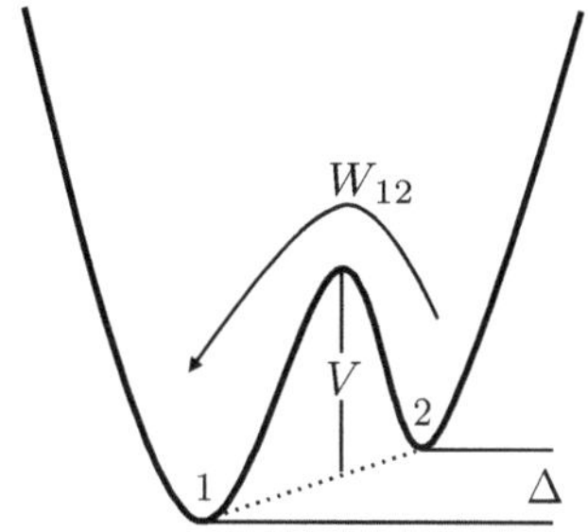

Fig. 1 Sketch of an asymmetric double well potential. The asymmetry is denoted by Δ and the average activation energy is V. The rate for a transition from state (orientation) 2 to 1 is given by W_{12}

In this case, the long-time behavior after equilibration within the wells can be properly described by a ME for the populations of the two wells, see e.g. [45]. This ME is then a discrete version of Eq. (6), which with the abbreviations $G_{kl}(t) = G(q_k, t|q_l)$ and $W_{kl} = W(q_k|q_l)$ is given by

$$\dot{G}_{kl}(t) = \sum_n \mathcal{W}_{kn} G_{nl}(t), \tag{14}$$

where the elements of the master-operator are $\mathcal{W}_{kl} = W_{kl} - \delta_{kl} \sum_n W_{nl}$, i.e. $\dot{G}_{kl}(t) = \sum_n [W_{kn} G_{nl}(t) - W_{nk} G_{kl}(t)]$.

For the ADWP model, the Green's functions in the field-free case are given by

$$G_{kl}(t) = p_k^{\mathrm{eq}} \left(1 - \mathrm{e}^{-t/\tau}\right) + \delta_{kl} \mathrm{e}^{-t/\tau} \quad \text{with} \quad \tau^{-1} = 2W \cosh(\beta\Delta/2) \quad \text{and} \quad p_k^{\mathrm{eq}} = \tau \cdot W_{kl}. \tag{15}$$

The time correlation function relevant for dielectric relaxation, $C_M(t)$ [Eq. (3)], is calculated assuming

$$M_k = M \cos(\theta_k) \quad \text{and therefore} \quad M_1 = M \cos(\theta) \quad ; \quad M_2 = -M \cos(\theta)$$

with M denoting the static molecular dipole moment. After performing the average over all orientations, $\langle \cos^n(\theta_k) \rangle = (n+1)^{-1}$ for n even and $\langle \cos^n(\theta_k) \rangle = 0$ for n odd, one finds

$$C_M(t) = \frac{M^2}{3}(1 - \delta^2)\mathrm{e}^{-t/\tau} \tag{16}$$

Here, the definition

$$\delta = \tanh(\beta\Delta/2) \tag{17}$$

has been used. The prefactor stems from $\langle M^2 \rangle = \sum_k M_k^2 p_k^{\mathrm{eq}} = (M^2/3)(1 - \delta^2)$ where the isotropic average has been performed (hence the factor $1/3$).

The Gaussian trap model

Also models that do not treat molecular reorientations explicitly can be used to compute response functions if the coupling of the relevant variables to the field are specified. In this chapter, we will consider the trap model with a Gaussian density of states [46, 47]. This type of model has been used quite successfully to understand certain aspects of glassy dynamics [48–50]. The main ingredients of the model can be summarized as follows. It is assumed that the minima of the free-energy landscape, ϵ, are distributed according to a Gaussian:

$$\rho(\epsilon) = \frac{1}{\sqrt{2\pi}\sigma} e^{-\epsilon^2/(2\sigma^2)} \tag{18}$$

with $\sigma = 1$. This assumption has been shown to be in reasonable agreement with the distributions of meta basin energies [51]. The transitions among the various 'states' ϵ are modeled as an activated escape out of the initial state with the destination state chosen at random, i.e., according to the density of states, Eq. (18):

$$W(\epsilon|\epsilon_0) = \rho(\epsilon) \cdot \kappa(\epsilon_0) \tag{19}$$

with the escape rate

$$\kappa(\epsilon) = \kappa_\infty e^{\beta\epsilon}. \tag{20}$$

Because $\rho(\epsilon)$ is normalized, $\int d\epsilon \rho(\epsilon) = 1$, the ME reads

$$\dot{G}(\epsilon, t|\epsilon_0) = -\kappa(\epsilon) G(\epsilon, t|\epsilon_0) + \rho(\epsilon) \int d\epsilon' \kappa(\epsilon') G(\epsilon', t|\epsilon_0). \tag{21}$$

The equilibrium populations are given by the long-time limit of the Green's function and are also Gaussian:

$$p^{eq}(\epsilon) = \lim_{t\to\infty} G(\epsilon, t|\epsilon_0) = \frac{1}{\sqrt{2\pi}\sigma} e^{-(\epsilon-\bar{\epsilon})^2/(2\sigma^2)} \quad ; \quad \bar{\epsilon} = -\beta\sigma^2 \tag{22}$$

The detailed calculation of the response functions and also the time correlation functions $C_M(t) = \langle M(t)M(0)\rangle$ requires a choice of the functional dependence of the 'moments' $M(t)$ on the trap energies ϵ. (Note that in the present discussion the trap energy ϵ plays the role of the generalized 'coordinate' q introduced in the Introduction.)

In order to calculate the response, one further has to quantify the dependence of the 'moment' $M(\epsilon)$ on the trap energy ϵ. The choice of this dependence represents a further assumption of the calculation and has a strong impact on the results for the response functions, as will be discussed below. According to Eqs. (2) and (3) the linear response is determined by the equilibrium auto-correlation function $C_M(t)$. In the present chapter, we will always make the following reasonable assumption:

$$\langle M(\epsilon)\rangle = 0 \quad \text{and} \quad \langle M(\epsilon)M(\epsilon_0)\rangle = \delta(\epsilon - \epsilon_0)\langle M(\epsilon)^2\rangle. \tag{23}$$

Using this, one finds for $C_M(t)$:

$$C_M(t) = \int d\epsilon \langle M(\epsilon)\rangle e^{-\kappa(\epsilon)t} p^{eq}(\epsilon). \tag{24}$$

We mention that for ϵ independent M, $\langle M(\epsilon)^2\rangle = \langle M^2\rangle$, this correlation function reduces to the 'jump-correlation function', which decays whenever a transition takes place and thus can be viewed as an intermediate scattering function for a large value of the modulus of the scattering vector [38].

3 Nonlinear Response Theory for Markov Processes

The response of the system to an external E field applied at time t_0 and measured by the observable $M(t)$ (the dipole moment) is given by Eq. (5), which in a discrete version reads as

$$\chi_M(t, t_0) = \langle M(t)\rangle = \sum_{kl} M_k G_{kl}^{(E)}(t, t_0) p_k(t_0). \tag{25}$$

Here, M_k is a shorthand notation for $M(q_k)$. The time-dependent perturbation theory for the Green's function $G_{kl}^{(E)}(t, t_0)$ is obtained in the following way. Starting from the ME in the discrete version of Eq. (6), one translates the ME into a matrix notation:

$$\partial_t \mathbf{G}(t, t_0) = \mathcal{W}(t)\mathbf{G}(t, t_0). \tag{26}$$

The matrix elements of the propagator are $\mathbf{G}(t, t_0)_{kl} = G_{kl}(t, t_0) = G(q_k, t|q_l, t_0)$ and those of the master—operator are given by $\mathcal{W}_{kl}(t) = \mathcal{W}(q_k, q_l, t)$. The formal solution of the ME can be written as $\mathbf{G}(t, t_0) = \mathcal{T} \exp\left(\int_{t_0}^t d\tau \mathcal{W}(\tau)\right)\mathbf{G}(t_0, t_0)$ where $\mathcal{T}$ denotes the time-ordering operator and $\mathbf{G}(t_0, t_0)_{kl} = \delta_{kl}$. Note that the same formalism is applicable to the solution of a FPE with the only difference that the master-operator is to be replaced by the FP-operator. Differences between the perturbation theory for a FPE and a ME appear, however, when dealing with the nonlinear response functions, see the discussion below. If an electric field is applied, the corresponding ME is written as $\partial_t \mathbf{G}^{(E)}(t, t_0) = \mathcal{W}^{(E)}(t)\mathbf{G}^{(E)}(t, t_0)$.

In order to set up the perturbation theory we decompose the master-operator into an unperturbed (field-free) part and a perturbation contribution,

$$\mathcal{W}^{(E)}(t) = \mathcal{W}(t) + \mathcal{V}(t). \tag{27}$$

Without specifying the expression for the coupling to the field, the elements of the individual terms $\mathcal{W}(t)$ and $\mathcal{V}(t)$ generally are obtained from a Taylor expansion of $\mathcal{W}^{(E)}(t)$ with respect to the electric field,

$$W_{kl}^{(E)}(t) = \sum_{n=0}^{\infty} \frac{1}{n!} W_{kl}^{(n)}(t) \cdot [\beta E(t)]^n \quad \text{with} \quad W_{kl}^{(n)}(t) = \frac{d^n}{d(\beta E)^n} W_{kl}^{(E)}(t)\bigg|_{E=0} \tag{28}$$

and accordingly we have the decomposition into the field-free part $W_{kl}(t) = W_{kl}^{(0)}(t)$ and the perturbation

$$\mathcal{V}(t) = \sum_{n=1}^{\infty} \mathcal{V}^{(n)}(t) \quad \text{with} \quad \mathcal{V}^{(n)}(t)_{kl} = \frac{[\beta E(t)]^n}{n!} \left[W_{kl}^{(n)}(t) - \delta_{kl} \sum_m W_{ml}^{(n)}(t) \right]. \tag{29}$$

The formal perturbation expansion for the propagator starts from the Dyson-like equation

$$\mathbf{G}^{(E)}(t, t_0) = \mathbf{G}(t, t_0) + \int_{t_0}^{t} dt' \mathbf{G}(t, t') \mathcal{V}(t') \mathbf{G}^{(E)}(t', t_0) \tag{30}$$

with $\mathbf{G}(t, t_0) \equiv \mathbf{G}^{(0)}(t, t_0)$. In an abbreviated form, we write for this equation

$$\mathbf{G}^{(E)} = \mathbf{G} + \mathbf{G} \otimes \mathcal{V} \otimes \mathbf{G}^{(E)} \tag{31}$$

where the time arguments are omitted and the convolution is abbreviated by the '$\otimes$' symbol. Using Eqs. (28) and (29), one finds the following general expression for the nth-order Green's function $\mathbf{G}^{(n)}(t, t_0)$:

$$\mathbf{G}^{(n)} = \sum_{m=0}^{n-1} \mathbf{G}^{(0)} \otimes \mathcal{V}^{(n-m)} \otimes \mathbf{G}^{(m)} \quad \text{with} \quad \mathbf{G}^{(0)} \equiv \mathbf{G} \tag{32}$$

In the next step, one uses the expression for the matrix elements of $\mathbf{G}^{(n)}(t, t_0)$, $G_{kl}^{(n)}(t, t_0)$, in Eq. (25) in order to compute the nth-order response,

$$\chi^{(n)}(t, t_0) = \sum_{kl} M_k G_{kl}^{(n)}(t, t_0) p_k(t_0). \tag{33}$$

Note that these response functions represent so-called integrated response functions. Of course, it is also possible to define the corresponding pulse-response functions, R, as in Eq. (2) and then the integrated response is obtained by multiplication of $R^{(n)}$ with the time-dependent fields and integrating over the internal times [27, 28].

The remaining problem in the formulation of the response theory is the determination of the field dependence of the transition rates $W_{kl}^{(E)}(t)$. We note that in general there is no definite recipe for this if the dynamics is described by a ME. Only in case of a diffusive dynamics represented by a FPE the field dependence of the FP-operator is determined by the force exerted by the applied field. Here, we will obtain the field dependence of the transition rates from the following considerations. One assumes that the field couples to a variable M, giving a contribution

$(-E \cdot M_k)$ to the energy and therefore the Boltzmann factor and the equilibrium population p_k^{eq} is altered accordingly. Together with the detailed balance condition $W_{kl}p_l^{\mathrm{eq}} = W_{lk}p_k^{\mathrm{eq}}$ this suggests an exponential dependence of the $W_{kl}^{(E)}(t)$ on the field. This is further substantiated by the following argument. One writes the discrete version of the FPE for the diffusion in a field, Eq. (13), on a grid of points $\{q_k\}$ with equal spacing $\Delta q = (q_{k+1} - q_k)$ and the definitions $\overline{D} = D/(\Delta q)^2$, $V_k = V(q_k)$, $p_k(t) = p(q_k, t)\Delta q$ in the form of a ME [33, 52]:

$$\dot{p}_k = \overline{W}_{k(k+1)}p_{k+1} + \overline{W}_{k(k-1)}p_{k-1} - \left(\overline{W}_{(k+1)k} + \overline{W}_{(k-1)k}\right)p_k$$

with

$$\overline{W}_{k(k\pm1)} = \overline{D}\left[1 - (\beta/2)(V_k - V_{k\pm1})\right] \simeq \overline{D}\mathrm{e}^{-(\beta/2)(V_k - V_{k\pm1})}$$

From this it is evident that the exponential field-dependence follows from $V_k \propto -EM_k$.

Using an exponential dependence, a quite general model is obtained using the following expression:

$$W_{kl}^{(E)}(t) = W_{kl}(t)\mathrm{e}^{\beta E(t)[\gamma M_k - \mu M_l]}. \tag{34}$$

Here, γ and μ can be chosen arbitrarily [53–55] and for $\gamma = 1 - \mu$, the model fulfills detailed balance. If $\mu = 1, \gamma = 0$, the field couples to the initial state of the transition and for $\mu = 0$, $\gamma = 1$, the coupling takes place via the destination state.

The above expressions for the $\mathbf{G}^{(n)}$ simplify considerably, if the stochastic dynamics is described by a FPE. This is because in that case there is no ambiguity in the choice of the coupling to the field. The coupling of the dipoles to the field adds a term $(-E \cdot M(q))$ to the potential $V(q)$ and this has to be incorporated into the FPE, yielding a strictly linear coupling [33]. Therefore, one has $\mathcal{V}^{(m)} = 0$ for $m > 1$ and Eq. (32) becomes $\mathbf{G}^{(n)} = \mathbf{G}^{(0)} \otimes \mathcal{V}^{(1)} \otimes \mathbf{G}^{(n-1)}$. If one uses Eq. (34) in the linearized form of the ME, one obtains $W_{(k+1)k}^{(E)}(t) = W_{(k+1)k}(t)(1 + \beta E(t)[\gamma M_{(k+1)} - \mu M_k])$ which for $\mu = \gamma = 1/2$ gives the force term in the FP-operator.

Using the expressions for the unperturbed Green's functions, one can calculate the corresponding nonlinear response functions for any model that is used to describe the molecular motion. In the following, we will focus on the response to sinusoidal fields of the form $H(t) = H_0 \cos(\omega t)$, for which the various response functions for times long compared to the initial transients can be written as

$$\begin{aligned}
\chi^{(1)}(t) &= \frac{H_0}{2}\left[\mathrm{e}^{-i\omega t}\chi_1(\omega) + \text{c.c.}\right] \\
\chi^{(3)}(t) &= \frac{H_0^3}{2}\left[\mathrm{e}^{-i\omega t}\chi_3^{(1)}(\omega) + \mathrm{e}^{-i3\omega t}\chi_3^{(3)}(\omega) + \text{c.c.}\right] \\
\chi^{(5)}(t) &= \frac{H_0^5}{2}\left[\mathrm{e}^{-i\omega t}\chi_5^{(1)}(\omega) + \mathrm{e}^{-i3\omega t}\chi_5^{(3)}(\omega) + \mathrm{e}^{-i5\omega t}\chi_5^{(5)}(\omega) + \text{c.c.}\right]
\end{aligned} \tag{35}$$

where c.c. denotes the complex conjugate.

4 Results for Simple Models

In this section, the results obtained for the linear and the nonlinear dielectric response using the stochastic models discussed above in Sect. 2 will be presented and discussed.

4.1 Reorientation Models

Here, we consider the rotational motion of molecules in terms of the simple models of rotational diffusion and of rotational random jumps. It is important to point out that no aspect of cooperative motion is incorporated in these models. In this sense, they do not allow for the treatment of so-called glassy correlations and constitute models that have been termed as trivial in the context of the nonlinear response of supercooled liquids [17, 26].

Rotational diffusion

The linear and nonlinear dielectric spectra for the model of isotropic rotational diffusion have been calculated and the corresponding expressions are repeated here for convenience [25]. We only mention that the method used in Ref. [25] is slightly different from the time-dependent perturbation theory as outlined above. The results, however, agree. The calculation using the perturbation theory starts from the FPE, Eq. (9), with the inclusion of a term $(-M\cos\theta\cdot E(t))$ coupling to the field. This gives rise to a decomposition of the FP-operator similar to Eq. (27), $\mathcal{L}^{(E)}(t)=\mathcal{L}_0+\mathcal{V}^{(1)}(t)$ with $\mathcal{V}^{(1)}(t)=\beta M E(t)(2\cos\theta+\sin\theta\partial_\theta)$.

For the linear response, one finds

$$\chi_{1,\mathrm{RD}}(\omega)=\Delta\chi_{1,\mathrm{RD}}\frac{1}{1-i\omega\tau_1}\quad\text{with}\quad\Delta\chi_{1,\mathrm{RD}}=\beta\frac{M^2}{3}.\tag{36}$$

For the cubic response, we consider the 3ω-component, which for this model is given by

$$\chi^{(3)}_{3,\mathrm{RD}}(\omega)=-\frac{1}{60}\beta^3M^4S_{\mathrm{RD}}(\omega\tau_1)\ ;\ S_{\mathrm{RD}}(x)=\frac{(3-17x^2)+ix(14-6x^2)}{(1+x^2)(9+4x^2)(1+9x^2)}\tag{37}$$

Here, $x=\omega\tau_1$ with $\tau_1=1/(2D_R)$.

Rotational random jumps

In this case, one has to consider a ME and thus one has to fix the values of μ and γ in Eq. (34). Here, we will choose these values according to the following consideration. The idea underlying the random jump model is that starting from a given orientation any other orientation can be reached in a single jump, thus completely randomizing the distribution of orientations. Therefore, we assume that the same holds for the dependence of the molecular dipole moment on the orientational degrees of freedom. Correspondingly, we start with a coupling to the initial orientation (say θ) according to $(-E\cdot M\cos\theta)$ and we average over the orientations that can be reached (say θ'), $\langle(-E\cdot M\cos\theta')\rangle$. Since $\langle\cos\theta'\rangle=0$, this corresponds to choosing $\mu=1$ and $\gamma=0$ and we will present results for this particular choice.

The linear response is given by the same expression as for the model of rotational diffusion, Eq. (36), with the replacement $\tau_1=1/\Gamma$. Also the third-order response can be written in a form that is very similar to Eq. (37). However, the spectral function is quite different and this gives rise to a different behavior.

$$\chi^{(3)}_{3,\mathrm{RJ}}(\omega)=-\frac{1}{60}\beta^3M^4S_{\mathrm{RJ}}(\omega/\Gamma)\ ;\ S_{\mathrm{RJ}}(x)=\frac{(2+62x^2-144x^4)-ix(1-167x^2+36x^4)}{6(1+x^2)(1+4x^2)(1+9x^2)} \tag{38}$$

with $x=\omega/\Gamma$. In the past, experimental results have either been presented in terms of real and imaginary part of the susceptibility or, alternatively, the modulus and the phase have been considered. In particular, it has proven meaningful to scale the modulus by the squared static linear response in the following way:

$$X_3(\omega)=\frac{T}{(\Delta\chi_1)^2}\left|\chi_3^{(3)}(\omega)\right| \tag{39}$$

This definition allows to get rid of the trivial temperature dependence of χ_3, $\chi_3\propto\beta^3$. Using this, one can write for the two models considered:

$$X_{3,Z}(\omega)=\frac{3}{20}\left|S_Z(\omega\tau_1)\right|\quad\text{with}\quad Z\in(\mathrm{RD},\mathrm{RJ}) \tag{40}$$

The limiting values are

$$X_{3,Z}(\omega\to0)=\frac{1}{20} \tag{41}$$

and for high frequencies both quantities vanish according to

$$X_{3,\mathrm{RD}}(\omega\to\infty)\simeq\frac{1}{40(\omega\tau_1)^3}\ ;\ X_{3,\mathrm{RJ}}(\omega\to\infty)\simeq\frac{1}{40(\omega\tau_1)}. \tag{42}$$

In Fig. 2 we show the real and the imaginary part of the cubic response for the two models considered. It is evident that the behavior of both quantities is somewhat different from the respective linear response. Apart from an irrelevant overall phase, the real part is not strictly monotonously changing from the low-frequency limit to the high frequency limit and also the imaginary part does not show the pure dissipative behavior of a Debye lineshape (Note that Kramers–Kronig relations do not exist in the nonlinear regime in general).

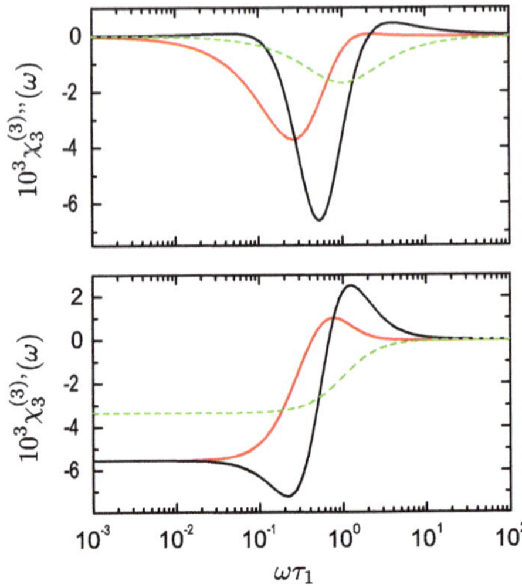

Fig. 2 Real and imaginary part of the cubic susceptibility $\chi_3^{(3)}(\omega)$ for the models of rotational diffusion (red lines) and rotational random jumps (black lines). The green dashed lines represent the linear response ($-10\chi_1(\omega)$)

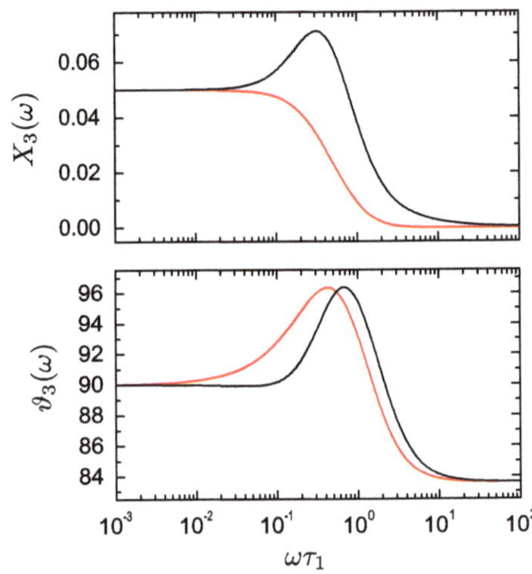

Fig. 3 Upper panel: $X_3(\omega)$ for the models of rotational diffusion (red lines) and rotational random jumps (black lines). Lower panel: Phase $\vartheta_3(\omega) = \mathrm{acos}(\chi_3^{(3)\prime}(\omega)/\chi_3^{(3)\prime}(\omega))$ (in deg.) as a function of frequency

The mentioned representation of the modulus is presented in Fig. 3. One can see that the model of rotational random jumps exhibits a hump located at a frequency somewhat smaller than the inverse relaxation time, whereas in case of rotational diffusion a monotonous decay is found from the low-frequency limit $X_3(0) = 1/20$ to the vanishing high-frequency limit. In the interpretation of Refs. [17, 18] the latter behavior is expected for so-called trivial dynamics without glassy correlations and the appearance of a hump is an indication of the occurance of such correlations. As indicated in the Introduction, also models without spatial correlations can give rise to a hump in $X_3(\omega)$. It is, however, interesting that also X_3 for a very simple model for the rotational motion of molecules shows a hump-like behavior. The phases $\vartheta_3(\omega)$

behave very similar and one cannot extract any significant difference between the two models.

4.2 ADWP Model

Performing the same calculation as for the reorientational models using Eq. (15) for the field-free Green's function, Eq. (15), one finds for the linear response:

$$\chi_{1,\mathrm{ADWP}}(\omega) = \Delta\chi_{1,\mathrm{ADWP}}\frac{1}{1-i\omega\tau} \quad \text{where} \quad \Delta\chi_{1,\mathrm{ADWP}} = \beta\langle\Delta M^2\rangle = \beta\frac{M^2}{3}\left(1-\delta^2\right) \tag{43}$$

where again $\delta = \tanh(\beta\Delta/2)$ and $\tau^{-1} = 2W\cosh(\beta\Delta/2)$ (Note that $\Delta\chi_{1,\mathrm{ADWP}}$ differs by a factor 1/2 from the definition of χ_{DWP} in Ref. [44]). $\Delta\chi_{1,\mathrm{ADWP}}$ is of course the same expectation value, $\langle\Delta M^2\rangle$, as in the expression for the two-time correlation function, Eq. (16). The spectral shape of $\chi_{1,\mathrm{ADWP}}(\omega)$ is identical to that of the reorientational models. The amplitude, however, shows an extra temperature dependence that is determined by the value of the asymmetry, cf. Fig. 4. For increasing Δ the low-temperature limit approaches zero. In case of vanishing asymmetry, the linear susceptibility is not distinguishable from the corresponding ones for the reorientational models.

For the higher order response functions one has to choose the values of γ and μ. However, as we have only two states and correspondingly there exists only a single destination state for each transition, this choice is irrelevant and the results all coincide provided that the detailed balance condition $\gamma + \mu = 1$ is fulfilled. In a straightforward calculation, one finds [27]:

$$\chi^{(3)}_{3,\mathrm{ADWP}}(\omega) = \frac{M^4}{20}\beta^3\left(1-\delta^2\right)S_{3,\mathrm{ADWP}}(\omega\tau) \tag{44}$$

Also in this case, the spectral function only depends on the product $x = \omega\tau$ and is given by

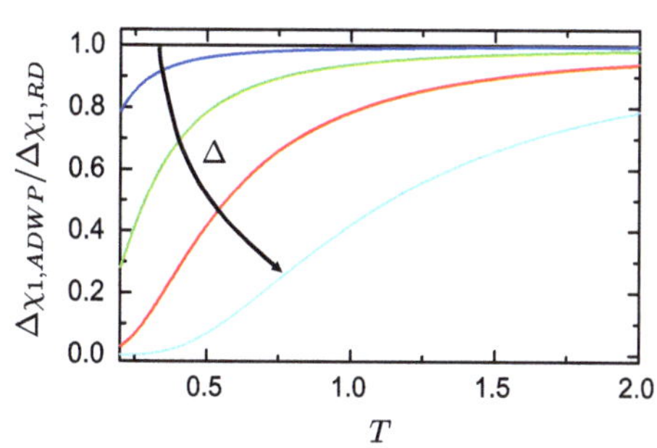

Fig. 4 Relative amplitude of the linear susceptibility, $\Delta\chi_{1,\mathrm{ADWP}}/\Delta\chi_{1,\mathrm{RD}}$ as a function of temperature for different values of the asymmetry; $\Delta = 0$: black, $\Delta = 0.2$: blue, $\Delta = 0.5$: green, $\Delta = 1.0$: red, $\Delta = 2.0$: cyan

$$S_{3,\mathrm{ADWP}}(x) = \delta^2 \frac{(1-11x^2)+i6x(1-x^2)}{(1+x^2)(1+4x^2)(1+9x^2)} + \frac{2(5x^2-1)+i3x(x^2-3)}{6(1+x^2)(1+9x^2)}$$

In the following, we will consider $X_3(\omega)$ according to Eq. (39). This quantity is given by, cf. Eqs. (43) and (44):

$$X_{3,\mathrm{ADWP}}(\omega) = \frac{9}{20} \frac{\left|S_{3,\mathrm{ADWP}}(\omega\tau)\right|}{\left(1-\delta^2\right)} \tag{45}$$

The limiting values for small and large frequencies are determined by the corresponding limits of $S_{3,\mathrm{ADWP}}(\omega\tau)$ and thus, one has

$$X_{3,\mathrm{ADWP}}(0) = \frac{3}{20} \frac{\left|3\delta^2-1\right|}{\left(1-\delta^2\right)} \quad ; \quad X_{3,\mathrm{ADWP}}(\omega \to \infty) \simeq \frac{1}{40\left(1-\delta^2\right)} \left|\frac{3\delta^2}{x^3} - \frac{1}{x}\right|. \tag{46}$$

Before discussing the frequency dependence of $X_{3,\mathrm{ADWP}}(\omega)$, it is instructive to consider $X_{3,\mathrm{ADWP}}(0)$. This quantity is plotted as a function of temperature in Fig. 5. The following features are evident immediately. For small Δ, $X_{3,\mathrm{ADWP}}(0) \to 3/20$ and for large Δ one has $X_{3,\mathrm{ADWP}}(0) \to \infty$. This behavior is reflected in Fig. 5 for small temperatures (large β and thus large $\beta\Delta$) where $X_{3,\mathrm{ADWP}}(0)$ becomes very large. Additionally, for finite values of the asymmetry, $X_{3,\mathrm{ADWP}}(0)$ approaches the limiting value 3/20 for high temperatures ($\beta \to 0$). The drop to zero of $X_{3,\mathrm{ADWP}}(0)$ is determined by the condition $\left|3\delta^2-1\right| = 0$ and therefore defines the characteristic temperature

$$T_3 = \frac{\Delta}{\ln\left[(\sqrt{3}+1)/(\sqrt{3}-1)\right]} \simeq 0.76\Delta. \tag{47}$$

The fact that $X_{3,\mathrm{ADWP}}(0)$ vanishes at T_3 has a strong impact on the behavior of its frequency dependence [27].

As a function of frequency, a hump-like behavior is observed in a certain temperature regime around T_3, as shown in Fig. 6. The existence of the hump-like structure is determined by $X_{3,\mathrm{ADWP}}(0)$ and for temperatures much lower or much higher than T_3 only trivial behavior as for the model of rotational diffusion is observed. This fact

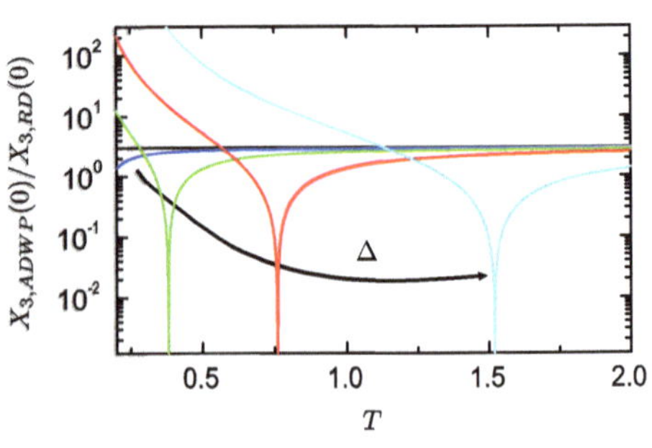

Fig. 5 Relative amplitude of the cubic susceptibility, $X_{3,\mathrm{ADWP}}(0)/X_{3,\mathrm{RD}}(0)$ as a function of temperature for different values of the asymmetry; $\Delta = 0$: black, $\Delta = 0.2$: blue, $\Delta = 0.5$: green, $\Delta = 1.0$: red, $\Delta = 2.0$: cyan

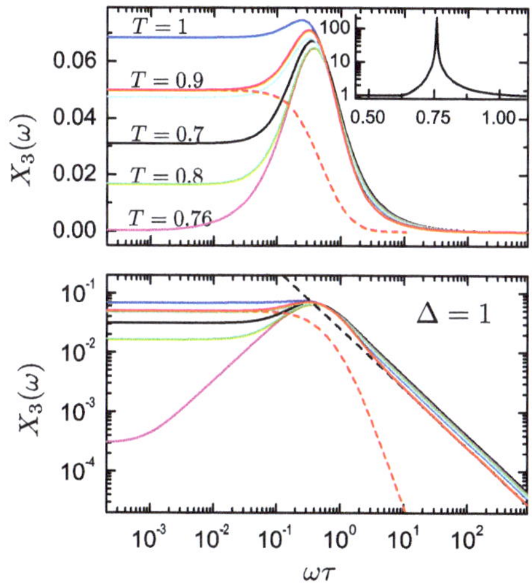

Fig. 6 $X_{3,\mathrm{ADWP}}(\omega)$ for $\Delta = 1$ as a function of frequency for different temperatures as indicated in the upper panel. The full red line represents $X_{3,\mathrm{RJ}}(\omega)$ and the dashed red line $X_{3,\mathrm{RD}}(\omega)$, cf. Fig. 3. The inset in the upper panel shows the ratio between the maximum value of $X_{3,\mathrm{ADWP}}(\omega)$ and its zero-frequency limit, $X^{\mathrm{max}}_{3,\mathrm{ADWP}}(\omega)/X_{3,\mathrm{ADWP}}(0)$ as a function of temperature. In the lower panel the same data are plotted on a logarithmic scale. The black dashed line has slope $1/\omega$

is quantified in the inset of the upper panel in Fig. 6, where $X^{\mathrm{max}}_{3,\mathrm{ADWP}}(\omega)/X_{3,\mathrm{ADWP}}(0)$ is plotted as a function of temperature. In supercooled liquids it is observed that the height of the peak decreases with increasing temperature [17]. In the ADWP model, this is true only for $T > T_3$, while for $T < T_3$ the height increases with temperature. For vanishing asymmetry, the hump disappears completely and $X_{3,\mathrm{ADWP}}(\omega)$ is very similar to the corresponding quantity for reorientational motions. At high frequencies, $X_{3,\mathrm{ADWP}}(\omega)$ behaves as that obtained for the model of rotational random jumps, cf. Eq. (46). This equation also shows that the high-frequency behavior depends on the value of Δ and the temperature, cf. Fig. 7. It is observed that for small values of $(\beta\Delta)$ the high frequency behavior is very similar to that obtained for the model of rotational random jumps. For large $(\beta\Delta)$ one finds a crossover from the behavior of $X_{3,\mathrm{RD}}(\omega)$ to $X_{3,\mathrm{RJ}}(\omega)$ around $\omega\tau \sim 1$. It should, however, be kept in mind that for $\delta \to 1$ the cubic susceptibility vanishes completely [and $X_{3,\mathrm{ADWP}}(\omega)$ diverges due to the denominator $(1-\delta^2)$ in Eq. (45)].

In the present chapter, we do not discuss the frequency dependence of the 1ω-component of the third-order response and only mention that this behaves very similar to the 3ω-component, for details we refer to Ref. [27].

Fifth-order response

For the ADWP model, in addition to the third-order response we present the results for the 5ω-component of the fifth-order response, $\chi^{(5)}_{5,\mathrm{ADWP}}(\omega)$. The calculation is performed in the same way as in case of the linear response and the cubic response

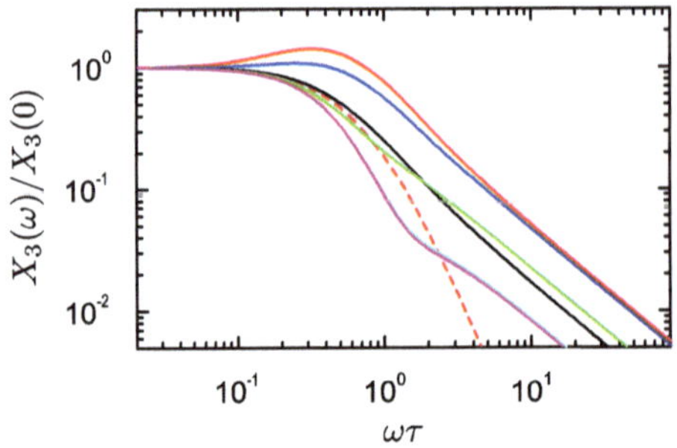

Fig. 7 $X_3(\omega)/X_3(0)$ as a function of frequency for different values of the asymmetry for $T = 1$. black: $\Delta = 0$, blue: $\Delta = 1$, green: $\Delta = 2$, cyan: $\Delta = 5$, magenta: $\Delta = 10$. The full red line represents $X_{3,\mathrm{RJ}}(\omega)/X_{3,\mathrm{RJ}}(0)$ and the dashed red line $X_{3,\mathrm{RD}}(\omega)X_{3,\mathrm{RD}}(0)$

with the result [28]:

$$\chi^{(5)}_{5,\mathrm{ADWP}}(\omega) = \left(\frac{M^6}{112}\right)\beta^5(1-\delta^2)S_{5,\mathrm{ADWP}}(\omega\tau) \tag{48}$$

where the isotropic average has been performed and the spectral function is given by

$$\begin{aligned}S_{5,\mathrm{ADWP}}(x) = \frac{1}{15N(x)}&\{(2-15\delta^2+15\delta^4)-5(6-155\delta^2+255\delta^4)x^2\\&+2(-612+3445\delta^2+2055\delta^4)x^4\\&-20(176+865\delta^2)x^6+3072x^8\}\\+\frac{ix}{8N(x)}&\{(11-104\delta^2+120\delta^4)+10(17+4\delta^2-180\delta^4)x^2\\&+(-293+9424\delta^2+960\delta^4)x^4\\&-20(157+160\delta^2)x^6+192x^8\}\end{aligned} \tag{49}$$

$$N(x) = (1+x^2)(1+4x^2)(1+9x^2)(1+16x^2)(1+25x^2). \tag{50}$$

Also in this case of the fifth-order response, the scaled modulus

$$X_{5,\mathrm{ADWP}}(\omega) = \frac{|\chi^{(5)}_{5,\mathrm{ADWP}}(\omega)|}{\beta^2(\Delta\chi_{1,\mathrm{ADWP}})^3} \tag{51}$$

can be considered in order to get rid of the trivial temperature dependence ($\chi_5^{(5)} \sim \beta^5$). We start the discussion of X_5 by considering the limits

$$X_{5,\mathrm{ADWP}}(0) = \frac{9}{560}\frac{\left|2-15\delta^2+15\delta^4\right|}{(1-\delta^2)^2} \quad ; \quad X_{5,\mathrm{ADWP}}(\omega\to\infty) = \frac{9}{22400(1-\delta^2)^2}\frac{1}{x}. \tag{52}$$

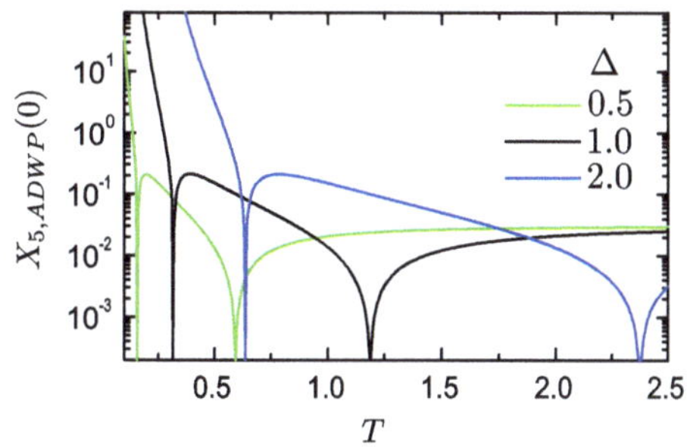

Fig. 8 $X_{5,\mathrm{ADWP}}(0)$ as a function of temperature for different values of the asymmetry; green: $\Delta = 0.5$, black: $\Delta = 1$, blue: $\Delta = 2$

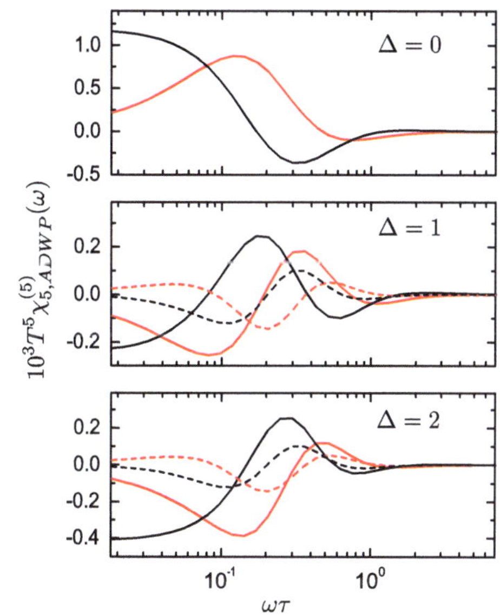

Fig. 9 $T^5 \chi^{(5)}_{5,\mathrm{ADWP}}(\omega)$ as a function of frequency for different values of the asymmetry as indicated. The black lines represent the real part and the red lines the imaginary part. Full lines correspond to $T = 1$ and dashed lines to $T = T_{5,a}$ (Figure adopted from Ref. [28])

The zero-frequency limit is shown in Fig. 8 for different values of the asymmetry as a function of temperature. It is obvious that $X_{5,\mathrm{ADWP}}(0)$ vanishes at two characteristic temperatures that are given by

$$T_{5;a/b} = \frac{\Delta}{\ln\left[(1+z_{a/b})/(1-z_{a/b})\right]} \quad \text{with} \quad z_{a/b} = \sqrt{\frac{15 \pm \sqrt{105}}{30}} \tag{53}$$

which yields

$$T_{5;a} \simeq 0.32\Delta \quad \text{with} \quad T_{5;b} \simeq 1.19\Delta. \tag{54}$$

The frequency dependence of the real and the imaginary part of $\chi^{(5)}_{5,\mathrm{ADWP}}(\omega)$ are displayed for some values of the asymmetry in Fig. 9. It is obvious that the overall behavior of both, the real and the imaginary part, is comparable to the corresponding third-order quantities, cf. Fig. 2. It is clear that also the fifth-order susceptibility for $\Delta = 0$ does not display a temperature dependence apart from the one of the relaxation time $\tau = 1/(2W)$, cf. Eq. (15). The modulus $X_{5,\mathrm{ADWP}}(\omega)$ is shown in Fig. 10 on a logarithmic scale. The temperature ranges encircled are centered around the two characteristic temperatures $T_{5;a} \simeq 0.32$ and $T_{5;b} \simeq 1.19$. It is obvious that for temperatures near to $T_{5;a/b}$ a hump-like behavior is found, but not for higher or lower temperatures. The behavior in the range of the characteristic temperatures is similar to what is observed for the third-order response around T_3. However, as is shown by the black line in Fig. 10, for $T = T_3$ only trivial behavior is observed for $X_{5,\mathrm{ADWP}}(\omega)$.

The temperature dependence of the relative maximum in $X_{5,\mathrm{ADWP}}(\omega)$ is shown in Fig. 11 and compared to that of $X_{3,\mathrm{ADWP}}(\omega)$. From this plot it can be seen that there is no hump in X_5 near T_3 and no hump in X_3 near $T_{5,a/b}$. This means, as a function of temperature one expects to observe a peak in X_5 at low temperatures (around $T_{5,a}$) the height of which first increases, then shows a maximum and decreases again. In X_3 only trivial behavior is observed in this temperature regime. Next, around T_3 this behavior is found in X_3 and no peak occurs in X_5. Only for still higher temperatures, X_5 exhibits a hump-like behavior in the regime around $T_{5,b}$.

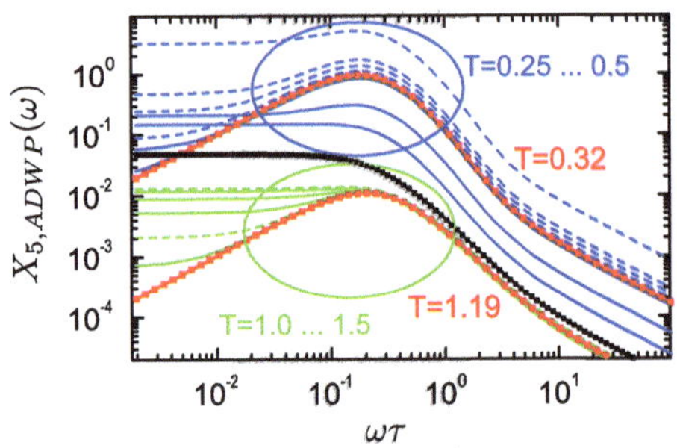

Fig. 10 $X_{5,\mathrm{ADWP}}(\omega)$ as a function of frequency for $\Delta = 1$. The circles indicate the temperature ranges around $T_{5;a}$ (blue) and $T_{5;b}$ (green). For these temperatures $X_{5,\mathrm{ADWP}}(\omega)$ is plotted in bold red. Full lines represent temperatures higher than $T_{5;a/b}$ and dashed lines lower temperatures. The bold black line is $X_{5,\mathrm{ADWP}}(\omega)$ for $T = T_3$ (Figure adopted from Ref. [28])

Fig. 11 Relative height of the hump, $X^{\max}_{k,\mathrm{ADWP}}(\omega)/X_{k,\mathrm{ADWP}}(0)$, as a function of temperature for $\Delta = 1$. $k = 3$ (blue line, $T_3 \simeq 0.76$) or $k = 5$ (red line $T_{5;a} \simeq 0.32$ and $T_{5;b} \simeq 1.19$) (Figure adopted from Ref. [28])

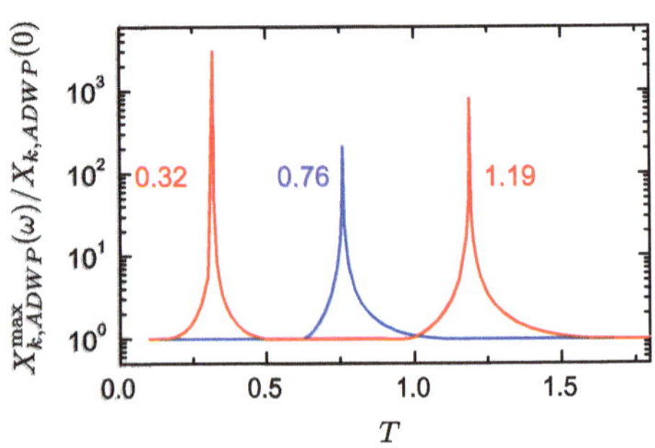

4.3 Gaussian Trap Model

In the calculation of the response for this model, it will be assumed that the coupling to the field takes place in the usual manner, i.e., via a reduction in the energy due to the alignment of the moments in the field, $\epsilon(E) = \epsilon - M(\epsilon)\cdot E(t)$. In addition, the explicit functional form of the variable $M(\epsilon)$ has to be fixed as this defines the particular version of the model. Here we use an Arrhenius-like dependence on the trap energies [56]:

$$\langle M(\epsilon)^2\rangle = \mathrm{e}^{-n\beta\epsilon} \tag{55}$$

where n is arbitrary and we set $M^2 = 1$. For $n = 0$, $\langle M^2\rangle$ is temperature-independent as in case of the models of reorientational motions discussed above.

One important reason for the particular choice (55) is that the spectral shape of the linear response is unaffected by this because it is given by (using $\gamma + \mu = 1$):

$$\chi_{1,GT}(\omega) = \beta\int d\epsilon p(\epsilon)^{\mathrm{eq}}\langle M(\epsilon)\rangle\frac{\kappa(\epsilon)}{\kappa(\epsilon) - i\omega} \tag{56}$$

If one now uses the relation

$$\int d\epsilon p(\epsilon)^{\mathrm{eq}}\mathrm{e}^{-n\beta\epsilon}\frac{\kappa(\epsilon)}{\kappa(\epsilon) - i\omega} = \mathrm{e}^{\frac{n(n+2)}{2}\beta^2\sigma^2}\int d\epsilon p(\epsilon)^{\mathrm{eq}}\frac{\kappa(\epsilon)}{\kappa(\epsilon) - i\omega_n}$$

with the scaled frequency $\omega_n = \omega\mathrm{e}^{n\beta^2\sigma^2}$ one finds

$$\chi_{1,GT}(\omega) = \Delta\chi_{1,GT}\int d\epsilon p(\epsilon)^{\mathrm{eq}}\frac{\kappa(\epsilon)}{\kappa(\epsilon) - i\omega_n} \quad \text{with} \quad \Delta\chi_{1,GT} = \beta\mathrm{e}^{\frac{n(n+2)}{2}\beta^2\sigma^2}. \tag{57}$$

Thus, the static susceptibility $\Delta\chi_{1,GT}$ strongly depends on the choice of n and the factor $\mathrm{e}^{\frac{n(n+2)}{2}\beta^2\sigma^2}$ becomes only temperature independent for $n = 0$ and for $n = -2$. For a further discussion of the properties of the linear susceptibility we refer to Ref. [27]. We only mention that $\chi_{1,GT}(\omega)$ is basically independent of the choice of n and that $\chi''_{1,GT}(\omega)$ broadens with decreasing temperature meaning that time–temperature superposition is not obeyed in the Gaussian trap model.

In the calculation of the nonlinear response functions, we have to fix the values of γ and μ in Eq. (34). Due to the fact that we consider a thermally activated escape from the initial trap of any transition with randomly chosen destination state, we assume $\mu = 1$ and $\gamma = 0$ as in the case of rotational random jumps. We note, however, that the results for the third-order response do not strongly depend on this particular choice [27].

We will not discuss the details of the calculations of the higher order response functions. Here, it suffices to mention that in the computation of $\chi^{(3)}_{3,GT}(\omega)$ the fourth moment $\langle M(\epsilon_1)M(\epsilon_2)M(\epsilon_3)M(\epsilon_4)\rangle$ and in case of $\chi^{(5)}_{5,GT}(\omega)$ the sixth moment $\langle M(\epsilon_1)M(\epsilon_2)M(\epsilon_3)M(\epsilon_4)M(\epsilon_5)M(\epsilon_6)\rangle$ have to be calculated. For these moments a

Gaussian factorization approximation was applied. This appears meaningful in terms of the physical properties of the model, it is however unclear how correlations will modify the results. For more details concerning the actual calculations and for the analytical expressions for the response functions we refer to Refs. [27, 28].

The nonlinear susceptibilities for $n = 0$ are shown as a function of frequency in Fig. 12. It is apparent that both, the real and the imaginary part of the susceptibilities vanish at low frequency. This can be understood from the analytic expressions for the corresponding limits. From Fig. 12 it is clear that a hump-like behavior is to be expected for the moduli. These quantities are plotted in Fig. 13, from which it is clear that there is a hump in both, the third-order and the fifth-order scaled modulus. The temperature dependence of the height of the peaks, however, is opposite to what is observed experimentally. The height increases with increasing temperature. Depending on the value of the model parameter n, different temperature dependencies for the maximum height of the hump are observed. This is shown for some examples in Fig. 14. It is obvious that for some values one observes a hump with a decreasing height as a function of temperature and for other values one has either a nearly temperature-independent behavior or an increase with temperature. This means, that calculations employing a simple mean-field like model like the Gaussian trap model considered here yield a rich scenario with very different results. A direct comparison between the height of the humps in X_3 and X_5 does not show the $X_5 \sim X_3^2$ behavior expected for the model of correlated domains [26].

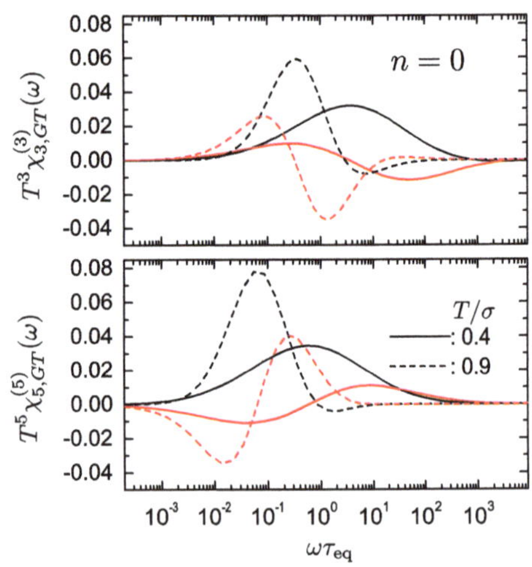

Fig. 12 $T^3\chi^{(3)}_{3,GT}(\omega)$ (upper panel) and $T^5\chi^{(5)}_{5,GT}(\omega)$ (lower panel) as a function of frequency for $n = 0$, i.e., energy-independent variables. The black lines represent the real part and the red lines the imaginary part. The relaxation time is given by $\tau_{eq} = \int dt C_M(t) = \kappa_\infty^{-1} e^{(3/2)\beta^2\sigma^2}$ (Figure adopted from Ref. [28])

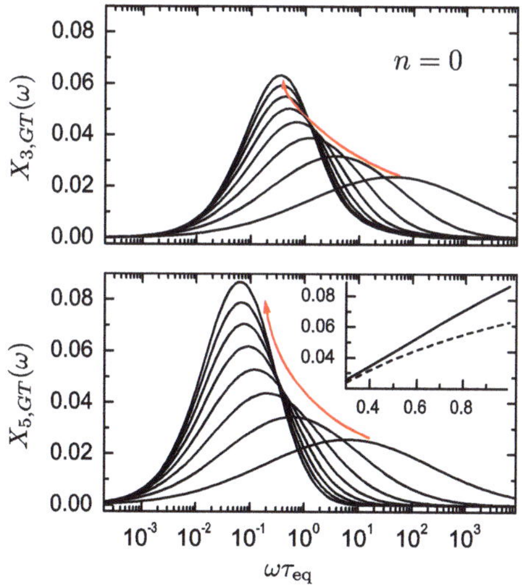

Fig. 13 $X_{3,GT}(\omega)$ (upper panel) and $X_{5,GT}(\omega)$ (lower panel) as a function of frequency for $n = 0$ and for $T = 0.3\sigma$ to $T = \sigma$ as indicated by the arrows. The inset in the lower panel shows the temperature dependence of the height of the hump with that of X_5 given by the full line and that of X_3 by the dashed lines (Figure adopted from Ref. [28])

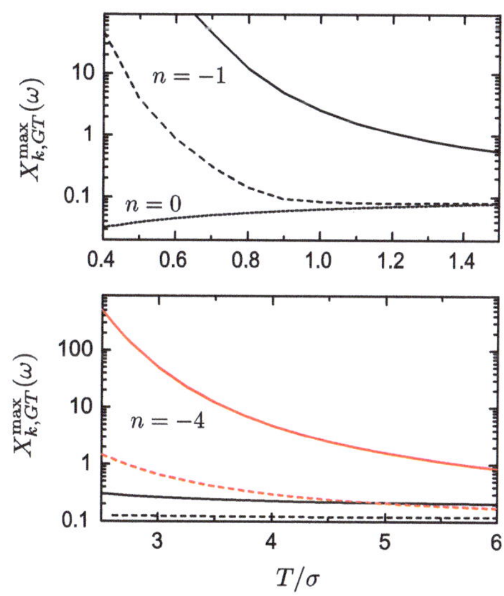

Fig. 14 $X^{\max}_{k,GT}$ for different values of n as a function of temperature. Full lines represent the fifth-order and dashed lines the third-order hump maxima. Upper panel: $n = -1$ and $X^{\max}_{3,GT}$ for comparison. Lower panel: $n = -4$ (red lines) and $n = 1$ (black lines)

5 Conclusions

In order to gain a deeper understanding of the information content of nonlinear dielectric response functions it is necessary to consider explicit models for the reorientational motion and the relaxation in the systems considered. The reason for this necessity lies in the fact that no analogue exists to the well-known fluctuation dissipation theorem holding for linear response functions. Therefore, nonlinear response functions cannot be related to equilibrium (multi-time) correlation functions in general and must be computed separately for each model considered.

In the present chapter, we reviewed the results of such calculations for models with a dynamics that can be viewed as Markovian. In this case, the time evolution of dynamic variables is governed by a ME for the corresponding probabilities. Response functions then are calculated as expectation values of the relevant (dipole) moments and time-dependent perturbation theory for the corresponding Green's functions is used to obtain the results in the desired order in the external electric field. In these calculations, the dependence of the transition rates on the field has to be fixed. A quite general model is provided by assuming an exponential dependence that can be motivated by the fact that the Boltzmann factors are modified due to the contribution of the dipole energy. In case of diffusive dynamics, the ME turns into a FPE and the exponential field dependence gives the corresponding force term in the FP-operator.

The important experimental observation of a hump in the nonlinear susceptibility of some glass-forming liquids has been interpreted in terms of the existence and the growth of amorphous order. The calculations using stochastic models for the reorientational motion of molecules presented in the present chapter show that it is possible to observe a hump-like behavior in case of rotational random jumps if it assumed that the field couples to the initial orientation of a transition. However, the height of the observed peak is temperature independent for this model. If on the other hand rotational diffusion is used as a model for molecular reorientations, only trivial behavior is observed.

Some further calculations have been performed that also exhibit a hump without any glassy correlations. If the model of reorientations in an asymmetric double-well potential is considered the observed decrease of the height of the hump with increasing temperature in X_3 is found for temperatures above T_3. However, taking into account X_5 one has to assume that the relevant temperature regime is above $T_{5;b}$. For these temperatures, however, X_3 only shows trivial behavior. Therefore, it appears that such a model cannot be used for the interpretation of experimental results. We have furthermore shown that also trap models that show some features of glassy relaxation can yield hump-like shapes for the third-order and fifth-order moduli for some values of the model parameters. It is possible to obtain different temperature dependencies of the heights, but in most cases the experimentally determined relations between X_3 and X_5 seems not to be observed.

In conclusion, it appears that measurements of different higher order nonlinear response functions are helpful to discriminate among various models for the relaxational processes in supercooled liquids and glasses.

Acknowledgements Useful discussions with Roland Böhmer, Gerald Hinze, Francois Ladieu, and Jeppe Dyre are gratefully acknowledged.

References

1. F. Kremer, A.E. Schönhals, *Broadband Dielectric Spectroscopy* (Springer, Berlin, 2002)
2. P. Lunkenheimer, U. Schneider, R. Brand, A. Loidl, Contemp. Phys. **41**, 15 (2000)
3. R. Richert, Adv. Chem. Phys. **156**, 101 (2014)
4. R. Böhmer, R. Chamberlin, G. Diezemann, B. Geil, A. Heuer, G. Hinze, S. Kübler, R. Richert, B. Schiener, H. Sillescu, H. Spiess, U. Tracht, M. Wilhelm, J. Non-Cryst, Solids **235–237**, 1 (1998)
5. M. Ediger, Annu. Rev. Phys. Chem. **51**, 99 (2000)
6. E.V. Russell, N.E. Israeloff, Nature **408**, 695 (2000)
7. R. Richert, J. Phys.: Cond. Matter **14**, R703 (2002)
8. K. Schmidt-Rohr, H. Spiess, Phys. Rev. Lett. **66**, 3020 (1991)
9. A. Heuer, M. Wilhelm, H. Zimmermann, H. Spiess, Phys. Rev. Lett. **75**, 2851 (1995)
10. R. Böhmer, G. Hinze, G. Diezemann, B. Geil, H. Sillescu, Europhys. Lett. **36**, 55 (1996)
11. B. Schiener, R. Böhmer, A. Loidl, R. Chamberlin, Science **274**, 752 (1996)
12. B. Schiener, R. Chamberlin, G. Diezemann, R. Böhmer, J. Chem. Phys. **107**, 7746 (1997)
13. R.V. Chamberlin, Phys. Rev. Lett. **83**, 5134 (1999)
14. X. Shi, G.B. McKenna, Phys. Rev. B **73**, 1 (2006)
15. J. Bouchaud, G. Biroli, Phys. Rev. B **72**, 064204 (2005)
16. R. Richert, J. Phys.: Cond. Matter **29**, 363001 (2017)
17. C. Crauste-Thibierge, C. Brun, F. Ladieu, D. L'Hote, G. Biroli, J.-P. Bouchaud, Phys. Rev. Lett. **104**, 165703 (2010)
18. C. Brun, F. Ladieu, D. L'Hote, M. Tarzia, G. Biroli, J.P. Bouchaud, Phys. Rev. B **84**, 104204 (2011)
19. R. Richert, S. Weinstein, Phys. Rev. Lett. **97**, 095703 (2006)
20. S. Weinstein, R. Richert, Phys. Rev. B **75**, 064302 (2007)
21. S. Rzoska, A. Drozd-Rzoska, J. Phys.: Cond. Matter **24**, 035101 (2012)
22. M. Tarzia, G. Biroli, A. Lefevre, J.-P. Bouchaud, J. Chem. Phys. **132**, 054501 (2010)
23. F. Ladieu, C. Brun, D. L'hôte, Phys. Rev. B **85**, 184207 (2012)
24. U. Buchenau, J. Chem. Phys. **146** (2017)
25. J. Dejardin, Y. Kalmykov, Phys. Rev. E **61**, 1211 (2000)
26. S. Albert, T. Bauer, M. Michl, G. Biroli, J.P. Bouchaud, A. Loidl, P. Lunkenheimer, R. Tourbot, C. Wiertel-Gasquet, F. Ladieu, Science **352**, 1308 (2016)
27. G. Diezemann, Phys. Rev. E **85**, 051502 (2012)
28. G. Diezemann, Phys. Rev. E **96**, 022150 (2017)
29. A. Morita, Phys. Rev. A **34**, 1499 (1986)
30. J. Dejardin, G. Debiais, Adv. Chem. Phys. **91**, 241 (1995)
31. Y. Kalmykov, Phys. Rev. E **65**, 021101 (2001)
32. P.M. Déjardin, F. Ladieu, J. Chem. Phys. **140**, 034506 (2014)
33. N. van Kampen, *Stochastic Processes in Physics and Chemistry* (North-Holland, Amsterdam, 1981)
34. H. Risken, *The Fokker-Planck Equation* (Springer, Berlin, 1989)
35. G. Gardiner, *Handbook of Statistical Methods for Physics, Chemistry and Natural Sciences* (Springer, Berlin, 1997)
36. D.T. Gillespie, *Markov Processes* (Academic Press, San Diego, 1992)
37. G. Diezemann, H. Sillescu, J. Chem. Phys. **111**, 1126 (1999)
38. R. Böhmer, G. Diezemann, G. Hinze, E.A. Roessler, Prog. Nucl. Magn. Reson. Spectrosc. **39**, 191 (2001)
39. A. Kivelson, S. Kivelson, J. Chem. Phys. **90**, 4464 (1989)
40. L. Alessi, L. Andreozzi, M. Faetti, D. Liporini, J. Chem. Phys. **114**, 3631 (2001)
41. G. Diezemann, J. Chem. Phys. **107**, 10112 (1997)
42. G. Diezemann, H. Sillescu, G. Hinze, R. Böhmer, Phys. Rev. E **57**, 4398 (1998)
43. G. Diezemann, R. Böhmer, G. Hinze, H. Sillescu, J. Non-Cryst, Solids **235–237**, 121 (1998)

44. R. Böhmer, G. Diezemann, *Broadband Dielectric Spectroscopy* (Springer, Berlin, 2002)
45. K. Schulten, Z. Schulten, A. Szabo, J. Chem. Phys. **74**, 4426 (1981)
46. J. Dyre, Phys. Rev. B **51**, 12276 (1995)
47. C. Monthus, J. Bouchaud, J. Phys. A: Math. Gen. **29**, 3847 (1996)
48. R. Denny, D. Reichman, J. Bouchaud, Phys. Rev. Lett. **90**, 025503 (2003)
49. G. Diezemann, J. Phys.: Cond. Mat. **19**, 205107 (2007)
50. C. Rehwald, N. Gnan, A. Heuer, T. Schrøder, J. Dyre, G. Diezemann. Phys. Rev. E **82**, 021503 (2010)
51. A. Heuer, J. Phys.: Cond. Matter **20**, 373101 (2008)
52. N. Agmon, J. Hopfield, J. Chem. Phys. **78**, 6947 (1983)
53. G. Diezemann, Europhys. Lett. **53**, 604 (2001)
54. A. Crisanti, F. Ritort, J. Phys. A: Math. Gen. **36**, R181 (2003)
55. G. Diezemann, Phys. Rev. E **72**, 011104 (2005)
56. S. Fielding, P. Sollich, Phys. Rev. Lett. **88**, 050603 (2002)

Effects of Strong Static Fields on the Dielectric Relaxation of Supercooled Liquids

Ranko Richert

Abstract When large DC-bias fields are applied to polar dielectric liquids, the orientational polarization of dipoles will lead to a considerable macroscopic dipole moment of the sample. In this situation, the dielectric relaxation behavior probed by a small amplitude AC-field superimposed onto the large DC-field will differ from the zero-bias field limit. This chapter summarizes the experimental approaches to dielectric spectroscopy in the presence of a large amplitude static field and the findings from such experiments. Only nonlinear effects that are completely reversible will be addressed, focusing on glass forming materials, as systems near their glass transition turn out to be particularly sensitive to external fields. The relation to third harmonic responses obtained from AC-fields is briefly discussed.

1 Introduction

Dielectric material properties are characterized by the dielectric constant ε or the dielectric susceptibility $\chi = \varepsilon - 1$. These quantities are defined by the respective constitutive equations, $D = \varepsilon\varepsilon_0 E$ or $P = \chi\varepsilon_0 E$, where E is the external electric field, D is the dielectric displacement, P is the polarization, and $\varepsilon_0 = 8.854 \times 10^{-12}$ $\mathrm{AsV^{-1}m^{-1}}$ represents the permittivity of vacuum [1]. Here, the susceptibility χ is meant to be a constant, implying that P (and likewise D) is proportional to the field E. For a considerable range of fields, this proportionality can be verified by experiment. However, if very large fields are applied or if polarization is detected with very high resolution, deviations from this linear behavior can be observed [2].

The above constitutive equations reflect steady-state quantities, equivalent to the limit of zero frequency. A much more complete characterization of a dielectric is obtained via broadband dielectric spectroscopy, i.e., with frequency as an additional

R. Richert (✉)
School of Molecular Sciences, Arizona State University, Tempe, AZ 85287-1604, USA
e-mail: ranko@asu.edu

R. Richert (ed.), *Nonlinear Dielectric Spectroscopy*, Advances in Dielectrics
https://doi.org/10.1007/978-3-319-77574-6_4

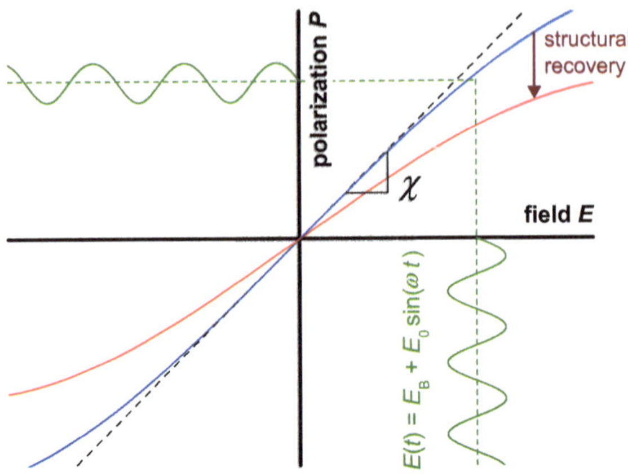

Fig. 1 Linear (dashed) and nonlinear (solid) relation in a P versus E representation. Oscillating curves represent fields $E(t)$ and the resulting responses $P(t)$ for the case of a high DC-bias field, E_B. Here, the slope that defines 'χ' represents the derivative $\partial P/\partial E = \varepsilon_0\chi(\omega, E_B, t_{HF})$ at a certain temperature and pressure, where the quantity t_{HF} is the time for which the system has been exposed to the high field. With increasing t_{HF}, structural recovery lets the system approach equilibrium with the high field

variable to obtain a spectrum $\hat{\varepsilon}(\omega) = \varepsilon'(\omega) - i\varepsilon''(\omega)$ [3, 4]. In order to measure a dielectric spectrum in the presence of a high DC-bias field, the typical approach is to superpose a large static field (E_B) and a small amplitude (E_0) oscillatory field, such that $E(t) = E_B + E_0 \sin(\omega t)$ with E_0 itself being within the linear response regime. This will lead to a polarization $P(t) = P_B + P_0(t)$, where the oscillating component, $P_0(t)$, can be used to obtain the permittivity $\hat{\varepsilon}(\omega)$ in the presence of the bias field. This situation is outlined schematically in a P versus E diagram in Fig. 1. Only for small DC-fields one can expect that the dielectric relaxation spectrum is independent of E_B.

The following section will outline the experimental techniques used to measure dielectric relaxation spectra for samples subject to a large DC-bias field. Experimental findings and a discussion of their explanations will be subdivided in the following manner. First, only those features will be addressed in which the system is in the steady state in the sense that the material is in equilibrium with the external electric field. This part is subdivided into nonlinear effects for which the dielectric relaxation amplitude is affected and those for which the relaxation time constant changes with the DC-field. This division is based upon the phenomenology of the effect and thus somewhat arbitrary, and the separate discussion is not meant to imply that these are independent effects. A subsequent section will then address the time scales the sample requires to achieve equilibrium with the field, a process we will refer to as structural recovery, due to its analogy to physical aging [5].

2 Experimental Approaches

2.1 Observations in the Stationary Limit

Both a high DC-bias voltage (V_B) and a small electrode separation (d) work in favor of a high field $E = V/d$. High fields that exceed the dielectric strength (breakdown field) of either the sample or the insulator used as electrode spacer can result in the destruction of the sample, the spacers, the electrodes, as well as the detection circuitry involved in the permittivity experiments. Therefore, protective measures should be in place to avoid the consequences of material failure. Typical spacer polymers used in disk geometries are polytetrafluorethylene (Teflon, PTFE) and polyimide (Kapton), with dielectric strengths of 180 and 220 kV/cm at ambient conditions, respectively [6].

Unless the capacitor electrode geometry is extremely rigid, the electrostrictive force resulting from a high DC-field can cause an apparent change in the permittivity [7, 8]. For a disk capacitor geometry with surface area A and separation d, the electrostrictive force amounts to $F = \varepsilon_s \varepsilon_0 A E_B^2 / 2$. Assuming electrode spacers with the total surface area a and Young's modulus Y, this force can lead to a relative distance change as large as $\Delta \ln d = -\varepsilon_s \varepsilon_0 A E_B^2 / (4aY)$. Accounting for the rigidity of the sample itself will reduce the effect further. This contribution to the field induced change of permittivity must be minimized or corrected for in order to obtain the true field effect on the sample. Under typical conditions, this effect is relatively small compared with other sources of nonlinear effects [9].

Common to practically all dielectric methods involving high DC-fields is the limited time for which the high field is applied. The main concerns associated with long time exposure to a DC-field are the effects of Joule heating and electrode polarization. Most samples studied by dielectric techniques display DC-conductivity quantified by σ_{dc}, which arises from the drift of unbound charges in the presence of an electric field. Exposure to a high field leads to the introduction of heat via the power density given by $p = jE = \sigma_{dc} E^2$. The amount q of heat added to the system then increases linearly with time, $q = p \times t$. The impact of Joule heating on the temperature of the sample greatly depends on the rate of heat loss to the electrodes, which often act as 'infinite' heat sinks relative to the small heat capacity of the sample. Assuming constant electrode temperatures, the average temperature increase within the sample amounts to $\Delta T_{avg} = pd^2/(12\kappa)$. Therefore, for a given power density p, the Joule heating effect on T is increased linearly with the power p, quadratically with electrode separation d, and reduced with thermal conductivity κ.

In addition to Joule heating, there is a transient heating effect that results from the field step when the electric field is changed from zero to a level of E_B for a polar material [10–12]. This electrocaloric effect originates from the entropy density change, $\Delta_E s$, that results from applying the field, which can be estimated using the relation $\Delta_E s = \varepsilon_0 (\partial \varepsilon_s / \partial T) E_B^2 / 2$ [1, 13], where the slope $(\partial \varepsilon_s / \partial T)$ is negative for most materials. In an adiabatic situation, or similarly for very short times, the temperature increase would amount to $\Delta T = -T \Delta_E s / (\rho C_p)$, where ρC_p is the volumetric heat capacity. This value for ΔT can serve as upper bound for nonadiabatic conditions. When the field is removed, the sample is cooled by the depolarization-induced entropy increase. Both power transients (field on, field off) will last for approximately the dielectric relaxation time of the system [12]. Obviously, this effect will not be influenced by the duration of how long the field is being applied.

In the long time limit, the current associated with DC-conductivity cannot be sustained because most electrodes (metals with electronic conduction mechanism) create blocking conditions with respect to the migration of ions. As a result, a DC-field will eventually lead to electrode polarization, i.e., a space charge build-up near the sample/electrode interfaces [14]. This space charge will counteract the charges that create the external electric field, so that the sample interior will be subject to a net field that is diminished relative to the externally applied field, $E_B = V_{\mathrm{B}}/d$, and the nonlinear effects will decline accordingly.

Regarding the timing of applying the bias field and collecting permittivity data from an additional small signal oscillating field, the bottom line of the above considerations is as follows. Subsequent to switching on the bias field, stationary conditions will not be achieved until the polarization response to the field step, the entropy heating transient, and the sample equilibrating with the new field all have completed. Data collection should begin only after steady state with respect to these processes has been achieved, but prior to changes that may result from Joule heating or electrode polarization. Several different experimental approaches to measuring DC-field nonlinear dielectric effects in the stationary limit have been described in the literature [15–21].

2.2 *Time-Resolved Experiments*

When a small field step is applied to a depolarized sample, the polarization builds up in a retarded fashion. If this polarization is within the regime of linear response, then the parameters characterizing the relaxation amplitude $\Delta\varepsilon = \varepsilon_s - \varepsilon_\infty$ and the relaxation time τ_{D} are not affected by the presence of the field. A sufficiently large electric DC-field, however, leads to changes in the relaxation parameters ($\Delta\varepsilon$, τ_{D}) and these quantities require some time to achieve their steady-state level. This process is analogous to the physical aging that occurs following a temperature step of considerable magnitude, and it is thus reasonable to adapt the term 'structural recovery' for the approach of nonlinear dielectric effects to their stationary levels [22].

Experimental techniques that are meant to record the evolution of nonlinear effects as a function of time generally fall into two categories. If one period of the applied frequency is very short compared with the desired time resolution, then permittivity can be obtained virtually in real time while the system is approaching the steady-state behavior. Equipment designed to achieve this with very high resolution has been described by Górny et al. [17].

An alternative technique is required when the aim is to obtain permittivity for each period of the applied field, i.e., with a time resolution of $2\pi/\omega$, where ω is the frequency of the oscillating field component [21]. In such a case, a smaller number of periods of the AC-signal fill the time, t_{HF}, that the high field is applied, with $t_{HF}/(2\pi/\omega) \approx 10$–1000. This situation facilitates recording the voltage (V) and current (I) traces as a function of time before and after the DC-field step-up or step-down, ideally with a density of at least a few 100 points per period. A schematic outline of the field protocol involved is presented in Fig. 2. Then, these curves are analyzed by subjecting the signal $\Psi(t)$ of interest, current $I(t)$, or voltage $V(t)$, to a period-by-period Fourier analysis, using

$$\Psi'(\omega) = \frac{\omega}{\pi} \int_{t}^{t+2\pi/\omega} \sin(\omega t')\, \Psi(t')\, \mathrm{d}t', \tag{1a}$$

$$\Psi''(\omega) = \frac{\omega}{\pi} \int_{t}^{t+2\pi/\omega} \cos(\omega t')\, \Psi(t')\, \mathrm{d}t', \tag{1b}$$

where t would progress in increments of $2\pi/\omega$. The quantities Ψ' and Ψ'' respectively denote the in-phase and out-of-phase amplitudes of the oscillating signal at the fundamental frequency. For each period, the signal amplitude is obtained by $A^{(\Psi)} = \sqrt{\Psi'^2 + \Psi''^2}$ and its phase by $\varphi^{(\Psi)} = \arctan\left(\Psi'/\Psi''\right)$. From these values for current ($\Psi = I$) and voltage ($\Psi = V$), the time-dependent analogs of ε' and ε'' can be obtained via

$$\varepsilon' = \frac{A^{(I)} \sin \Delta\varphi}{\omega A^{(V)} C_{\mathrm{geo}}}, \quad \varepsilon'' = \frac{A^{(I)} \cos \Delta\varphi}{\omega A^{(V)} C_{\mathrm{geo}}}, \tag{2}$$

where $\Delta\varphi = \left|\varphi^{(I)} - \varphi^{(V)}\right|$ and with $C_{\mathrm{geo}} = \varepsilon_0 A/d$ representing the geometric capacitance. Strictly speaking, susceptibility χ as well as related quantities such as ε' and ε'' are correctly defined only for stationary conditions.

As indicated in Fig. 2, both field steps, on and off, lead to a current response that hampers the Fourier analysis of the first periods after a field step. Since the field step response lasts for about the structural relaxation time of the system, the periods affected by the step response are those during which the time dependence of interest occurs. A remedy to this problem is to record the signals again, but with the polarity of the DC-field inverted, and then compute the average of the two results. In that manner, the entire dataset can be subjected to Fourier analysis.

3 Nonlinear Effects in the Stationary Limit

This section deals with results obtained in the stationary limit, i.e., those for which the sample is in equilibrium with both the small amplitude AC-field component and, more importantly, with the strong DC-field. As will be discussed in more detail further

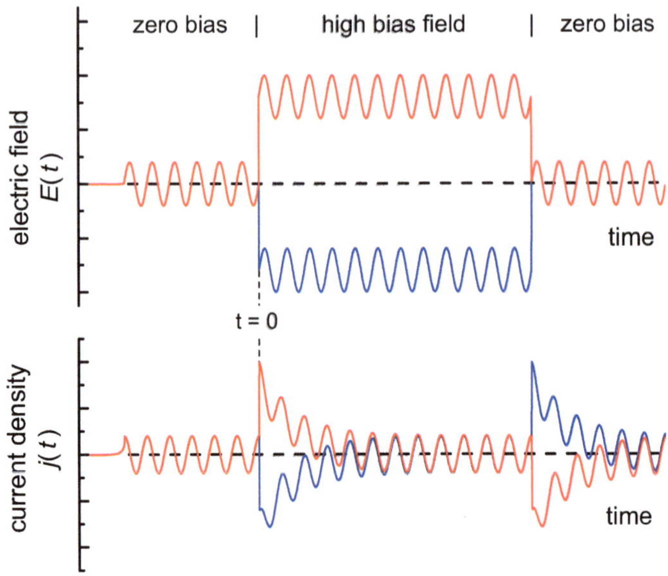

Fig. 2 Field protocols for the time-resolved experimental approaches to nonlinear effects. Low amplitude AC-fields before and after the positive or negative DC-bias interval facilitate probing the low field limit of χ and the decay of the nonlinear effect. Averaging over positive and negative going DC-field steps facilitates the elimination of the step response, so that the odd Fourier components can be obtained even for times for which the step response would otherwise distort the oscillatory signal

below, it can take a time that is approximately equal to the structural relaxation time before the nonlinear effect has approached its steady-state value. A schematic and simplified view of how the permittivity is expected to be modified by a large DC-field is depicted in Fig. 3, albeit using exaggerated changes in both the amplitude and the time constant between the low and high field cases. While this approach is one out of many different routes to quantifying nonlinear dielectric behavior, the basis for Fig. 3 is to approximate the field-induced changes in the dielectric relaxation in terms of the two parameters ($\Delta\varepsilon$, τ_α) that are most sensitive to a DC electric field in the case of polar liquids. For a more detailed description, changes in the shape parameters of the loss profile and higher Fourier components of the dielectric response may be required.

3.1 Field-Induced Changes in Relaxation Amplitude

In a polar liquid, a considerable contribution to polarization can be due to orientational polarizability (α_{or}) of molecular dipoles, while the electronic polarizability (α_{el}) is small by comparison. For noninteracting dipoles with moments μ, the polarization

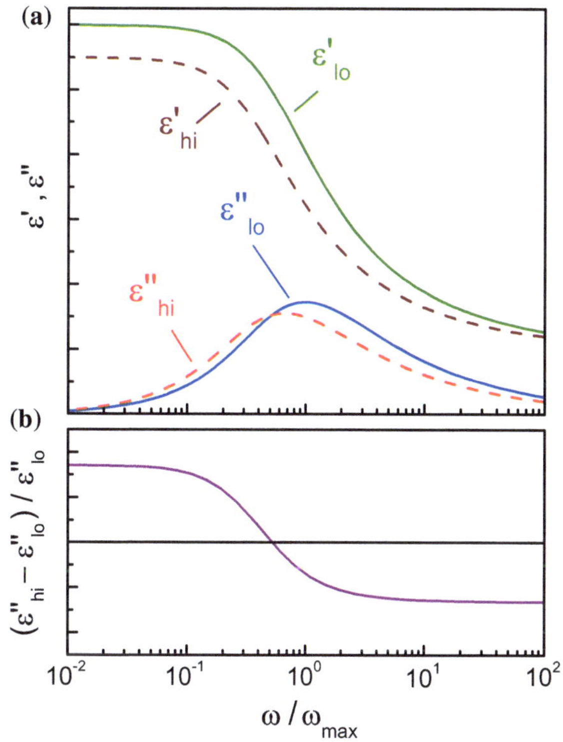

Fig. 3 **a** Calculated dielectric storage (ε'_{lo}) and loss ($\varepsilon''_{\text{lo}}$) components for a HN-type system (τ_{HN} = 1 s, $\alpha_{\text{HN}} = 1$, and $\gamma_{\text{HN}} = 0.5$) representing the low field limit ('lo'). The high field counterparts (ε'_{hi}, $\varepsilon''_{\text{hi}}$) are based on the same parameters but with smaller $\Delta\varepsilon$ and larger τ_{HN} to model the saturation and electrorheological effect, respectively. Curves are shown on a reduced frequency scale, $\omega/\omega_{\max}$, where $\omega_{\max}$ is the loss peak frequency in the low field limit. **b** The field-induced change of the dielectric loss, $(\varepsilon''_{\text{hi}} - \varepsilon''_{\text{lo}})/\varepsilon''_{\text{lo}}$, relative to the low field limit, ($\varepsilon''_{\text{lo}}$). The curve is based on the loss data shown in panel (**a**), where the changes are strongly exaggerated compared with typical experiments

$P \propto (\alpha_{\text{el}} + \alpha_{\text{or}})\, E$ can be expressed in terms of the average of the cosine of the angle θ between dipole moment and field direction, $P \propto \alpha_{\text{el}} E + \mu \langle \cos\theta \rangle$ [3, 4]. At very small fields, the value of $\langle \cos\theta \rangle$ increases linearly with the field, but it can not exceed unity, because $\langle \cos\theta \rangle = 1$ corresponds to the situation in which all dipoles are perfectly aligned with the field, and a further increase of E will not lead to additional polarization. This implies that the practically linear relation between P and E will have to break down for sufficiently high fields, leading to a reduction of the susceptibility χ in the relation $P = \chi \varepsilon_0 E$.

This above notion of nonlinear behavior can be quantified by requiring that the equilibrium distribution of dipole orientations corresponds to maximal entropy at a given temperature T. For noninteracting dipoles, the value of $\langle \cos\theta \rangle$ as a function of field E and temperature T is given by [23, 24]

$$\langle\cos\theta\rangle = \frac{\iint\limits_{4\pi} \cos\theta \, e^{\mu E \cos\theta / kT} d\Omega}{\iint\limits_{4\pi} e^{\mu E \cos\theta / kT} d\Omega} = \mathrm{cotanh}\,(a) - \frac{1}{a} = L\,(a)\,. \tag{3}$$

Here, $L(a)$ represents the Langevin function and its argument a is defined as $a = \mu E/k_{\mathrm{B}}T$. The dependence of $\langle\cos\theta\rangle$ on the field can be approximated by

$$L\,(a) \approx \frac{1}{3}a - \frac{1}{45}a^3 + \frac{2}{945}a^5 - \frac{2}{9450}a^7 + \cdots. \tag{4}$$

The recognition of this feature has led Debye to state: "Here we see that the mean moment is not a linear function for large values of the argument, and for such values the dielectric constant would not be a true constant but would depend upon the field intensity" [25]. It is important to realize that the quantitative treatment of the saturation effect of Eq. (4) is valid only for noninteracting dipoles.

For the more general case of a polar liquid, a description of steady-state polarization that includes nonlinear terms would read

$$\frac{P}{\varepsilon_0} = \chi E + \chi^{(3)} E^3 + \chi^{(5)} E^5 + \chi^{(7)} E^7 + \cdots, \tag{5}$$

where E represents a static field. In practice, the susceptibility in the presence of a high DC-bias field would be determined by a small amplitude oscillatory field, i.e., using $E\,(t) = E_B + E_0 \sin\,(\omega t)$ with $E_0 \ll E_B$. Provided that the amplitude E_0 is sufficiently small, the polarization would display two Fourier components, a static level (P_0) and another component (P_1) that oscillates at a frequency ω [26]:

$$\frac{\hat{P}_0}{\varepsilon_0 E_B} = \hat{\chi}_0^{(3)} E_B^2, \tag{6a}$$

$$\frac{\hat{P}_1\,(\omega)}{\varepsilon_0 E_0} = \hat{\chi}_1^{(1)}\,(\omega) + \hat{\chi}_1^{(3,dc)}\,(\omega)\, E_B^2. \tag{6b}$$

Higher order Fourier components will emerge only when E_0 is large enough to pick up the curvature of $P(E)$ at the high DC-fields. The quantities $\chi_k^{(n)}$ represent contributions from the nth power of the field in the kth Fourier component, where $\chi_1^{(1)}$ is usually written as χ. There is no simple connection among the values of $\chi_k^{(n)}$ for a given material, and for higher fields additional terms such as $\chi_k^{(5)}$ may be required.

In order to study the impact of a large DC-bias electric field on the dielectric constant (ε_s) or the relaxation amplitude ($\Delta\varepsilon$), it is sufficient to measure the effect of E_B at a single frequency ω positioned within the low-frequency plateau of $\varepsilon'(\omega)$. Because electronic polarizability and thus ε_∞ ($\approx n^2$) is very insensitive to the field magnitude, the field effect on $\Delta\varepsilon$ and on ε_s are usually equivalent. In a first approximation, deviations from the low field limit depend quadratically on the field, so that the Piekara factor, $\Delta_E \varepsilon_s / E^2$ or $(\varepsilon_{\mathrm{hi}} - \varepsilon_{\mathrm{lo}}) / E^2$, is a useful metric of nonlinear dielectric effects [27], where $\varepsilon_{\mathrm{hi}}$ and $\varepsilon_{\mathrm{lo}}$ denote the high and low field values, respec-

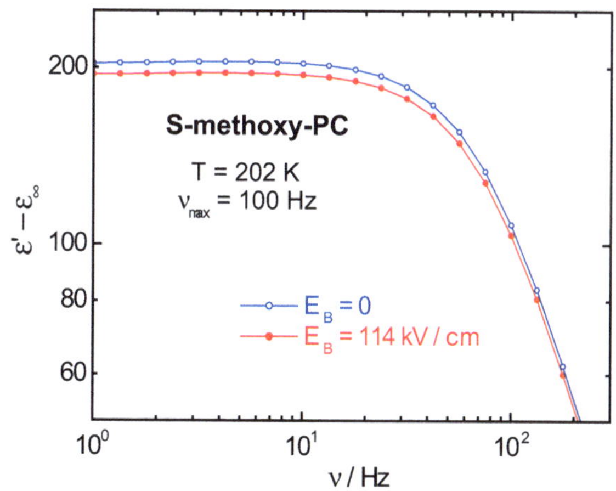

Fig. 4 Frequency dependent real part of permittivity, $\varepsilon' - \varepsilon_\infty$, for (*S*)-(–)-methoxy-PC at $T = 202$ K for a high bias field of $E_B = 114$ kV/cm (solid symbols) and for zero bias (open symbols). At low frequencies, the field-induced reduction of $\Delta\varepsilon$ amounts to 5%. Data from Ref. [57]

tively. Relative changes of the dielectric relaxation amplitude have also be used to report field effects, derived either from the real or imaginary contribution to permittivity at a given frequency, $\left(\varepsilon'_{hi} - \varepsilon'_{lo}\right) / \left(\varepsilon'_{lo} - \varepsilon_\infty\right)$ or $\left(\varepsilon''_{hi} - \varepsilon''_{lo}\right) / \varepsilon''_{lo}$, respectively. As long as E_0 is within the low field limit regime and no other source of nonlinear behavior is present, a frequency dependence is not expected.

This saturation effect has been verified experimentally for numerous liquids, including diethyl ether, for which Herweg has reported a reduction of the dielectric constant with the field as early as 1920 [2]. Typical field induced relative changes of $\Delta\varepsilon$ amount to about 1% for polar liquids at fields of the order of few hundred kV/cm. A large number of materials has been characterized in terms of the dielectric saturation effect [2, 28–32] and various theories have been developed to rationalize saturation in condensed dipole systems such as liquids [33–40]. An example of a strong saturation effect is shown for a highly polar supercooled liquid in Fig. 4, revealing a reduction of 5% of $\Delta\varepsilon$ at a field only $E_B = 114$ kV/cm.

In the 1930s, it has been discovered by Piekara and Piekara that the dielectric constant can *increase* by virtue of a large electric field [28]. This was observed for a composition series involving benzene and nitrobenzene. While the benzene-rich mixture displayed the negative values for $\Delta_E\varepsilon$ that are typical of saturation, the sign of $\Delta_E\varepsilon$ changed to positive when the mixture was high in nitrobenzene content. This nonlinear dielectric effect has been labeled as 'negative saturation', but the present term 'chemical effect' is more appropriate, as it reflects better the origin of this increase of the dielectric constant [41]. Very generally, whenever molecular or supramolecular structures with different net dipole moments coexist in thermodynamic equilibrium in the liquid, an electric field will generate a preference for the more polar species [42]. In order to lead to a net positive change in $\Delta\varepsilon$, the field must increase the dielectric constant by more than the saturation that will occur inevitably at the same time.

A variety of mechanisms can be envisioned that give rise to such chemical effects. The structure of molecules with intramolecular degrees of freedom can be altered toward a higher molecular dipole moment, or two isomers with different dipole moment may coexist in dynamic equilibrium. Small effects could result from aligning the axes of highest electronic polarizability with the electric field, analogous to the Kerr effect. Larger field induced effects are possible when different supramolecular structures are in thermodynamic equilibrium. Even if the molecular dipole moment remains virtually unaltered, distinct supramolecular structures can differ in their effective dipole moment via differences in their Kirkwood correlation factors [43–46].

Correlation in the orientation of dipoles can influence the dielectric constant considerably, even at a constant dipole density. This feature is captured in the Kirkwood–Fröhlich equation [1, 3, 5, 8],

$$\frac{(\varepsilon_s - \varepsilon_\infty)(2\varepsilon_s + \varepsilon_\infty)}{\varepsilon_s (\varepsilon_\infty + 2)^2} = \frac{\rho N_\mathrm{A} \mu^2}{9 k_\mathrm{B} T \varepsilon_0 M} \times g_\mathrm{K}, \tag{7}$$

where ε_s and ε_∞ are the dielectric constants in the limits of low and high frequency, respectively. In Eq. (7), N_A is Avogadro's constant, ρ is the density, M is the molar mass, k_B is Boltzmann's constant, and ε_0 represents the permittivity of vacuum. The correlation factor g_K in Eq. (7) is often expressed in terms of the average dipole orientation of the z neighboring dipoles, $g_\mathrm{K} = 1 + z \langle \cos\theta \rangle$, and it can be viewed as rescaling the squared molecular dipole moment, μ^2, to an effective value, $\mu_\mathrm{eff}^2 = \mu^2 g_\mathrm{K}$. Depending on the supramolecular structure of the system, μ_eff can turn out larger or smaller than μ itself, depending on whether g_K is larger or smaller than unity. Values of $g_\mathrm{K} > 1$ indicate more parallel alignment of dipoles, whereas $g_\mathrm{K} < 1$ reflects antiparallel dipoles [47].

Electrostatic interactions among dipoles can lead to departures of g_K from unity, but stronger effects can be expected when orientational correlations result from covalent bonds (as in polymers) or from hydrogen bonds as in alcohols. In particular, monohydric alcohols can display a considerable range of g_K values [48], from 0.1 to 4.2 within a series of octanol isomers [45–47]. The case $g_\mathrm{K} = 0.1$ is understood as indicating mainly ring-like structures stabilized by hydrogen bonds, in which dipoles cancel effectively. The other extreme, $g_\mathrm{K} = 4.2$, would suggest a preference for hydrogen-bonded chains in which dipole moments are enhanced by a more parallel alignment. In the context of the chemical effect as a nonlinear dielectric feature, of particular interest are those monohydric alcohols in which ring and chain-like structures coexist in a dynamic equilibrium [45–47].

The compound 5-methyl-3-heptanol (5M3H) is a monohydric alcohol case where the value of g_K changes from 1.5 to 0.5 in a matter of a 20 K temperature increase in the viscous regime above the glass transition temperature, T_g. In this situation $g_\mathrm{K} = 1$ must be understood as indicating the coexistence of chain-like ($g_\mathrm{K} > 1$) and ring-like ($g_\mathrm{K} < 1$) structures, rather than uncorrelated dipole orientation. Near $g_\mathrm{K} \approx 1$, 5M3H is very sensitive to changes in external parameters, and only a moderate electric field of 50 kV/cm is needed to enhance the relaxation amplitude, $\Delta\varepsilon$, by

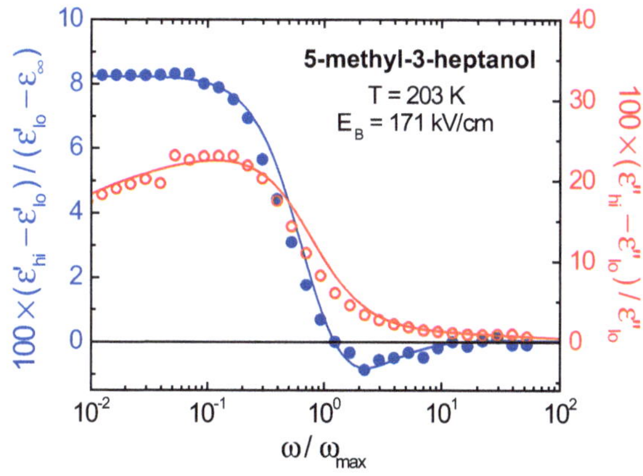

Fig. 5 Chemical effect for 5-methyl-3-heptanol at $T = 203$ using a dc-field of $E_B = 171$ kV cm^{-1}. Solid symbols represent the experimental field induced steady state spectra of the relative change of the real part of permittivity, $(\varepsilon'_{\text{hi}} - \varepsilon'_{\text{lo}})/(\varepsilon'_{\text{lo}} - \varepsilon_\infty)$, and open symbols are for the loss, $(\varepsilon''_{\text{hi}} - \varepsilon''_{\text{lo}})/\varepsilon''_{\text{lo}}$. The lines are fits reflecting the 11% increase in the Debye peak amplitude. Data from Ref. [49]

1% [49]. An example of such a pronounced chemical effect is shown for 5M3H in Fig. 5, where the amplitude of the Debye peak is increased by 11% at a field of E_B = 171 kV/cm.

3.2 *Field-Induced Changes in Relaxation Times*

Complex liquids can be designed to change their viscosity as a matter of an external electric field [50]. For simple single-component molecular liquids, such an electrorheological effect is not immediately expected. However, several thermodynamic arguments exist that the glass transition temperature T_g of a polar liquid should change with the application of an external electric field, which is the equivalent of a field-dependent structural relaxation time at a given temperature. Moynihan predicted a shift of the glass transition, $\Delta_E T_g$, when a static field (E_B) is applied to a system. For two distinct assumptions, a relaxation invariant volume ($\Delta_{\text{rel}} V = 0$) and or a relaxation invariant entropy ($\Delta_{\text{rel}} S = 0$), the effect has been quantified via the respective relations [51],

$$\Delta_E T_g = \varepsilon_0 \frac{\Delta_{\text{rel}} \left[\partial\left(\chi V\right) / \partial p \right]}{2V \Delta_{\text{rel}} \alpha} E_B^2, \quad \text{for } \Delta_{\text{rel}} V = 0, \tag{8a}$$

$$\Delta_E T_g = -\varepsilon_0 \frac{T \Delta_{\text{rel}} \left[\partial\left(\chi V\right) / \partial T \right]}{2\Delta_{\text{rel}} C_p} E_B^2, \quad \text{for } \Delta_{\text{rel}} S = 0. \tag{8b}$$

Here, χ denotes the static dielectric susceptibility, V the volume, p the pressure, α the volume thermal expansion coefficient, and C_p the heat capacity. The differences labeled as 'Δ_{rel}' refer to changes between the relaxed liquid and the unrelaxed glassy state. For a typical polar glass-forming liquid (glycerol at $T = 218$ K) at a field of $E_B = 100$ kV/cm, values of $\Delta T_g = 2$ mK and $\Delta T_g = 40$ mK were derived from for the respective conditions $\Delta_{rel} V = 0$ and $\Delta_{rel} S = 0$.

More recently, Johari reported a model leading to a field-induced shift of $\Delta_E T_g \approx 23$ mK at a static field of $E_B = 100$ kV/cm for glycerol [52], i.e., not far from the magnitude of effects expected on the basis of Moynihan's approach. The basic idea is to realize that the thermodynamic entropy is field dependent according to the relation given by Fröhlich [1],

$$\Delta_E S = \upsilon \frac{\varepsilon_0}{2} \left(\frac{\partial \varepsilon_s}{\partial T} \right)_V E_B^2, \tag{9}$$

where υ is the molar volume. According to this model, this change, $\Delta_E S$, should be added to the configurational entropy, S_{cfg}, that impacts the average structural relaxation time (τ_α) via the Adam–Gibbs (AG) approach [53],

$$\log_{10} \left(\tau_\alpha / s \right) = A + \frac{C}{T \times S_{cfg} (T)}. \tag{10}$$

Comparable field-induced shifts of the Kauzmann temperature, T_K, at which the configurational entropy vanishes [54], have been derived by Matyushov [55]. Shifts of T_K and T_g have a similar impact on the relaxation dynamics.

Common to the above three approaches to how dynamics change with the application of an external electric field is that relaxation times (and perhaps likewise viscosity) increase with the magnitude of a static field. These predictions are obtained under isothermal conditions, meaning that the field effects are not due to temperature changes that would result from the electrocaloric effects that modify temperature under adiabatic conditions. It has been demonstrated only recently that such small changes are accessible to experiment. Presently, a resolution of about $\Delta_E T_g = 0.3$ mK at a static field of $E_B = 10$ kV/cm for glycerol is possible, as reported by L'Hôte et al. [20], which is equivalent to $\Delta_E T_g = 38$ mK at $E_B = 100$ kV/cm.

Experimentally, conclusive evidence of a change in relaxation time is obtained by recording the permittivity spectrum for frequencies that cover the loss peak range, e.g., by impedance spectroscopy in the presence of a DC-bias field of sufficient magnitude. Because the slope $\partial \lg \varepsilon'' / \partial \lg \omega$ vanishes at the loss peak at ω_{max}, the value of $\varepsilon''_{max} = \varepsilon''(\omega_{max})$ will not change as a result of small changes in the relaxation time. Instead, ε''_{max} is only affected by the dielectric relaxation amplitude, i.e., saturation and/or chemical effects. The signature of a loss peak shift would be the occurrence of elevated values of $\varepsilon''(\omega < \omega_{max})$ and reduced values of $\varepsilon''(\omega > \omega_{max})$ relative to the high field level of ε''_{max}. Such behavior has been observed for numerous glass forming materials: poly(vinyla cetate) (PVAc), phenyl salicylate (SAL), 2-methyltetrahydrofuran (MTHF), cresolphthalein dimethylether (CPDE), glycerol

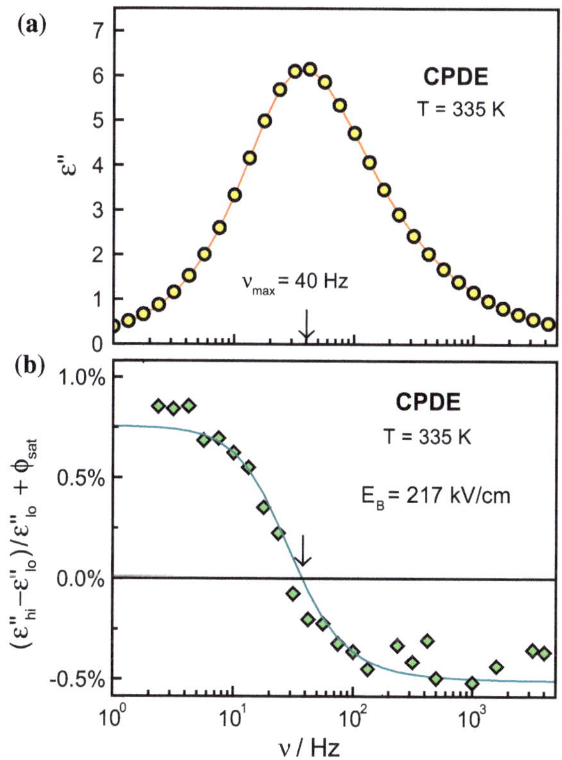

Fig. 6 **a** Dielectric loss spectrum of CPDE at $T = 335$ K with peak frequency positioned at $\nu_{\max} = 40$ Hz. The solid line represents a Cole–Davidson fit with the parameters $\Delta\varepsilon = 15.0$, $\tau_{\mathrm{CD}} = 5.8$ ms, and $\gamma_{\mathrm{CD}} = 0.67$. **b** Quasi steady-state values of the field induced relative changes of the dielectric loss for CPDE at $T = 335$ K. Symbols depict the nonlinear effect, $(\varepsilon''_{\mathrm{hi}} - \varepsilon''_{\mathrm{lo}})/\varepsilon''_{\mathrm{lo}}$, after correcting for the frequency invariant saturation effect, $\phi_{\mathrm{sat}} = 0.72\%$. The subscripts 'hi' and 'lo' refer to bias electric fields of $E_B = 217$ kV/cm and $E_B = 0$, respectively. The line is based on the Cole–Davidson fit of (**a**), with $\Delta\varepsilon$ reduced by $\phi_{\mathrm{sat}} = 0.72\%$ and τ_{CD} increased by 0.75% for the $\varepsilon''_{\mathrm{hi}}$ case. Data from Ref. [62]

(GLY), propylene carbonate (PC), propylene glycol (PG), *N*-methyl-ε-caprolactam (NMEC), [56], and 4-vinyl-1,3-dioxolan-2-one (vinyl-PC) [57]. These systems differ considerably in their dielectric relaxation amplitudes ($\Delta\varepsilon = 3$–100), glass transition temperature ($T_{\mathrm{g}} = 91$–340 K), and chemical constitution. If the results are normalized to a common field of $E_B = 100$ kV/cm, shifts of T_{g} for these eight compounds range from $\Delta_E T_{\mathrm{g}} = 3$ to 28 mK, and time constant elevations vary between $\Delta_E \ln\tau = 0.14$ and 1.65%. An example of the field-induced change of the loss spectrum associated with an increased relaxation time is given in Fig. 6 for the liquid CPDE near its glass transition.

For a number of the glass-formers listed above, not only are the $\Delta_E \ln\tau$ values available from high-field dielectric studies, but calorimetric data has also been reported, so that the Adam–Gibbs parameter C in Eq. (10) can be evaluated [56, 58]. This parameter quantifies the extent to which the average structural relaxation time depends on changes of the excess entropy S_{exc}, i.e., the difference between the entropies of liquid and crystal. Therefore, it appears that all information is available to test whether the AG relation also holds for field-induced entropy changes. However, this is not as straightforward as indicated by Johari. First, Eq. (9) requires as input the slope of $\partial\varepsilon_s/\partial T$ at constant volume, $dV = 0$, whereas most experimental results for this slope will refer to isobaric conditions, $dp = 0$. Typically, the discrepancy between $(\partial\varepsilon_s/\partial T)_V$ and $(\partial\varepsilon_s/\partial T)_p$ amounts to not more than 20% for these liquids [55]. Moreover, the entropy change calculated via Eq. (9) refers to that of the total thermodynamic entropy, i.e., not to the excess entropy ($S_{exc} = S^{liquid} - S^{crystal}$) that is accessible to adiabatic calorimetry, nor to the configurational entropy (S_{cfg}) that is meant to enter the AG relation of Eq. (10). In tests of the AG relation where the temperature is used to tune entropy, it is usually assumed that S_{cfg} and S_{exc} are connected by a temperature invariant factor, $f_S = S_{cfg}/S_{exc}$, and most estimates of f_S are in the range from 0.5 to 0.9 [59, 60]. Quantitative results for several liquids indicate that the Adam–Gibbs model fails to predict the effect of DC-field on the relaxation time, unless unrealistic discrepancies between $(\partial\varepsilon_s/\partial T)_V$ and $(\partial\varepsilon_s/\partial T)_p$ or between S_{cfg} and S_{exc} are accepted [56, 61]. Nevertheless, entropy may be related to this electrorheological effect, as a correlation between $\Delta_E \ln\tau$ and $\Delta_E S$ has been observed for some liquids [62]. On the other hand, this correlation may simply imply that dynamics (τ, η) are generally more sensitive to a field whenever the dielectric constant (ε_s) is more sensitive to temperature. At present, there is no straightforward approach to determine the magnitudes of the effects for a given system in the steady-state limit.

4 Field-Induced Structural Recovery

The discussion of the above subchapter was limited to stationary effects of a DC-bias field. Naturally, it is not expected that these steady-state field effects are established the instant that the field is applied. Analogous to physical aging and related features, driving a system beyond the regime of linear response initiates structural recovery [5]. On the basis of this analogy, one would expect the nonlinear effects to approach their equilibrium values on a time scale that is reminiscent of that of the primary structural relaxation [63, 64]. However, two important differences to the aging phenomenology are worth pointing out: (i) field-induced changes of the logarithmic relaxation times, $\ln\tau$, usually remain very small (a few percent), so that the time scale of structural recovery remains practically constant while the system approaches equilibrium; (ii) the typical experimental quantities used to gauge the deviation from the zero-field state are quadratic in the field. The latter feature is the result of the symmetry of the problem, as the polarity (sign) of the applied field does not impact the nonlinear field effect. The resulting quadratic field dependence has consequences for the temporal pattern of the structural recovery process [12], which will be outlined in what follows.

Consider a depolarized dielectric to which a field is applied at a time $t = t_{on}$ and subsequently removed at a time $t = t_{off} > t_{on}$, i.e., $E(t) = E_B$ for $t_{on} \leq t \leq t_{off}$ and $E(t) = 0$ otherwise. For a simple Debye-type system, the normalized polarization response following t_{on} would follow $R_{rise}(t) = 1 - \exp(-t/\tau)$, where t_{on} is now set to zero. The polarization $P(t)$ is normalized using $R_{rise}(t) = (P(t) - P_\infty)/\Delta P$, with P_∞ and ΔP representing the instantaneous polarization response and the time-dependent polarization step magnitude, respectively. By analogy, the decay of the normalized polarization is given by $R_{decay}(t) = \exp(-t/\tau)$, now with t_{off} set to zero. Even for a high field E_B, the above time dependences would still be very good approximations to a nonlinear polarization process, as the deviations from the linearity $P \propto E$ rarely exceed a few percents. However, the time dependence expected for quantities that gauge nonlinear effects ($\Delta_E \varepsilon_s$, $\varepsilon''_{hi} - \varepsilon''_{lo}$, third harmonic signal amplitudes $\chi_3 E^2$) is different, as their steady-state levels depend quadratically on the field (in a first approximation).

It has been argued that the generalization of the quadratic field dependence in the static limit is the quadratic dependence on the time-dependent polarization expressed as a fictive field, $E_{fic}(t) = \Delta P(t) / (\varepsilon_0 \Delta\varepsilon)$, which approaches E_B in the long time limit [12, 62]. As a result, quantities whose steady-state level scale with E_B^2 are expected to approach that level with a time pattern that follows $P^2(t)$. For the Debye-type examples mentioned above, this leads to $R_{rise}^2(t) = 1 - 2\exp(-t/\tau) - \exp(-2t/\tau)$ and $R_{decay}^2(t) = \exp(-2t/\tau)$. Clearly, the symmetry of the rise and decay patterns is lost in the nonlinear regime. For the simple case of a Debye-type system, this feature is illustrated in Fig. 7.

When the field-induced effects are gauged via changes of the parameters $\ln\tau(t)$ or $\Delta\varepsilon(t)$, their time dependence would follow

$$\ln\tau(t) = \ln\tau^0 + \Delta_E \ln\tau \times \left(\frac{\Delta P(t)}{\varepsilon_0 \Delta\varepsilon^0 E_B}\right)^2, \tag{11a}$$

$$\Delta\varepsilon(t) = \Delta\varepsilon^0 + \Delta_E \Delta\varepsilon \times \left(\frac{\Delta P(t)}{\varepsilon_0 \Delta\varepsilon^0 E_B}\right)^2, \tag{11b}$$

where the superscript '0' identifies a quantity evaluated at $E_B = 0$. The steady-state levels of the field-induced changes, $\Delta_E \ln\tau$ and $\Delta_E \Delta\varepsilon$, will depend quadratically on the field E_B, and the terms in parentheses are normalized such that $0 \leq \Delta P(t)/(\varepsilon_0 \Delta\varepsilon^0 E_B) \leq 1$.

The situation regarding the dependence on $P^2(t)$ becomes more complicated for systems with dispersive dynamics, i.e., those for which the loss spectrum is widened compared with the Debye-type case used above for illustration purposes. Analogous to approaches to structural recovery in the context of physical aging [5], one needs to decide whether a model with a single or with multiple fictive fields is the more appropriate description of the problem at hand [65–67]. As dispersive heterogeneous dynamics involve different modes (labeled 'i') with specific time constants (τ_i) [68, 69], the question of the number of fictive fields amounts to deciding whether structural recovery of each mode 'i' progresses according to its own polarization state,

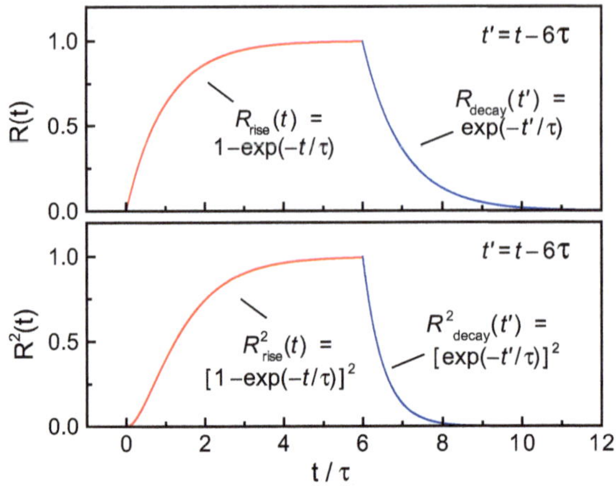

Fig. 7 Schematic representation of a system characterized by a polarization step response function, $\chi(t) = \exp(-t/\tau)$ with a single time constant τ, subject to a field $E(t)$ that is constant at the level E_0 for times $0 \leq t \leq 6$ ms and zero otherwise. In the linear response regime, the rise and decay responses, $R(t)$, are symmetric exponentials, see top panel. In the nonlinear regime, a quantity that depends quadratically on E_0 in the steady state case is expected to follow $R^2(t)$, resulting in the rise/decay asymmetry shown in the bottom panel. Relative to the linear response, the rise of $R^2(t)$ is retarded while its decay is accelerated. Adapted from Ref. [62]

$\Delta P_i(t)/\left(\varepsilon_0 \Delta\varepsilon_i^0\right)$, or governed by the average polarization, $\Delta P(t)/\left(\varepsilon_0 \Delta\varepsilon^0\right)$. Here, $\Delta\varepsilon_i$ refers to the contribution of mode 'i' to the total relaxation amplitude $\Delta\varepsilon$, with $\Delta\varepsilon = \sum_i \Delta\varepsilon_i$ and similarly $\Delta P(t) = \sum_i \Delta P_i(t)$. A single fictive field implies that even those modes that have a very small time constant (τ_i) relative to the average (τ_α) will not approach equilibrium any faster than the modes with larger τ_i.

For various nonlinear dielectric effects [62], the time-resolved structural recovery that has been initiated by a high static electric field displays the rise/decay asymmetry mentioned above: saturation [12, 56–58], chemical effect [12, 49], and electrorheological effect [12, 56–58]. An example for which the time-dependent change of ε'' is due to both saturation and the electrorheological effect is shown in Fig. 8, again for CPDE. The same asymmetry is observed for other time-resolved changes that are quadratic in the field: nonlinearities arising from energy absorbed from time-dependent fields [70, 71], and the birefringence observed in studies of the electro-optical Kerr effect (EOKE) [72–76]. The analogy between nonlinear dielectric effects and EOKE has been emphasized in many treatments of dipolar systems subject to high fields [77–80]. Quantitative analyses of these patterns have shown that the rise and decay curves can be explained by the same time constant, provided that the quadratic polarization dependence is accounted for. The present explanation for the apparent rise/decay asymmetry has been validated by more complex field patterns, where expectedly the transitions among two high field levels show almost symmetric rise/decay behavior [12, 56].

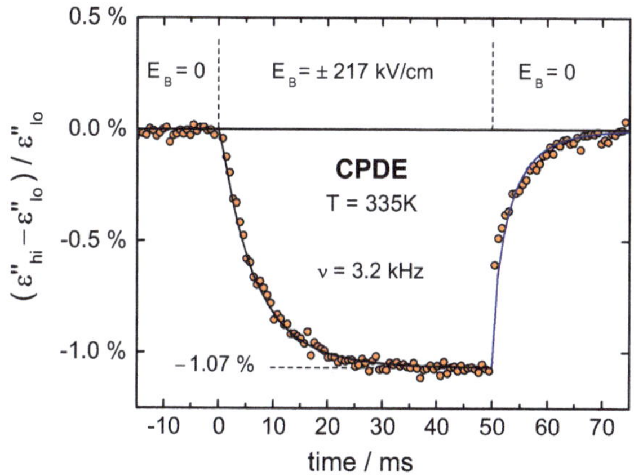

Fig. 8 Field-induced relative change of the 'dielectric loss' component, ε'', for CPDE at a temperature of $T = 335$ K, versus time with a resolution of one period. The ac-field, $E(t) = E_0 \sin(2\pi \nu t)$, is characterized by $E_0 = 43$ kV/cm and $\nu = 3.2$ kHz, the dc-field pattern is indicated at the top. Solid circles represent the values corrected for energy absorption and are thus assumed to reflect the nonlinear effect associated with the high bias field. Lines are squared KWW fits to the rise and fall behavior, respectively using $\phi_0[1 - \phi(t)]^2$ and $\phi_0[\phi(t)]^2$ with $\phi(t) = \exp[-(t/\tau_0)^\beta]$ and $\phi_0 = -1.07\%$. Data from Ref. [62]

5 Relation to Cubic Susceptibilities

Cubic susceptibilities or third harmonic responses are usually not measured using a high amplitude DC-field, but rather with a sufficiently strong alternating field, $E(t) = E_0 \sin(\omega t)$, and without DC-bias [81]. Therefore, a detailed discussion of nonlinear effects measured via higher harmonics is outside the scope of this chapter. Nevertheless, it is worthwhile pointing out the connection between what has been established as nonlinear features from high DC-field experiments and the third harmonic signals, with the latter often reported in terms of the dimensionless quantity $|\chi_3|E_0^2$ versus frequency, defined via

$$\frac{\hat{P}_3(\omega)}{\varepsilon_0 E_0} = \hat{\chi}_3(\omega) E_0^2, \tag{12}$$

where $\hat{P}_3(\omega)$ denotes the frequency domain polarization signal at 3ω.

A model has been proposed that links the changes of the peak amplitude of $|\chi_3|E_0^2$ spectra with temperature to a change in a nontrivial length scale via the number, N_{corr}, of dynamically correlated particles [82–84]. Because the model does not predict absolute values for $|\chi_3|$, only relative changes for $N_{corr}(T)$ can be obtained within the framework of this model, which pertains to systems close to a critical point [82]. The interest in $N_{corr}(T)$ has prompted numerous measurements of cubic susceptibilities on supercooled liquids in recent years [85–90].

An alternative approach to rationalizing cubic susceptibilities rests on knowing the steady-state levels of the field-induced change in the amplitude ($\Delta_E \Delta\varepsilon$) and in the relaxation time ($\Delta_E \ln\tau$) from experiments, as well as their time-dependent structural recovery as expressed in Eq. (11a, 11b) [91]. In order to quantify polarization under these nonlinear conditions, we begin with a relation that provides the time-dependent polarization, $P(t)$, for a Debye-type mode for any time-dependent field $E(t)$,

$$\frac{\mathrm{d}P_i(t)}{\mathrm{d}t} = \frac{\varepsilon_0 \Delta\varepsilon_i(t)\, E(t)}{\tau_i(t)} - \frac{P_i(t)}{\tau_i(t)}, \tag{13}$$

but with the parameters characterizing amplitude ($\Delta\varepsilon$) and time constant (τ) depending on time according to their link to $P(t)$ as outlined in Eq. (11a, 11b). For the time-dependent fields of present interest, an extra term is needed in Eq. (11a) to account for the reduction of τ as a result of the sample absorbing energy from the field [92–95], but well-tested models are available to quantify this contribution [68, 69, 96]. If the values of $\Delta\varepsilon$ and τ in Eq. (13) were constant, then the linear response polarization would be obtained for a given $E(t)$. Thus, the nonlinear features are accounted for by the changes of $\Delta\varepsilon$ and τ, in perfect analogy to the common models employed to capture physical aging and related nonlinear phenomena in response to changes in temperature [5].

The relaxation time dispersion observed for supercooled liquids can be accounted for by expressing the frequency dependent part of the permittivity as a sum of Debye modes, $\hat{\varepsilon}(\omega) - \varepsilon_\infty = \sum_i \Delta\varepsilon_i / (1 + i\omega\tau_i)$. Then, Eq. (13) would be applied to each Debye contribution and solved separately for each mode 'i' using a field $E(t) = E_0 \sin(\omega t)$. The total polarization is obtained from the sum of the individual $P_i(t)$, plus the instantaneous response, $P(t) = P_\infty + \Delta P(t)$ with $\Delta P(t) = \sum_i P_i(t)$. For each frequency ω, the numeric calculation of $P(t)$ is continued until stationary conditions are achieved, and subsequent periods of the oscillating $P(t)$ curve can be subjected to Fourier analysis to determine the desired susceptibility, $\chi_3(\omega)$ [89, 97]. While this approach conforms to the heterogeneous nature of structural relaxation [68, 69], it should not be concluded that these nonlinear effects can only be explained on the basis of heterogeneous dynamics.

For viscous glycerol, the cubic susceptibilities in terms of $\chi_3 E^2$ have been computed on the basis of the approach outlined above. Interestingly, the results reflect the experimental counterpart in an almost quantitative fashion [95], see the comparison provided in Fig. 9a. Moreover, the model can be used to separate the three different contributions to the cubic susceptibility: saturation, electrorheological effect, and energy absorption, and the distinct contributions are included in Fig. 9b. Within the framework of this model, it turns out that the electrorheological effect constitutes the main contribution to the 'hump' in $\chi_3 E^2$, while the low-frequency plateau level is determined solely by the saturation effect. One result of this agreement between model and experimental data is that the main features of the $|\chi_3(\omega)|$ spectrum can be explained without explicit involvement of a length scale. Other models have also demonstrated that dynamical correlations need not be assumed to explain maxima in cubic susceptibility spectra [98–100].

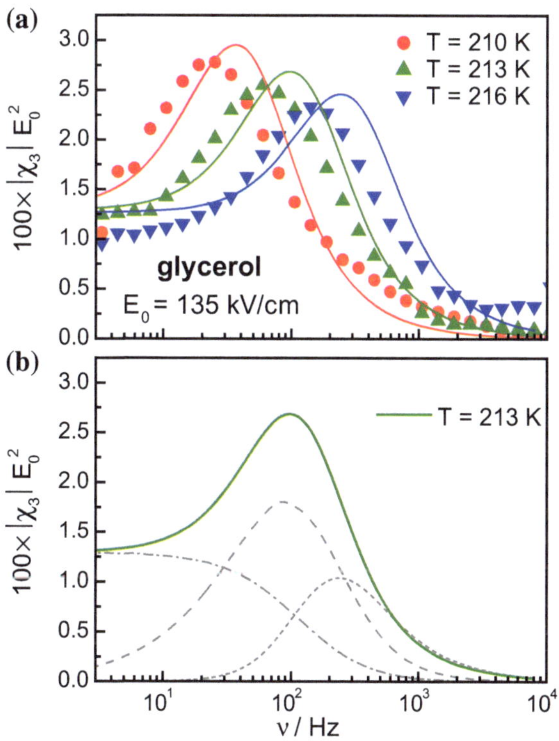

Fig. 9 **a** Symbols represent experimental spectra of the third harmonic susceptibility for glycerol at the three temperatures indicated, reported in terms of the quantity $|\chi_3|\,E_0^2$ and using peak fields of $E_0 = 135$ kV cm^{-1}. Solid lines are calculated steady state spectra of $|\chi_3|\,E_0^2$ at $E_0 = 135$ kV cm^{-1} using the model outlined in Sect. 5 with the parameters selected to represent glycerol at the three temperatures indicated. **b** The solid curve reproduces the model calculation for $T = 213$ K from panel (**a**), while dashed curves represent the distinct contributions to the $T = 213$ K case: saturation (dash-dot), electrorheological effect (dash), and energy absorption (short dash). Adapted from Ref. [62]

A very different and model-free way of demonstrating a close connection between the cubic susceptibility and the dc-field induced change in permittivity at the fundamental frequency rests on representing both the quantities in the same fashion, i.e., as modulus of the complex susceptibility versus frequency. The AC-field results are then quantified by $|\chi_3|E_0^2$, while the DC-field results are cast into the form $|\chi_1^{(3,\mathrm{dc})}|E_B^2$, see Eq. (6b). Based on permittivity data obtained at high ($\varepsilon_{\mathrm{hi}}$) and zero ($\varepsilon_{\mathrm{lo}}$) DC-field, this quantity can be obtained using

$$\left|\hat{\chi}_1^{(3,dc)}(\omega)\right| E_B^2 = \sqrt{\left(\varepsilon'_{hi}(\omega) - \varepsilon'_{lo}(\omega)\right)^2 + \left(\varepsilon''_{hi}(\omega) - \varepsilon''_{lo}(\omega)\right)^2}. \qquad (14)$$

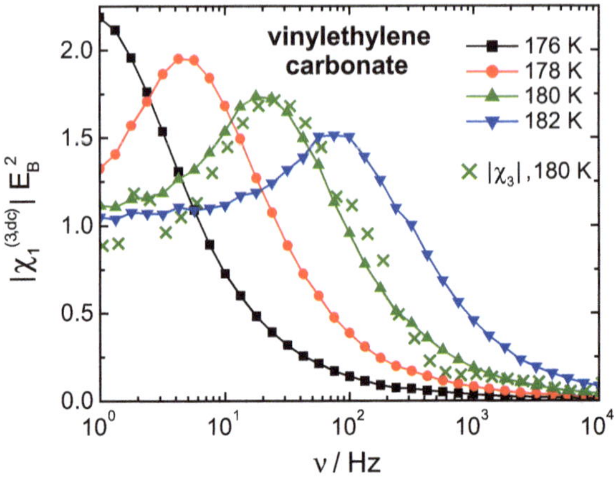

Fig. 10 Field-induced susceptibility change at the fundamental frequency, shown as $|\chi_1^{(3,\mathrm{dc})}|E_B^2$ versus fundamental frequency ν for vinyl ethylene carbonate (VEC) at the temperatures indicated. The curves are measured using a small AC-field with peak value $E_0 = 35$ kV/cm superimposed onto a DC-bias field of $E_B = 250$ kV/cm. The $|\chi_3|$ spectrum for vinyl-PC at $T = 180$ K is shown as crosses, rescaled to match the amplitude of $|\chi_1^{(3,\mathrm{dc})}|E_B^2$. Data from Ref. [57]

Spectra of $|\chi_1^{(3,\mathrm{dc})}|E_B^2$ are shown for 4-vinyl-1,3-dioxolan-2-one (vinyl ethylene carbonate, VEC) in Fig. 10, and their overall appearance is reminiscent of the $|\chi_3|E_0^2$ spectra of Fig. 9a obtained for glycerol. A direct comparison with $|\chi_3|E_0^2$ data is provided for VEC at $T = 180$ K [57], shown as crosses in Fig. 10, supporting the idea that third harmonic data obtained with high AC-fields can be modeled on the basis of permittivity data measured in the presence of a strong DC-field.

6 Concluding Remarks

Dielectric saturation is a well-known phenomenon that occurs when high electric fields are applied to a polar liquid. It is an immediate consequence of polarization possessing an upper bound, which corresponds to an orientational distribution with $\langle\cos\theta\rangle \leq 1$. Much more recently, it has been recognized that a DC-field of sufficient magnitude also modifies the relaxation times, thereby changing the glass transition temperature and possibly viscosity. Dipole correlations that are important in polar condensed systems prohibit a straightforward determination of the magnitude of these effects for a given material and field. As a general trend, however, both features become more pronounced with increasing dielectric constant of the material.

When the high DC-field is applied or removed, parameters characterizing the relaxation amplitude ($\Delta\varepsilon$) and time constants (τ) gradually approach new equilibrium values. As in the case of physical aging, this structural recovery process is governed by the structural relaxation time of the system, but following $P^2(t)$ rather than $P(t)$ itself regarding the time dependence. This quadratic dependence on the time-dependent fictive field, $E_{\mathrm{fic}}(t) = \Delta P(t) \big/ (\varepsilon_0 \Delta\varepsilon)$, originates from the quadratic dependence of the steady-state levels on the field. It explains the rise/decay asymmetry observed for all nonlinear dielectric effects, which is also seen in the birefringence traces of electro-optical Kerr effect studies. This time-dependent fictive field is also a critical input to a model that links third-order susceptibilities measured with AC-fields to the changes in permittivity obtained from DC-field experiments. The stationary levels of the DC-field induced changes remain relatively small, but the high resolution of dielectric spectroscopy facilitates recording the structural recovery process with good resolution. Therefore, the changes that occur in response to applying a large field can serve as physical aging experiments with fast time resolution, because applying or removing a field can be performed in a matter of microseconds.

Acknowledgments This work is partly supported by the National Science Foundation under Grant No. CHE-1564663.

References

1. H. Fröhlich, *Theory of Dielectrics* (Clarendon, Oxford, 1958)
2. J. Herweg, Die elektrischen dipole in flüssigen Dielektricis. Z Physik **3**, 36 (1920)
3. F. Kremer, A. Schönhals (eds.), *Broadband Dielectric Spectroscopy* (Springer, Berlin, 2002)
4. R. Richert, Supercooled liquids and glasses by dielectric relaxation spectroscopy. Adv. Chem. Phys. **156**, 101 (2014)
5. I.M. Hodge, Enthalpy relaxation and recovery in amorphous materials. J. Non-Cryst. Solids **169**, 211 (1994)
6. J. Brandrup, E.H. Immergut (eds.), *Polymer Handbook*, 2nd edn. (Wiley, New York, 1975)
7. D.G. Lahoz, G. Walker, An experimental analysis of electromagnetic forces in liquids. J. Phys. D Appl. Phys. **8**, 1994 (1975)
8. C.J.F. Böttcher, *Theory of Electric Polarization*, vol. 1 (Elsevier, Amsterdam, 1973)
9. S. Weinstein, R. Richert, Nonlinear features in the dielectric behavior of propylene glycol. Phys. Rev. B **75**, 064302 (2007)
10. J.A. Schellman, Dielectric saturation. J. Chem. Phys. **24**, 912 (1956)
11. G.G. Wiseman, J.K. Kuebler, Electrocaloric effect in ferroelectric Rochelle salt. Phys. Rev. **131**, 2023 (1963)
12. A.R. Young-Gonzales, S. Samanta, R. Richert, Dynamics of glass-forming liquids. XIX. Rise and decay of field induced anisotropy in the non-linear regime. J. Chem. Phys. **143**, 104504 (2015)
13. L.D. Landau, E.M. Lifshitz, *Electrodynamics of Continuous Media* (Pergamon, Oxford, 1984)
14. P. Ben Ishai, M.S. Talary, A. Caduff, E. Levy, Y. Feldman, Electrode polarization in dielectric measurements: a review. Meas. Sci. Technol. **24**, 102001 (2013)
15. P.A. Bradley, G. Parry Jones, A system for the investigation of nonlinear dielectric effects using digital techniques. J. Phys. E: Sci. Instrum. **7**, 449 (1974)
16. A.E. Davies, M.J. van der Sluijs, G. Parry Jones, Notes on a system for the investigation of nonlinear dielectric effects. J. Phys. E: Sci. Instrum. **11**, 737 (1978)
17. M. Górny, J. Zioło, S.J. Rzoska, A new application of the nonlinear dielectric method for studying relaxation processes in liquids. Rev. Sci. Instrum. **67**, 4290 (1996)
18. S.J. Rzoska, V.P. Zhelezny (eds.), *Nonlinear Dielectric Phenomena in Complex Liquids* (Kluwer Academic Publishers, Dordrecht, 2004)

19. S.J. Rzoska, A. Drozd-Rzoska, Dual field nonlinear dielectric spectroscopy in a glass forming EPON 828 epoxy resin. J. Phys.: Condens. Matter **24**, 035101 (2012)
20. D. L′Hôte, R. Tourbot, F. Ladieu, P. Gadige, Control parameter for the glass transition of glycerol evidenced by the static-field-induced nonlinear response. Phys. Rev. B **90**, 104202 (2014)
21. S. Samanta, R. Richert, Dynamics of glass-forming liquids. XVIII. Does entropy control structural relaxation times? J. Chem. Phys. **142**, 044504 (2015)
22. S. Samanta, R. Richert, Non-linear dielectric behavior of a secondary relaxation: glassy D-sorbitol. J. Phys. Chem. B **119**, 8909 (2015)
23. P. Langevin, Sur la théorie du magnétisme. J. Phys. Theor. Appl. **4**, 678 (1905)
24. P. Debye, Der Rotationszustand von Molekülen in Flüssigkeiten. Phys. Z. **36**, 100 (1935)
25. P. Debye, *Polar Molecules* (Chemical Catalog Company, New York, 1929)
26. R. Richert, Frequency dependence of dielectric saturation. Phys. Rev. E **88**, 062313 (2013)
27. G.P. Jones, in *Non-Linear Dielectric Effects: Dielectric and Related Molecular Processes*, specialist periodical reports vol. 2, ed. by M. Davies (The Chemical Society, London, 1975)
28. A. Piekara, B. Piekara, Saturation électrique dans les liquides purs et leurs mélanges. Compt. Rend. Acad. Sci. (Paris) **203**, 852 (1936)
29. A. Piekara, Dielectric saturation and hydrogen bonding. J. Chem. Phys. **36**, 2145 (1962)
30. J. Małecki, Dielectric saturation in aliphatic alcohols. J. Chem. Phys. **36**, 2144 (1962)
31. A. Piekara, A. Chelkowski, New experiments on dielectric saturation in polar liquids. J. Chem. Phys. **25**, 794 (1956)
32. I. Danielewicz-Ferchmin, On the non-linear dielectric effect in some non-polar liquids and nitrobenzene. Chem. Phys. Lett. **155**, 539 (1989)
33. J.H. van Vleck, On the role of dipole-dipole coupling in dielectric media. J. Chem. Phys. **5**, 556 (1937)
34. S. Kielich, Semi-macroscopic treatment of the theory of non-linear phenomena in dielectric liquids submitted to strong electric and magnetic fields. Acta Phys. Polon. **17**, 239 (1958)
35. R.L. Fulton, The theory of nonlinear dielectric. Polar, polarizable molecules. J. Chem. Phys. **78**, 6877 (1983)
36. R.L. Fulton, On the theory of nonlinear dielectrics. J. Chem. Phys. **78**, 6865 (1983)
37. J.L. Déjardin, Y.P. Kalmykov, P.M. Déjardin, Birefringence and dielectric relaxation in strong electric fields. Adv. Chem. Phys. **117**, 275 (2001)
38. I. Szalai, S. Nagy, S. Dietrich, Nonlinear dielectric effect of dipolar fluids. J. Chem. Phys. **131**, 154905 (2009)
39. S. Buyukdagli, Dielectric anisotropy in polar solvents under external fields. J. Stat. Mech. **2015**, P08022 (2015)
40. D.V. Matyushov, Nonlinear dielectric response of polar liquids. J. Chem. Phys. **142**, 244502 (2015)
41. J. Małecki, The relaxation of the nonlinear dielectric effect. J. Mol. Struct. **436–437**, 595 (1997)
42. J. Małecki, Non-linear dielectric behaviour and chemical equilibria in liquids. Electrochim. Acta **33**, 1235 (1988)
43. J. Małecki, Investigations of hexanol-1 multimers and complexes by the method of dielectric polarization in weak and strong electric fields. J. Chem. Phys. **43**, 1351 (1965)
44. J.A. Małecki, Study of self-association of 2-methyl-2-butanol based on non-linear dielectric effect. Chem. Phys. Lett. **297**, 29 (1998)
45. L.P. Singh, R. Richert, Watching hydrogen bonded structures in an alcohol convert from rings to chains. Phys. Rev. Lett. **109**, 167802 (2012)
46. L.P. Singh, C. Alba-Simionesco, R. Richert, Dynamics of glass-forming liquids. XVII. Dielectric relaxation and intermolecular association in a series of isomeric octyl alcohols. J. Chem. Phys. **139**, 144503 (2013)
47. W. Dannhauser, Dielectric study of intermolecular association in isomeric octyl alcohols. J. Chem. Phys. **48**, 1911 (1968)

48. R. Böhmer, C. Gainaru, R. Richert, Structure and dynamics of monohydroxy alcohols—milestones towards their microscopic understanding, 100 years after Debye. Phys. Rep. **545**, 125 (2014)
49. A.R. Young-Gonzales, R. Richert, Field induced changes in the ring/chain equilibrium of hydrogen bonded structures: 5-methyl-3-heptanol. J. Chem. Phys. **145**, 074503 (2016)
50. W.M. Winslow, Induced fibration of suspensions. J. Appl. Phys. **20**, 1137 (1949)
51. C.T. Moynihan, A.V. Lesikar, Comparison and analysis of relaxation processes at the glass transition temperature. Ann. New York Acad. Sci. **371**, 151 (1981)
52. G.P. Johari, Effects of electric field on the entropy, viscosity, relaxation time, and glass-formation. J. Chem. Phys. **138**, 154503 (2013)
53. G. Adam, J.H. Gibbs, On the temperature dependence of cooperative relaxation properties in glass-forming liquids. J. Chem. Phys. **43**, 139 (1965)
54. W. Kauzmann, The nature of the glassy state and the behavior of liquids at low temperatures. Chem. Rev. **43**, 219 (1948)
55. D.V. Matyushov, Configurational entropy of polar glass formers and the effect of electric field on glass transition. J. Chem. Phys. **145**, 034504 (2016)
56. S. Samanta, R. Richert, Electrorheological source of nonlinear dielectric effects in molecular glass-forming liquids. J. Phys. Chem. B **120**, 7737 (2016)
57. A.R. Young-Gonzales, K. Adrjanowicz, M. Paluch, R. Richert, Nonlinear dielectric features of highly polar glass formers: derivatives of propylene carbonate. J. Chem. Phys. **147**, 224501 (2017)
58. S. Samanta, O. Yamamuro, R. Richert, Connecting thermodynamics and dynamics in a supercooled liquid: cresolphthalein-dimethylether. Thermochim. Acta **636**, 57 (2016)
59. M. Goldstein, Comparing landscape calculations with calorimetric data on ortho-terphenyl, and the question of the configurational fraction of the excess entropy. J. Chem. Phys. **123**, 244511 (2005)
60. L.-M. Wang, R. Richert, Measuring the configurational heat capacity of liquids. Phys. Rev. Lett. **99**, 185701 (2007)
61. R. Richert, Relaxation time and excess entropy in viscous liquids: electric field versus temperature as control parameter. J. Chem. Phys. **146**, 064501 (2017)
62. R. Richert, Nonlinear dielectric effects in liquids: a guided tour. J. Phys.: Condens. Matter **29**, 363001 (2017)
63. P. Lunkenheimer, R. Wehn, U. Schneider, A. Loidl, Glassy aging dynamics. Phys. Rev. Lett. **95**, 055702 (2005)
64. R. Richert, P. Lunkenheimer, S. Kastner, A. Loidl, On the derivation of equilibrium relaxation times from aging experiments. J. Phys. Chem. B **117**, 12689 (2013)
65. A.J. Kovacs, J.J. Aklonis, J.M. Hutchinson, A.R. Ramos, Isobaric volume and enthalpy recovery of glasses. II. A transparent multiparameter theory. J. Polym. Sci. B: Polym Phys. **34**, 2467 (1996)
66. R. Richert, Physical aging and heterogeneous dynamics. Phys. Rev. Lett. **104**, 085702 (2010)
67. S. Samanta, R. Richert, Limitations of heterogeneous models of liquid dynamics: very slow rate exchange in the excess wing. J. Chem. Phys. **140**, 054503 (2014)
68. M.D. Ediger, Spatially heterogeneous dynamics in supercooled liquids. Annu. Rev. Phys. Chem. **51**, 99 (2000)
69. R. Richert, Heterogeneous dynamics in liquids: fluctuations in space and time. J. Phys.: Condens. Matter **14**, R703 (2002)
70. W. Huang, R. Richert, Dynamics of glass-forming liquids. XIII. Microwave heating in slow motion. J. Chem. Phys. **130**, 194509 (2009)
71. R. Richert, Reverse calorimetry of a supercooled liquid: propylene carbonate. Thermochim. Acta **522**, 28 (2011)
72. R. Coelho, D. Khac Manh, Utilisation de la biréfringence électro-optique pour l'étude de la relaxation dipolaire dans les liquides polaires faiblement conducteurs. C R Acad. Sc Paris—Serie C **264**, 641 (1967)

73. M.S. Beevers, J. Crossley, D.C. Garrington, G. Williams, Dielectric and dynamic Kerr-effect studies in liquid systems. Faraday Symp. Chem. Soc. **11**, 38 (1977)
74. M.S. Beevers, D.A. Elliott, G. Williams, Static and dynamic Kerr-effect studies of glycerol in its highly viscous state. J. Chem. Soc. Faraday Trans. **2**(76), 112 (1980)
75. J. Crossley, G. Williams, Structural relaxation in 2-methyl-2,4-pentanediol studied by dielectric and Kerr-effect techniques. J. Chem. Soc. Faraday Trans. **2**(73), 1651 (1977)
76. J. Crossley, G. Williams, Relaxation in hydrogen-bonded liquids studied by dielectric and Kerr-effect techniques. J. Chem. Soc., Faraday Trans. 2 **73**, 1906 (1977)
77. W.T. Coffey, B.V. Paranjape, Dielectric and Kerr effect relaxation in alternating electric fields. Proc. R. Ir. Acad. **78**, 17 (1978)
78. J.L. Déjardin, P.M. Déjardin, Y.P. Kalmykov, Nonlinear electro-optical response. I. Steady state Kerr effect relaxation arising from a weak ac electric field superimposed on a strong dc bias field. J. Chem. Phys. **106**, 5824 (1997)
79. W.T. Coffey, Y.P. Kalmykov, S.V. Titov, Anomalous nonlinear dielectric and Kerr effect relaxation steady state responses in superimposed ac and dc electric fields. J. Chem. Phys. **126**, 084502 (2007)
80. M.S. Beevers, J. Crossley, D.C. Garrington, G. Williams, Consideration of dielectric relaxation and the Kerr-effect relaxation in relation to the reorientational motions of molecules. J. Chem. Soc. Faraday Trans. 2 **72**, 1482 (1976)
81. C. Thibierge, D. L'Hôte, F. Ladieu, R. Tourbot, A method for measuring the nonlinear response in dielectric spectroscopy through third harmonics detection. Rev. Sci. Instrum. **79**, 103905 (2008)
82. J.-P. Bouchaud, G. Biroli, Nonlinear susceptibility in glassy systems: a probe for cooperative dynamical length scales. Phys. Rev. B **72**, 064204 (2005)
83. M. Tarzia, G. Biroli, A. Lefèvre, J.-P. Bouchaud, Anomalous nonlinear response of glassy liquids: general arguments and a mode-coupling approach. J. Chem. Phys. **132**, 054501 (2010)
84. F. Ladieu, C. Brun, D. L'Hôte, Nonlinear dielectric susceptibilities in supercooled liquids: a toy model. Phys. Rev. B **85**, 184207 (2012)
85. C. Crauste-Thibierge, C. Brun, F. Ladieu, D. L'Hôte, G. Biroli, J.-P. Bouchaud, Evidence of growing spatial correlations at the glass transition from nonlinear response experiments. Phys. Rev. Lett. **104**, 165703 (2010)
86. C. Brun, F. Ladieu, D. L'Hôte, M. Tarzia, G. Biroli, J.-P. Bouchaud, Nonlinear dielectric susceptibilities: accurate determination of the growing correlation volume in a supercooled liquid. Phys. Rev. B **84**, 104204 (2011)
87. C. Crauste-Thibierge, C. Brun, F. Ladieu, D. L'Hôte, G. Biroli, J.-P. Bouchaud, Nonlinear susceptibility measurements in a supercooled liquid close to T_g: growth of the correlation length and possible critical behavior. J. Non-Cryst. Solids **357**, 279 (2011)
88. T. Bauer, P. Lunkenheimer, A. Loidl, Cooperativity and the freezing of molecular motion at the glass transition. Phys. Rev. Lett. **111**, 225702 (2013)
89. R. Casalini, D. Fragiadakis, C.M. Roland, Dynamic correlation length scales under isochronal conditions. J. Chem. Phys. **142**, 064504 (2015)
90. S. Albert, T. Bauer, M. Michl, G. Biroli, J.-P. Bouchaud, A. Loidl, P. Lunkenheimer, R. Tourbot, C. Wiertel-Gasquet, F. Ladieu, Fifth-order susceptibility unveils growth of thermodynamic amorphous order in glass-formers. Science **352**, 1308 (2016)
91. R. Richert, Nonlinear dielectric signatures of entropy changes in liquids subject to time-dependent electric fields. J. Chem. Phys. **144**, 114501 (2016)
92. B. Schiener, R. Böhmer, A. Loidl, R.V. Chamberlin, Nonresonant spectral hole burning in the slow dielectric response of supercooled liquids. Science **274**, 752 (1996)
93. R. Richert, S. Weinstein, Nonlinear dielectric response and thermodynamic heterogeneity in liquids. Phys. Rev. Lett. **97**, 095703 (2006)
94. W. Huang, R. Richert, The physics of heating by time-dependent fields: microwaves and water revisited. J. Phys. Chem. B **112**, 9909 (2008)
95. T. Bauer, P. Lunkenheimer, S. Kastner, A. Loidl, Nonlinear dielectric response at the excess wing of glass-forming liquids. Phys. Rev. Lett. **110**, 107603 (2013)

96. K.R. Jeffrey, R. Richert, K. Duvvuri, Dielectric hole burning: signature of dielectric and thermal relaxation time heterogeneity. J. Chem. Phys. **119**, 6150 (2003)
97. P. Kim, A.R. Young-Gonzales, R. Richert, Dynamics of glass-forming liquids. XX. Third harmonic experiments of non-linear dielectric effects versus a phenomenological model. J. Chem. Phys. **145**, 064510 (2016)
98. G. Diezemann, Nonlinear response theory for Markov processes: simple models for glassy relaxation. Phys. Rev. E **85**, 051502 (2012)
99. G. Diezemann, Higher-order correlation functions and nonlinear response functions in a Gaussian trap model. J. Chem. Phys. **138**, 12A505 (2013)
100. G. Diezemann, Nonlinear response functions in an exponential trap model. J. Non-Cryst. Solids **407**, 61 (2015)

Nonresonant Spectral Hole Burning in Liquids and Solids

Ralph V. Chamberlin, Roland Böhmer and Ranko Richert

Abstract A review of nonresonant spectral hole burning (NHB) is given. NHB utilizes a large-amplitude, low-frequency pump oscillation in an externally applied field to modify the response of a sample nonlinearly, then a small probe step is applied to measure its modified response. When combined with other techniques, NHB indicates that the non-exponential relaxation in most substances comes from an ensemble of independently relaxing regions, with length scales on the order of nanometers. Various models are presented, focusing on a "box" model that gives excellent agreement with NHB measurements, often with no adjustable parameters. The box model is based on energy absorption that changes the local "fictive" temperature of slow degrees of freedom in spectrally selected regions, with a return to equilibrium only after this excess energy flows into the heat bath. A physical foundation for such thermodynamic heterogeneity is presented, based on concepts from nanothermodynamics. Guided by this approach, a Landau-like theory and Ising-spin model are described that yield several features found in glassforming liquids. Examples of results from NHB are shown, with special emphasis on dielectric hole burning (DHB) of liquids and magnetic hole burning (MHB) of solids.

1 Introduction

Although viscoelastic creep was already known to Robert Hooke in 1678 [1], Wilhelm Weber was arguably the first person to attempt a physical interpretation of measured non-exponential response in materials when he analyzed the mechanical relaxation of silk fibers in 1835 [2]. In any case, it was Weber's doctoral student Friedrich Kohlrausch who first recognized that the measured relaxation

R. V. Chamberlin (✉)
Department of Physics, Arizona State University, Tempe, AZ 85287, USA
e-mail: ralph.chamberlin@asu.edu

R. Böhmer
Fakultät Physik, Technische Universität Dortmund, 44221 Dortmund, Germany

R. Richert
School of Molecular Sciences, Arizona State University, Tempe, AZ 85287, USA

R. Richert (ed.), *Nonlinear Dielectric Spectroscopy*, Advances in Dielectrics
https://doi.org/10.1007/978-3-319-77574-6_5

of several types of materials, some crystalline and others amorphous, could be characterized by an empirical formula [3, 4], now known as the Kohlrausch or Kohlrausch–Williams–Watts law [5], or the stretched exponential function [6, 7]. Rudolph Kohlrausch, Friedrich's father, had introduced this function in 1854 to describe his measurements of electric charge as a function of time in a Leyden jar (a glass-filled capacitor) [8], but Friedrich seems to have been the first to recognize its ubiquity. The function can be written as

$$\Phi(t) = \Phi_0 \exp[-(t/\tau)^{\beta}]. \tag{1}$$

Here Φ_0 is an initial amplitude, τ a relaxation time, and β a parameter that governs the effective width of the decay function. Empirically, it is often found that β varies between 0.5–0.7 for liquids, and 0.2–0.4 for solids. Physically, it has long been a goal to understand the fundamental mechanism that could cause such common response in a wide range of diverse materials.

At least, since Newton introduced his law of cooling in 1701 [9], single-exponential relaxation ($\beta = 1$) has been known to come from the simple differential equation, $d\Phi(t)/dt = -\Phi(t)/\tau$. Two distinct scenarios that yield net stretched exponential relaxation are: a "homogeneous" picture, where the relaxation time itself is time-dependent, $\tau(t)$, or a "heterogeneous" picture with a distribution of relaxation times that are location-dependent, $\tau(x)$. The original interpretation used by Rudolph Kohlrausch was to assume a homogeneous time dependence, $\tau(t) \propto t^{1-\beta}$. This scenario remained a common picture for interpreting the behavior through the 1970s and 80s, when it was suggested that the dielectric correlation function may have [10] "a natural non-exponential dependence upon time due to the cooperative nature of the process, and may not be regarded … as arising from a distribution of relaxation times, as is usually supposed," a theme that also underlies Jonscher's well-known book on "Dielectric relaxation in solids" [11]. A contrasting picture with a history that is equally prominent is that any measured response can be represented by a sum of simple exponentials [12]. Indeed, a heterogeneous distribution of relaxation times provides a versatile paradigm for interpreting stretched exponential relaxation and other empirical formulas for the slow response of materials [13–25]. Figure 1 depicts these two contrasting scenarios.

Over the past few decades, direct evidence to distinguish between these scenarios has finally become available. Specific approaches include measurements on nanoscale systems [26], multidimensional nuclear magnetic resonance (NMR) [27], a combination of optical absorbance and modulated calorimetry [28], which builds on earlier work [29], time-dependent solvation [30] and photobleaching [31], as well as scanning probe microscopy [32] and particle scattering techniques [33–35]. Indeed, spectrally selective methods capable of identifying specific sub-ensembles in the net response have shown that most materials contain a heterogeneous distribution of independently relaxing degrees of freedom [36–43]. Several of these techniques demonstrate that the independent degrees of freedom are spatially localized in independently relaxing "regions," often on a length scale of 1–3 nm involving 10–1000 molecules, spins, or monomer units. Each region is defined by having a

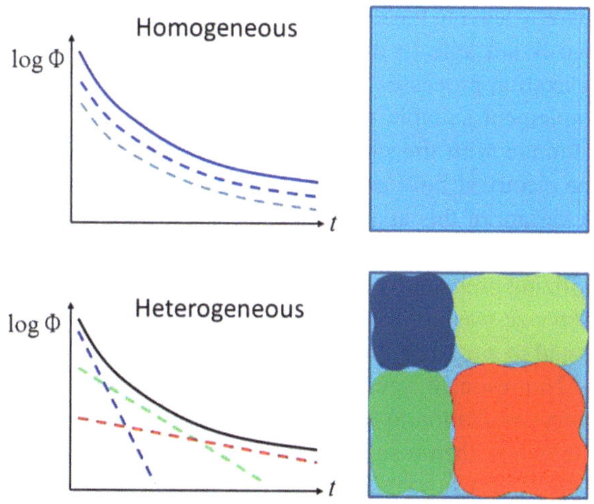

Fig. 1 Solid lines on the left side show identical net responses for the logarithm of a relaxing quantity, log Φ, as a function of time. Dashed lines indicate that the net response may come from homogeneous response throughout the sample (upper frame) or a heterogeneous distribution of independently relaxing regions (lower frame). Right side shows a cartoon sketch of how each type of response may occur inside a bulk sample

distinct relaxation time for its local response, $\tau(x)$, typically yielding effectively single-exponential relaxation.

Some measurement techniques establish that this dynamic heterogeneity is also in the thermodynamics. The focus of the present chapter, nonresonant spectral hole burning (NHB), is one of these techniques. A related approach is based on dynamic specific-heat measurements [44, 45], which involves monitoring the temperature of the thermal bath as heat flows slowly into the slow degrees of freedom. Thermodynamics requires that net heat flow can occur only if the effective temperatures of the slow degrees of freedom differ from that of the heat bath. NHB is essentially the inverse process, monitoring the slow degrees of freedom as their excess energy flows slowly into the heat bath. The versatility and power of NHB have facilitated characterizing thermodynamic heterogeneity in many types of materials, see Sect. 1.2. Indeed, NHB has shown that thermodynamic heterogeneity involves transient but distinct variations in several thermodynamic quantities including energy, temperature, dielectric polarization, magnetic alignment, and/or mechanical stress. NHB has found numerous applications for a wide range of materials including liquids, glasses, spin glasses, polymers, and even high-quality single crystals.

1.1 Background to Development

In order to emphasize different parts of a presumed spectrum of different dielectric relaxation times, a technique called pulsed dielectric spectroscopy was devised [46]. It is based on sequences of (relatively small) electrical fields and field steps. When presenting first results from this technique at a meeting in Denmark, in early 1994, Jeppe Dyre reminded us that without entering the nonlinear response regime, and thereby leaving the range in which Boltzmann's superposition principle [47] applies, one is not able to directly distinguish between homogeneous and heterogeneous relaxation processes. Immediately thereafter, some of us started to discuss how to implement suitable nonlinear excitation and detection schemes. One idea was to continue with the process exploited in Ref. [46], but to increase the step size of the electrical field until the response function would depend on the field strength. A variant of this approach based on the nonlinear generalization of the Boltzmann principle, the so-called Wiener–Volterra series [48], was eventually implemented utilizing pseudo-stochastic multiple-pulse excitation [49]. The idea of increasing the step size was not tested at the time, mainly because the basic NHB technique was found to work beautifully [50, 51].

Nonresonant spectral hole burning utilizes a large-amplitude, low-frequency pump oscillation to modify the dynamics of a sample on the timescale of the pump frequency, then a small probe step is applied to measure the modified spectrum as a function of time. The pump oscillation and probe step are often separated by a variable recovery time, as sketched in Fig. 2. The pump and probe may be in any applied field—e.g., electric, magnetic, or strain field—whenever sufficient energy from this field can be absorbed by the sample. The original inspiration for NHB came from the standard technique of spectral hole burning (SHB) that is normally used to investigate broadened resonances. Although originally applied to NMR [52], now SHB most often involves optical spectroscopy using narrow-bandwidth lasers [53]. The usual goal of SHB is to determine the fundamental source of broadening if the spectrum is broader than a simple Lorentzian. Typically, a sample is irradiated with an intense laser beam at a chosen wavelength of light. The intense beam causes "bleaching" of any response that absorbs significant energy from the light. Bleaching involves eliminating the response from the spectrum by either saturating it or displacing it to distant frequencies. If the spectrum is uniformly bleached at all frequencies, the spectrum is homogeneously broadened. However, if the laser bleaches a narrow "hole" in the spectrum that is roughly the width of a single Lorentzian, the spectrum is inhomogeneously broadened. Most samples show this kind of heterogeneously broadened spectra so that resonance-based SHB facilitates investigating the fundamental features of individual contributions to the full spectrum. Nonresonant spectral hole burning relies on the same idea, but with several key differences:

(1) The spectra studied come from relaxation phenomena, not from resonances.
(2) The pump field frequencies are usually about 15 orders of magnitude slower, e.g., 1 Hz instead of 500 THz.
(3) The mechanism usually involves shifting the response rate, not reducing the magnitude of response.
(4) Because this response-rate-shift mechanism is found to be short-lived, it is crucial to measure the entire modified spectrum rapidly, most suitably in a single step using time-domain spectroscopy.

Another set of experiments that inspired the initial conception of NHB involves dynamic specific heat. Although frequency-dependent heat capacity measurements showed that slow degrees of freedom near the glass transition involve slow heat flow

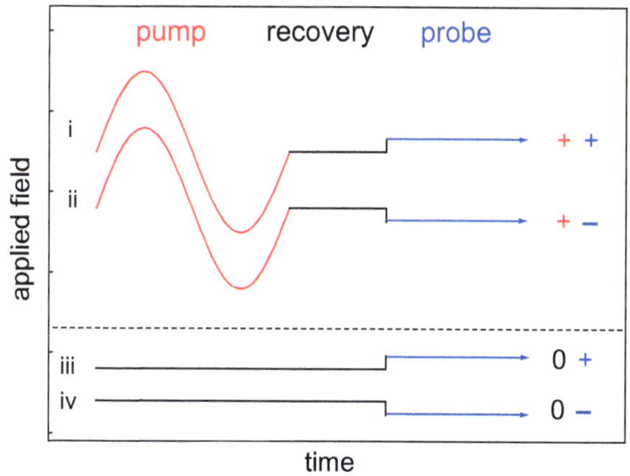

Fig. 2 Sketch of the general protocol of applied fields as a function of time for typical nonresonant spectral hole burning experiments. The field pattern shown consists of a sequence of pump, recovery, and probe periods. Inspired by phase cycles exploited in nuclear magnetic resonance (NMR), undesired signal contributions (such as, e.g., the linear after effect generated by the pump cycle) can be suppressed by suitable linear combination of the responses generated using the numbered traces

[54, 55], it was the low-temperature time-dependent measurements of Meissner and coworkers that provided a more direct picture of the universality of this phenomenon [44, 56–58]. In these time-dependent measurements, a short heat pulse (~1 μs) is applied to one side of a sample, while the temperature, $T(t)$, is monitored as a function of time on the other side. Over several orders of magnitude in time, from the end of the heat pulse until the excess energy flows out through the wires that connect the sample to the cryostat, this measurement yields the specific heat, with heat capacity inversely proportional to $T(t)$. This $T(t)$ is found to first increase rapidly as the heat from the pulse travels across the sample and reaches the thermometer, then the sample cools slowly in a non-exponential manner over intermediate times. Because the sample is adiabatic until the timescale when heat starts to flow out of the sample, such slow non-exponential relaxation must be due to the slow transfer of heat from the heat bath to the slow degrees of freedom in the sample. According to the laws of thermodynamics, this net heat flow can occur only if there is a temperature difference, with the slow degrees of freedom having what is often called an "effective temperature" that distinguishes it from the temperature of the heat bath, similar to the concept of spin temperature known from magnetic resonance [59] or fictive temperature from aging in glasses. Thus, such time-dependent specific-heat measurements imply thermodynamic heterogeneity inside bulk samples. Because the heat is injected into the heat bath, heat flow out of the bath could be into a single slow system that is homogenous (but separate from the bath), or into an ensemble of slow systems that are themselves heterogeneous. The inspiration leading to NHB was to invert the process, so that extra energy is imparted directly into selected slow degrees of freedom, chosen by their response frequency.

An advantage of NHB is that it can yield direct evidence for thermodynamic heterogeneity in the response, even using just the raw data, with no need for analysis, modeling, or interpretation. Another advantage is that detailed agreement with the experiment can be obtained with a simple "box" model, often with no adjustable parameters. The basic idea of this model is that the sample contains independently relaxing regions (the boxes), and that each box has its own time-dependent effective temperature that is governed by a timescale for response of the box, the energy it absorbs from the field, its heat capacity, and timescale for recovery by thermal connection to the heat bath. It is assumed that the timescale for response and recovery are identical, i.e., energy goes into and out of the box at the same rate, a feature that has since been found to usually be the case [67]. For a quantitative connection between the energy absorbed and change in effective temperature, it is assumed that all boxes have the same specific heat, consistent with various measurements over at least four orders of magnitude in frequency [60, 61]. This specific heat can often be deduced from the change in heat capacity of the sample due to the slow degrees of freedom that freeze at the glass transition. Additional details about this box model, including improvements that make it extremely versatile and accurate, will be discussed in Sect. 2.

The theoretical model for NHB in dielectric *and* magnetic response (a simplified box model) was conceived in the summer of 1994 [50]. This model predicted not only general NHB behavior but also several details that were confirmed by measurements— some at a quantitative level. The dielectric variant of the experiment was implemented during the summer of 1995 in the laboratory of Alois Loidl, in Darmstadt, Germany. After Bernd Schiener, a talented student then in his lab, had learned how to prepare samples that did not (always!) break through when large electrical fields were applied, extensive measurements of dielectric hole burning (DHB) in two glassforming liquids were made [51]. The remarkable similarity of the results to the predictions confirmed, in particular, the model's key ingredient: thermodynamic heterogeneity. Some years after the development of DHB, magnetic hole burning (MHB) experiments were successfully implemented [61]. NHB and the ideas on which this method is based have since been applied to a variety of new techniques to identify and investigate heterogeneous relaxation inside bulk materials.

1.2 *Versatility of Nonresonant Hole Burning*

Following the initial results showing heterogeneous response in the α-relaxation of glassforming liquids, similar NHB techniques have been applied to a wide array of liquids and solids with the goal of probing different relaxation processes involving not only dielectric but also magnetic and shear fields. The DHB technique has been successfully applied to investigations of primary and secondary relaxations as well as the excess wing of neat and binary glassforming liquids [50, 60, 62–67], the molecular reorientation in a supercooled plastic crystal [68], the structurally decoupled motion in an ion conductor [69, 70], and the dielectric response of relaxor ferroelectrics

[71–74]. MHB has been applied to the study of spin glasses [61] and ferromagnets. Mechanical hole burning has been applied to investigations of polymers [75, 76] and polymer solutions [77, 78]. In all of these studies, evidence for spectral selectivity was found, and thus based on a phenomenological criterion [37], it can be concluded that heterogeneity exists in the dynamics of the studied liquids and solids. The relative merits of mechanical hole burning as compared to other nonlinear techniques, e.g., those based on the application of large-amplitude oscillatory shear (LAOS) were recently reviewed [78]. Therefore, mechanical hole burning is not further considered in the present chapter. Details regarding LAOS experiments can be found in the chapter by Wilhelm and Hyun in this volume [79].

About 10 years after the conception of the original pump-and-probe scheme inherent in the early DHB experiment, it was suggested that a more easily implemented sequence of small, large, and again small ac fields can serve a similar purpose, which is to deposit energy and probe its consequences in a frequency selective manner [80]. A detailed overview of the results emerging from these insights was given recently [81]. Questions regarding nonlinear response in supercooled liquids [82] can be addressed particularly well using this alternative approach [67], with heterogeneity deduced from the box model.

Another example of the versatility of the pump-and-probe scheme used for NHB would be to generalize the technique to pump with one type of field, and then probe with another. Specifically, one could exploit rheology for the nonlinear spectrally selective excitation, and then apply dielectric spectroscopy for broadband linear-response detection. Such an approach would generalize a related cross-technique, rheodielectric spectroscopy, see for instance Refs. [83–85].

2 Box Model and Other Approaches for Characterizing Nonresonant Hole Burning

2.1 *Introduction*

This section outlines a model that aims at providing a quantitative account of the results expected for dielectric hole burning, with similar models applicable to related measurements such as magnetic hole burning. For DHB, the fundamental idea is that a polar liquid will absorb energy from a time-dependent external electric field [86]. The task is to quantify the influx of energy and determine how it might modify the dielectric relaxation spectrum. Furthermore, it is important to clarify whether adiabatic or isothermal conditions provide a more realistic approach to the experimental situation. First, consider the hypothetical case of a dynamically homogeneous system subject to a sinusoidal field of high amplitude. In this case, the qualitative outcome expected for steady state is relatively straightforward. The energy irreversibly transferred from the electric field (E) to the sample is proportional to E^2. This energy will eventually be converted to heat, which uniformly raises the bath temperature of the sample [87]. The field-induced changes of the dielectric relaxation behavior will then be equivalent to heating the whole sample, yielding a reduction of dielectric relaxation times, τ_D, and a miniscule change in the relaxation amplitude, $\Delta\varepsilon$. In the more common case of heterogeneity for the dynamics and thermodynamics [27, 39, 40], fast and slow modes are independent and thus differ in their energy uptake [51], which is the situation addressed by the model outlined below.

It is important to realize that this model is not meant to capture all contributions to nonlinear dielectric responses. Instead, the scope of this model is limited to rationalizing how the energy that is irreversibly transferred from the field to the sample modifies the dielectric relaxation behavior.

2.2 General Relations

We start by assuming dynamic heterogeneity, implying that the net non-exponential polarization response originates from a superposition of contributions from independently relaxing regions. While other models can be envisioned, we will adopt the case for which each independently relaxing region contributes with a Debye type (single exponential) behavior and that the overall response can be expressed as the sum of these distinct local responses from the regions, weighted by their volume fractions. For the net dielectric permittivity function, this results in the complex permittivity

$$\hat{\varepsilon}(\omega) - \varepsilon_\infty = \Delta\varepsilon \int_0^\infty \frac{g(\tau)}{1 + i\omega\tau} \mathrm{d}\tau \approx \sum_{i=1}^{n} \frac{\Delta\varepsilon_i}{1 + i\omega\tau_i}, \tag{2}$$

where $\Delta\varepsilon_i \approx \Delta\varepsilon g(\tau)\,\mathrm{d}\tau$, with $g(\tau)$ being the normalized probability density of finding a mode with time constant τ in the interval $\mathrm{d}\tau$. While the integral over an effectively continuous $g(\tau)$ containing a very large number of time constants τ is appropriate for describing bulk liquid behavior, the summation over a relatively small number of regions n (labeled by subscript 'i') is convenient for numerical calculations. This model neglects all time-constant fluctuations, referred to as rate exchange [88] or "lifetime of heterogeneity," so that the values of the time constants τ are assumed to remain time-invariant at low-field amplitudes. Because these fluctuations are found to occur on timescales that are usually as slow (or slower) than the average relaxation time, their omission is expected to be inconsequential for all but the very slowest modes.

Similarly, as suggested by dynamic specific-heat measurements, the dynamic heat capacity spectrum, $\hat{C}_p(\omega)$, is modeled as the sum of individual contributions with Debye character

$$\hat{C}_p(\omega) - C_{p,\infty} = \Delta C_p \int_0^\infty \frac{g(\tau)}{1 + i\omega\tau} \mathrm{d}\tau \approx \sum_{i=1}^{n} \frac{\Delta C_{p,i}}{1 + i\omega\tau_i}, \tag{3}$$

where $\Delta C_{p,i} = \Delta C_p g(\tau)\,\mathrm{d}\tau$. The same probability density $g(\tau)$ is employed for describing both permittivity and dynamic heat capacity spectra [89]. This approach to the dynamic heat capacity of the sample implies the assumption that $C_{p,i}$ is proportional to the volume fraction of each independently relaxing region, labeled 'i', equivalent to a constant heat capacity per molecule. Consequently, each region is assigned its individual enthalpy density increase, Δh_i, and thus its own increase in fictive temperature, $\Delta T_i = \Delta h_i/(\rho \Delta C_{p,i})$. Direct enthalpy exchange among different modes is neglected, and each mode is assigned the same time constant τ for both dielectric and enthalpy response. This assumption is supported by evidence against different time constants for dielectric relaxation and enthalpy recovery, as discussed in the literature [90–92]. This picture implies that excess energy remains localized for a relatively long time within a mode, consistent with the slow enthalpy relaxation observed in calorimetry and aging experiments [93]. Because this energy is localized for such long times it must be very weakly coupled to the heat bath, so that the energy may have time to equilibrate among all localized degrees of freedom in the region, thereby mimicking a local fictive temperature that is different from the bath temperature. Thus, $\Delta T_i \neq 0$ has physical meaning.

In order to reflect reproducible initial conditions, it is assumed that each region 'i' is depolarized and at zero field prior to applying an external field at $t = 0$:

$$P_i\,(t \leq 0) = 0, \quad E\,(t \leq 0) = 0$$

The subsequent polarization response of each mode is governed by the differential equation

$$\frac{\mathrm{d}P_i\,(t)}{\mathrm{d}t} = \frac{\varepsilon_0 \Delta\varepsilon_i\,(t)\,E\,(t)}{\tau_i\,(t)} - \frac{P_i\,(t)}{\tau_i\,(t)}. \tag{4}$$

The overall polarization $P(t)$ involves the sum of an instantaneous (P_∞) and a retarded (ΔP) contribution with respect to the field E, as given by

$$P\,(t) = \varepsilon_0\,(\varepsilon_\infty - 1)\,E\,(t) + \sum_i P_i\,(t) = P_\infty\,(t) + \Delta P\,(t)\,. \tag{5}$$

This relation yields the ordinary linear-response result for $P(t)$ for a given time-dependent field $E(t)$ in the special case that all $\Delta\varepsilon_i$ and τ_i remain time-invariant. Allowing these quantities to change with fictive temperature T_i, and thus with time represents the box model manner of incorporating nonlinear effects, whereas their changes are negligible for sufficiently small fields.

For a given region 'i', the current density $j_i(t) = \mathrm{d}P_i(t)/\mathrm{d}t$ creates an additional enthalpy density per unit time, $p_i\,(t) = \mathrm{d}h_i\,(t)\big/\mathrm{d}t$, according to the power absorbed $p_i\,(t) = j_i^2\,(t)\,\tau_i\,(t)\big/\,(\varepsilon_0 \Delta\varepsilon_i)$, while the recovery of h_i toward equilibrium is governed by the recovery time τ_i of that particular region, leading to

$$\frac{\mathrm{d}h_i(t)}{\mathrm{d}t} = p_i(t) - \frac{h_i(t) - h_i^{\mathrm{eq}}}{\tau_i(t)} = \left(\frac{\mathrm{d}P_i(t)}{\mathrm{d}t}\right)^2 \frac{\tau_i(t)}{\varepsilon_0 \Delta\varepsilon_i(t)} - \frac{h_i(t) - h_i^{\mathrm{eq}}}{\tau_i(t)}. \quad (6)$$

All enthalpy lost by a mode due to its recovery is added as heat to the bath with heat capacity $C_{p,\infty}$, i.e., it is assumed that no enthalpy is passed directly from one mode to another. These enthalpy densities can be converted to fictive temperatures T_i via $\mathrm{d}h_i(t) = \rho\Delta C_{p,i}\,\mathrm{d}T_i(t)$ and $h_i^{\mathrm{eq}} = \rho\Delta C_{p,i} T$. Similar to the situation encountered in the context of physical aging, the concept of a fictive or configurational temperature is also used here to characterize the state of a nonequilibrium system in terms of the value of T_i that would give rise to the same state if the system were in equilibrium at a temperature $T = T_i$.

Finally, it needs to be specified how the fictive temperatures T_i impact the parameters that determine the relaxation behavior. Because field-induced changes in T_i are small, only the relaxation time τ_i is considered to be modified by external fields, as the sensitivity of the amplitudes $\Delta\varepsilon_i$ to temperature is negligible by comparison. The dependence of τ_i on T_i is quantified by the nontrivial assumption that $\tau_i(T_i)$ tracks the overall equilibrium activation behavior $\tau_\alpha(T)$ via

$$\frac{\mathrm{d}\tau_i}{\mathrm{d}T_i} = -\frac{\tau_\alpha}{T^2}\left(\frac{\mathrm{d}\ln\tau_\alpha}{\mathrm{d}(1/T)}\right). \quad (7)$$

This relation couples Eqs. (4) and (6), i.e., the polarization response is modified by the changes in mode-specific fictive temperatures.

Regarding the heat exchange with the surroundings, limited average power will ensure that heat is transported effectively away from the sample toward a thermal reservoir. In the situation in which a sample of thickness d is sandwiched between two metal electrodes that can be considered isothermal, the field-induced temperature increase can be estimated. The steady-state result for a spatially uniform average power density, $p = \mathrm{d}q/\mathrm{d}t$, is obtained by solving $\partial^2\Delta T/\partial z^2 = -p/\kappa$, where z is the spatial coordinate normal to the electrode surface and κ the thermal conductivity of the sample [94, 95]. The average temperature increase is thus $\Delta T_{\mathrm{avg}} = pd^2/(12\kappa)$, and the maximal effect at the center ($z = d/2$) amounts to $\Delta T_{\mathrm{max}} = 1.5\Delta T_{\mathrm{avg}}$. If p is sufficiently small and ΔT_{max} is negligible compared with the fictive temperature changes measured in DHB, isothermal conditions may be assumed for the sample with the bath temperature T kept as a constant. The adiabatic limit, with no heat transfer to the surroundings, is accounted for by having the enthalpy recovery from all regions combined into an enthalpy increase of the bath, i.e.,

$$\frac{\mathrm{d}T}{\mathrm{d}t} = \frac{1}{\rho C_{p,\infty}} \sum_{i=1}^{n} \frac{h_i(t) - h_i^{\mathrm{eq}}}{\tau_i(t)} = \sum_{i=1}^{n} \frac{T_i(t) - T}{\tau_i(t)} \frac{\rho\Delta C_{p,i}}{\rho C_{p,\infty}}. \quad (8)$$

If the high-amplitude field is applied for a longer time, this may result in a significant increase of the bath temperature T of the sample [92].

2.3 Dielectric Hole Burning Protocol

The protocol of the pump-probe experiment is designed to demonstrate heterogeneity regarding the dynamics by modifying the response of a subset of modes, while the other modes remain unaffected. Accordingly, the narrowest possible power spectrum of the field is desired, so that a sinusoidal field with frequency Ω is the pattern of choice for the duration of the pump, usually an integer number n_p of complete cycles during the pump time, t_p, cf. Fig. 2. For a subsequent recovery time (t_r) that separates the pump from the probe cycle, the field is zero (or very small), allowing for the recovery of the pump-induced effects over the recovery time. Thus, the overall field sequence can be represented by

$$E\left(t_p\right) = \begin{cases} 0 & , t_p < 0 \\ E_0 \sin\left(\Omega t_p\right) & , 0 \le t_p \le n_p 2\pi/\Omega \\ 0 & , n_p 2\pi/\Omega \le t_p \le n_p 2\pi/\Omega + t_r \end{cases} . \tag{9}$$

This field generates an influx of energy density per unit time, $p_i(t)$, to modes with relaxation time τ_i and dielectric amplitude $\Delta\varepsilon_i$, according to [82, 96]

$$p_i\left(t_p\right) = \frac{\varepsilon_0 E_0^2 \Delta\varepsilon_i \Omega^2 \tau_i}{\left(1+\Omega^2\tau_i^2\right)^2} \times \begin{cases} \left[\Omega\tau_i \sin\left(\Omega t_p\right) + \cos\left(\Omega t_p\right) - \exp\left(-t_p/\tau_i\right)\right]^2 & , 0 \le t_p \le n_p 2\pi/\Omega \\ \left[1 - \exp\left(n_p 2\pi/\Omega\tau_i\right)\right]^2 \exp\left(-2t_p/\tau_i\right) & , t_p > n_p 2\pi/\Omega \end{cases} . \tag{10}$$

An example of the time-dependent power for $n_p = 2$ is depicted in Fig. 3. The impact of this power on the fictive temperature of mode 'i' is then determined by

$$\frac{\mathrm{d}T_i(t)}{\mathrm{d}t} = \frac{p_i(t)}{\rho\Delta C_{p,i}} - \frac{T_i(t) - T}{\tau_i(t)} . \tag{11}$$

This $T_i(t)$ then modifies the relaxation parameters τ_i (and possibly $\Delta\varepsilon_i$) as outlined above.

The amplitude of the probe step field, $E_{\mathrm{probe}} \ll E_0$, applied at $t = 0$, where $t_p = n_p 2\pi/\Omega + t_r$, is considered to be in the regime of linear response, so no additional power is absorbed from this field and the polarization response to this field step is proportional to E_{probe}. However, unless t_r is very long, the enhancement of fictive temperatures will persist for times $t > 0$ and thus modify the polarization response to this probe field. Because the pump field amplitude is much larger than E_{probe}, much of the polarization for $t > 0$ may originate from the pump field, while the response to the probe step is small. In the experiment, a "phase cycle" is used to remove the linear response to the pump field by taking the average of two polarization

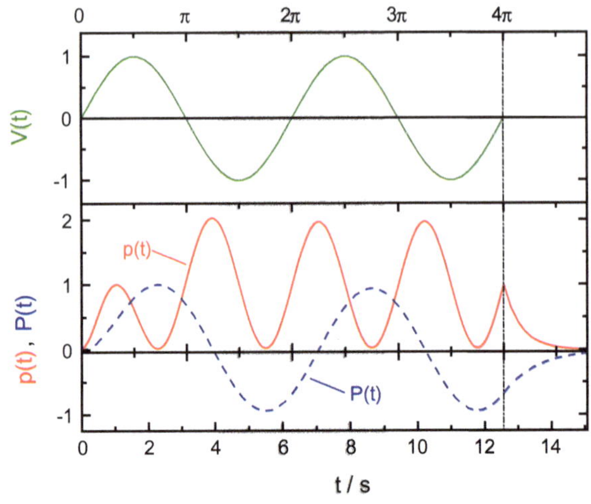

Fig. 3 Applied sinusoidal voltage $V(t) = E(t)/d$ and the resulting polarization $P(t)$ and dissipated power $p(t)$ in a single mode with $\tau = 1$ s and $\Omega = 1\ \mathrm{s}^{-1}$. The curve for $V(t)$ follows Eq. (9) for $n_p = 2$, $P(t)$ is derived from Eq. (4), $p(t)$ is calculated according to Eq. (10). Note that some power remains after $V(t)$ has been turned off, i.e., for $t > 4\pi/\Omega$, because a significant amount of polarization P is present at $t = 4\pi/\Omega$. The curves are scaled arbitrarily. The vertical line marks the end of the pump process. Adapted from Ref. [96]

signals that differ only in the sign of E_0, equivalent to a 180° phase shift of the pump field, cf. Fig. 2. Within the model calculation, this "phase cycle" can be accounted for by setting $P_i(t) \equiv 0$ for all times $t < 0$, i.e., disregarding all polarization prior to the probe step. The final goal is to compare the overall step polarization response following the pump process, $P^*(t)$, with that in the absence of the pump event, $P(t)$. The result is presented either as vertical difference, $\Delta_\mathrm{V}P(t) = P^*(t) - P(t)$, or its horizontal analog, $\Delta_\mathrm{H}P(t)$, which for sufficiently small differences can be approximated by [60]

$$\Delta_\mathrm{H}P(t) = -\frac{\Delta_\mathrm{V}P(t)}{\mathrm{d}P(t)/d\ \ln(t/s)}. \tag{12}$$

Because the probe field E_probe is time-invariant for times $t > 0$, the polarization response to the probe step is linearly related to permittivity $\varepsilon(t)$ in the time domain. An alternative to probing the step response in a constant-field situation is to use a displacement step, D_probe, to polarize the system and then monitor the field as a function of time [60, 80, 97, 98]. In this constant charge case, polarization is linearly related to the field $E(t)$ and to the electric modulus $M(t)$, defined via the steady-state relation $\hat{M}(\omega) = 1/\hat{\varepsilon}(\omega)$. For a given system, the relaxation, $M(t)$, is always associated with a smaller time constant relative to the retardation, $\varepsilon(t)$, with the relation between the linear averages obeying $\langle\tau_M\rangle / \langle\tau_\varepsilon\rangle = \varepsilon_\infty/\varepsilon_\mathrm{s}$ [99]. Within the framework of the above model, this constant charge case can be realized by an initial

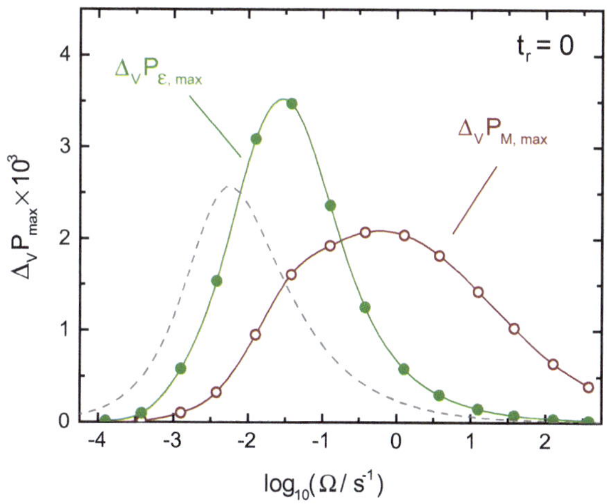

Fig. 4 Dependence of the peak amplitude of vertical hole signals (at zero waiting time) on the pump frequency Ω. The curve labeled $\Delta_V P_{\varepsilon,\max}$ refers to constant voltage detection with $P(t) \sim \varepsilon(t)$ while $\Delta_V P_{M,\max}$ refers to constant charge detection with $P(t) \sim M(t)$. The comparison indicates the advantage of using modulus type detection $M(t)$ at high pump frequencies, here with $\Omega > 1\ \mathrm{s}^{-1}$. The calculation is for glycerol at $T = 187.3$ K, with the arbitrarily scaled loss profile, $\varepsilon''(\omega)$, shown as dashed line. Adapted from Ref. [80]

probe field set to $E(0) = D_{\text{probe}}/(\varepsilon_0\varepsilon_\infty)$, which will change with time according to $\mathrm{d}E(t)/\mathrm{d}t = -(\mathrm{d}\Delta P/\mathrm{d}t)/(\varepsilon_0\varepsilon_\infty)$. Constant charge conditions require $D(t = 0) = D(t \to \infty)$, thus implying a finite steady-state field $E(t \to \infty) = (\varepsilon_\infty/\varepsilon_s)E(0)$ for a dielectric sample, and a subsequent decay to zero only in the case of dc conductivity [98]. A model-based comparison of the two types of probing the polarization in time domain, constant-field versus constant charge, is depicted in Fig. 4, which shows the peak vertical difference $\Delta_V P_{\varepsilon,\max}$ and $\Delta_V P_{M,\max}$ versus pump frequency Ω, respectively. These data demonstrate that the constant charge case provides increased sensitivity to the pump-induced modification at the higher pump frequencies, a direct consequence of the $\langle\tau_M\rangle < \langle\tau_\varepsilon\rangle$ feature.

Comparisons of predictions of the model discussed in the present section (or variants thereof) with experimental data can be found in various reports [62, 64, 66, 69, 90–92, 94].

2.4 Other Applications of the Box Model

The general version of the box model, as outlined via Eq. (2) through Eq. (8), has been applied to field protocols other than those exploited for dielectric hole burning. Changes in fictive temperatures and the concomitant modifications of polarization responses should be expected whenever a time-dependent field of sufficient magnitude is applied to a sample with considerable dielectric loss, ε''. The model reveals that a high-field step modifies the polarization response relative to the low-field counterpart, and the effects are larger than the DHB results for the same field amplitude as the temporal separation between pump and probe is eliminated [80]. However, the power spectrum of a field step is much wider than that of the DHB pump field, and the spectral selectivity is diminished accordingly. For a field step of magnitude E applied at time $t = 0$, the power density for a single mode 'i' is given by

$$p_i(t) = \frac{\varepsilon_0 \Delta\varepsilon_i(t) E^2}{\tau_i(t)} \exp\left(-\frac{2t}{\tau_i(t)}\right). \tag{13}$$

This term has been employed successfully to capture how field steps modify the dielectric behavior of liquids [100].

Sinusoidal fields of high amplitudes E_0 can lead to very prominent changes in the dielectric behavior. For isothermal conditions and a field given by $E(t) = E_0 \sin(\omega t)$, the steady-state fictive temperature for mode 'i' follows:

$$T_i = T + \frac{\varepsilon_0 E_0^2 \Delta\varepsilon}{2\rho \Delta C_p} \times \frac{\omega^2 \tau_i^2}{1 + \omega^2 \tau_i^2}, \tag{14}$$

where the frequency-dependent term approaches unity for $\omega \gg \tau_i$ [67]. For an entire typical loss spectrum, $\varepsilon''(\omega)$, this modifies the high-frequency wing ($\omega > 10\omega_{\max}$) as if the temperature had increased by the amount $\varepsilon_0 E_0^2 \Delta\varepsilon / (2\rho \Delta C_p)$. For frequencies not too far above the loss peak frequency $\omega_{\max}$, high-field impedance experiments have demonstrated agreement with these predictions [90, 92, 101–103]. At very high frequencies relative to $\omega_{\max}$, the magnitude of this steady-state effect may be reduced due to excess wing or secondary processes.

In order to explore how the fictive temperature approaches its steady-state value as a function of time, the sinusoidal field technique has been supplemented with amplitude steps. A typical protocol could consist of 32 periods of the sine wave at low fields (within the linear response regime), followed by 64 periods with high-field amplitude, after which another 32 periods of the low field are appended. Fourier analysis of the voltage and current traces for each period facilitates studying the field-induced modifications as a function of time after applying or removing the high field. This protocol is best modeled with the general approach, Eq. (2) through Eq. (8), with $E(t)$ representing the amplitude-modulated sinusoidal field. For frequencies not too far above $\omega_{\max}$, the model captures the experimentally observed time dependence with high fidelity [91, 92, 102, 104]. At higher frequencies, very fast modes adjust their time constants or fictive temperatures on the timescale of the average structural relaxation time (τ_α) rather than τ_i. Therefore, in the range $\omega > 100\omega_{\max}$, it can take many periods of a high field before steady state is reached [102, 103].

So far each mode 'i' is assumed to be single-exponential in time, or Debye-like as a function of frequency, cf. Eq. (2). Thus, the shapes of the spectral holes are relatively narrow. Similar assumptions underlie the "excitation profile" that was used in the initial applications of the box model [50] and most of [105] its refinements in studies dealing primarily with supercooled liquids [51, 62]. The latter typically display relatively narrow dielectric loss spectra. However, for disordered systems featuring very broad loss spectra—such as relaxor ferroelectrics [73] and some binary mixtures of glass formers [66]—the effects of intrinsically non- or multi-exponential responses are sometimes relevant, as was considered for the description of molecular dynamics in other contexts [37, 106–108]. On a phenomenological basis, rather than starting from exponential behavior, intrinsic nonexponentialities were typically written akin to Eq. (1), for instance as

$$P_i \propto \exp[-(t/\tau_i)^{\beta_{\mathrm{in}}}]. \tag{15}$$

If intrinsic nonexponentialities are present, it was suggested to expand the box model [66]: In particular, the expression describing the change in field-energy density per unit time, that is generated by the application of an external electrical (pump) field, then comprises two parts. The one referring to the dissipative contribution is given by that considered in Eq. (10). In addition, in the presence of intrinsic nonexponentialities, a "reactive" contribution was found to arise which corresponds to power "temporarily stored in the system" [66]. For further details and implications of that approach the reader is referred to Ref. [66].

2.5 Asymmetric Double Wells and Other Approaches

Various other approaches can yield net behavior analogous to the box model. A particularly appealing framework for many workers in the field is based on models involving an asymmetric double-well potential (ADWP) [97, 105, 109]. The double well may be viewed to refer to the potential describing two molecular orientations or to two minima in a slice of a potential energy surface, with asymmetry needed to lift the energetic degeneracy between the minima, as depicted in Fig. 5. The population numbers of the two minima are given by q_1 and q_2 so that for a total of N dipoles with moment μ_{el}, the (linear) polarization $P(t) = \mu_{\mathrm{el}} \cos\theta\; n(t)$ arising in response to a time-dependent electrical field $E(t)$ can be obtained from the population difference

$$n(t) \equiv N\,[q_1(t) - q_2(t)]\,. \tag{16}$$

The nonlinear response of ADWPs has been treated for external fields with an arbitrary time dependence, see Eq. (14.21) in Ref. [68]. Its frequency-domain response is further considered elsewhere in this volume [110]. Therefore, here we just summarize the main results of the calculations insofar as they refer to the NHB protocol sketched in Fig. 2. Changes in the population difference are assumed to be governed by the master equation, see Eq. (14.18) in Ref. [68]. In lowest order ($|e(t)| \ll 1$) in the reduced electric field $e(t) \equiv \mu_{\mathrm{el}} E(t) \cos\theta/(2k_{\mathrm{B}}T)$ one has

$$\dot{n}(t) = \tau^{-1}[e(t) + \delta]N - \tau^{-1}[1 + e(t)\,\delta]n(t)\,. \tag{17}$$

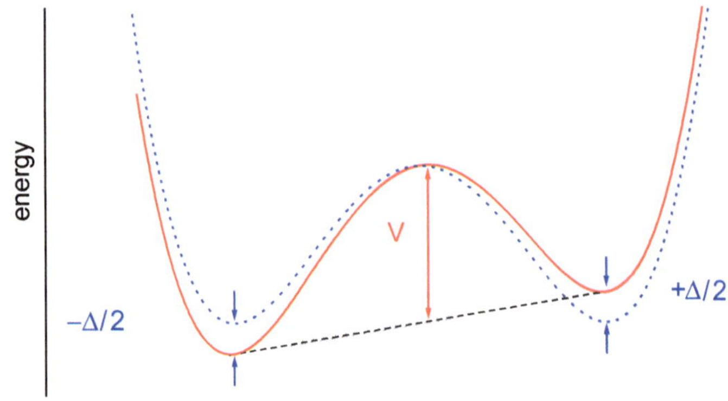

Fig. 5 The solid line sketches an ADWP characterized by an asymmetry Δ. The dotted line depicts a symmetric double-well potential with a barrier height V, e.g., with respect to some rotational coordinate

Here, use was made of the reduced asymmetry $\delta \equiv \tanh[\Delta/(2k_\mathrm{B}T)]$ and the relaxation rate

$$\tau^{-1} \equiv 2W\cosh[\Delta/(2k_\mathrm{B}T)] \tag{18}$$

with W denoting the transition rate for vanishing asymmetry, $\Delta = 0$. The rate τ^{-1}, via $W = \nu_0\exp[-V/(k_\mathrm{B}T)]$, can in turn be related to the barrier height V and the attempt frequency ν_0 (typically on the order of about 10^{13} s^{-1}).

It is important to point out that modeling of heterogeneous response requires choosing suitable distributions for both the barrier heights V_i and asymmetries Δ_i. However, the distributions of V_i and Δ_i cannot be chosen independent of each other, but only in such a way that the resulting distribution of relaxation times $\tau_i = \tau(V_i,\delta_i)$ faithfully describes the response $\Sigma_i P_i(t)$ induced by a small step in the external field.

In order to obtain the contributions of various polarization modes $P_i(t)$ one has to solve Eq. (17). Let us, therefore, start from an initial condition $n_i(0) = N\delta_i$, which simply refers to the equilibrium value for the population difference of a single ADWP $n_i(t \to \infty) = N\delta_i$. Then, the linear response of an ADWP to a step field E_probe (after integration over the entire solid angle) gives $P_i(t) = \chi_{\mathrm{ADWP},i}\, E_\mathrm{probe}\, [1 - \exp(-t/\tau_i)]$ with the static susceptibility $\chi_{\mathrm{ADWP},i} = \frac{1}{6}\, N\mu_\mathrm{el}^2(1 - \delta_i^2)\,/\,(k_\mathrm{B}T)$.

Next, let us assume that the sequence represented by Eq. (10) was implemented, i.e., after n_p cycles of a sinusoidal pump field, $E(t) = E_0\sin(\Omega t_p)$, and after a recovery time t_r, a probe field E_probe is finally applied. For this sequence of fields and, like before, using $n_i(0) = N\delta_i$ as the initial condition, Eq. (17) was solved to quadratic order in $e(t)$ during the pump period, and in linear order thereafter. Using the NHB protocol including the phase cycling sketched in Fig. 2 for the vertical difference one then obtains

$$\Delta_V P_{\mathrm{ADWP,i}}(t) = C_{\mathrm{ADWP,i}}(\delta_i, E_{\mathrm{probe}}) X_{\mathrm{SIN}}(\Omega) \frac{t}{\tau_i} \exp\left(-\frac{t+t_r}{t_i}\right). \quad (19)$$

Here, $C_{\mathrm{ADWP},i}(\delta_i, E_{\mathrm{probe}}) \equiv E_{\mathrm{probe}}\ N\mu_{\mathrm{el}}^4 \delta_i^2 (1 - \delta_i^2)/[40(k_B T)^3] = \chi_{\mathrm{ADWP},i}\ E_{\mathrm{probe}} \frac{3}{20}[\mu_{\mathrm{el}}\delta_i/(k_B T)]^2$ is a prefactor that tends to vanish for asymmetries $\delta_i \to 0$. The so-called excitation profile referring to n_p pump cycles is given by $X_{\mathrm{SIN}}(\Omega) = \frac{3}{2} E_0^2 \varepsilon''(\Omega t_p)\varepsilon''(2\Omega t_p)\{1-\exp[-2\pi n_p/(\Omega t_p)]\}$ and $\varepsilon''(k\Omega t_p) = k\Omega t_p/[1+(k\Omega t_p)^2]$ with $k = 1, 2$. The profile is peaked at $\Omega\tau_i \approx 1$, which means that like for the box model, a modification of the polarization is achieved in a frequency selective manner.

Note that the Eq. (19) can be interpreted to imply that the pump process has shortened the relaxation time of polarization contribution 'i', so that $\Delta\ln(\tau_i/\mathrm{s}) = -\frac{3}{20}[\mu_{\mathrm{el}}\delta_i/(k_B T)]^2 X(\Omega)$. By virtue of Eq. (18) the reduced relaxation time may be ascribed, e.g., to an increase in the local *effective* temperature due to deviations from the equilibrium population difference. In the latter case, the pump process has changed the effective occupation of the two wells of the ADWP, analogous to the spin temperature in NMR.

If desired, based on Eq. (19) the horizontal difference $\Delta_H P_{\mathrm{ADWP},i}(t)$ can be calculated simply by applying the general relation given by Eq. (12). A representation of this function specific for the ADWP was given in Appendix 2 of [68].

The ADWP is one of the simplest examples of a multi-state model. While it is straightforward to formulate generalizations to more complex models [68], it is interesting to point out that rather involved perturbative approaches were developed and applied in connection with the NHB protocol as well. Such approaches include descriptions of nonlinear dipolar response in terms of stochastic dynamics [105, 110] and of spin models [111].

Yet another approach to address the issues of homogeneous versus heterogeneous dynamics starts from Brownian dipole oscillators in harmonic potentials [112]. By expanding, e.g., the dipole moments in terms of normal vibrational coordinates, and retaining only terms referring to the lowest nontrivial order, the cubic response of an ensemble of Brownian oscillators was calculated [112]. An interesting feature of the results is that they can be applied to resonance as well as to relaxation phenomena. In the underdamped case, these calculations may be useful for application in the field of terahertz spectroscopy. In the overdamped case, response functions akin to those referring to a relaxational scenario are recovered [112].

When expanding the polarizability α_{el} rather than the dipole moment μ_{el} in terms of the normal coordinates, applications in the areas of Raman or Kerr effect spectroscopies are conceivable. Then, the coupling of the responding system to the external electrical field is described by a Hamiltonian $H = -\mu_{\mathrm{el}}\ E(t)\ P_1(\cos\theta) - \alpha_{\mathrm{el}}\ E^2(t)\ P_2(\cos\theta)$, where $P_l(\cos\theta)$ denotes a Legendre polynomial of rank l [113]. With the goal to discriminate homogenous from heterogeneous scenarios, the $P_2(\cos\theta)$ response was studied in the framework of the Brownian oscillators (also including anharmonic potentials as a source of nonlinearity) [114] as well as in terms of a rotational diffusion model for application to the slow dynamics of supercooled liquids [113].

3 Nanothermodynamics

3.1 Introduction

When combined with information from other experimental techniques, the basic picture that emerges from NHB is that the primary response of most materials involves thermodynamic heterogeneity from an ensemble of independently relaxing nanometer-sized regions. As described in Sect. 2, various models based on this picture are used to characterize NHB, including the box model. Despite its basic simplicity, the box model provides excellent agreement with all essential features found in NHB, and other measurements involving nonlinear response and dynamic heterogeneity, often with no adjustable parameters. Indeed, by an accidental coincidence of names, the box model is a prime example of the aphorism introduced by George Box [115], and used by statisticians: "All models are wrong, but some are useful." Here, we seek a more deeply-rooted foundation for the box model, not only with respect to NHB experiments but also for a general understanding of how dynamic heterogeneity can occur in the equilibrium response of ostensibly homogenous liquids and solids. Therefore, in the following, we will summarize some insights provided by the theory of small-system thermodynamics, which was originally developed by Terrell Hill to describe finite-size effects in ensembles of nanometer-sized systems [116–118]. By analogy, this "nanothermodynamics" has been adapted to provide a general framework for the description of heterogeneous samples that contain an ensemble of independently relaxing nanometer-sized regions [23, 119, 120], as studied using NHB and other techniques. Nanothermodynamics provides a fundamental physical foundation for the box model, ADWPs, Brownian dipole oscillators, and other models proposed to describe the thermal and dynamic heterogeneity inside bulk samples [24, 121–123].

3.2 Thermodynamic Heterogeneity in Bulk Samples

We start by assuming dynamic heterogeneity due to distinct spatial regions that are effectively uncorrelated during their response. This assumption is supported by several experimental techniques that identify dynamical correlation lengths of 1–3 nm [26, 27, 32], including some that indicate decorrelation occurs abruptly across subnanometer interfaces in glassforming liquids [102, 124] and in crystals [33–35]. Furthermore, computer simulations have shown that correlations between regions due to interactions cause serious deviations from the standard fluctuation–dissipation relation for uniform specific heat [125, 126], so that decorrelation may be necessary to justify using Boltzmann's factor ($\mathrm{e}^{-\Delta U/k_\mathrm{B}T}$) for the probability that a fluctuation increases the internal energy by an amount ΔU. Empirical evidence for uniform specific heat comes from measurements of NHB, especially in the high-frequency wing of glassforming liquids showing that the specific heat is constant over two orders of magnitude in amplitude and four orders of magnitude in frequency [60], unlike the size-dependent energy fluctuations found in simulations of interacting regions. However, we also recognize that over long enough times, the dynamic heterogeneity exhibited by many samples becomes homogeneous, which generally involves exchange of energy and particles, thereby mixing the relaxation rates. Our assumption is that significant correlation between regions occurs only after the primary response is essentially complete, consistent with most measurements, and models used to describe them. Thus, we adopt one of the requirements given by Feynman for Boltzmann's factor to be valid [127]: "...if all the 'fast' things have happened and all the 'slow' things not, the system is said to be in thermal equilibrium."

In general, dynamic heterogeneity yields statistical independence, so that the net probability from uncorrelated regions that fluctuate and relax, independently is the product of their separate probabilities. Specifically, if system 1 has W_1 ways to yield its response, and system 2 has W_2 ways to yield its response, statistical independence yields $W_1 W_2$ for the total number of ways that the combined system can give its net response. Thus, dynamic heterogeneity implies thermodynamic heterogeneity, where the entropy of the combined system is the sum of the two separate entropies: $S_{1+2} = k_B \ln(W_1 W_2) = k_B [\ln(W_1) + \ln(W_2)] = S_1 + S_2$. NHB is one of the techniques that can measure this thermodynamic heterogeneity directly. Because standard thermodynamics is based on the assumption that each system is homogeneous, with a total entropy that cannot be subdivided, thermodynamic heterogeneity places strict constraints on the foundation of any theory proposed to explain the heterogeneity inside bulk samples. The box model meets these criteria by simply assuming that the sample can be modeled by an ensemble of independent boxes, each with its own temperature that is governed by energy absorption, heat capacity, and heat flow to the bath. As we will show, nanothermodynamics provides the theoretical foundation for justifying such models based on thermodynamic heterogeneity. First, however, we review some classical treatments of thermal fluctuations inside bulk materials, and one way to make them heterogeneous.

Theoretical treatment of thermal fluctuations was pioneered by Einstein, culminating in his theory of critical opalescence inhomogeneous liquids [128]. Einstein's basic idea was to include a second-order term in the Taylor-series expansion in the probability of finding specific states: $p \propto e^{\Delta S/k_B}$, where to second order, the change in entropy is $\Delta S = (\partial S'/\partial E')\Delta E' + \frac{1}{2}\lambda^2(\partial^2 S/\partial \lambda^2)$. Here, S is the total entropy, S' and E' are the entropy and energy of the heat bath, and λ is an internal order parameter of the system. Note that Boltzmann's factor comes from the thermodynamic definition $\partial S'/\partial E' = 1/T$, while $\partial^2 S/\partial \lambda^2 < 0$ is needed for stable fluctuations. Extending this idea, Landau added a quartic term in the Taylor-series expansion to obtain a unified theory of phase transitions [129]. Both of these theories assume that λ is uniform throughout the sample so that the theories are unable to accommodate heterogeneity. The Ginzburg–Landau theory for phase transitions improves on these ideas by allowing local variations in λ. One way to implement this theory is to start with separate cells that have distinct values of λ, then couple the cells via an interface term to obtain a homogeneous phase [130, 131]. Thermodynamic heterogeneity can be achieved by simply eliminating the coupling between cells, yielding independently relaxing regions [132, 133]. However, such spatial heterogeneity is incompatible with standard thermodynamics, where it is assumed that each system is macroscopically homogeneous in all of its thermal properties. Below, we use the thermodynamics of small systems to provide a theoretical foundation for independently relaxing regions, but first, a brief review of standard thermodynamics is provided.

The fundamental equation of thermodynamics, also known as Gibbs' equation, is the combined first and second laws. It is often written in terms of a differential in the internal energy as

$$\mathrm{d}U = T\mathrm{d}S - p\,\mathrm{d}V + \mu\,\mathrm{d}N. \tag{20}$$

Here, the intensive environmental variables are temperature T, pressure p, and chemical potential μ; while the extensive parameters for the system are entropy S, volume V, and number of particles N. Equation (20) is based on several assumptions. Basically, the system must be linear, homogeneous, and large. Large implies the thermodynamic limit, where $N \to \infty$ with V/N finite, which may apply to equilibrium properties of bulk samples when averaged over long enough times, but not to the primary response from independently relaxing regions that usually have length scales on the order of nanometers. Homogeneous implies all thermodynamic variables are uniform throughout the sample, unlike systems exhibiting thermodynamic heterogeneity investigated by NHB and other techniques. Linear implies that the intensive environmental variables are the linear (first-order) derivatives of the internal energy with respect to the extensive parameters; higher-order derivatives are neglected. Neglected terms include fluctuations involving the heat capacity, e.g., $\partial^2(U - TS) / (\partial T^2)|_{V,N} = -C_V/T$, as well as the quadratic and quartic contributions introduced by Einstein and Landau. The usual argument is that for macroscopic systems, fluctuations are negligible compared to average properties, at least when far from any phase transitions. For example, relative fluctuations in energy, $\langle \Delta U^2 \rangle / \langle U \rangle^2 = k_B T^2 C_v / \langle U \rangle^2$, are proportional to $1/N$ and hence negligible as $N \to \infty$. However, energy changes due to finite-size effects cannot be neglected in nanometer-sized systems, including heterogeneous systems having local fluctuations that are independent of neighboring fluctuations. Thus, nanothermodynamics provides the theoretical foundation necessary for treating systems that show thermal and dynamic heterogeneity on the scale of nanometers.

Finite-size effects alter the energy of nanometer-sized systems. In general, the resulting internal energy can be written as

$$U = Nu + N^{2/3}a_0 + N^{1/3}b_0 + c_0 + \cdots. \tag{21}$$

Here u is the bulk energy per particle, a_0 governs the surface energy, b_0 is a length-dependent factor, c_0 a fluctuation term, etc. Non-extensive contributions are always present, but they are negligible if $N \to \infty$. However, for finite-sized systems (or fluctuations) the surface term is about 1% of the total energy when $N \sim 10^6$ particles, while the fluctuation term is also significant for $N \leq 1000$. Including finite-size effects in U alters the left side of Eq. (20), so that if energy is to be conserved, something must change on the right side. In 1962, Terrell Hill introduced the theory of small-system thermodynamics. His theory restores conservation of energy for finite-sized systems by adding a new pair of conjugate variables, a subdivision potential ε and the number of subdivisions η, so that Eq. (20) becomes [116–118]

$$dU = T\,dS - p\,dV + \mu\,dN + \varepsilon\,d\eta. \tag{22}$$

For bulk samples containing thermodynamic heterogeneity, the number of subdivisions (η) is essentially the number of independently relaxing regions, while the subdivision potential ($\varepsilon = U - TS + pV - \mu N$) yields the residual free energy from these regions after all linear and homogeneous contributions are removed.

Hill's theory was originally developed to treat separate small systems, such as individual molecules or independent nanoparticles, which he conceptually combined into a macroscopic ensemble that is assumed to obey standard thermodynamics. Using a "different approach to nanothermodynamics" [118] Hill's ideas should also apply to independently relaxing nanometer-sized regions, which naturally form a macroscopic ensemble of small systems inside bulk samples. Indeed, nanothermodynamics facilitates a unique feature that is needed for conservation of energy in independently relaxing regions. Specifically, although separate interface (a_0) and/or length-scale terms (b_0) from static structures inside bulk samples could be added to any Hamiltonian, free energy changes due to independently-fluctuating nanometer-sized regions, e.g., c_0 in Eq. (21), cannot be accommodated by Eq. (20). Other non-Hamiltonian contributions to free energy come from configurational entropy and single-particle effects. In other words, although Eq. (20) is adequate for $\eta = 1$ ($d\eta = 0$) in the thermodynamic limit ($N \to \infty$) of a homogeneous system with uniform correlations, Eq. (22) provides a systematic way of treating independently relaxing regions that have contributions from configurational entropy, and unrestricted fluctuations. These corrections become increasingly important for small regions ($N \ll 1000$), where η is very large, approaching the total number of particles in the sample. Thus, small-system thermal effects are crucial for conservation of energy in theoretical treatments of independently relaxing regions.

Because $\varepsilon = U - TS + pV - \mu N$ yields the residual free energy after all macroscopic contributions have been removed, $\varepsilon \equiv 0$ is required by the Gibbs–Duhem relation for linear and homogeneous systems in the thermodynamic limit—but ε is often nonzero for finite-sized systems. The subdivision potential can be understood by comparison to the chemical potential. μ is the partial differential of free energy with respect to the number of particles in an infinite system, while ε involves discrete differences when a single particle is added to finite-sized systems. Values of ε for several models in their relevant ensembles are given on pages 101–102 in part II of [117]. To emphasize various unique features in nanothermodynamics, we will review key steps in obtaining ε for an ideal gas of noninteracting point-like particles in two distinct ensembles.

3.3 Finite-Size Thermal Effects in Ideal Gases

First, consider an ideal gas in the canonical ensemble. Using the environmental variables (N, V, and T) and thermal de Broglic wavelength $\Lambda \propto 1/\sqrt{k_B T}$, the canonical partition function is $Q_N = V^N/(N!\Lambda^{3N})$, see Eq. 15–59 in [117], and the Helmholtz free energy is $F_N = -k_B T \ln(Q_N)$. In standard thermodynamics of large systems, the chemical potential comes from the derivative of F_N with respect to N so that $-\mu/k_B T = \partial \ln(Q_N)/\partial N|_{T,V} = \ln(V/N\Lambda^3) - 1/(2N) + 1/(12N^2) + \cdots$, see Eq. (15–62) in [117]. In nanothermodynamics of small systems, the derivative is replaced by a finite difference, which improves accuracy by treating N as an integer, and by avoiding Stirling's approximation for the factorial, yielding $-\mu_N/k_B T = \ln(Q_{N+1}) - \ln(Q_N) = \ln(V/N\Lambda^3) - 1/N + 1/(2N^2) + \cdots$. Thus, to lowest order, finite-size effects increase the chemical potential by $1/(2N)$. Adding similar finite-size effects to the pressure, the subdivision potential is found from $\varepsilon_N/k_B T = -\ln(Q_N) + N(1 - \mu_N/k_B T) = N + \ln[N!/(1+N)^N]$ (Eq. (15–64) in [117]). Indeed, even with no interactions, finite-size effects in an ideal gas of N particles yield $\varepsilon_N > 0$, so that the free energy increases when the sample is subdivided into independently relaxing regions. Thus, subdividing a bulk sample into an ensemble of smaller systems is energetically unfavorable, at least in the canonical ensemble, where the number of particles in each fluctuation is fixed at N; an unrealistic constraint that is often assumed in standard theoretical treatments.

Now consider the same ideal gas in the generalized (μ, p, T) ensemble. Note that this is the only ensemble that allows equilibrium fluctuations inside bulk samples, where nanometer-sized fluctuations should be unconstrained by fixed N, V, or U. Hence it might be called the nanocanonical ensemble. Assuming independently-fluctuating regions with an average size $\langle N \rangle$, Eqs. (10–88) and (10–89) in [117] yield the subdivision potential $\varepsilon_{\langle N \rangle}/k_B T = -\ln(1 + \langle N \rangle)$. Now subdividing the sample into independently relaxing regions decreases the free energy, opposite to the canonical ensemble. This reduction in free energy is due to the increased entropy from allowing unconstrained fluctuations in region size, which is unique to nanothermodynamics. For a quantitative estimate, we integrate the subdivision potential over $\eta(\approx 1/\langle N \rangle)$, from the thermodynamic limit ($\eta = 0$) to $1/\langle N \rangle$, giving $\Delta\varepsilon_{\langle N \rangle}/k_B T = -\int_0^{1/\langle N \rangle} \ln(1 + 1/\eta)\, d\eta = -(1 + 1/\langle N \rangle)\,[\ln(\langle N \rangle) + \ln(1 + 1/\langle N \rangle)]$. Dividing this total (integrated) change in free energy by the macroscopic kinetic energy of the ideal gas $(3\langle N \rangle k_B T/2)$, the free energy per particle is decreased by 3.1% for $\langle N \rangle = 100$, and decreased by 17% for $\langle N \rangle = 10$. Thus, even for a system of ideal gas particles with no interactions, due to the increased configurational entropy from regions with different sizes, the free energy is decreased by subdividing the system. Because this configurational entropy is not in the Hamiltonian, it is usually neglected in standard statistical mechanics. Furthermore, unlike standard thermodynamics, where all ensembles are equivalent, if finite-size effects from fluctuations are included, the free energy depends strikingly on the ensemble, with $\varepsilon_N > 0$ for a fixed number of particles, while $\varepsilon_{\langle N \rangle} < 0$ if the number of particles is allowed to fluctuate. Standard thermodynamics does not include this significant reduction in free energy, which drives the formation of independently relaxing regions.

From this simplified picture of an ideal gas in the nanocanonical ensemble, the free energy decreases monotonically with decreasing system size, favoring individual particles that are uncorrelated. For more realistic particles, inter-particle interactions cause correlations that tend to favor larger regions due to the cost of interface energies. Balancing these factors using the appropriate model should yield an equilibrium distribution about an average size that is consistent with experiments. A simpler model that also yields an equilibrium distribution of finite-sized regions is for noninteracting binary degrees of freedom (e.g., Ising-like spins or a binary alloy), where it is found that the average size of independently relaxing regions in thermal equilibrium is $\langle N \rangle \approx 11.07$ [134]. Although this value is suggestively close to $\langle N \rangle \approx 10$ for the number of molecules in slow domains of glycerol as measured by multidimensional NMR, it is much smaller than $\langle N \rangle \approx 76$ molecules for ortho-terphenyl and $\langle N \rangle \approx 390$ monomer units for poly(vinyl acetate) [41]. Furthermore, because this model treats only noninteracting particles, it is too simple to capture details such as the expected temperature dependence of $\langle N \rangle$ [135]. Nevertheless, we emphasize that nanothermodynamics provide a general paradigm for an equilibrium distribution of independently relaxing regions, where the entropy of neighboring regions is additive so that the dynamics is truly uncorrelated, and free energies of small regions can be lowered by more than 10% due solely to increased configurational entropy. Thus, nanothermodynamics provide a fundamental foundation for thermodynamic heterogeneity and a significant addition to any model that is proposed for describing NHB and other measurements showing heterogeneity.

3.4 Landau-like Theory for Phase Transitions in Finite-Sized Systems

If treated in the nanocanonical ensemble, Landau's unified theory of phase transitions provides a common explanation for several properties of supercooled liquids [133]. The theory predicts a second-order phase transition at a critical temperature T_c, super-Arrhenius activation rates that mimic the Vogel–Fulcher–Tammann (VFT) law, and a distribution of system sizes yielding net non-exponential relaxation that mimics the stretched exponential. The VFT law for a characteristic relaxation time can be written as $\tau \propto \exp[DT_0/(T - T_0)]$, where T_0 is the Vogel temperature and D the strength parameter. Mathematically, the VFT law can be attributed to activation energies that obey the Curie–Weiss law, which is a common feature of Landau's unified theory. Empirically, T_c is usually close to the Kauzmann temperature (T_K), where the entropy of the liquid would become less than that of the solid if not for a transition. Specifically for glycerol, $T_K = 135$ K [136] and $T_c = 131$ K [133], both far below the glass temperature $T_g = 193$ K. Thus, equilibrium behavior near T_c is impossible to measure directly, at least in such strong glassforming liquids, where

$D > 10$ (e.g. $D \approx 16$ in glycerol [137]), so that $T_c \ll T_g$. However, various characteristics of the transition can be deduced by extrapolating behavior from $T > T_g$. In contrast, fragile glassforming liquids have $T_c \approx T_g$ (e.g. $D \approx 4.2$ in propylene carbonate), so that clear deviations from the VFT law around T_c can be attributed to finite-size effects around the phase transition [23].

Here, we outline a Landau-like theory for phase transitions in small systems. Because of nonlinear size dependences, it is useful to expand the free energy *per particle* in a Taylor series, $f(\lambda) = f_0 + f_2\lambda^2 + f_4\lambda^4$, where λ is the intensive (per particle) order parameter. As in the original Landau theory, the system is assumed to have sufficient symmetry that the expansion has only even coefficients f_0, f_2, and f_4. For binary degrees of freedom, the quadratic term can be written as $f_2 = \frac{1}{2}(k_B T - \varepsilon_2)$, which contains a contribution from the entropy $-\frac{1}{2}\, k_B \lambda^2$, and a mean-field interaction term $-\frac{1}{2}\, \varepsilon_2 \lambda^2$. A similar contribution from entropy for the quartic term yields $f_4 = \frac{1}{12}\, k_B T$. The canonical ensemble partition function for a system of N particles is $Z_N = \int_{-\infty}^{+\infty} e^{-Nf(\lambda)/k_B T} d\lambda$. Here, the integral is extended to $\lambda = \pm\infty$, valid at high temperatures ($T \gg T_c$), where $\lambda \approx 0$, as most systems are relatively disordered (small systems often fluctuate to $\lambda \approx \pm 1$ near T_c). At $T \gg T_c$ the quartic term is also negligible, yielding Gaussian integrals that can be evaluated to give an average energy per particle of $\langle \varepsilon_N \rangle = -\int_{-\infty}^{+\infty} \frac{1}{2}\varepsilon_2 \lambda^2 e^{-Nf(\lambda)/k_B T} d\lambda / Z_N \approx \frac{1}{2}\varepsilon_2 k_B T / [N\,(k_B T - \varepsilon_2)]$. Note the $1/N$ dependence, which explains why this energy is neglected above T_c in standard Landau theory of large systems, and why subdividing the sample into small systems lowers the energy per particle. Using the magnitude of the total energy $|N \langle \varepsilon_N \rangle|$ as an activation energy in the Arrhenius law yields the VFT law, with $T_0 = \varepsilon_2 / k_B$ as the Vogel temperature and $D = \frac{1}{2}$ as the strength parameters—but this D is too small for real substances.

Realistic strength parameters can be obtained using the nanocanonical ensemble [133]. Specifically, assuming unrestricted sizes for the systems $1 \le N < \infty$ yields the partition function $\Gamma = \sum_{N=1}^{\infty} Z_N e^{N\mu/k_B T}$, then unrestricted numbers of indistinguishable systems in a supersystem $1 \le n < \infty$ yields $\Upsilon = \sum_{n=1}^{\infty} \Gamma^n / n!$. Now using the average energy of a supersystem ($|\langle E \rangle| = k_B T^2 \partial \ln(\Upsilon) / \partial T$) in the Arrhenius law gives VFT-like behavior that matches measurements from various glassforming liquids, as shown by the solid lines in Fig. 6. Including the quartic term to obtain best fits to the glycerol and propylene glycol data yields constant values $(f_0 - \mu)/k_B T = 0.0349$ and 0.0323, with $\varepsilon_2/k_B = 131$ and 112 K, respectively, which we now use to deduce the average system size. Returning to the Gaussian approximation for the integrals, the canonical partition function becomes $Z_N \approx e^{-Nf_0/k_B T} \sqrt{\pi k_B T / (N f_2)}$. The average size is $\langle N \rangle \approx \sum_{N=1}^{\infty} \sqrt{N}\, e^{-N(f_0-\mu)/k_B T} \,/\, \sum_{N=1}^{\infty} [\, e^{-N(f_0-\mu)/k_B T} / \sqrt{N}\,] = \mathrm{Li}_{-1/2}(e^{-(f_0-\mu)/k_B T}) / \mathrm{Li}_{1/2}(e^{-(f_0-\mu)/k_B T})$, where $\mathrm{Li}_s(z)$ is the polylogarithm. Using the constant values deduced from the data for $(f_0 - \mu)/k_B T$, and evaluating the functions numerically, yields $\langle N \rangle \approx 16.9$ and 18.1 particles for glycerol and propylene glycol, respectively. $\langle N \rangle \approx 16.9$ is already too high for the measured value from multidimensional NMR (10 molecules), and the mathematical approximations become even worse near T_g, where the measurements are made. Thus, this general theory does not accurately describe such specific details. Nevertheless, a Landau-like theory for

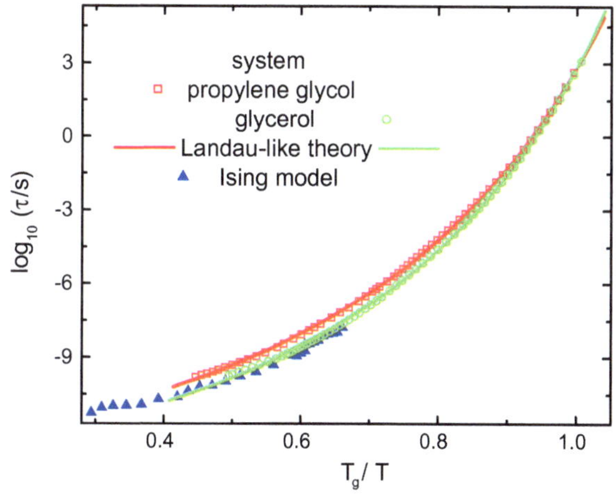

Fig. 6 Angell plot of logarithm of characteristic relaxation times versus inverse temperature from measurements, theory, and computer simulations. Arrhenius behavior would be a straight line on this plot, whereas curvature is indicative of the VFT law. Open symbols are from measurements of the inverse frequency of the peak dielectric loss in propylene glycol and glycerol [139]. Lines come from the activation energies of a Landau-like theory for phase transitions in finite-sized systems. Solid triangles show the net relaxation time from Monte Carlo simulations of the Ising model, with a local configurational entropy term based on nanothermodynamics. Adapted from Ref. [133]

finite-sized systems exhibits several features characteristic of glassforming liquids, including: average system sizes that are within a factor of two of measured sizes, a phase transition near the Kauzmann temperature, VFT-like activation energies, and energy reduction from subdividing into small systems that yields an equilibrium distribution of system sizes that mimics measured dielectric loss spectra [133]. Furthermore, the Landau-like theory also predicts entropy changes that agree with measured nonlinear response, at least within a factor of two [138].

3.5 Toward a Microscopic Model for the Heterogeneous Response in Complex Systems

The Landau-like theory for small systems in the nanocanonical ensemble yields behavior that is consistent with several features in glassforming liquids, but the theory is not microscopic so that its parameters are most accurately determined by measurements. Still, because the theory is based on thermodynamics, including finite-size effects, these parameters are few and relatively fundamental, e.g., the chemical potential and mean-field interaction energy. Models used to quantitatively simulate NHB also require input from measurements, e.g., the VFT parameters, excess specific heat, and distribution of relaxation times in the box model. Although the notion of ADWPs or Brownian dipole oscillators might suggest microscopic systems, these models require at least as many empirical parameters to obtain agreement with experiments. Furthermore, because NHB tells us nothing about the physical size of the systems, they could be individual molecules, independent plane waves, or anything in-between. Therefore, we rely on various other techniques to learn that typical independently relaxing systems are spatial regions, with length scales of 1–3 nm [26, 27, 32]. Thus, a typical region studied by NHB contains at least ten interacting particles, with relaxation behavior that is effectively independent of neighboring regions. Nanothermodynamics provides the fundamental foundation for allowing thermal and dynamic heterogeneity from such nanometer-sized systems of particles. Moreover, nanothermodynamics guides the development of microscopic models, with the goal of predicting observed behavior without having to use input from measurements.

The ferromagnetic Ising model for interacting binary degrees of freedom ("spins") is a simplistic microscopic model that yields a thermodynamic phase transition. By adding the configurational entropy from local regions in the sample, the Ising model gives behavior similar to that found in glassforming liquids and shows evidence that neighboring regions become de-correlated [35]. Although the Ising model was originally developed for uniaxial magnetic spins, it maps directly to several other interacting systems, including uniaxial electric dipoles, lattice gases, and binary alloys. Furthermore, regions of Ising spins have two ground states separated by a potential barrier, which may mimic an asymmetric double-well potential when acted on by surrounding parts of the sample. In fact, when finite-size thermal effects from nanothermodynamics are included, this Ising model provides accurate agreement with the thermal and dynamic properties of many substances, including ferromagnetic materials and critical fluids [119, 126, 140], and the ubiquitous low-frequency fluctuations that yield $1/f$-like noise [134, 141, 142].

Here, we present some results from Monte Carlo (MC) simulations of a large system of Ising spins on a simple cubic lattice. The Hamiltonian is $U = -J \sum_{<i,j>} \sigma_i \sigma_j$, where σ_i and σ_j are spin alignments that may be up (+1) or down (–1), J is an interaction energy, and the sum is over all six nearest-neighbor spins on the lattice. This model is usually simulated in the canonical ensemble using the Metropolis algorithm. Specifically, each MC step involves choosing a spin at random, calculating the change in interaction energy to invert the spin (ΔU), then accepting the new configuration only if Boltzmann's factor is greater than a random number uniformly distributed between 0 and 1:

$$e^{-\Delta U/k_B T} > [0, 1)\,. \tag{23}$$

Thus, steps that decrease the energy are always accepted, whereas steps that increase the energy are unlikely to occur at low temperatures. This standard Ising model on a cubic lattice has a ferromagnetic transition at $k_B T/J = 4.51$. However, the Metropolis algorithm does not include energy changes from configurational entropy, which come from finite-size effects in the thermodynamics, similar to the contributions to the subdivision potential that cause the ideal gas to favor subdividing into independent particles.

Configurational entropy has long been recognized as an important contributor to the behavior of glassforming liquids [121]. Configurational entropy for a system of n noninteracting Ising-like spins, $S_\lambda = k_B \ln(W_\lambda)$, is easily calculated from their multiplicity given by the binomial coefficient: $W_\lambda = n!/\{[\frac{1}{2}n(1+\lambda)]!\,[\frac{1}{2}n(1-\lambda)]!\}$. Here, the intensive order parameter is $\lambda = M/n$, where the net alignment (M) equals the number of up spins minus the number of down spins. Expanding in a Taylor series of λ to quartic order gives $S_\lambda = k_B n[\ln(2) - \frac{1}{2}\lambda^2 - \frac{1}{12}\lambda^4]$. (Note these same entropy terms were used in the free energy per particle for the Landau-like theory.) In general, thermal averages of the intensive quantities have a size dependence of $\langle\lambda^2\rangle \propto 1/n$ and $\langle\lambda^4\rangle \propto 1/n^2$, so that these fluctuation terms yield non-extensive contributions to the energy that must be accommodated by the subdivision potential. For spins in a field H, the nanocanonical ensemble has intensive environmental variable μ, H, and T, so that the conjugate variables fluctuate about average values $\langle n\rangle$, $\langle M\rangle$, and $\langle U\rangle$. Now assume that n fluctuates more slowly than M and U, consistent with NHB and other measurements, at least above the α-peak in glassforming liquids. Next, separate the system variables into time-averaged quantities $\langle S\rangle$ and $\langle M\rangle$ that depend on the interaction energy at fixed n, and time-dependent quantities S_λ and M_λ that vary with the configurational entropy, but are independent of the interaction energy under these conditions. Then, the subdivision potential for small systems may be written as $\varepsilon = \langle U\rangle - T(S_\lambda + \langle S\rangle) - H(M_\lambda + \langle M\rangle) - \mu\langle n\rangle$. Recall that $\varepsilon = 0$ in the nanocanonical ensemble when the sample contains a thermal equilibrium distribution of independently relaxing regions, similar to how $\mu = 0$ when a system contains a thermal equilibrium distribution of phonons or photons. Balancing the quantities that depend on interaction energy gives $\langle U\rangle - T\langle S\rangle - H\langle M\rangle - \mu\langle n\rangle = 0$, yielding the Gibbs–Duhem relation for macroscopic (time-averaged) behavior. The remaining (time-varying) term is: $HM_\lambda + TS_\lambda = 0$. In zero applied field there is no external work, but still, something must balance changes in S_λ during equilibrium fluctuations if energy is to be conserved. Two possible mechanisms are: work done against an internal field from neighboring regions, or work done on the thermal bath when the configurational entropy changes. The second mechanism is analogous to the work done on the thermal bath when an ideal polymer of noninteracting links is straitened, which can be measured as an increase in temperature when a rubber band is stretched [143]. In any case, some sort of internal work must be done during equilibrium fluctuations to balance the change in entropy. The Metropolis criterion, Eq. (23), comes from conservation of interaction energy. However, to ensure conservation of total energy, including contributions from configurational entropy, a second criterion is required. The second criterion can be written as

$$e^{(S_\lambda - S_0)/k_B} > [0, 1), \tag{24}$$

where $S_0 = k_B \ln\{n!/[(n/2)!]^2\}$ is the maximum configurational entropy in the region. Note that to facilitate faster dynamics in practical simulations, a Kronecker delta function from the local energy is included in the exponent of Eq. (24) [35].

Equation (24) accommodates local configurational entropy in the Ising model and yields VFT-like activation energies shown by the solid triangles in Fig. 6 that mimic

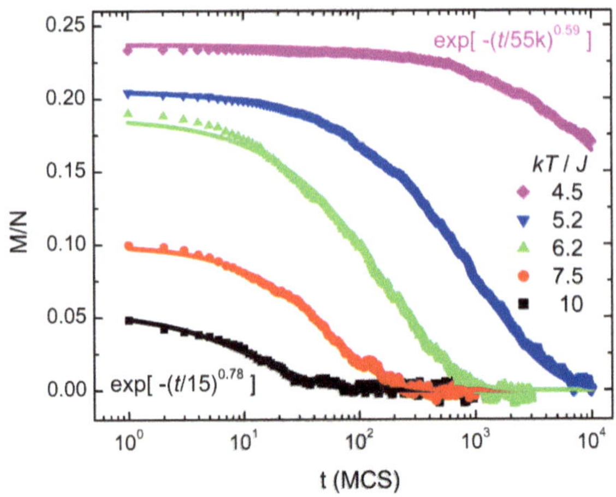

Fig. 7 Semilogarithmic plot of magnetization as a function of time after removing an applied field, at five temperatures given in the legend. Solid symbols come from MC simulations of the Ising model with entropy from cube-shaped regions containing $n = 4096$ spins. Time is given in MC sweeps (MCS), the unit of time for N attempted MC steps in a simulation of N spins. Solid lines are fits to the simulations using the stretched exponential function, yielding stretching exponents that range between $\beta = 0.50$ and 0.78

the behavior of glycerol. Although these simulations use the entropy from cube-shaped regions containing $n = 4096$ spins, much larger than the value of $\langle N \rangle \approx 10$ measured by NMR, this difference could be due to the inadequacy of binary spins to simulate classical dipoles. As a function of time, Fig. 7 shows that this model yields net relaxation that mimics the stretched exponential, with stretching exponents of $\beta = 0.5$–0.8 that encompass the usual range found for supercooled liquids. Moreover, Fig. 8 shows that individual regions in the microscopic model exhibit large jumps when the configurational entropy is near its maximum value $S_\lambda \approx S_0$, interspersed by small steps due to entropic trapping when $S_\lambda \ll S_0$. Similar jumps and steps for the rotation of individual dipoles in supercooled liquids are deduced from multidimensional NMR [144], as shown in the inset of Fig. 8. The Ising model, with configurational entropy needed to conserve total energy, provides a simplified but microscopic picture for measured thermal and dynamic response of glassforming liquids. Indeed, using only microscopic parameters as input, simulations of this model exhibit VFT-like activation energies as a function of temperature, and net stretched exponential relaxation as a function of time, including jumps and steps in the relaxation of individual regions that mimic measurements.

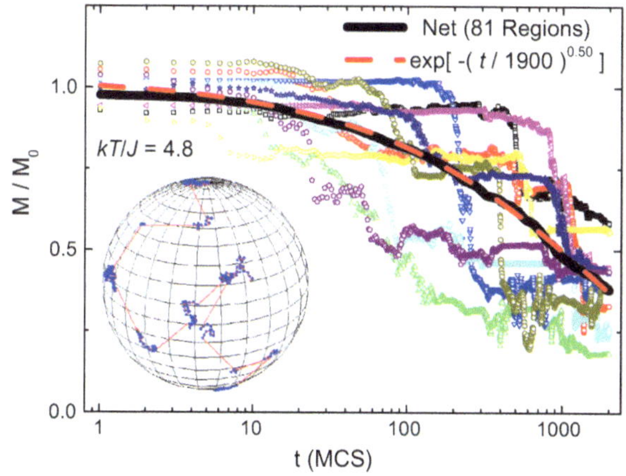

Fig. 8 Semilogarithmic plot of normalized magnetization as a function of time after removing an applied field, from simulations similar to those in Fig. 7, but at a single temperature of $k_B T/J = 4.8$. The solid line is the net response from 81 regions, each containing 4096 particles. The dashed line is a fit to the net response yielding the stretched-exponential function given in the legend. The symbols show the response from 10 of the 81 regions, showing that the net response hides many details in the response of individual regions. The inset shows the behavior deduced from multidimensional NMR for individual molecules in glassforming liquids [144], which also exhibit jumps and steps, reminiscent of the simulations

4 Experimental Details

4.1 Dielectric Hole Burning

Although nonlinear response is required to distinguish between homogeneous and heterogeneous scenarios, strong nonlinear effects can cause unwanted deviations from equilibrium response. Hence, early in the development of NHB, it was recognized that the technique should utilize minimal nonlinearity, involving the smallest-order nonlinear terms in a suitable expansion of the response in powers of the pump field. For DHB, the nonlinear dielectric relaxation of small-molecule liquids was expected to be small [145] and thus for them, this constraint is easier to fulfill than, e.g., for the highly polar relaxor ferroelectrics. Using typical values of dipole moments, $\mu_{el} = 1$ D, temperatures, $T = 150$ K, and electrical field strengths, $E = 1$ kV/50 μm = 200 kV/cm, the ratio of dipolar to thermal energy, $\mu_{el}E/k_B T$, is only of the order of 0.032. Thus for supercooled liquids, only relatively small DHB effects are to be expected. The considerations in Sect. 2 suggest that under these challenging conditions, samples suitable to perform the dielectric variant of this experiment should fulfill a number of conditions, which are as follows:

(i) In the studied temperature range, the dielectric loss, more precisely, the product of loss and the square of the applied field, should be large to maximize the energy absorption.

(ii) Relaxation times should vary rapidly with temperature to achieve large changes in the dynamics from the moderate effective temperature change of the degrees of freedom addressed by selective excitation.
(iii) The specific heat of those degrees of freedom should be small.

Thus, (i) high-loss and (ii) fragile supercooled liquids, such as propylene carbonate, were chosen for early implementations of DHB. For typical glass formers, (iii) does not provide a particularly useful criterion for optimization, since near the glass transition the configurational specific heat does not differ all that much from liquid to liquid. Of course, for decoupled degrees of freedom such as those related to β-relaxations or ionic motions that are typically addressed below the glass transition temperature, the relevant specific heat will be much smaller. Furthermore, the application of large electric fields can easily lead to effects of dielectric breakdown. Thus, high-voltage-induced sample failure is often a limiting factor, even when using thoroughly polished electrodes and with edge effects carefully avoided. If only relatively small nonlinear signals are available a number of issues become important not just regarding the sample but also regarding the apparatus. For a successful implementation of DHB, careful consideration needs to be given to items such as

(i) The excitation profile of the pump should be substantially narrower than any single-exponential contribution to the relaxation peak that has a full width at half maximum of 1.14 decades.
(ii) Short persistence times of spectral holes require fast detection schemes.
(iii) The small nonlinear response will be overlaid by a large unwanted linear aftereffect of the pump.
(iv) A small increase in effective relaxation rate can be viewed to correspond to a small increase in effective temperature. Yet an overall dielectric heating of the sample is to be avoided.
(v) Large electrical fields can lead to an electrode spacing that due to effects of electrostriction can change with time.

To address item (i), let us note that the sinusoidal excitation sketched in Fig. 2 provides optimum selectivity for a given pump length. If (ii), the persistence time of the spectral holes is short, then a fast probing scheme is required. This can be implemented by the field step illustrated in Fig. 2, which allows one to record the full spectrum in single sweep. To eliminate unwanted polarization signals (iii), a phase cycle inspired by NMR has proven indispensable. The field sequence and phase cycling can be implemented using an arbitrary waveform generator in conjunction with a sufficiently fast high-voltage amplifier. On the detection side, which was implemented using a modified Sawyer-Tower circuit, it is mandatory to employ a high-impedance electrometer amplifier.

Item (iv) is particularly demanding. It implies that one should keep the temperatures stable over the extended timescales required to perform a full phase cycle. With fictive temperature changes of the order of only a few millikelvin, a temperature stability on the order of ~1 mK is advisable, which can be maintained, e.g., using a well-regulated closed-cycle refrigerator. To minimize overall sample heating during the (local) energy input from the pump process , it is necessary to keep the electrode separation relatively small (usually 10–50 μm) [94]. As emphasized early on [50, 62], see also Sect. 1.1, it is important to ascertain that the local energy input does *not* lead to a spatial modulation of the heat-bath temperatures within the sample. Otherwise, subsequent to the pump process, effects of thermal conductivity would lead to an equilibration on timescales much shorter than typical hole persistence times. Conversely, one may argue that successful observation of spectral holes in the nonresonant dielectric response implies that such effects are not necessary to consider in the present context [62].

The required small electrode spacing necessary to avoid extraneous heating effects and to facilitate large pump fields is usually enforced by using rigid inserts, such as rod-like fibers, micron-sized spheres, or thin insulator rings. Otherwise, in the presence of large electric fields the electrodes, if not sufficiently rigid, can "breath" or simply squeeze soft sample materials, during the ac excitation. Given the finite mechanical modulus of the sandwiched sample, seemingly nonlinear contributions to the polarization response may arise, which can be magnified by electrostrictive or piezoelectric samples [62, 81].

4.2 Magnetic Hole Burning

Magnetic hole burning (MHB) provides another example of the versatility and power of the NHB technique [61]. Like DHB, MHB involves a large-amplitude, low-frequency pump oscillation followed by a small probe step, but the field is magnetic instead of electric. A sketch depicting a typical set of field sequences is shown in Fig. 9, which includes the phase cycling and background-removal processes. Magnetization as a function of time after the probe step is usually measured using a high-speed SQUID magnetometer. To minimize background drift, it is best to have the magnetization measured in zero applied field, so that the probe step is usually from a small value to zero field, different from the situation in Fig. 2, where the step is from zero to a small value. Specifically, each sequence in Fig. 9 includes a large pump oscillation with the same amplitude and phase, but one has a small positive initial offset, and the other has a small negative initial offset. The offset is abruptly removed following a recovery time (t_r) after the pump oscillation, so that subtracting the two responses yields the linear response of a sample that had been modified nonlinearly by the pump oscillation, then aged for a controlled recovery time.

One advantage of studying MHB is that the spin degrees of freedom in spin glasses are found to have specific heats that are about a million times smaller than those for dielectric degrees of freedom in glassforming liquids. Thus, a single pump oscillation of 100 Oe can cause the local effective (spin) temperature to change by 2 K. Furthermore, spin-glass transition temperatures are often an order of magnitude lower than liquid-glass freezing temperatures so that such large changes in spin temperature have an even larger effect on the dynamics. Another advantage is that

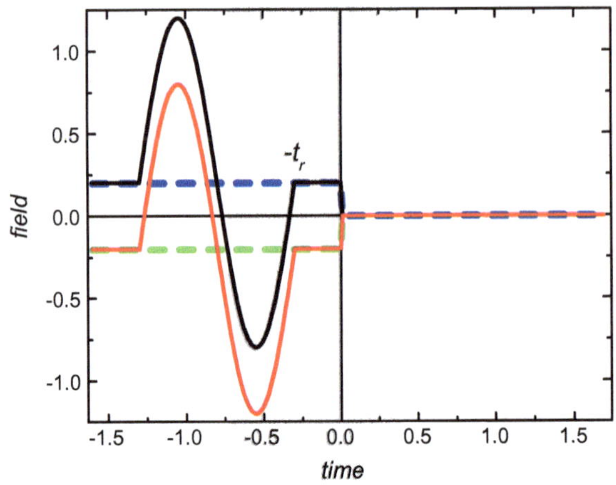

Fig. 9 Sketch of applied magnetic field as a function of time for magnetic hole burning. First, the sample is equilibrated at the measurement temperature in a small field that is positive (upper oscillation and constant offset) or negative (lower oscillation and constant offset). Solid lines show a large pump oscillation, followed by a recovery time t_r, then a small step to zero field. Dashed lines show the field for measuring the response with no pump oscillation. Magnetization is measured as a function of time for up to 9 orders of magnitude, from about 10^{-5} to 10^4 s after removing the small field, using a high-speed SQUID magnetometer

there is no magnetic analog of dielectric breakdown so that much larger magnetic fields can be applied without concern for sample survival.

4.3 Modulus Technique

For pump frequencies Ω that exceed the loss peak frequency ($\omega_{\max}$) of the system by more than a factor of about 100, the sensitivity of a polarization response to a field step, i.e., of $\varepsilon(t)$, to pump-induced modifications is poor, because $\varepsilon(t)$ is nearly time-invariant for times that are short compared with the average response time. A remedy to the resulting small hole amplitudes is to probe the dielectric behavior by a charge step rather than using a field step, as relaxation $M(t)$ after a charge step approaches steady state more rapidly than retardation $\varepsilon(t)$ after a field step [97, 98, 146]. Regarding the linear averages of the relaxation (τ_{M}) and retardation (τ_ε) times, in Sect. 2.3 it was already pointed out that the effect amounts to $\langle \tau_M \rangle \,/\, \langle \tau_\varepsilon \rangle = \varepsilon_\infty / \varepsilon_{\mathrm{s}}$, and is thus more pronounced for polar materials with $\varepsilon_{\mathrm{s}} \gg \varepsilon_\infty$ [99]. The resulting polarization under constant charge conditions is linearly related to the electric modulus, $\hat{M}(\omega) = 1/\hat{\varepsilon}(\omega)$, which can be measured directly in terms of the time-dependent field while the charge or dielectric displacement remains unchanged. As shown in Fig. 4, this detection method provides enhanced sensitivity to pump-induced modifications at the higher pump frequencies [80].

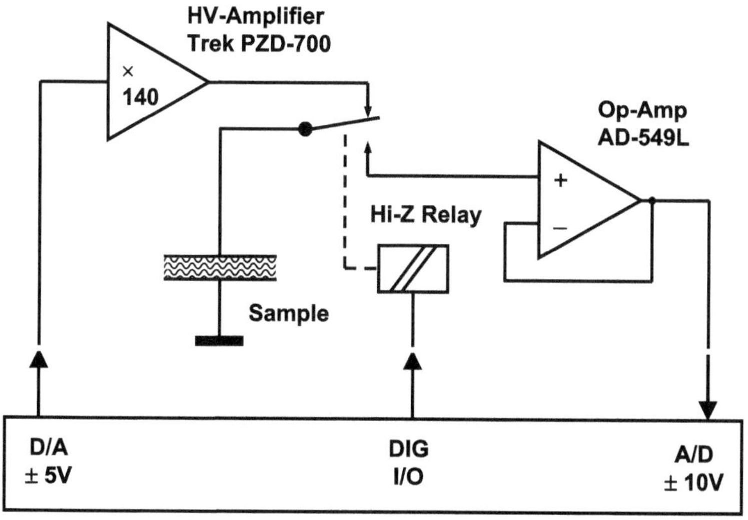

Fig. 10 Schematic representation of the circuit used for measuring DHB with the modulus type detection of dielectric polarization. The relay connects the sample to the voltage source for applying the burn voltage, for the duration of the waiting time, and for creating the charge step. Subsequently, the sample is connected to the electrometer circuit which measures the voltage as a function of time, ideally without loss of charge. Adapted from Ref. [60]

In order to realize measuring the time-domain modulus subsequent to a charge step, a high-impedance relay is employed to connect the sample capacitor to the voltage source during the pump and recovery cycle, see Fig. 10. Subsequently, a voltage step is applied to the sample and the relay immediately disconnects the sample from the voltage source and connects it to a high input impedance electrometer circuit. Due to the input characteristics of the operational amplifier ($R_{\mathrm{in}} = 10^{15}\ \Omega$, $I_{\mathrm{bias}} \leq 70$ fA) and the high relay impedance ($R_{\mathrm{off}} > 10^{14}\ \Omega$), the capacitor charge remains virtually constant for the duration of detecting the voltage across the sample capacitor, 1 ms to 100 s [60, 82]. Voltage signal generation, relay action, as well as recording the step response in quasi-logarithmic time steps is under software control, such that data for entire phase cycles can be acquired in an automated fashion.

5 Dielectric Hole Burning

An extensive list of references describing experimental studies of DHB is given in Sect. 1.2. Below, we focus on the primary dielectric relaxation and high-frequency wing in two small-molecule glassforming liquids, propylene carbonate, and glycerol. Figure 11a shows that the primary response of glycerol is relatively narrow, and the same holds for many other glass formers such as propylene carbonate or 2-picoline. As an example of a broader response, Fig. 11b features the dielectric loss spectrum

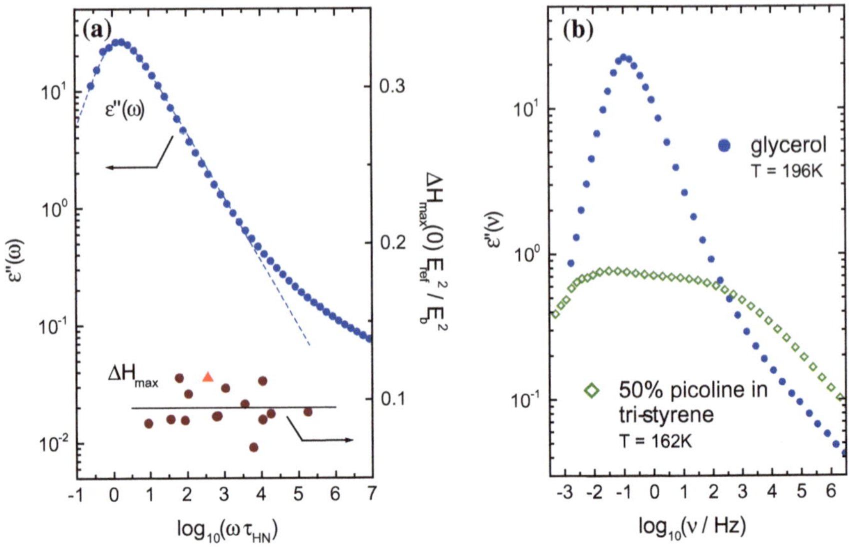

Fig. 11 **a** Dielectric loss $\varepsilon''(\omega)$ and field normalized peak amplitudes of horizontal holes, $\Delta H_{\max}(t_r = 0)$, plotted against the reduced logarithmic frequency scale, $\log_{10}(\omega\tau_{HN})$. The dashed line is a Havriliak–Negami (HN) fit with exponents $\alpha = 0.95$ and $\gamma = 0.58$. Horizontal hole amplitudes $\Delta H_{\max}(0)$ are normalized to a common reference field. The horizontal line represents an amplitude of 0.095, the $\Delta H_{\max}(0)$ data are characterized by 0.095 ± 0.013 across the relative burn frequency range $1 \leq \log_{10}(\omega\tau_{HN}) \leq 5.3$. **b** Double-logarithmic representation of dielectric loss spectra of glycerol [60] and of 50% picoline in tri-styrene. Adapted from Ref. [66]

of a binary mixture of 2-picoline ($T_g = 132.7$ K) in tri-styrene (molecular weight $M_n = 370$ g/mol, $T_g = 233$ K) [66]. One recognizes that in comparison with pure substances, the mixture is not only much broader but also characterized by a lower amplitude of the maximum loss. It should be emphasized that the dielectric response of this mixture is dominated by that of picoline ($\mu_{el} = 2.1$ D) with the maximum loss in pure tri-styrene about two decades smaller than in picoline [66]. The challenge posed in DHB studies by low dielectric loss levels, and hence, low pump-induced energy input, has to be dealt with not only in binary glass formers of the type referred to in Fig. 11b but also when performing experiments far off the dielectric loss peak in the regime of the so-called excess wing, as shown for glycerol in Fig. 11a.

The present section is structured with respect to various phenomena in order to facilitate their comparisons in different glass formers or with respect to different relaxation phenomena: In Sect. 5.1 results for vertical and horizontal spectral holes are presented, then in Sect. 5.2 the amplitudes and the frequency positions the spectral holes are dealt with, and in Sect. 5.3 their recovery is discussed.

5.1 Horizontal and Vertical Spectral Holes

The supercooled liquids glycerol and propylene carbonate fulfill many of the requirements to achieve large DHB signals, see Sect. 4.1, and were the first liquids studied using this technique [51]. Figure 12 depicts experimental results obtained for propylene carbonate slightly above its calorimetric glass transition. In panel (a) we compare the equilibrium step response together with a phase-cycled response recorded subsequent to applying a single pump oscillation for three different pump frequencies, Ω. The phase cycling scheme is necessary to cancel the undesired large linear polarization aftereffect due to the pump so that at short and long measuring times the equilibrium and modified responses agree. However, differences arise for suitably chosen pump frequencies at intermediate times. The inset in Fig. 12a illustrates how the horizontal and vertical differences are obtained, as displayed in Fig. 12b, c, respectively. In panel (b), the amplitude of the vertical difference $\Delta_V\varepsilon(t)$ is seen to be strongly dependent on Ω while its position is less sensitive to the pump frequency. This difference can be viewed as a direct demonstration of frequency selective input of energy. Even more impressive are the horizontal holes, $\Delta_H\varepsilon(t)$. One recognizes that the spectrally modified responses are most pronounced for low pump frequencies at long times and for high pump frequencies at short times. Thus, these data readily reveal one of the basic hallmarks expected for heterogeneous response. Conversely, in the presence of a homogeneous scenario a pump-induced energy input is expected to shift the entire relaxation function $\varepsilon(t)$ uniformly along the time axis to the left, i.e., to shorter times. Hence, for homogeneous relaxations a $\Delta_H\varepsilon(t)$ pattern is expected that is in contrast to the experimental observations. The $\Delta_H\varepsilon(t)$ representation is unique in enabling one to distinguish homogeneous from heterogeneous responses directly on the basis of the raw data for a single pump frequency Ω. With the response probed at *several* Ω such a model-free distinction is possible using $\Delta_V\varepsilon(t)$ data as well.

According to Eqs. (7) and (12) horizontal differences $\Delta_H\varepsilon(t)$ may be expressed in terms of changes in fictive temperature T_i. The maximum change in T_i that may be inferred from the data in Fig. 12c is on the order of 60 mK which is much larger than the typical temperature stability of a few millikelvin that was maintained in these experiments. It has been emphasized that changes in T_i should not be confused with effects of local heating in the thermal bath [94]. Estimates show that enhanced local bath temperatures, implied within such a picture on nanoscopic length scales, would readily equilibrate within timescales much shorter than microseconds [64]. None of the NHB experiments carried out in the last 25 years has accessed this time regime. In other words, the observed NHB effects cannot be due to local heating in the thermal bath but must be associated with extra energy in the slow degrees of freedom. This notion is compatible with quantitative descriptions of the data such as those shown in Fig. 12, in terms of a local fictive temperature for the box model [51] and other approaches [105].

DHB in the range of the α-relaxation peak has also been performed for glycerol and yielded data similar to those presented in Fig. 12 for propylene carbonate.

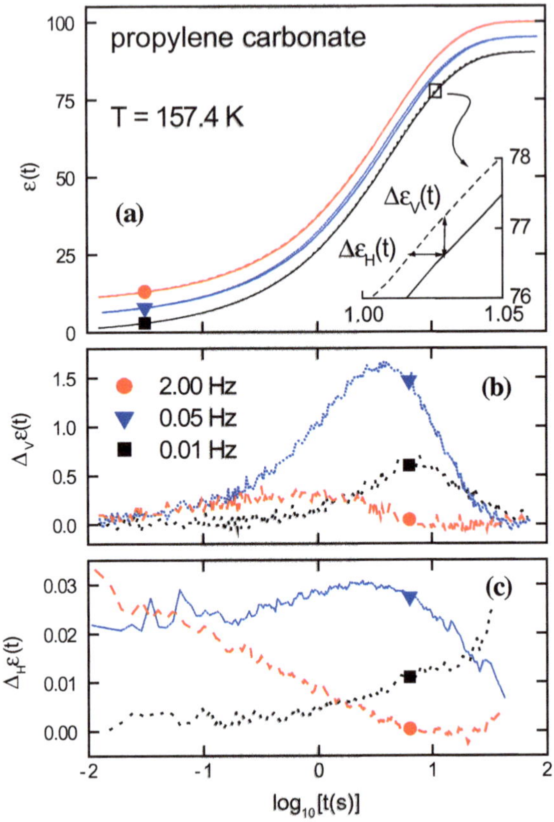

Fig. 12 Dielectric function $\varepsilon(t)$ of propylene carbonate: **a** Solid lines show the equilibrium response, while dashed lines show the responses modified by a single pump oscillation, with the lower and upper pair of curves offset (by ±5) for clarity. Symbol shape identifies the frequency (given in the legend) of each pump oscillation of 900 V across a 50 μm thick sample. The inset in (**a**) illustrates how the vertical and horizontal holes are calculated. The vertical holes, $\Delta\varepsilon_V(t)$ shown in panel (**b**), are obtained from the modified amplitude minus the equilibrium amplitude at each time. The horizontal holes, $\Delta\varepsilon_H(t)$ shown in panel (**c**), are obtained from the logarithmic difference between the equilibrium and modified response times at each amplitude. Adapted from Ref. [51]

Rather than reproducing these data for glycerol here, we will now turn to the frequency range above the α peak in which a pronounced excess wing is observed, see Fig. 11a. In the corresponding frequency range, the dielectric losses are small and in order to achieve a sufficient signal-to-noise ratio, the modulus technique detailed in Sect. 4.3 was employed. An example for such a DHB measurement is depicted in Fig. 13 for glycerol at $T = 187.30$ K, where the characteristic time constant of the dielectric retardation is $\tau_{HN} = 285$ s, as obtained from a Havriliak-Negami fit to the low-field loss profile, $\varepsilon''(\omega)$. The results are represented as both the vertical, $\Delta M(t)$, and the horizontal, $\Delta H(t)$, hole for a pump frequency of $\Omega = 1.26\,\mathrm{s}^{-1}$, which

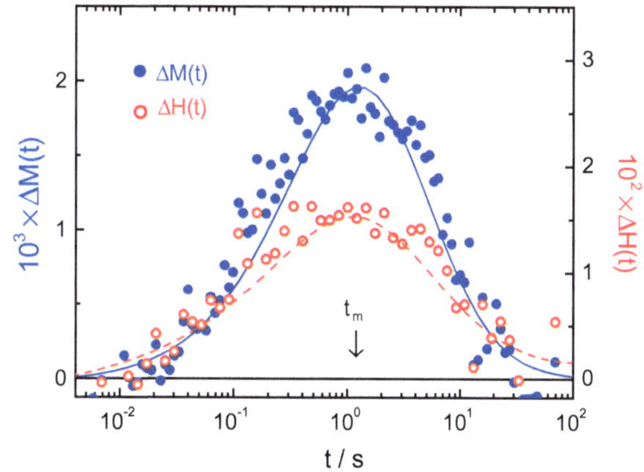

Fig. 13 Vertical and horizontal dielectric holes in glycerol at $T = 187.30$ K, with the ΔH data obtained from the $\Delta M(t)$ measurements using Eq. (12). The data were obtained using a pump field $E_{\text{pump}} = 90$ V/6.4 μm, a pump frequency $\Omega = 2\pi \times 50.2$ Hz, $n_p = 6$ pump cycles, and a recovery time $t_r = 1$ s. Adapted from Ref. [82]

is about 2.5 decades higher than the loss peak position, see red triangle in Fig. 11a positioned at $\log_{10}(\Omega\tau_{\text{HN}}) = 2.55$. In this situation, Fig. 4 suggests that the effect in terms of $M(t)$ exceeds that of $\varepsilon(t)$ by a factor of about 5. Clearly, the horizontal 'hole' signal $\Delta H(t)$ becomes very small at short and long times relative to the peak at $t = 1$ s. This is a clear indication of spectral selectivity, as this feature had not been forced by normalization. The lines in Fig. 13 demonstrate the favorable agreement between experiment and the box model, as these have been calculated as outlined in Sect. 2 with no adjustable parameter [82].

For the binary mixtures with their small dielectric losses, cf. Figure 11b, sufficient signal-to-noise ratio was achieved by applying large fields of up to $E_{\text{pump}} = 550$ kV/cm without experiencing dielectric breakdown [66]. Following the standard phase cycle and adapting the voltage to optimize the signal quality, the spectral holes shown in Fig. 14 were obtained. Pump frequencies were incremented in steps of about half a decade and one recognizes that the times of maximum modification t_m, at which the minima in the spectral holes appear, follow this step size in Ω. This directly demonstrates spectral selectivity and dynamic heterogeneity and was also to be expected on the basis of the rather flat loss spectrum shown in Fig. 11.

A closer look at the hole widths reveals that they are broader than those apparent for neat liquids (cf. Figs. 12 and 13), indicating that these spectra contain a component that is homogeneously broadened. This impression is confirmed quantitatively by the data shown in Fig. 15. In addition to the experimental data, this figure features calculations based on the assumption that each relaxation mode 'i' is characterized (i) by an exponential relaxation $P_i \propto \exp(-t/\tau_i)$ which does *not* fit the experimental data or (ii) by an intrinsically stretched function $P_i \propto \exp[-(t/\tau_i)^{\beta_{\text{in}}}]$, cf. Eq. (15),

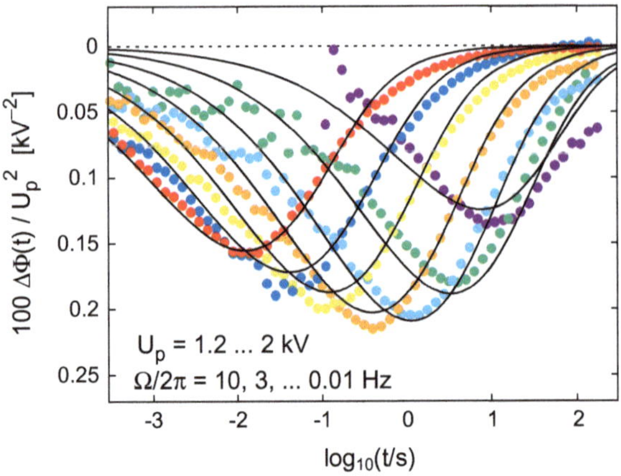

Fig. 14 Vertical spectral holes for a mixture of 50% 2-picoline in tri-styrene measured at $T = 161$ K after $n_p = 1$ pump cycle with frequencies (from left to right) of $\Omega/2\pi = 10, 3, 1, 0.3, 0.1, 0.03$, and 0.01 Hz. Here the data are inverted to make them look like "holes." Lines represent calculations employing the box model (assuming intrinsically exponential behavior). Adapted from Ref. [66]. Courtesy of T. Blochowicz

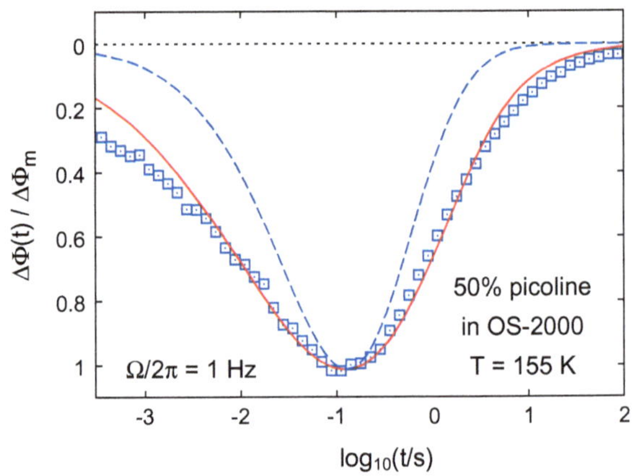

Fig. 15 Vertical hole of a mixture of 2-picoline in oligo-styrene (OS-2000: $M_n = 2140$ g/mol, $T_g = 325$ K). Calculations employing the box model are shown as dashed line (assuming intrinsically exponential behavior, cf. Eq. (15) with $\beta_{\text{in}} = 1$) and as solid line ($\beta_{\text{in}} = 0.62$). Adapted from Ref. [66]. Courtesy of T. Blochowicz

with a Kohlrausch exponent $\beta_{\text{in}} = 0.62$. The calculations based on the latter approach are seen to capture the observed hole broadening much better [66].

5.2 Frequency-Dependent Amplitudes and Positions

After having dealt with the width of the spectral holes, let us now discuss their amplitudes as a function of the pump frequency Ω. Horizontal holes referring to the primary relaxation are shown in Fig. 12c and one recognize that their depth $\Delta_H \varepsilon_m(\Omega)$

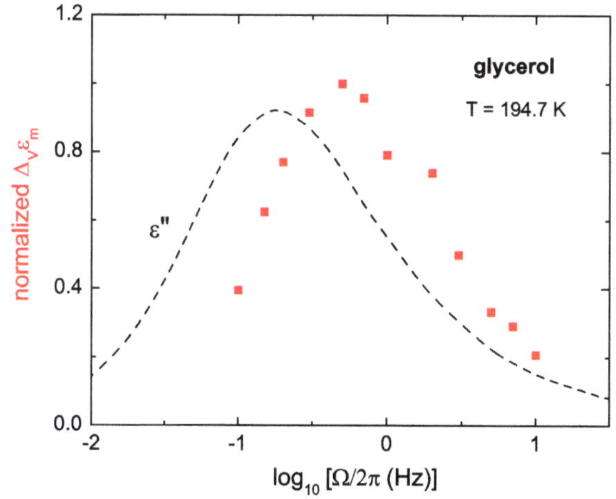

Fig. 16 Symbols represent pump-frequency-dependent vertical hole amplitudes $\Delta_V\varepsilon_m(\Omega)$ normalized by $\Delta_V\varepsilon_m(\Omega/2\pi = 0.5\ \text{Hz}) = 4.8 \times 10^{-3}$ of glycerol as measured at 194.7 K with $n_p = 1$ and $E_{\text{pump}} = 1.2$ kV/50 μm. The dashed line represents the dielectric loss spectrum $\varepsilon''(\omega)$ (in arbitrary units) measured at the same temperature [147]

does not vary much with Ω. This feature, which is in accord with the box model, may be assessed even better from, e.g., Fig. 10 in Ref. [64] and Fig. 3b in Ref. [92]. In any case, near the maximum of the dielectric loss peak, the vertical hole amplitudes are much more sensitive to Ω, as is evident for propylene carbonate from Fig. 12b. For glycerol the hole amplitudes $\Delta_V\varepsilon_m(\Omega)$ as measured near 195 K are shown in Fig. 16 [147]. When comparing with the conventional dielectric loss spectrum $\varepsilon''(\nu)$ measured at the same temperature, one recognizes that $\Delta_V\varepsilon_m(\Omega/2\pi)$ is peaked at a frequency that is somewhat higher than that of $\varepsilon''(\nu)$. Alternatively, one could say the pattern somewhat resembles the later nonlinear work [67] which demonstrated that under strong ac irradiation the resulting (nonlinear) dielectric loss is significantly enhanced only on the high-frequency flank of the (linear) loss spectrum, ε''. An advantage of this experiment done purely in the frequency domain [67], which is now highly developed [81], is that the modification is probed directly *during* irradiation, while the early measurements of $\Delta_V\varepsilon_m(\Omega)$ capture its effect only after a varying amount of recovery has occurred. Quantitative amplitudes for $\Delta_V\varepsilon_m(\Omega)$, including all aftereffects, are still fully captured by the box model. In fact, in Fig. 4 corresponding calculations are shown that refer to glycerol, but at a slightly lower temperature than the one used for Fig. 16. A similar but broader $\Delta_V\varepsilon_m(\Omega)$ pattern than the one shown in Fig. 16 may be inferred from the data in Fig. 14.

Regarding the excess wing, the amplitudes of the horizontal holes (for an example see the glycerol data in Fig. 13) are independent of Ω, see Fig. 11a. Intuition might suggest that the amount of fictive temperature change (and thus the horizontal hole amplitude) should decrease with increasing Ω, because the loss and power absorbed are reduced. Over the frequency range of Fig. 11a, $\varepsilon''(\omega)$ diminishes by two orders of magnitude while Ω increases by four orders of magnitude, yet no systematic changes in $\Delta H_{\max}$ are observed. The explanation of this feature is part of the box model, namely that $C_p(\omega)$ traces the frequency dependence of the dielectric loss, leading to an increase of the fictive temperature that is largely independent of Ω. This observation provides a strong validation of the box model assumption of a constant heat capacity per molecule (in other words, each molecule contributes the same amount to the local heat capacity of the slow degrees of freedom). By contrast, vertical hole amplitudes will become smaller at higher pump frequency because the slope $\mathrm{dlog}\varepsilon''/\mathrm{dlog}\omega$ will eventually decline, see the dashed extrapolation of the power-law behavior indicated in Fig. 11a.

Comparing the positions of the holes in Figs. 12b and 14, major differences become apparent. In particular, this concerns the pump frequency variation of the times of maximum modification, t_m. To allow for a better comparison of the Ω-dependent hole shifts, Fig. 17 shows the time of maximum modification t_m as a function of the pump time $t_p = 2\pi/\Omega$ for several glass formers. The very small shift of t_m with t_p seen there for propylene carbonate follows a power law, $t_m \propto t_p^{\alpha}$, with an exponent of only $\alpha = 0.3$. The other end of the scale is set by the binary liquids containing 50% picoline for which $\alpha = 1$ was reported [66]. The same behavior, $t_m = t_p$, is also found to describe the positions of the vertical holes pumped in the excess wing of glycerol [60], while $\alpha = 0.4$ characterizes the regime of the structural relaxation in that glass former.

In fact, these differing exponents reflect the differing widths in the distributions of relaxation times of the addressed degrees of freedom. If there is no distribution (e.g., a Debye-type or homogeneous process prevails), the hole positions will not depend on Ω at all. With the degrees of freedom responding (or being modified) on the timescale on which they are perturbed, the exponent $\alpha = 1$ becomes plausible in the limit of very broad distribution widths.

An analogous $t_m(t_p)$ pattern is evident also from DHB studies of solids. For lanthanum-modified lead zirconate titanate (PLZT) the exponent is $\alpha = 0.5$ [74]. For lead magnesium niobate based relaxor ferroelectric (PMN) $\alpha = 1$ [71] and from the data for Ca^{2+} doped strontium titanate ($Sr_{0.998}Ca_{0.002}TiO_3$), a diluted relaxor [72], a similar exponent can be inferred. An even more interesting behavior is displayed by the calcium–potassium nitrate glass $2Ca(NO_3)_2.3KNO_3$ at $T = 300$ K, i.e., about 33 K below its glass transition temperature [69]. Here, a DHB study employing the modulus technique yielded a power law with $\alpha = 0.5$ on the high-frequency flank of the dielectric loss peak, while for frequencies below the peak (a region mostly masked by dc conductivity) the hole positions were found to be independent of Ω (or t_p) so that $\alpha = 0$. This latter homogenous behavior was rationalized in terms of a spatial averaging over the heterogeneity of local ion diffusivities that occurs if the pump period is very long with respect to the ion hopping time. The latter can be estimated to be close to the inverse loss peak frequency [69].

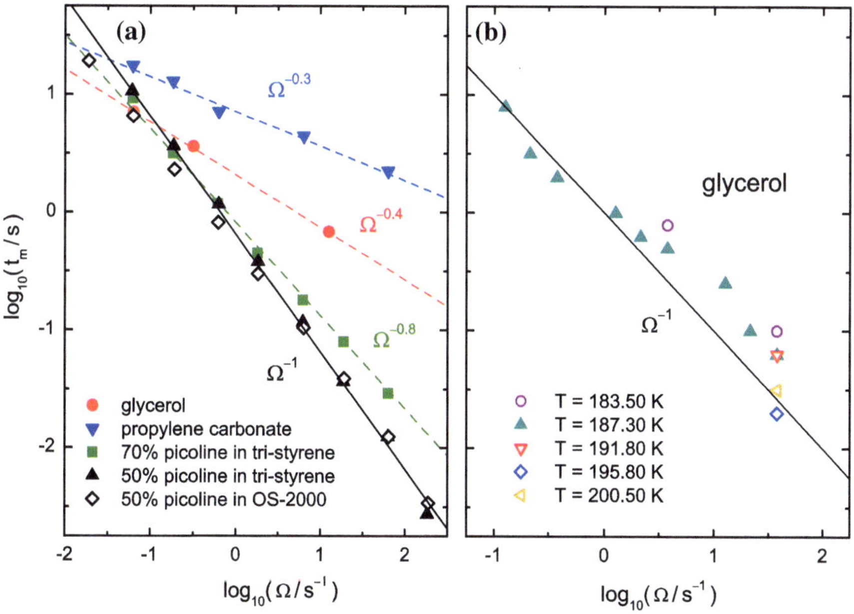

Fig. 17 Times of maximum modification t_m at which vertical hole positions occur as a function of the pump time $t_p = 2\pi/\Omega$. Examples include (**a**) the primary relaxation of propylene carbonate [51], the main processes of glycerol and several binary mixtures [66], as well as (**b**) the high-frequency wing of glycerol [60]. The lines represent $t_m \propto \Omega^{-\alpha}$ behaviors. In both frames the solid lines indicate $t_m = \Omega^{-1}$. Figure adapted from Refs. [60] and [66]

5.3 *Hole Recovery*

By increasing the time interleaved between pump and probe, cf. Fig. 2, during which the external field is constant (typically it is set to zero), the spectral holes will refill. As an example in Fig. 18a we present horizontal spectral holes measured for propylene carbonate using different recovery times t_r. One recognizes that within experimental uncertainty, the shape of the spectral holes does not change significantly during t_r. Analogous data were obtained for other pump frequencies and for other temperatures [62, 64]. The hole depths were analyzed, with results collected in Fig. 19a, which allow for several remarkable observations: (i) The recovery is pump frequency independent which is plausible in view of the rather narrow distribution of relaxation times characterizing this van der Waals liquid. (ii) The recovery takes place on the timescale of the primary (α) relaxation (corresponding data are added to Fig. 19 as dashed line). This finding is plausible as well since the peak relaxation sets the longest timescale relevant in (most) supercooled liquids and certainly in propylene carbonate. (iii) The recovery of the horizontal holes is compatible with a single exponential (solid lines), $\Delta_H\varepsilon_m(t_r) \propto \exp(-t_r/\tau_r)$, where τ_r denotes a characteristic recovery time, sometimes also called a refilling time or hole lifetime.

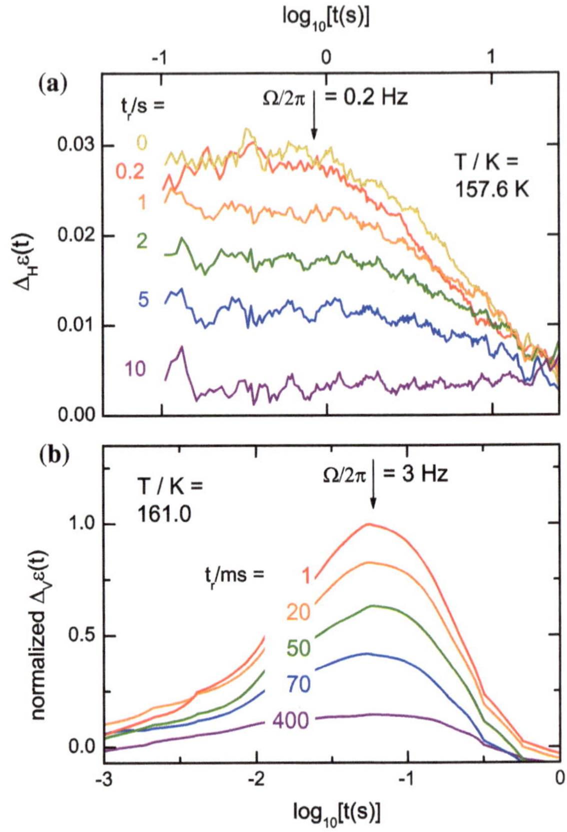

Fig. 18 Time-dependent modifications, $\Delta_{\mathrm{H}}\varepsilon(t)$, as measured for different recovery times t_r in propylene carbonate after a single pump oscillation of (**a**) 0.2 Hz at 157.6 K (adapted from Ref. [51]) and (**b**) 3.0 Hz at 161.0 K [147]. Arrows mark the inverse pump frequencies $1/\Omega$. The amplitudes in frame (**b**) are normalized to that of the hole measured at the shortest t_r

A similar analysis was performed also for the vertical holes that are presented in Fig. 18b and the t_r dependence of the hole amplitudes is summarized in Fig. 19b. Essentially, most observations for the vertical holes mimic those obtained from the horizontal holes. In particular, there is no significant variation in the shape of the holes during recovery. A difference in the behavior of the horizontal holes seems to be that the vertical holes do not recover in an exponential fashion. The t_r-dependent hole depths $\Delta_{\mathrm{V}}\varepsilon_m(t_r)$ follow more closely the time dependence of the properly scaled linear polarization response function.

For another quantitative comparison of hole recovery data and box model prediction, we turn again to the DHB results obtained for glycerol using the $M(t)$ technique with pump frequencies that exceed the loss peak frequency considerably. Figure 20a shows hole recovery data for various pump frequencies, and the timescale of recovery clearly changes with Ω. The dependence of the characteristic recovery time τ_r on the reciprocal pump frequency $(1/\Omega)$ is compiled in Fig. 20b, showing that $\tau_r \approx 10t_p$. In both panels of Fig. 20, solid lines represent predictions of the box model

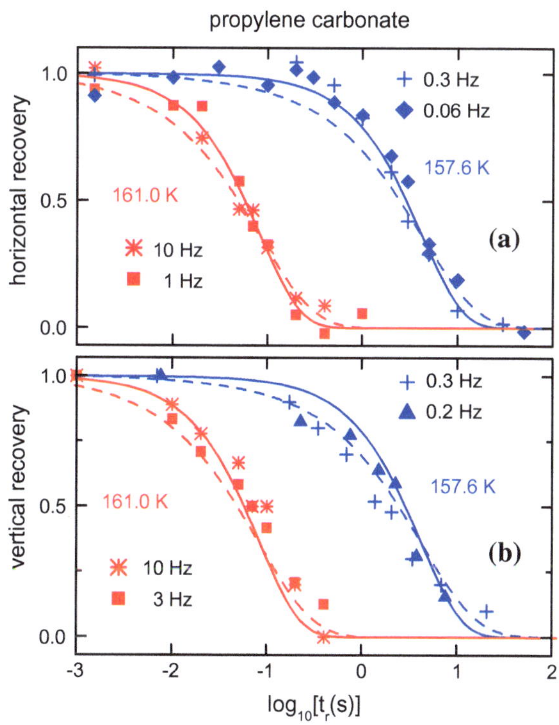

Fig. 19 Comparison of normalized linear dielectric response $\Phi(t)$ (dashed line) and recovery (symbols) for several pump frequencies. Panel (**a**) shows the refilling of the horizontal holes $\Delta_{\mathrm{H}}\varepsilon_m(t_r)$ (adapted from Ref. [51]) and panel (**b**) shows the refilling of the vertical holes $\Delta_{\mathrm{V}}\varepsilon_m(t_r)$ [147]. The solid lines are best single-exponential fits to the recovery data in panel (**a**). Solid and dashed lines from panel (**a**) are reproduced in panel (**b**) for comparison with the experimental data

with no adjustable parameters. Consistent with the model, hole recovery is governed by modes with time constants somewhat longer than $1/\Omega$.

It is instructive to compare these recovery results with those for liquids displaying very broad dielectric loss peaks for which the binary mixture of 50% picoline in tri-styrene is an example [66]. Figure 21 summarizes vertical holes for this liquid that were recorded for a range of recovery times. One recognizes how, for increasing t_r, the position of the minima shifts to longer times. This behavior is expected on the basis of the box model considering that modes appearing at shorter times should recover faster. In fact, the solid lines in Fig. 21 were calculated using a variant of this model in which intrinsically non-exponential behavior was included [66]. Furthermore, in that reference, it was demonstrated explicitly that an additional significant hole broadening does not emerge during recovery. Thus, it was concluded that mechanisms such as spectral diffusion are not operative in these experiments.

The t_r-dependent depths of the vertical holes obtained from experiments carried out at three different pump frequencies are summarized in Fig. 22. The experimental data were found to be compatible with calculations using the box model including intrinsic nonexponentiality (Eq. (15) with $\beta_{\mathrm{in}} = 0.65$) [66]. The hole depths themselves were described using a suitably adapted Kohlrausch function

$$\Delta_{\mathrm{V}} P_m(t_r, \Omega) \propto \exp[-(t_r/\tau_r)^{\beta}], \tag{25}$$

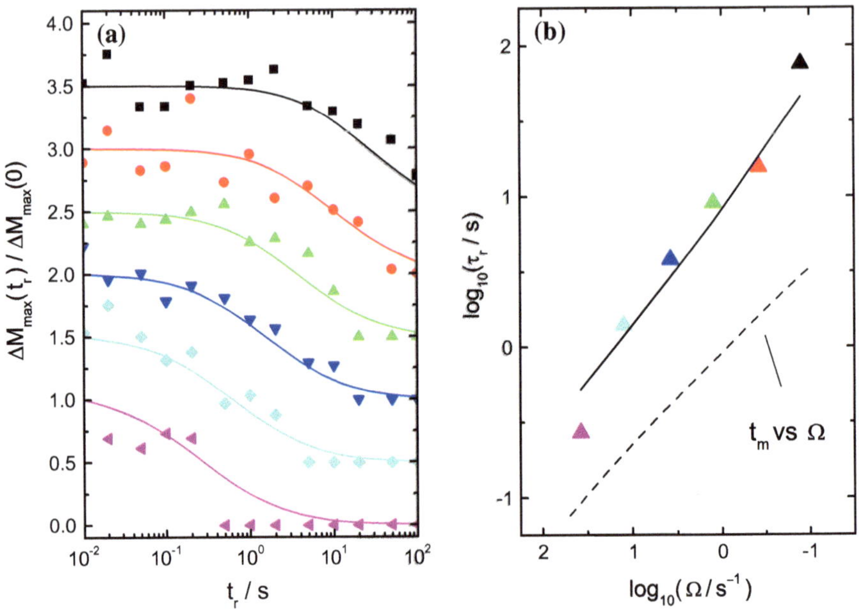

Fig. 20 **a** Normalized peak hole amplitude, $\Delta M_{\mathrm{max}}(t_{\mathrm{w}})/\Delta M_{\mathrm{max}}(0)$, as a function of the recovery time t_r measured for glycerol at $T = 187.30$ K. In the order from top to bottom, different symbols refer to different pump frequencies: $\Omega/2\pi = 0.02, 0.06, 0.2, 0.6, 2$, and 6 Hz. The lines are calculated results from the box model. To improve clarity, symbols and lines are incrementally offset by +0.5. **b** Hole lifetimes τ_r determined as the '1/e' times from data in (**a**), plotted versus the pump frequency Ω. The symbols are the experimental results, the solid line reflects the box model calculation. The dashed line shows t_m versus Ω to illustrate the relation $\tau_r \approx 10\, t_m$. Adapted from Ref. [82]

Fig. 21 Recovery of vertical holes in a binary glass former. The solid lines are calculations using the box model including intrinsic nonexponentiality. Adapted from Ref. [66]. Courtesy of T. Blochowicz

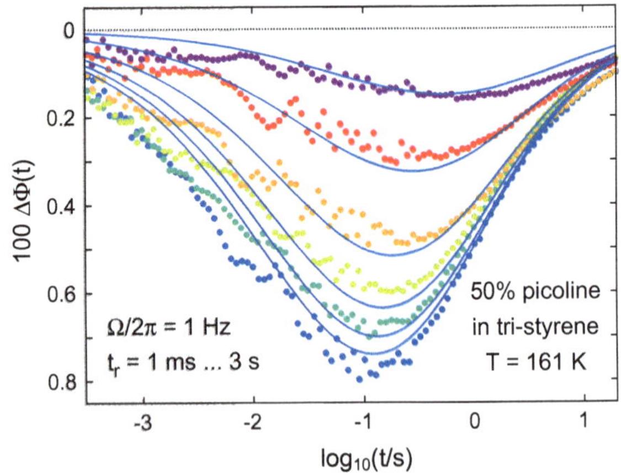

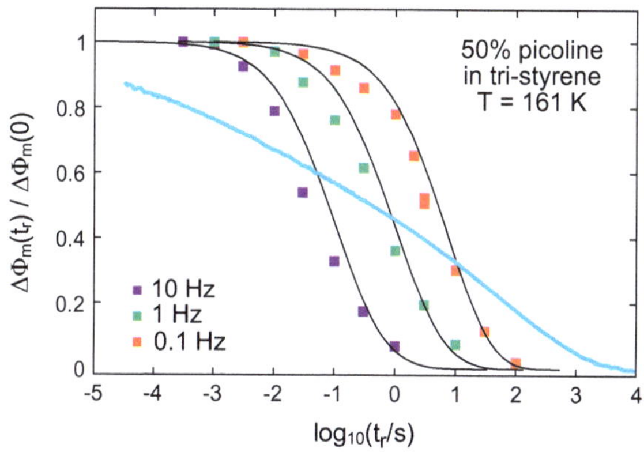

Fig. 22 Recovery of vertical holes pumped at 0.1, 1, and 10 Hz in the response of a binary glass former. Symbols were determined at the time of maximum modification. Thin dashed lines represent recoveries as calculated using the box model with $\beta_{in} = 0.65$, cf. Eq. (15). The calculations result in stretched-exponential recovery functions with a Kohlrausch exponent of 0.56. For comparison the very broad linear step response function is added as thick solid line. Adapted from Ref. [66]. Courtesy of T. Blochowicz

with a characteristic recovery time $\tau_r \approx 6/\Omega$ and $\beta = 0.56$ (so that $\Omega \langle \tau_r \rangle \approx 10$). Also included in Fig. 22 is the almost logarithmically decaying linear response of the binary glass former [66]. Interestingly, this comparison of broad linear response and narrower recovery shows the finite excitation width of the pump process and that the subsequent recovery only concerns the degrees of freedom that were selected by the pump irradiation.

For the DHB data shown in Fig. 22 the hole depths were read out at their actual minimum which is simple to determine experimentally. Alternatively, one may argue the degrees of freedom modified most by a specific pump frequency are those which give rise to the vertical holes at $t_r \to 0$ at their initial position $t_m = t_{m,0}$. In the absence of a coupling of degrees of freedom referring to different spectral positions—or put in terms of the box model: an absence of a cross-coupling of the boxes [64] which in all modeling performed so far is found to be consistent with experimental data—it may be preferable to read out the hole amplitude at $t_{m,0}$ for *all* recovery times. For liquids with a relatively narrow primary relaxation, this may not be an issue at all since the holes do not show a significant shift with recovery time. This is obviously different for the vertical holes depicted in Fig. 21 and was analogously observed for some relaxor ferroelectrics [73, 74]. It appears, however, that $\Delta_V P_m(t_r,t_m)$ and $\Delta_V P(t_r,t_{m,0})$ were quantitatively compared only for the PLZT [74] and PMN [73] relaxor materials that both exhibit very broad polarization responses. For PLZT it was found that the recovery measured (i) at constant time $t_{m,0}$ proceeds about a factor of two faster than when (ii) reading out at $t_m(t_r)$; for PNM a slightly larger factor was found [73]. On a qualitative level, such differences may be inferred directly from the data in Fig. 21. A shift of the hole pattern to longer detection times together with hole shapes that do not change with t_r necessarily implies faster recovery in case (ii).

Recovering *horizontal* holes were not reported for the binary mixtures, but were studied for the relaxor PMN that is also characterized by a very broad response [73]. Interestingly, for this example, a shift of the time t_m of maximum modification did *not* occur as t_r progresses. Nevertheless, the characteristic recovery time τ_r assessed on the basis of $\Delta_H P(t_r, t_{m,0})$ data was found to agree with τ_r determined from $\Delta_V P_m(t_r, t_m)$. This observation seems to provide some justification for analyzing recoveries also in terms of vertical holes.

Finally, it may be asked on which factors the characteristic recovery time may depend. On the one hand, for propylene carbonate and glycerol, it is obvious from Fig. 19 that t_r is essentially given by the relaxation time corresponding to that of the α peak. This consideration appears to apply to the supercooled plastic crystal cyclooctanol as well [68]. On the other hand, for the binary glass former, below Eq. (25) it was stated that $\langle\tau_r\rangle \approx 10/\Omega$ or, expressed in terms of the pump time, $\langle\tau_r\rangle \approx Rt_p$ with $R \approx 10$. Figure 23 shows that more or less the same factor characterizes the recovery of holes pumped in the excess wing of glycerol [60]. At frequencies that exceed the loss peak position by at least an order of magnitude, the probability density $g(\tau)$ of time constants becomes relatively flat. Furthermore, the actual $g(\tau)$ profile is less important in this high-frequency regime. Then, the characteristic hole recovery time is dominated by the time constants of those modes that are associated with the largest changes in fictive temperature after a few pump cycles. While the power spectrum peaks at Ω, the modes with τ somewhat larger than $1/\Omega$ will absorb less but retain the added energy for times longer than $1/\Omega$, implying that subsequent cycles will add to the fictive temperature, specifically for modes slower than $1/\Omega$. According to Fig. 23, this leads to a factor R of ≈10 for glycerol at $n_p = 3$, as predicted by the box model.

The characteristic recovery times obtained for the relaxors were found to be relatively long [71, 74]; R values even much larger than 10 were reported [73]. However, these values can significantly depend on temperature and on the amplitude of the pump field. These observations were interpreted in terms of a scenario involving the motion and depinning of domain walls. Mechanisms of this kind are obviously not applicable to supercooled liquids which are in the focus of this section.

6 Magnetic Hole Burning

The technique of magnetic hole burning (MHB) facilitates investigations of intrinsic heterogeneity in the nonresonant relaxation of magnetic degrees of freedom in bulk samples. Although the earliest application of resonant SHB investigated heterogeneity in (underdamped) NMR [52], with subsequent studies of spectral selection in magnetic response involving resonances at low frequencies [148–151], MHB allows investigations of nonresonant (overdamped) magnetic relaxation. Some general features regarding the procedure are given in Sect. 4.2.

The upper panel of Fig. 24 shows time-dependent magnetization from a 5% Au:Fe spin-glass sample. The measurements were made at a temperature of 18.8 K, some-

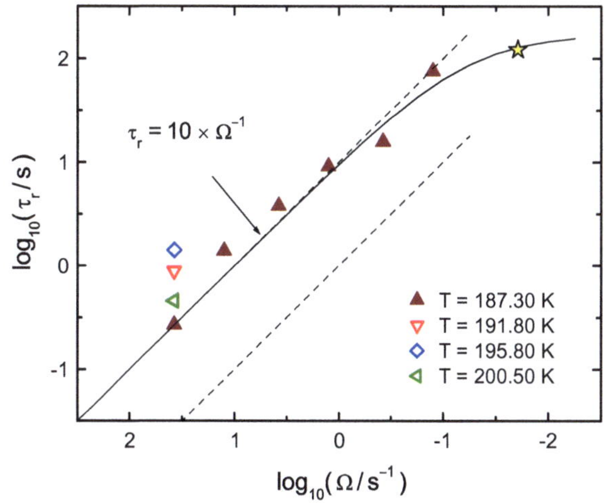

Fig. 23 Characteristic recovery times (also called hole lifetimes) τ_r as a function of the pump time t_p. Different symbols are for different temperatures as indicated. The $T = 187.30$ K data set spans 2.5 decades in Ω and is well approximated by the upper dashed line indicating $\tau_r = 10\Omega^{-1}$. The star is the result of Schiener et al. [51], after shifting that $T = 194.7$ K result to match the present $T = 187.30$ K case. The solid line is given by $\tau_r^{-1} = \Omega/10 + \langle\tau_\alpha\rangle^{-1}$ with $\langle\tau_\alpha\rangle = 170$ s. Adapted from Ref. [60]

what below the spin-glass transition temperature of $T_{sg} = 21.6$ K. The procedure was to first cool the sample from 25 to 18.8 K in a small field of magnitude $|h| = 8$ Oe, stabilize the temperature for about two minutes, modify the sample using a pump oscillation of amplitude $H_0 = 96$ Oe, then remove h and measure magnetization as a function of time. Similar measurements were made with no pump oscillation to characterize the equilibrium response. Data shown in Fig. 24 were obtained by subtracting the magnetizations measured with $h = -8$ Oe from those with $h = +8$ Oe, and subtracting a similar set of measurements with the sample in the other coil of a first-order gradiometer flux transformer. Different symbols show the response with no pump oscillation (black circles), and after a single pump oscillation at a frequency of 30 Hz (blue), 1 Hz (green), or 30 MHz (red). The lower panel of Fig. 24 shows amplitude differences between equilibrium and modified responses, using the same symbols as in the upper panel. Note that the peak modification occurs near the timescale corresponding to each pump frequency, as given by the arrows.

The solid line in the upper panel of Fig. 24, which mimics the measured equilibrium response, comes from a model for independently relaxing regions with a Gaussian distribution of energy levels, similar to that used for supercooled liquids [20]. Solid lines in the lower panel of Fig. 24 come from simulations based on a simplified box model, where the Gaussian distribution is modified by the average rate of energy absorbed in each region integrated over the pump cycle. The only adjustable parameter is the excess specific heat per spin, ΔC_p, an amplitude factor. The solid lines in Fig. 24 yield $\Delta C_p/k_B = 0.5 \times 10^{-6}$, six orders of magnitude smaller than the

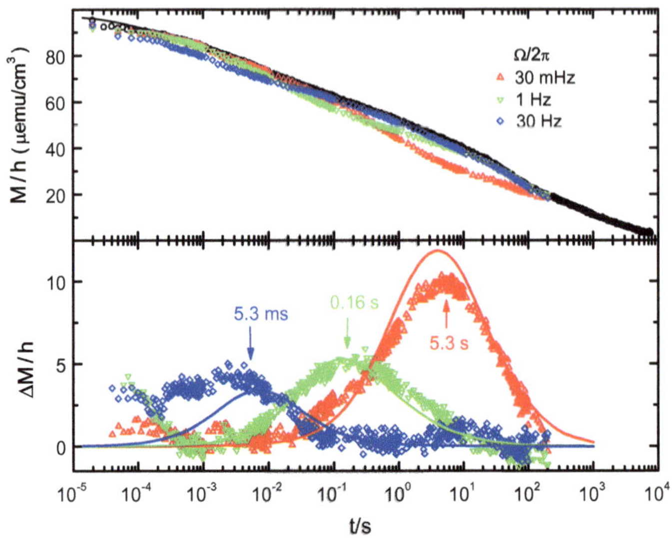

Fig. 24 Magnetization (upper panel) and its difference (lower panel) as a function of logarithm of time after removing a small field, from measurements of magnetic hole burning on a 5% Au:Fe spin-glass sample. The measurements are made at 18.8 K, about 3 K below the spin-glass transition temperature of 21.6 K. Black circles show equilibrium response with no pump oscillation. Colored symbols show response after a single pump oscillation of 30 Hz (blue), 1 Hz (green), and 30 MHz (red), with differences between equilibrium and modified response shown in the bottom panel. The arrows indicate the timescale corresponding to each pump frequency. The solid lines come from simulations using a simplified box model, where average power is integrated over the pump cycle, with specific heat (an amplitude factor) as the only adjustable parameter. Adapted from Ref. [61]

specific heat of the Au lattice at the spin-glass transition: $C_p/k_B = 8.35 \times 10^{-5} T_{sg} + 5.3 \times 10^{-5} T_{sg}^3 \approx 0.54$ [152]. This relatively small value of ΔC_p is consistent with the very small changes in C_p measured at the spin-glass transition, which can usually be seen only by taking differences [153]. It is interesting to note that the relatively large signal in MHB is a consequence of this small value for $\Delta C_p/k_B$. Indeed, smaller values of $\Delta C_p/k_B$ yield larger effective temperature changes, and hence larger hole burning signals. Thus, NHB is an ideal technique for investigating the time- and temperature-dependent specific heats of slow degrees of freedom that are difficult to measure directly by other techniques.

Figure 25 shows MHB difference spectra from 5% Au:Fe at several recovery times after a single pump oscillation of 1 Hz (upper panel) and 0.1 Hz (lower panel). Note that the recovery rate is proportional to the pump frequency and that the modified spectrum becomes narrower as the response recovers. The behavior is consistent with the simplified box model (solid lines), where the recovery rate of each region is assumed to match its response rate. The inset shows the normalized peak of the spectral hole as a function of scaled recovery time for four different pump oscillation

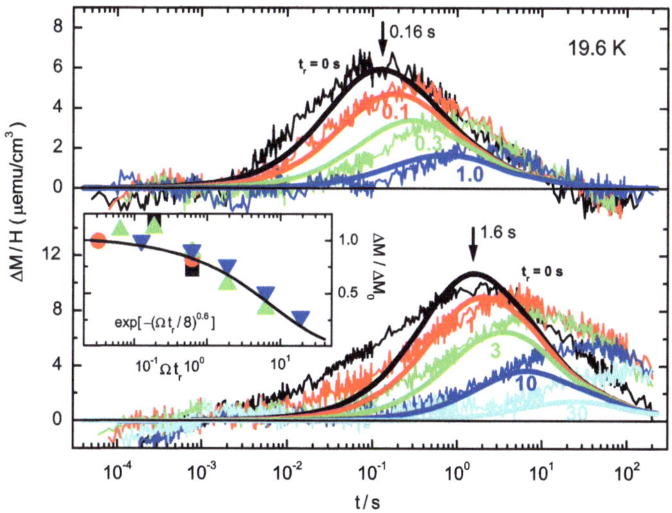

Fig. 25 Difference spectra from MHB of 5% Au:Fe at 19.6 K as a function of recovery time after a pump oscillation of 1 Hz (upper panel) and 0.1 Hz (lower panel). Arrows mark the characteristic time for each pump frequency. The solid curves are from the simplified box model, with thermal-coupling rates equal to response rates. Note that the spectral holes recover first at the short-time side, becoming narrower with increasing recovery time, consistent with the data. The inset shows the normalized peak in the spectral hole as a function of scaled recovery time for four different pump oscillation frequencies. A stretched exponential function (solid curve) characterizes the recovery. Adapted from Ref. [61]

frequencies. A stretched-exponential function $\Delta M \propto \exp[-(t_r \Omega/8)^{0.6}]$ (solid curve in the inset of Fig. 25), can be used to characterize the recovery.

Figure 26 shows magnetization as a function of time from the 5% Au:Fe sample at two measurement temperatures after various pump amplitudes. Note that at the lower temperature and smaller pump amplitudes the response exhibits a spectral hole, while at the higher temperature or larger pump amplitudes the response shows a "spectral step." Spectral hole burning is identified by accelerated response over 1–2 decades in time, without changing the total magnitude of response (both the initial and final magnetizations match). In contrast, spectral step burning has significant response that is missing from the measurement window, so that the initial remanent magnetization is reduced due to the pump oscillation. For materials that are deep within their frozen phase (see Ref. [154]), this missing magnitude can be attributed to high-field saturation of the response. However, a different mechanism is indicated for Au:Fe, because of the sharp onset of spectral step burning with increasing pump amplitude combined with the sharp onset with increasing temperature close to T_{sg}. The mechanism can be attributed to selected degrees of freedom that are heated by the pump oscillation to above T_{sg}, into the paramagnetic phase, where they have no remanent magnetization and hence relax to zero before the start of the measurement window. From Fig. 26 it can be deduced that a pump oscillation of 96 Oe causes

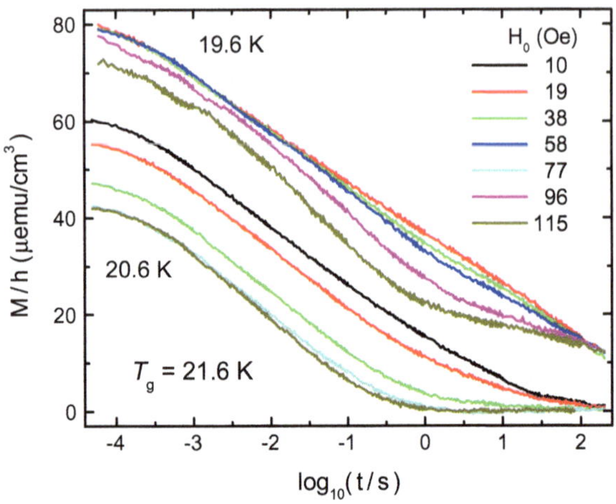

Fig. 26 Time dependence of magnetization from the 5% Au:Fe spin-glass sample after a pump oscillation of 0.1 Hz with several amplitudes H_0 (given in legend), measured at two temperatures. Note that at low temperatures and low pump amplitudes these data exhibit a spectral hole, where the magnetization is shifted to shorter times over 1–2 decades in time while maintaining the total magnitude of response. However, higher pump amplitudes and higher temperatures yield a spectral step, where some of the response is completely absent from the time window. Adapted from Ref. [61]

some degrees of freedom to respond as if they have an effective temperature that is at least 2 K above the bath temperature, while a pump oscillation of 19 Oe causes some degrees of freedom to respond as if they are at least 1 K above the bath temperature.

Figure 27 shows difference spectra from measurements of magnetization as a function of time from a single-crystal whisker of iron. The difference spectra come from the time-dependent remanent magnetization after a single pump oscillation of 5 Oe, minus the magnetization after no pump. The measurements are made at 4.2 K, far below the Curie temperature of 1043 K. The differences exhibit pump-frequency-dependent spectral steps, not spectral holes, indicating that the mechanism involves selective saturation of the response due to the pump oscillation, not response-rate acceleration.

7 Conclusions

Nonresonant spectral hole burning has become a versatile and powerful technique for investigating the intrinsic response inside bulk materials. The original purpose of NHB was to distinguish between the homogeneous and heterogeneous scenarios for non-exponential relaxation behavior. The heterogeneous scenario is found for the dielectric, magnetic, and mechanical response of all systems studied so far, which include liquids, glasses [155], polymers, relaxor ferroelectrics, and spin glasses.

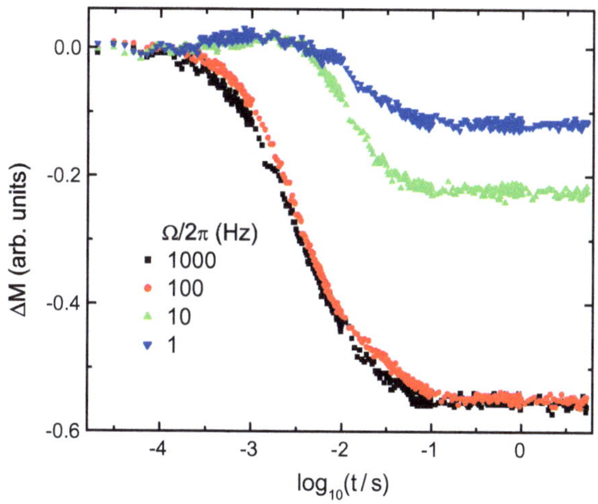

Fig. 27 Difference spectra from a single-crystal whisker of iron at 4.2 K after a single pump oscillation of 5 Oe at four different frequencies, as given in the legend

Moreover, this dynamic heterogeneity is found to be thermodynamic in nature, coming from a heterogeneous ensemble of independently relaxing regions that have distinct local energies as well as distinct fluctuations regarding temperatures, dielectric polarizations, magnetic moments, and/or mechanical strains or stresses.

Several phenomenological models have been developed to describe NHB experiments, as outlined in Sect. 2. The box model is one such model that shows excellent agreement with most NHB measurements. The parameters in this model can usually be determined by separate measurements so that this level of agreement is often achieved with no adjustable parameters. In other cases, the box model has a single adjustable parameter that yields the specific heat of the slow degrees of freedom, some of them too small to be measured using other techniques. The box model is based on independently relaxing regions (the boxes), each with its own local fictive temperature that is selectively modified by a spatially uniform external field. The box model describes several features in measurements of NHB, including both the amplitude- and frequency dependence of the spectral holes, their dependence on the number and amplitude of the pump oscillations, and their recovery as a function of time after the end of the pump. In general, the response and recovery rates of each region are found to be similar, indicating that energy is absorbed from the external field at the same rate at which it flows from the local regions into the bath.

The theory of small-system thermodynamics developed in Sect. 3 provides a fundamental foundation for the heterogeneous distribution of independently relaxing nanometer-sized regions. These regions are probed by NHB and several other techniques, and characterized by the box model and other approaches for treating heterogeneity. The basic idea of nanothermodynamics is that the macroscopic ensemble of independently relaxing regions inside bulk samples must obey the laws of thermodynamics for bulk systems, but the individual regions have behavior that is unique to small systems. Specifically, because the independently relaxing regions are statistically independent, their entropies should be additive. Similarly, their energies must be strictly conserved, including contributions from fluctuations and discrete-particle effects that are neglected in the usual thermodynamic limit. For example, nanothermodynamics includes a subdivision potential (ε) that is identically zero in the thermodynamic limit, but often nonzero for finite-sized systems. In fact, for the classical ideal gas of point-like particles it is found that $\varepsilon_N > 0$ in the canonical ensemble with fixed number of particles for all fluctuations, while $\varepsilon_{\langle N \rangle} < 0$ in the nanocanonical ensemble with variable N. Furthermore, nanothermodynamics facilitates the treatment of this nanocanonical ensemble, where the number of particles, volume, and energy of individual regions may fluctuate without artificial constraints. Because the nanocanonical ensemble lowers the total energy of the ideal gas, it provides a basic mechanism favoring the formation of an ensemble of independently relaxing regions, even inside bulk samples. Using nanothermodynamics as a guide, a Landau-like theory and Ising-spin model are described that mimic several features in the behavior of supercooled liquids.

Measurements of dielectric hole burning permit investigations of the intrinsic dielectric response inside bulk samples. Systems studied to date using DHB include supercooled liquids, relaxor ferroelectrics, a plastic crystal, and an ion conductor. Dynamics investigated include the primary (α) response, both near the peak and in the excess wing, secondary (β) relaxations, and the structurally decoupled motion of ions in the ionic conductor. Similarly, measurements of magnetic hole burning facilitate investigations of intrinsic magnetic response inside bulk samples. MHB has been used to investigate the slow magnetic relaxation of spin glasses and single-crystal ferromagnets. Furthermore, measurements of rheological hole burning facilitate investigations of intrinsic mechanical responses inside bulk samples, such as the slow relaxation of stress or strain in polymers.

In summary, these many measurements of nonresonant spectral hole burning indicate that the net non-exponential response of most materials involves dynamic and thermodynamic heterogeneity. Although NHB gives no direct information about the length scale of this heterogeneity, which in-principle could involve any length scale between individual molecules and independent plane waves, other measurements show that this heterogeneity usually involves a distribution of independently relaxing regions with characteristic length scales on the order 1–3 nm. The box model provides a way to quantitatively characterize the linear and low-order nonlinear parts of the measured response, sometimes with no adjustable parameters. While the box model is a phenomenological approach, nanothermodynamics provides a fundamental physical foundation for the measured thermodynamics heterogeneity, as well as for the box model and other models proposed to describe net relaxation from an ensemble of independently relaxing regions. Thus, nonresonant spectral hole burning represents a versatile, powerful, and direct technique for studying the thermodynamic heterogeneity that is found in the response of most materials.

Acknowledgements Current work in the general area covered in this article is supported by the Deutsche Forschungsgemeinschaft under Grant No. BO1301/14-1. We thank Thomas Blochowicz for kindly sharing data and figures from Ref. [66].

References

1. H. Scher, M.F. Shlesinger, J.T. Bendler, Time-scale invariance in transport and relaxation. Phys. Today **44**, 26 (1991)
2. W. Weber, Ueber die Elasticität der Seidenfäden. Pogg. Ann. Phys. **24**, 711 (1835)
3. F. Kohlrausch, Ueber die elastische Nachwirkung bei der Torsion. Pogg. Ann. Phys. **114**, 337 (1863)
4. F. Kohlrausch, Beiträge zur Kenntniß der elastischen Nachwirkung. Pogg. Ann. Phys. **128**, 1 (1866)
5. G. Williams, D.C. Watts, Non-symmetrical dielectric relaxation behaviour arising from a simple empirical decay function. Trans. Faraday Soc. **66**, 80 (1970)
6. R.V. Chamberlin, G. Mozurkewich, R. Orbach, Time decay of the remanent magnetization in spin-glasses. Phys. Rev. Lett. **52**, 867 (1984)
7. M. Cardona, R.V. Chamberlin, W. Marx, The history of the stretched exponential function. Ann. Phys. (Leipzig) **16**, 842 (2007)
8. R. Kohlrausch, Theorie des elektrischen Rückstandes in der Leidener Flasche. Pogg. Ann. Phys. **91**, 56 (1854)
9. U. Grigull, Newton's temperature scale and the law of cooling. Wärme- und Stoffübertragung. **18**, 195 (1984)
10. G. Williams, M. Cook, P.J. Hains, Molecular motion in amorphous polymers consideration of the mechanism for α, β and ($\alpha\beta$) dielectric relaxations. J. Chem. Soc. Faraday Trans. II **2**, 1045 (1972)
11. A.K. Jonscher, *Dielectric Relaxation in Solids* (Chelsea Dielectrics Press, London, 1983)
12. E. von Schweidler, Studien über die Anomalien im Verhalten der Dielektrika. Ann. Phys. **24**, 711 (1907)
13. K.W. Wagner, Zur Theorie der unvollkommenen Dielektrika. Ann. Phys. **345**, 817 (1913)
14. K.S. Cole, R.H. Cole, Dispersion and absorption in dielectrics I. Alternating current characteristics. J. Chem. Phys. **9**, 341 (1941)
15. D.W. Davidson, R.H. Cole, Dielectric relaxation in glycerol, propylene glycol, and n-propanol. J. Chem. Phys. **19**, 1484 (1951)
16. C.J.F. Böttcher, P. Bordewijk, *Theory of Electric Polarization*, vol. II (Elsevier, Amsterdam, 1978)
17. E. Donth, The size of cooperatively rearranging regions at the glass transition. J. Non-Cryst. Solids **53**, 325 (1982)
18. R.V. Chamberlin, D.N. Haines, Percolation model for relaxation in random systems. Phys. Rev. Lett. **65**, 2197 (1990)
19. V.I. Yukalov, Phase transitions and heterophase fluctuations. Phys. Rep. **208**, 395 (1991)
20. R.V. Chamberlin, R. Böhmer, E. Sanchez, C.A. Angell, Signature of ergodicity in the dynamic response of amorphous systems. Phys. Rev. B **46**, 5787 (1992)
21. R.V. Chamberlin, M.R. Scheinfein, Slow magnetic relaxation in iron: a ferromagnetic liquid. Science **260**, 1098 (1993)
22. C. Hansen, R. Richert, E.W. Fischer, Dielectric loss spectra of organic glass formers and Chamberlin cluster model. J. Non-Cryst. Solids **215**, 293 (1997)
23. R.V. Chamberlin, Mesoscopic mean-field theory for supercooled liquids and the glass transition. Phys. Rev. Lett. **82**, 2520 (1999)
24. G. Biroli, J.P. Bouchaud, K. Miyazaki, D.R. Reichman, Inhomogeneous mode-coupling theory and growing dynamic length in supercooled liquids. Phys. Rev. Lett. **97**, 195701 (2006)
25. A. Heuer, Exploring the potential energy landscape of glass-forming systems: from inherent structures via metabasins to macroscopic transport. J. Phys.: Condens. Matter **20**, 373101 (2008)
26. C.T. Rogers, R.A. Buhrman, Composition of $1/f$ noise in metal-insulator-metal tunnel junctions. Phys. Rev. Lett. **53**, 1272 (1984)
27. K. Schmidt-Rohr, H.W. Spiess, Nature of nonexponential loss of correlation above the glass transition investigated by multidimensional NMR. Phys. Rev. Lett. **66**, 3020 (1991)

28. R. Böhmer, E. Sanchez, C.A. Angell, AC technique for simultaneous study of local and global linear responses near the glass transition: the case of doped $Ca^{2+}/K^{+}/NO_3^{-}$. J. Phys. Chem. **96**, 9089 (1992)
29. A. Barkatt, C.A. Angell, Use of structural probe ions for relaxation studies in glasses. 2. Temperature-jump and temperature-ramp studies of cobalt(II) in nitrate glasses. J. Phys. Chem. **82**, 1972 (1978)
30. R. Richert, Origin of dispersion in dipolar relaxations of glasses. Chem. Phys. Lett. **216**, 223 (1993)
31. M.T. Cicerone, M.D. Ediger, Relaxation of spatially heterogeneous dynamic domains in supercooled ortho-terphenyl. J. Chem. Phys. **103**, 5684 (1995)
32. E. Vidal Russell, N.E. Israeloff, Direct observation of molecular cooperativity near the glass transition. Nature **408**, 695 (2000)
33. X. Qiu, T. Proffen, J.F. Mitchell, S.J.L. Billinge, Orbital correlations in the pseudocubic *O* and rhombohedral *R* phases of $LaMnO_3$. Phys. Rev. Lett. **94**:177203 (2005). Data from this study using neutron scattering to investigate correlations in $LaMnO_3$ are presented in Ref. [35]. It is believed that abrupt loss of dynamical correlation across interatomic distances could be a general phenomenon in crystals, Billinge SJL, private communication
34. Discussion regarding the use of the pair distribution function to study static and dynamic correlations in a variety of systems is given in C. A. Young, A. L. Goodwin, Applications of pair distribution function methods to contemporary problems in materials chemistry. J. Mater. Chem. **21**, 6464 (2011)
35. R.V. Chamberlin, Monte Carlo simulations including energy from an entropic force. Phys. A **391**, 5384 (2012)
36. R. Böhmer, Nanoscale heterogeneity in glass-forming liquids: experimental advances. Curr. Opin. Solid State Mater. Sci. **3**, 378 (1998)
37. R. Böhmer, R.V. Chamberlin, G. Diezemann, B. Geil, A. Heuer, G. Hinze, S.C. Kuebler, R. Richert, B. Schiener, H. Sillescu, H.W. Spiess, U. Tracht, M. Wilhelm, Nature of the non-exponential primary relaxation in structural glass-formers probed by dynamically selective experiments. J. Non-Cryst. Solids **235–237**, 1 (1998)
38. H. Sillescu, Heterogeneity at the glass transition: a review. J. Non-Cryst. Solids **243**, 81 (1999)
39. M.D. Ediger, Spatially heterogeneous dynamics in supercooled liquids. Annu. Rev. Phys. Chem. **51**, 99 (2000)
40. R. Richert, Heterogeneous dynamics in liquids: fluctuations in space and time. J. Phys.: Condens. Matter **14**, R703 (2002)
41. S.A. Reinsberg, A. Heuer, B. Doliwa, H. Zimmermann, H.W. Spiess, Comparative study of the NMR length scale of dynamic heterogeneities of three different glass-formers. J. Non-Cryst. Solids **307**, 208 (2002)
42. L. Hong, V.N. Novikov, A.P. Sokolov, Dynamic heterogeneities, boson peak, and activation volume in glass-forming liquids. Phys. Rev. E **83**, 061508 (2011)
43. L.J. Kaufman, Heterogeneity in single-molecule observables in the study of supercooled liquids. Annu. Rev. Phys. Chem. **64**, 177 (2013)
44. M. Meissner, K. Spitzmann, Experimental evidence on time-dependent specific heat in vitreous silica. Phys. Rev. Lett. **46**, 265 (1981)
45. P.K. Dixon, S.R. Nagel, Frequency-dependent specific heat and thermal conductivity at the glass transition in o-terphenyl mixtures. Phys. Rev. Lett. **61**, 341 (1988)
46. R. Böhmer, B. Schiener, J. Hemberger, R.V. Chamberlin, Pulsed dielectric spectroscopy of supercooled liquids. Z. Phys. B Condensed Matter. **99**, 91 (1995); R. Böhmer, B. Schiener, J. Hemberger, R.V. Chamberlin, Erratum. Z Phys. B Condensed Matter. **99**, 624 (1996)
47. L. Boltzmann, Zur Theorie der elastischen Nachwirkung. Wiener Sitzungsber **70**, 275 (1874). See, e.g., page 622 of the collection of papers available under https://phaidra.univie.ac.at/view/o:63647
48. R.R. Ernst, G. Bodenhausen, A. Wokaun, *Principles of Nuclear Magnetic Resonance in One and Two Dimensions* (Clarendon, Oxford, 1987)

49. O. Kircher, R. Böhmer, G. Hinze, Pseudo-stochastic multiple-pulse excitation in dielectric spectroscopy: application to a relaxor ferroelectric. J. Phys.: Condens. Matter. **15**, S1069 (2003)
50. R.V. Chamberlin, Heterogeneity in the primary response of amorphous and crystalline materials. Proposal to the National Science Foundation (1994)
51. B. Schiener, R. Böhmer, A. Loidl, R.V. Chamberlin, Nonresonant spectral hole burning in the slow dielectric response of supercooled liquids. Science **274**, 752 (1996)
52. N. Bloembergen, E.M. Purcell, R.V. Pound, Relaxation effects in nuclear magnetic resonance absorption. Phys. Rev. **73**, 679 (1948)
53. S. Völker, Hole-burning spectroscopy. Annu. Rev. Phys. Chem. **40**, 499 (1989)
54. N.O. Birge, S.R. Nagel, Specific heat spectroscopy of the glass transition. Phys. Rev. Lett. **54**, 2674 (1985)
55. T. Christensen, The frequency-dependence of the specific-heat at the glass transition. J. Phys. (Paris) **46**, C8–635 (1985)
56. W. Knaak, M. Meissner, Time-dependent specific heat of vitreous silica between 0.1 and 1 K, in *Phonon Scattering in Condensed Matter*, ed. by W. Laßmann, S. Döttinger (Springer, Berlin, 1984), pp. 416–418
57. B. Mertz, J.F. Berret, R. Böhmer, A. Loidl, M. Meissner, W. Knaak, Calorimetric investigations of $(NaCN)_{1-x}(KCN)_x$ orientational glasses. Phys. Rev. B **42**, 7596 (1990)
58. N. Sampat, M. Meissner, Time-dependent specific heat of crystals and glasses at low temperatures, in *Die Kunst of Phonons*, ed. by T. Paszkiewicz, K. Rapcewicz (Springer, Boston, 1994), pp. 105–112
59. see e.g. C.P. Slichter, *Principles of Magnetic Resonance* (Chap. 6), 3rd edn. (Springer, Berlin, 1990)
60. K. Duvvuri, R. Richert, Dielectric hole burning in the high frequency wing of supercooled glycerol. J. Chem. Phys. **118**, 1356–1363 (2003)
61. R.V. Chamberlin, Nonresonant spectral hole burning in a spin glass. Phys. Rev. Lett. **83**, 5134 (1999)
62. R.V. Chamberlin, B. Schiener, R. Böhmer, Slow dielectric relaxation of supercooled liquids investigated by nonresonant spectral hole burning. Mater. Res. Soc. Symp. Proc. **455**, 117 (1997)
63. R.V. Chamberlin, Experiments and theory of the nonexponential relaxation in liquids, glasses, polymers and crystals. Phase Transitions **65**, 169 (1998)
64. B. Schiener, R.V. Chamberlin, G. Diezemann, R. Böhmer, Nonresonant dielectric hole burning spectroscopy of supercooled liquids. J. Chem. Phys. **107**, 7746 (1997)
65. R. Richert, Spectral selectivity in the slow β-relaxation of a molecular glass. Europhys. Lett. **54**, 767 (2001)
66. T. Blochowicz, E.A. Rössler, Nonresonant dielectric hole burning in neat and binary organic glass formers. J. Chem. Phys. **122**, 224511 (2005); see also T. Blochowicz, Broadband Dielectric Spectroscopy in Neat and Binary Molecular Glass Formers: frequency and Time Domain Spectroscopy, Non-Resonant Spectral Hole Burning. Dissertation, Universität Bayreuth (2003)
67. R. Richert, S. Weinstein, Nonlinear dielectric response and thermodynamic heterogeneity. Phys. Rev. Lett. **97**, 095703 (2006)
68. R. Böhmer, G. Diezemann, Principles and applications of pulsed dielectric spectroscopy and nonresonant dielectric hole burning, in *Broadband dielectric spectroscopy*, ed. by F. Kremer, A. Schönhals (Springer, Berlin, 2002), pp. 523–569
69. R. Richert, R. Böhmer, Heterogeneous and homogeneous diffusivity in an ion-conducting glass. Phys. Rev. Lett. **83**, 4337 (1999)
70. R. Richert, The modulus of dielectric and conductive materials and its modification by high electric fields. J. Non-Cryst. Solids **305**, 29 (2002)
71. O. Kircher, B. Schiener, R. Böhmer, Long-lived dynamic heterogeneity in a relaxor ferroelectric. Phys. Rev. Lett. **81**, 4520 (1998)

72. W. Kleemann, V. Bobnar, J. Dec, P. Lehnen, R. Pankrath, S.A. Prosandeev, Relaxor properties of dilute and concentrated polar solid solutions. Ferroelectrics **261**, 43 (2001)
73. O. Kircher, G. Diezemann, R. Böhmer, Nonresonant dielectric hole-burning spectroscopy on a titanium modified lead magnesium niobate ceramic. Phys. Rev. B **64**, 054103 (2001)
74. T. El Goresy, O. Kircher, R. Böhmer, Nonresonant hole burning spectroscopy of the relaxor ferroelectric PLZT. Solid State Commun. **121**, 485 (2002)
75. X. Shi, G.B. McKenna, Mechanical hole burning spectroscopy: evidence for heterogeneous dynamics in polymer systems. Phys. Rev. Lett. **94**, 157801 (2005)
76. X.F. Shi, G.B. McKenna, Mechanical hole-burning spectroscopy: demonstration of hole burning in the terminal relaxation regime. Phys. Rev. B **73**, 014203 (2006)
77. Q. Qin, H. Doen, G.B. McKenna, Mechanical Spectral Hole Burning in polymer solutions. J. Polym. Sci. Part B: Polym. Phys. **47**, 2047 (2009)
78. N. Shamim, G.B. McKenna, Mechanical spectral hole burning in polymer solutions: comparison with large amplitude oscillatory shear fingerprinting. J. Rheol. **58**, 43 (2014)
79. M. Wilhelm, K. Hyun, Nonlinear oscillatory shear mechanical response, in *Nonlinear Dielectric Spectroscopy*, ed. by R. Richert (Springer, *this book*, 2018)
80. S. Weinstein, R. Richert, Heterogeneous thermal excitation and relaxation in supercooled liquids. J. Chem. Phys. **123**, 224506 (2005)
81. R. Richert, Nonlinear dielectric effects in liquids: a guided tour. J. Phys.: Condens. Matter **29**, 363001 (2017)
82. K.R. Jeffrey, R. Richert, K. Duvvuri, Dielectric hole burning: signature of dielectric and thermal relaxation time heterogeneity. J. Chem. Phys. **119**, 6150 (2003)
83. S. Capaccioli, D. Prevosto, A. Best, A. Hanewald, T. Pakula, Applications of the rheo-dielectric technique. J. Non-Cryst. Solids **353**, 4267 (2007)
84. T. Uneyama, Y. Masubuchi, K. Horio, Y. Matsumiya, H. Watanabe, J.A. Pathak, C.M. Roland, A theoretical analysis of rheodielectric response of type-A polymer chains. J. Polym. Sci. Part B: Polym. Phys. **47**, 1039 (2009)
85. K. Horio, T. Uneyama, Y. Matsumiya, Y. Masubuchi, H. Watanabe, Rheo-dielectric responses of entangled cis-polyisoprene under uniform steady shear and LAOS. Macromolecules **47**, 246 (2014)
86. L.D. Landau, E.M. Lifshitz, *Electrodynamics of Continuous Media* (Pergamon, Oxford, 1984)
87. T. Matsuo, H. Suga, S. Seki, Dielectric loss measurement by differential thermal analysis (DTA). Bull. Chem. Soc. Jpn. **39**, 1827 (1966)
88. A. Heuer, Information content of multitime correlation functions for the interpretation of structural relaxation in glass-forming systems. Phys. Rev. E **56**, 730 (1997)
89. K. Schröter, E. Donth, Viscosity and shear response at the dynamic glass transition of glycerol. J. Chem. Phys. **113**, 9101 (2000)
90. L.-M. Wang, R. Richert, Measuring the configurational heat capacity of liquids. Phys. Rev. Lett. **99**, 185701 (2007)
91. L.-M. Wang, R. Richert, Reply to comment on "Measuring the configurational heat capacity of liquids". Phys. Rev. Lett. **104**, 239603 (2010)
92. R. Richert, Reverse calorimetry of a supercooled liquid: propylene carbonate. Thermochim. Acta **522**, 28 (2011)
93. I.M. Hodge, Enthalpy relaxation and recovery in amorphous materials. J. Non-Cryst. Solids **169**, 211 (1994)
94. A. Khalife, U. Pathak, R. Richert, Heating liquid dielectrics by time dependent fields. Eur. Phys. J. B **83**, 429 (2011)
95. C. Crauste-Thibierge, C. Brun, F. Ladieu, D. L'Hôte, G. Biroli, J.-P. Bouchaud, Evidence of growing spatial correlations at the glass transition from nonlinear response experiments. Phys. Rev. Lett. **104**, 165703 (2010)
96. R. Richert, Dielectric hole burning in an electrical circuit analog of a dynamically heterogeneous system. Phys. A **322**, 143 (2003)
97. H. Fröhlich, *Theory of Dielectrics* (Clarendon, Oxford, 1958)

98. H. Wagner, R. Richert, Dielectric relaxation of the electric field in poly(vinylacetate): a time domain study in the range 10^{-3} s to 10^6 s. Polymer **38**, 255 (1997)
99. J. Jäckle, R. Richert, Why retardation takes more time than relaxation in a linear system. Phys. Rev. E **77**, 031201 (2008)
100. A.R. Young-Gonzales, S. Samanta, R. Richert, Dynamics of glass-forming liquids. XIX. Rise and decay of field induced anisotropy in the non-linear regime. J. Chem. Phys. **143**, 104504 (2015)
101. S. Weinstein, R. Richert, Nonlinear features in the dielectric behavior of propylene glycol. Phys. Rev. B **75**, 064302 (2007)
102. W. Huang, R. Richert, Dynamics of glass-forming liquids. XIII. Microwave heating in slow motion. J. Chem. Phys. **130**, 194509 (2009)
103. S. Samanta, R. Richert, Limitations of heterogeneous models of liquid dynamics: very slow rate exchange in the excess wing. J. Chem. Phys. **140**, 054503 (2014)
104. W. Huang, R. Richert, Response to "Comment on 'Dynamics of glass-forming liquids. XIII. Microwave heating in slow motion'". [J. Chem. Phys. **137**:027101 (2012)] J. Chem. Phys. **137**:027102
105. G. Diezemann, Response theory for nonresonant hole burning: stochastic dynamics. Europhys. Lett. **53**, 604 (2001)
106. R. Richert, Molecular dynamics analysed in terms of continuous measures of dynamic heterogeneity. J. Non-Cryst. Solids **235–237**, 41 (1998)
107. M. Winterlich, G. Diezemann, H. Zimmermann, R. Böhmer, Microscopic origin of the non-exponential dynamics in a glassy crystal. Phys. Rev. Lett. **91**, 235504 (2003)
108. M. Storek, J. Tilly, K.R. Jeffrey, R. Böhmer, Four-time ^{7}Li stimulated-echo spectroscopy for the study of dynamic heterogeneities: application to lithium borate glass. J. Magn. Reson. **282**, 1 (2017)
109. A. Wagner, H. Kliem, A comment on dielectric hole burning. J. Chem. Phys. **111**, 1043 (1999)
110. G. Diezemann, Stochastic models of higher-order dielectric responses, in *Nonlinear Dielectric Spectroscopy*, ed. by R. Richert (Springer, *this book,* 2018)
111. G. Diezemann, Dynamic heterogeneities in the out-of-equilibrium dynamics of simple spherical spin models. Phys. Rev. E **68**, 021105 (2003)
112. U. Häberle, G. Diezemann, Nonresonant holeburning in the Terahertz range: Brownian oscillator model. J. Chem. Phys. **120**, 1466 (2004)
113. U. Häberle, G. Diezemann, Kerr effect as a tool for the investigation of dynamic heterogeneities. J. Chem. Phys. **124**, 044501 (2006)
114. U. Häberle, G. Diezemann, Dynamic Kerr effect responses in the Terahertz-range. J. Chem. Phys. **122**, 184517 (2005)
115. G.E.P. Box, Robustness in the strategy of scientific model building, in *Robustness in Statistics*, ed. by R.L. Launer, G.N. Wilkerson (Academic Press, New York, 1979), pp. 201–236
116. T.L. Hill, Thermodynamics of small systems. J. Chem. Phys. **36**, 3182 (1962)
117. T.L. Hill, *Thermodynamics of Small Systems (parts I and II)* (Dover, Mineola, NY, 1994)
118. T.L. Hill, A different approach to nanothermodynamics. Nano Lett. **1**, 273 (2001)
119. R.V. Chamberlin, Mean-field cluster model for the critical behaviour of ferromagnets. Nature **408**, 337 (2000)
120. R.V. Chamberlin, The big world of nanothermodynamics. Entropy **17**, 52 (2015)
121. G. Adam, J.H. Gibbs, On the temperature dependence of cooperative relaxation properties in glass-forming liquids. J. Chem. Phys. **43**, 139 (1965)
122. G. Tarjus, S.A. Kivelson, Z. Nussinov, P. Viot, The frustration-based approach of supercooled liquids and the glass transition: a review and critical assessment. J. Phys.: Condens. Matter **17**, R1143 (2005)
123. A. Wisitsorasak, P.G. Wolynes, Dynamical heterogeneity of the glassy state. J. Phys. Chem. **118**, 7835 (2014)
124. U. Tracht, M. Wilhelm, A. Heuer, H. Feng, K. Schmidt-Rohr, H.W. Spiess, Length scale of dynamic heterogeneity at the glass transition determined by multidimensional nuclear magnetic resonance. Phys. Rev. Lett. **81**, 2727 (1998)

125. K. Binder, Finite size scaling analysis of Ising model block distribution functions. Z. Phys. B **43**, 119 (1981)
126. R.V. Chamberlin, G.H. Wolf, Fluctuation-theory constraint for extensive entropy in Monte-Carlo simulations. Eur. Phys. J. B **67**, 495 (2009)
127. R.P. Feynman, *Statistical Mechanics: A Set of Lectures* (Westview Press, Boulder, CO, 1998), p. 1
128. A. Einstein, Theorie der Opaleszenz von homogenen Flüssigkeiten und Flüssigkeitsgemischen in der Nähe des kritischen Zustandes. Ann. Phys. **33**, 1275 (1910)
129. L. Landau, The theory of phase transitions. Nature **138**, 840 (1936)
130. M.J. Klein, L. Tisza, Theory of critical fluctuations. Phys. Rev. **76**, 1841 (1949)
131. S.K. Ma, *Modern theory of critical phenomena* (W.A. Benjamin, Reading, MA, 1976)
132. R.V. Chamberlin, Critical behavior from Landau theory in nanothermodynamic equilibrium. Phys. Lett. A **315**, 312 (2003)
133. M.R.H. Javaheri, R.V. Chamberlin, A free-energy landscape picture and Landau theory for the dynamics of disordered materials. J. Chem. Phys. **125**, 154503 (2006)
134. R.V. Chamberlin, Reducing low-frequency noise during reversible fluctuations. Eur. Phys. J. Spec. Topics **226**, 365 (2017)
135. X.H. Qiu, M.D. Ediger, Length scale of dynamic heterogeneity in supercooled D-Sorbitol: comparison to model predictions. J. Phys. Chem. B **107**, 459 (2003)
136. C.A. Angell, D.L. Smith, Test of the entropy basis of the Vogel-Tammann-Fulcher equation. Dielectric relaxation of polyalcohols near T_g. J. Phys. Chem. **86**, 3845 (1982)
137. R. Böhmer, K.L. Ngai, C.A. Angell, D.J. Plazek, Nonexponential relaxations in strong and fragile glass formers. J. Chem. Phys. **99**, 4201 (1993)
138. S. Samanta, R. Richert, Dynamics of glass-forming liquids. XVII. Does entropy control structural relaxation times. J. Chem. Phys. **142**, 044504 (2015)
139. F. Stickel, E.W. Fischer, R. Richert, Dynamics of glass-forming liquids. II. Detailed comparison of dielectric relaxation, dc-conductivity and viscosity data. J. Chem. Phys. **104**, 2043 (1996)
140. R.V. Chamberlin, J.V. Vermaas, G.H. Wolf, Beyond the Boltzmann factor for corrections to scaling in ferromagnetic materials and critical fluids. Eur. Phys. J. B **71**, 1 (2009)
141. R.V. Chamberlin, D.M. Nasir, $1/f$ noise from the laws of thermodynamics for finite-size fluctuations. Phys. Rev. E **90**, 012142 (2014)
142. R.V. Chamberlin, S. Abe, B.F. Davis, P.E. Greenwood, A.S.H. Shevchuk, Fluctuation theorems and $1/f$ noise from a simple matrix. Eur. Phys. J. B **89**, 185 (2016)
143. R.P. Feynman, R.B. Leighton, M. Sands, *The Feynman Lectures on Physics* (Chap. 44-1) (Addison Wesley, Reading, MA, 1963)
144. R. Böhmer, G. Diezemann, G. Hinze, E. Rössler, Dynamics of supercooled liquids and glassy solids. Prog. Nucl. Magn. Reson. Spectrosc. **39**, 191 (2001)
145. L. Wu, Relaxation mechanisms in a benzyl chloride–toluene glass. Phys. Rev. B **43**, 9906 (1991)
146. N.G. McCrum, B.E. Read, G. Williams, *Anelastic and Dielectric Effects in Polymeric Solids* (Dover, New York, 1991)
147. B. Schiener et al (1995) unpublished
148. R.A. Webb, New technique for improved low-temperature SQUID NMR measurements. Rev. Sci. Instr. **48**, 1585 (1977)
149. R.V. Chamberlin, L.A. Moberly, O.G. Symko, High-sensitivity magnetic-resonance by SQUID detection. J. Low Temp. Phys. **35**, 337 (1979)
150. B.T. Saam, M.S. Conradi, Low-frequency NMR polarimeter for hyperpolarized gases. J. Magn. Reson. **134**, 67 (1998)
151. S. Ghosh, R. Parthasarathy, T.F. Rosenbaum, G. Aeppli, Coherent spin oscillations in a disordered magnet. Science **296**, 2195 (2002)
152. A. Berton, J. Chaussy, J. Odin, J. Peyrard, J.J. Prejean, J. Souletie, Apparent specific heat of a spin glass (Au Fe 6 at %) in presence of a remanent magnetization and associated energy and magnetization relaxations. Solid State Commun. **37**, 241 (1981)

153. W.E. Fogle, J.D. Boyer, N.E. Phillips, J. Van Curen, Calorimetric investigation of spin-glass ordering in CuMn. Phys. Rev. Lett. **47**, 352 (1981)
154. R. Richert, Effects of strong static fields on the dielectric relaxation of supercooled liquids, in *Nonlinear Dielectric Spectroscopy*, ed. by R. Richert (Springer, *this book*, 2018)
155. However, homogeneous behavior was found from low-frequency NHB measurements on an ion conductor, see Ref. [69]

Nonlinear Dielectric Effect in Critical Liquids

Sylwester J. Rzoska, Aleksandra Drozd-Rzoska and Szymon Starzonek

Abstract The nonlinear dielectric effect (*NDE*) describes changes of dielectric constant under the strong electric field, with the metric defined $(\varepsilon(E) - \varepsilon(E \to 0))/E^2$. This contribution discusses (i) miscibility, (ii) the isotropic phase of nematic liquid crystals, and (iii) the supercooled nitrobenzene. For comparison, the precritical evolution of $\varepsilon = \varepsilon(E \to 0)$ is also presented. The discussion is extended for dynamic issues related to the "linear" and "nonlinear" relaxation times. Finally, basic problems of the dual-field *NDE* experimental technique are briefly presented.

1 Introduction

At the beginning of the twentieth century, a number of new and revolutionary technologies and inventions appeared, changing the world around. The emergence of the radio technique associated with discoveries of Marconi and Armstrong is of particular importance for this work [1]. New technologies make it also possible to reach very low temperatures, very high pressures, strong magnetic or electric fields, etc. These extraordinary conditions often led to great fundamental discoveries [1]. All these could influence Herweg's decision to study the influence of the strong electric field on dielectric constant (ε) in liquids, particularly in diethyl ether (DEE) [2, 3]. He discovered that $\varepsilon(E) < \varepsilon(E \to 0)$ and introduced the magnitude $\Delta\varepsilon^E/E^2 = \left[\varepsilon(E) - \varepsilon(E \to 0)/E^2\right] < 0$ as the metric for the new phenomenon: it was later called dielectric saturation or nonlinear dielectric effect (*NDE*) [2, 3]. It turned out that this result can be interpreted within the Herweg–Debye–Langevin model describing the orientation of noninteracting or weakly interacting permanent dipole moments coupled to DEE molecules. Its key output is

S. J. Rzoska (✉) · A. Drozd-Rzoska · S. Starzonek
Institute of High Pressure Physics, Polish Academy of Sciences, Ul. Sokolowska 29/37, 01-142 Warsaw, Poland
e-mail: sylwester.rzoska@unipress.waw.pl

R. Richert (ed.), *Nonlinear Dielectric Spectroscopy*, Advances in Dielectrics
https://doi.org/10.1007/978-3-319-77574-6_6

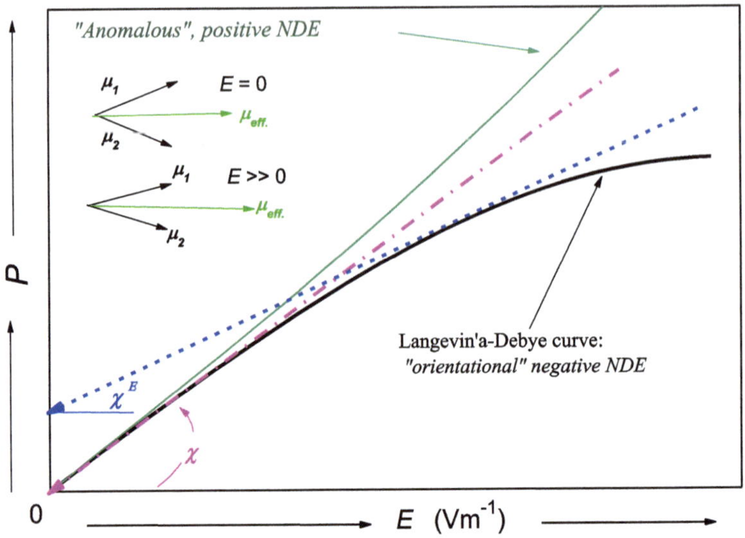

Fig. 1 The evolution of dielectric polarization on increasing the intensity of the electric field in isotropic liquid dielectrics. The thick (black) solid curve is for Eq. (1) with correlation factors $R_P = 1$ and $R_S = 1$ Eq. (1), leading to *NDE* < 0. The thin and green curve is for the "anomalous" *NDE* > 0. The dielectric constant is defined as $\chi = \varepsilon - 1 = \mathrm{d}P(E)/\mathrm{d}E$: it is shown as the dashed-dotted (magenta) line for $E \to 0$ and via the slope of the dashed (blue) line for $E >> 0$. The concept of Piekara's "scissors" [7–10] to explain the positive *NDE* in nitrobenzene is also sketched

the relation describing the behavior of the dielectric polarization under the strong electric field [3, 4]:

$$P(E) = F_1 \frac{N\mu}{3k_B T} R_P E - F_3 \frac{N\mu^4}{45k_B T^3} R_S E^3 + \cdots \tag{1}$$

where F_1 and F_3 are the local field factors, N is the number of permanent dipole moments in a unit volume, μ is the permanent dipole moment, E denotes the intensity of the electric field, R_P and R_S stand for correlation factors. For liquids and isotropic dielectrics, the term $\sim E^2$ is absent. Dielectric constant is for the real part of dielectric permittivity $\varepsilon^*(f) = \varepsilon'(f) - i\varepsilon''(f)$ in the "static" frequency domain where $\varepsilon'(f)$ value is (almost) frequency independent: for DEE, this is (kHz–MHz) region. For Eq. (1) and Fig. 1, $\varepsilon' - 1 = \varepsilon - 1 = \chi = \mathrm{d}P/\mathrm{d}E$, where χ denotes the electric susceptibility.

The factor R_S in Eq. (1) was introduced by Piekara in the mid of the 30s to explain the anomalous positive *NDE* ($\varepsilon(E) > \varepsilon(E > 0)$, which he discovered in nitrobenzene [5–10]. Piekara linked the phenomenon to the intermolecular coupling of nitrobenzene molecules causing the dipole–dipole "scissor-like" arrangement of permanent dipole moments, as shown in Fig. 1. Under the strong electric field, this leads to the increase of the effective dipole moment [7–10].

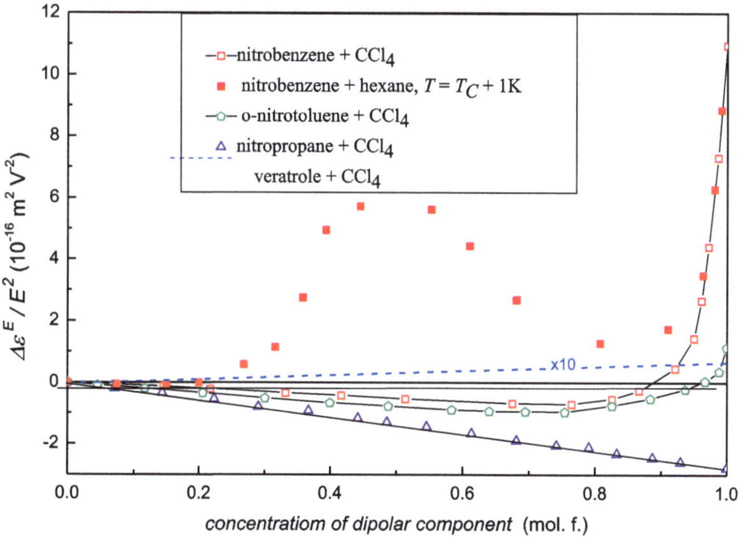

Fig. 2 The *NDE* in mixtures of a dipolar component in a non-dipolar solvent (CCl_4, hexane): the isothermal, concentration-related dependence. Results are for the near room temperature (20 °C). For nitrobenzene–hexane mixture, results are related to the homogeneous phase just above the critical consolute temperature $T_C \approx 19.5\,°C$ [11]

Figure 2 shows the sensitivity of *NDE* to different molecular mechanisms (for the review see refs. [10, 12]): (i) in nitropropane mixtures, the behavior predicted by Eq. (1) for $R_S \approx 1$ (no relevant intermolecular coupling) and $-\Delta\varepsilon^E/E^2 \propto N$ takes place; (ii) the same occurs in nitrobenzene and nitrotoluene mixtures in CCl_4 for small concentrations of dipolar components; when increasing their concentrations, the intermolecular coupling and $\Delta\varepsilon^E/E^2 > 0$ appears; (iii) in mixtures of veratrole, *NDE* detects the intramolecular rotation—this contribution is small, "positive" and $\Delta\varepsilon^E/E^2 \sim x$ $(x \sim N)$; (iv) the unique case constitutes the nitrobenzene–hexane mixture where the additional strong and positive *NDE* contribution appears in the vicinity of the critical concentration, $x \approx 0.43$ mole fraction of nitrobenzene.

In the 30s, Piekara discovered two more unique phenomena [5–8]. He carried out temperature-related measurements of dielectric constant and its strong electric field related counterpart—*NDE* on approaching the critical consolute temperature in nitrobenzene–hexane mixture, the system of limited miscibility for $T < T_C \approx 19.5\,°C$ (Fig. 3). In the homogeneous phase well above the critical consolute temperature (T_C), the dielectric constant first increased linearly on cooling but in the immediate vicinity of T_C it slightly bent down (decreases) below such extrapolated dependence [12]. In similar studies for *NDE,* Piekara reported a very strong increase toward positive values [5, 6]. Surprisingly, attempts to describe evolutions of ε' $(T \to T_C)$ and $\Delta\varepsilon^E/E^2$ $(T \to T_C)$ remained puzzling for the next five decades [10, 15–23].

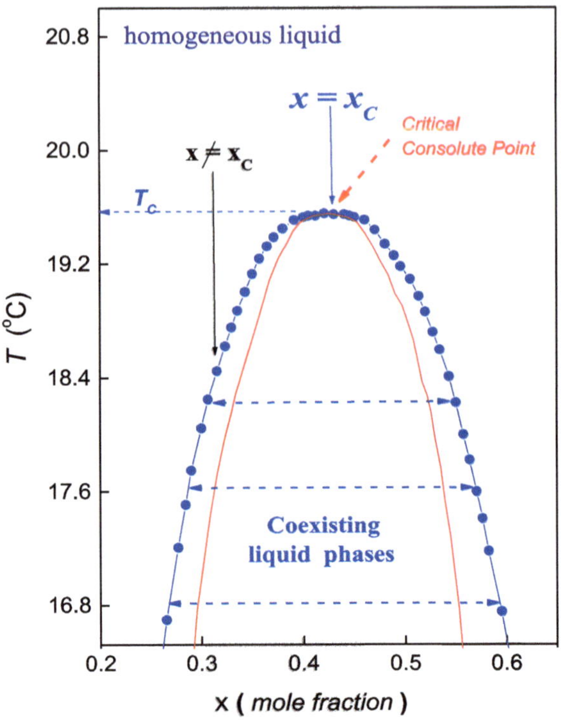

Fig. 3 The coexistence curve (binodal) in the nitrobenzene–hexane mixture of limited miscibility (solid circles and the curve in blue) [11]. It has been determined by the authors using the "cathetometric method" described in ref. [13]. The red curve below the binodal shows the spinodal curve determined via *NDE* measurements in mixtures of noncritical concentration, using the "pseudospinodal" analysis as in ref. [14]. The critical concentration: $x_C = 0.43$ mole fraction of nitrobenzene and the critical consolute temperature $T_C = 19.5\,°\mathrm{C}$

This contribution first resumes the problem of dielectric constant in critical mixtures. Subsequently, the behavior of the nonlinear dielectric effect (*NDE*) in critical mixtures and in liquid crystals is discussed. This is supplemented by the evidence for the possible critical origins of the anomalous positive *NDE* in nitrobenzene. Some issues related to *NDE*-related dynamics, i.e., the *NDE* extension toward the nonlinear dielectric spectroscopy (*NDS*), are also discussed. Basic problems of the dual-field nonlinear dielectric technique are given. The contribution is based on the authors earlier but reanalyzed results and on yet unpublished new evidence.

2 Dielectric Constant in Critical Mixtures

Studies of dielectric permittivity in binary critical mixtures of limited miscibility, which followed Piekara's findings [12], were continued from the 50s [15–23], but results were puzzling. Most often no pretransitional anomaly or the pretransitional bending up or even a very strong pretransitional increase from the linear behavior remote from the critical temperature were observed [15, 23]. Only in the late 80s, Thoen et al. [24] discovered the key reason of such scatter: the necessity of taking into account the applied measurement frequency. For benzonitrile–isooctane, nitroethane–cyclohexane, and nitrobenzene–isooctane critical mixtures the pretransitional "bending down" for $\varepsilon\,(T \to T_C)$ was observed for $f > 100\,\mathrm{kHz}$ and the obtained precritical "anomalies" were portrayed via the model related to Eq. (2). For lower frequencies, the parasitic impact of the ionic Maxwell–Wagner effect caused the recalled above problems. Such behavior illustrates Fig. 4 for the critical behavior of dielectric constant in nitrobenzene–hexane critical mixture for few frequencies [11]. The model for the parameterization of the anomalous, pretransitional behavior of dielectric constant was proposed by Goulon, Greffe and Oxtoby (GGO, 1979) [25] using the "critical" droplet model and Sengers et al. (1980, [26]) basing on the thermodynamic scaling analysis within the theory of critical phenomena:

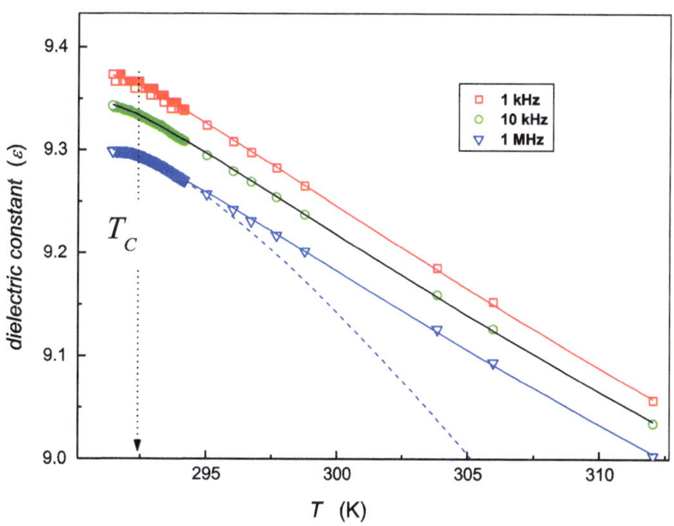

Fig. 4 The temperature behavior of dielectric constant on approaching the critical point in nitrobenzene–hexane mixture of critical concentration. Note the bending down in the immediate vicinity of the critical temperature for the measurement frequency $f = 1\,\mathrm{MHz}$. This is related to the precritical effect, portrayed via Eq. (2). The dashed curve is for Eq. (2) when neglecting the correction-to-scaling term. Already for $f = 1\,\mathrm{kHz}$ no pretransitional is visible, due to the Maxwell–Wagner effect [11]

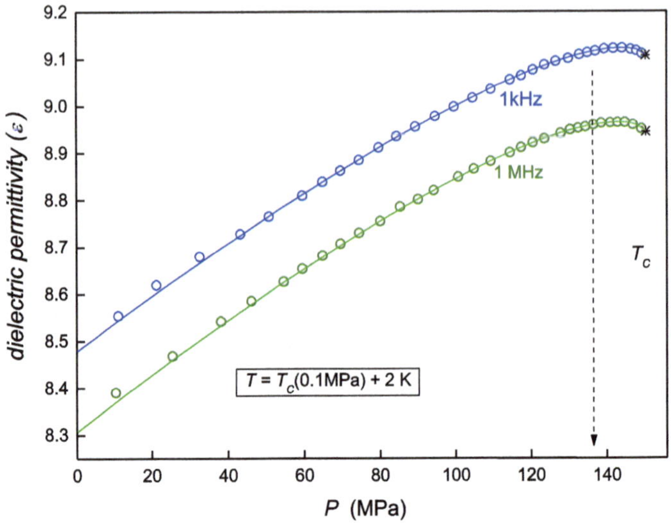

Fig. 5 The evolution of dielectric constant in nitrobenzene–decane critical mixture for the isothermal, pressure-related approach to the critical point. Solid curves are for Eq. (3). Note that for the pressure paths correction-to-scaling terms can be neglected and the MW ionic-related parasitic contribution is absent, even for $f = 100$ Hz (!). Note that $\mathrm{d}T_C/\mathrm{d}P > 0$ for this mixture. The plot prepared basing on results from ref. [27]

$$\varepsilon(T) = \varepsilon_C + a(T - T_C) + A(T - T_C)^{1-\alpha}\left[1 + (T - T_C)^{\Delta} + \cdots\right], \quad P = \mathrm{const} \tag{2}$$

where the critical exponent $\alpha \approx 0.115$ is for the specific heat critical anomaly: the value is for $d = 3$ (the dimension of space) and $n = 1$ (the dimension of the order parameter) universality class. The exponent $\Delta \approx 0.5$ is for the first correction-to-scaling term, important when moving away from T_C [25, 26, 29].

In ref. [27], the pressure counterpart of the above relation was introduced:

$$\varepsilon(P) = \varepsilon_C + a_P|P - P_C| + A_P|P - P_C|^{1-\alpha} \tag{3}$$

For the pressure path, there are no correction-to-scaling terms, and no impact of the MW effect even for as low frequency as $f = 100$ Hz on the critical effect is observed, as shown in Fig. 5. Moreover, the pretransitional anomaly is much "stronger" than for the isobaric, temperature path: compare results presented in Figs. 4 and 5. All these show notable advantages of high-pressure studies of the pretransitional anomaly of dielectric constant. The form of the pretransitional anomaly and the sign of the amplitude A_P in Eq. (3) depend on the path of approaching the critical point and the sign of $\mathrm{d}T_C/\mathrm{d}P$, as shown in Fig. 6.

Regarding possible reasons of the long-standing puzzles for the pretransitional anomaly $\varepsilon(T \to T_C)$ [15–23], it is worth stressing that they were obtained mainly

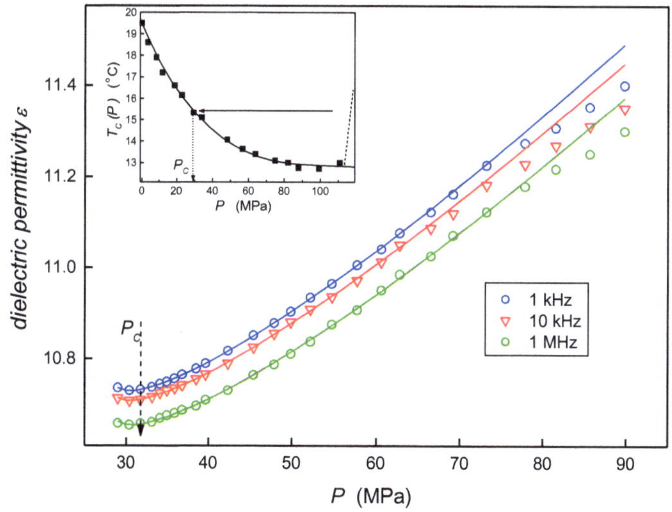

Fig. 6 The pressure-related, isothermal behavior of dielectric constant for nitrobenzene–hexane critical mixture when approaching the critical consolute point under atmospheric pressure: $T_C = 19.5\,°C$, $P = 0.1$ MPa. Solid curves are portrayed via Eq. (3). Note that for this system $dT_C/dP < 0$, as shown in the inset where the pressure evolution of the critical temperature is presented. Based on results from ref. [28]

using the standard Wheatstone bridge apparatus for frequencies between DC and few kHz. In the 80s, new HP impedance analyzers appeared: they enabled the frequency scan up to few MHz and introduced the sampling way of measurements. This was the key for the success in explaining $\varepsilon(T)$ mystery in critical mixtures. The question arises, in which way Piekara [12] obtained the "correct" pattern of the critical anomaly for $\varepsilon(T)$ already in the early 30s? This can be associated with the way of measurements he applied: Piekara used the resonant circuit which was switched on for a short period of time ("by hand"). The system operated in the near—MHz domain. In fact all these resembled the "modern" way of measurements.

3 Nonlinear Dielectric Effect in Critical Mixtures of Limited Miscibility

The nonlinear dielectric effect (*NDE*) can be recognized as the "nonlinear" counterpart of dielectric constant [see Fig. 1; Eq. (1)]. It was clear from the late 70s that in the homogeneous phase of critical, binary, mixtures *NDE* follows the pattern characteristic for critical phenomena [10, 30 and refs therein]:

$$\frac{\Delta\varepsilon^E}{E^2} \propto (T - T_C)^{-\psi} \tag{4}$$

Following the physics of critical phenomena one can expect the same value of the *NDE*-related critical exponent for any critical mixture: surprisingly, the scatter $0.2 < \psi < 0.8$ was observed [10 and refs. therein]. Such behavior is unusual for strong precritical anomalies [29]. Only in 1986, it was shown that qualitative discrepancies between experimental results were associated with poor estimations of the noncritical (molecular) background effect, namely [30]:

$$\frac{\Delta\varepsilon^E}{E^2} = \left(\frac{\Delta\varepsilon^E}{E^2}\right)_{\text{bckg}} + \left(\frac{\Delta\varepsilon^E}{E^2}\right)_{\text{Crit.}} = \left(\frac{\Delta\varepsilon^E}{E^2}\right)_{\text{bckg}} + A'_{NDE}\,(T - T_C)^{-\psi} \tag{5}$$

Examples of "background" contributions are shown in Fig. 2. For mixtures of limited miscibility composed of a dipolar component and a non-dipolar solvent, the optimal method is *NDE* measurements in a reference solution of unlimited miscibility [30]: taking into account the concentration in the volume fraction and the properly selecting the non-dipolar solvent [30]. In refs. [30, 31] the universal value $\psi \approx 0.37$ for $T \to T_C$ and $\psi \approx 0.4$ when tests are for the extended temperature range $T - T_C >$ 10 K [30, 31] were obtained. It was suggested that the latter is associated with the lack of correction-to-scaling terms. In 1979, Oxtoby et al. [25] developed so-called droplet model and linked the critical effect in binary mixtures of limited miscibility to the elongation under the strong electric field of initially spherical fluctuations—droplets. The following "universal" relation was obtained:

$$\left(\frac{\Delta\varepsilon^E}{E^2}\right)^{\text{elong.}}_{\text{crit.}} = A_{NDE}\,(T - T_C)^{\gamma-2\beta}, \quad \text{i.e. the critical exponent } \psi = \gamma - 2\beta \tag{6}$$

where the critical amplitude $A_{NDE} \propto (\varepsilon_1 - \varepsilon_2)^4/\varepsilon_S^2$, ε_1 and ε_2 are dielectric constants of component of the binary mixture, and ε_S is for the mixture (solution). The exponent $\gamma \approx 1.23$ is for the order parameter-related susceptibility: $\chi_T = \chi_0\,(T - T_C)^{-\gamma}$; the exponent $\beta \approx 0.325$ is for the order parameter: $M = B_0\,(T_C - T)^{\beta}$. Values of exponents are for the $d = 3$ and $n = 1$ universality class [29].

It was also indicated that the "universal" electrostriction contribution, i.e., the change of the volume of fluctuations droplets due to the strong electric field, also appears [25]:

$$\left(\frac{\Delta\varepsilon^E}{E^2}\right)^{\text{el.strict.}}_{\text{crit.}} \propto (T - T_C)^{2\beta-1} \tag{7}$$

However, this contribution is qualitatively smaller than the one related to the elongation of critical fluctuations. Hence, the precritical anomaly of *NDE* should be described only by the critical exponent $\psi = \gamma - 2\beta \approx 0.59$ [25]. The same functional form for the pretransitional behavior was obtained by Hoye and Stell [32] when studying the strong electric field-induced distortion on the orientationally averaged particle–particle correlation function and by Onuki and Doi [33] when considering the structure factor and dipolar interactions induced in critical fluctuations under the strong electric field. All mentioned model predicted the same behavior for *NDE* and the electro-optic Kerr effect (*EKE*). Hence, theoretical models led to the following conclusion [25, 32, 33]:

$$\psi_{NDE}\,(\exp.) = \psi\,(\text{theor.}) = \psi_{EKE}\,(\exp.) \tag{8}$$

However, existing experimental evidences yielded [30, 31, 34]:

$$\psi_{NDE}\,(\exp.) \approx 0.37 < \psi\,(\text{theor.}) < \psi_{EKE}\,(\exp.) \approx 0.85 \tag{9}$$

It was shown in ref. [35] that both Eqs. (8) and (9) can be valid (!). The proposed explanation took into account changes of the form of the correlation length under the strong electric field:

$$\xi\,(E \to 0) = \xi_o\,(T - T_C)^{-\nu} \to \xi\,(E) = \left(\xi_{||}, \xi_{\perp}, \xi_{\perp}\right) \tag{10}$$

where the correlation length $\xi\,(T) = \xi_0\,(T - T_C)^{-\nu}$, the critical exponent $\nu = \nu_{||} = \nu\,(\text{nonclassical}) \approx 0.63$ [the value for ($d = 3$, $n = 1$) universality class], and the exponent $\nu_{\perp} = \nu\,(\text{classical}) = 1/2$, i.e., $\xi_{\perp}$ is described within the mean-field approximation which takes place for the dimensionality $d > 4$ or equivalently for the large enough range of interactions [29]. The latter is possible for $\xi\,(E)$ due to the rod-like elongation which can increase the number of neighbors. Using scaling relations introduced by Fisher [37], one can show that the basic output relation of models [25, 32, 33] for *NDE* and *EKE* can be presented as follows [35]:

$$\begin{aligned}\frac{\Delta\varepsilon^E}{E^2}, \frac{\Delta n}{E^2} &= A_C\chi_T\,(T)\left\langle\Delta M^2\right\rangle_V \\ &\propto A_C\chi_0\,(T - T_C)^{-\gamma}\,\Delta M_0\,(T - T_C)^{2\beta} \\ &= C\,(T - T_C)^{\gamma - 2\beta}\end{aligned} \tag{11}$$

and alternatively

$$\frac{\Delta\varepsilon^E}{E^2}, \frac{\Delta n}{E^2} = \chi_T\,(T)\left\langle\Delta M^2\right\rangle_V = \xi^d\chi^2 \tag{12}$$

Hence $\psi = \gamma - 2\beta = \mathrm{d}\nu - 2\gamma$. Regarding amplitudes, for *NDE*: $A_C \propto (\varepsilon_1 - \varepsilon_2)^4/\varepsilon_S^2$ and for *EKE*: $A_C \propto (\varepsilon_1 - \varepsilon_2)^2\,(n_1 - n_2)/\varepsilon_S n_S$, where n stands for the refractive index [36]. Recalling the classical–non-classical asymmetry of the correlation length, taking into account difference in definitions of *NDE* $\left(\Delta\varepsilon^E/E^2 = (\varepsilon\,(E) - \varepsilon)/E^2 = \left(\varepsilon_{||} - \varepsilon\right)/E^2\right)$, and $\Delta n^E/E^2 = \left(n_{|||||} - n_{\perp}\right)/E^2$ for *EKE* and taking into account the Fisher's scaling [37], one obtains [35]:

$$\psi\,(EKE) = 2\gamma - \left(\nu_{||} + 2\nu_{\perp}\right) \approx 2 \times 1.24 - (0.63 + 2 \times 0.5) = 0.85 \qquad (13)$$

$$\psi\,(NDE) = \gamma^{\text{nonclassical}} + \gamma^{\text{classical}} - \mathrm{d}\nu = 1.24 + 1.02 - 3 \times 0.63 = 0.37 \qquad (14)$$

or alternatively

$$\psi\,(NDE) = \gamma^{classical} - 2\beta = 1.02 - 2 \times 0.325 = 0.37 \qquad (15)$$

where the critical exponent $\gamma \approx 1.02$ denotes the susceptibility exponent within the mean-field approximation but with the logarithmic correction (0.02), important near the classical–non-classical crossover [29]. Values in relations (13–15) are in fair agreement with *NDE* experimental results for $T \to T_C$ [31]. For *NDE,* local (critical) fluctuations of the order parameter $\langle \Delta M^2 \rangle_V$ in the homogeneous phase are associated with one of the susceptibilities (compressibility χ_T) in Eq. (11) and with the volume of critical fluctuations ξ^d. Critical exponents describing their behavior are non-classical, as in the "normal" critical mixture, i.e., $\beta \approx 0.325$ (order parameter), $\gamma \approx 1.24$ (susceptibility related to the order parameter changes), and $\nu \approx 0.63$ (the correlation length). Under the strong electric field, critical fluctuations are elongated and oriented along lines of the strong electric field, for $\varepsilon_{\text{fluct.}} > \varepsilon_{\text{Solution}}$. On fluctuations acts the force, which is perpendicular to the lines of the electric field. This process is described the second susceptibility in Eq. (11), related to the perpendicular direction, which shows the classical, mean-field, behavior: with power exponents for the susceptibility $\gamma = 1$, for the order parameter $\beta = 1/2$, and for the correlation length $\nu = 1/2$. One can conclude that for critical anomalies of *NDE* and *EKE* essential is the non-symmetric deformation of fluctuations leading to the "mixed" criticality [35].

The important issue in the analysis of the critical effect of *NDE* is the precise estimation of the noncritical, purely molecular, background effect. The "reference solution" method, described above, is possible only if the background effect is associated with one of the components of the critical mixture, for instance, the dipolar one in nitrobenzene–hexane mixture. This method cannot be applied if both components significantly contribute to the background effect or for the gas–liquid critical point. In studies on critical systems, the background effect is often determined via the extrapolation of the behavior from the high-temperature region, well above T_C [29]. However, such approach cannot be applied for the critical anomaly of *NDE*, since for this method the impact of critical fluctuations can be significant even 50 K above T_C[30]. In ref. [38], the derivative-based analysis for estimating the critical contribution without a knowledge of the background effect in prior was proposed. The molecular background generally [see Eq. (11); Fig. 2] can be well approximated via the linear temperature dependence in the limited range of temperatures occurring for experimental data.

Hence, taking into account the experimental error, one can obtain the following relations:

$$\left(\Delta\varepsilon^E/E^2\right)(T) \approx \left[\frac{A_{NDE}}{T - T_C}\right]_{\text{critical}} + [(a + bT)]_{\text{background}} \tag{16}$$

$$\frac{\mathrm{d}\left(\Delta\varepsilon^E/E^2\right)}{\mathrm{d}T} = -\phi\, A\,(T - T_C)^{-\psi-1} + b \tag{17}$$

and subsequently

$$\begin{aligned}\frac{\mathrm{d}^2\left(\Delta\varepsilon^E/E^2\right)}{\mathrm{d}T^2} &= -\psi\,(-\psi - 1))\,A_{NDE}\,(T - T_C)^{-\psi-2}\\ &= \psi\,(\psi + 1)\,A_{NDE}\,(T - T_C)^{-\psi}\,(T - T_C)^{-2}\end{aligned} \tag{18}$$

After the rearrangement, one defines the plot via $(T - T^*)^2 \log_{10}\left(\mathrm{d}^2\left(\Delta\varepsilon^E/E^2\right)/\mathrm{d}T^2\right)$ versus $\log_{10}(T - T_C)$ at which for *NDE* critical effect a linear dependence with the slope $b = -\psi$ and the interception $a = \log_{10}[\psi\,(\psi + 1)\,L_0]$: $\psi\,(\psi + 1)\,L_0 = 10^a$ should appear. The critical contribution and the total measured *NDE* obtained in this way make it possible to estimate (calculate) also the background effect. Such analysis was applied for results presented in Fig. 7, showing the behavior of the critical effect in the homogeneous phase of a critical mixture. Figure 7 shows also the impact of the measurement frequency, which is discussed at the end of this paragraph. Figure 8 presents that the form of the critical *NDE* pretransitional effect in a critical mixture is the same (isomorphic) for the pressure and temperature paths. However, Fig. 8 shows the unique case of nitrobenzene–hexane critical mixture tested above the critical consolute temperature under atmospheric pressure: in the formally exclusively homogeneous region. The observed anomaly is due to the critical point hidden in the negative pressures domain [40] and located on the extension of $T_C(P)$ curve into this region. The critical effect is well portrayed by the relation $\left(\Delta\varepsilon^E/E^2\right)_{crit} = A_{NDE}/\left(P - P_C\right)^{\psi}$: in this case $P_C < 0$, yielding the possibility of estimating hidden $T_C\,(P)$ location.

Piekara discovered the strong pretransitional increase of *NDE* when cooling toward the critical consolute temperature. He also observed the impact of the vicinity of the critical consolute point in isothermal, concentrational studies (Fig. 2) [5–7]. Figure 9 presents first results of such behavior for few isotherms, including the path located in immediate vicinity of the critical consolute temperature, from $T = T_C + 10\,\mathrm{K}$ to $T = T_C + 0.02\,\mathrm{K}$ [11].

Figure 10 presents the analysis of the critical behavior for the isotherm $T = T_C + 0.02\,\mathrm{K}$, basing on data from Fig. 9. Linking relations for the temperature-related *NDE* critical anomaly $\Delta\varepsilon^E/E^2 = A\,(T - T_C)^{-\psi}$ and for the order parameter $M = B_o\,|x_C - x|$, where x denotes the concentration along the coexistence curve for $x > x_C$ and $x < x_C$ branches, one obtains

$$\frac{\Delta\varepsilon^E}{E^2} \propto |x_C - x|^{-\psi/\beta} \tag{19}$$

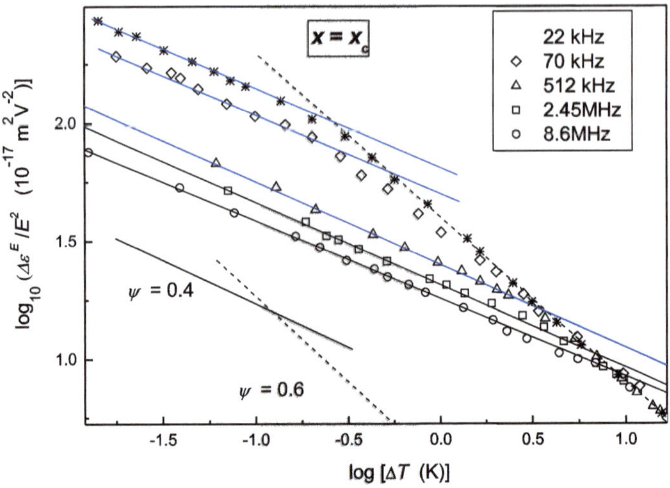

Fig. 7 The temperature evolution of *NDE* on approaching the critical consolute point in the homogeneous phase of nitrobenzene–hexane mixture. Results are for few selected measuring frequencies, indicated in the figure. For the applied scale, the critical behavior is visualized via the linear dependence: the slope is related to the critical exponent. Based on reanalyzed results from ref. [11]; see also ref. [39]

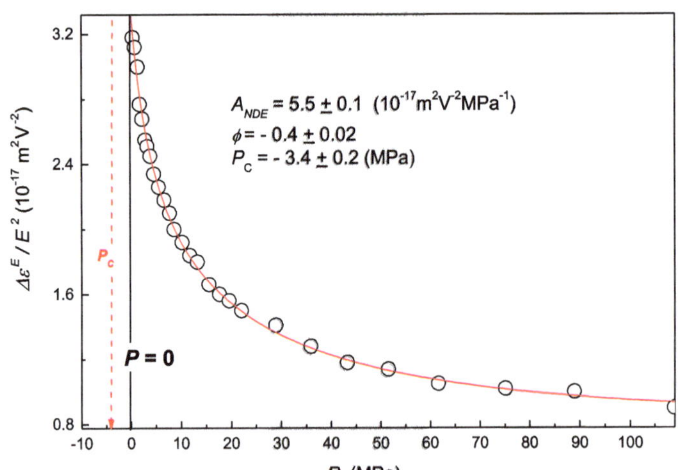

Fig. 8 The *NDE* pretransitional effect in the homogeneous phase of nitrobenzene–hexane critical mixture for the isotherm $T = T_C + 0.5$ K. Results are for the pressure path. The pressure counterpart of Eq. (5) was used for the parameterization. Critical exponents are the same for temperature and pressure paths of approaching the critical point, according to the isomorphism of critical phenomena [29]. Note that for the given mixture $\mathrm{d}T_C/\mathrm{d}P < 0$. The pretransitional effect is due to the critical point hidden in negative pressure's domain. Based on results from ref. [28]

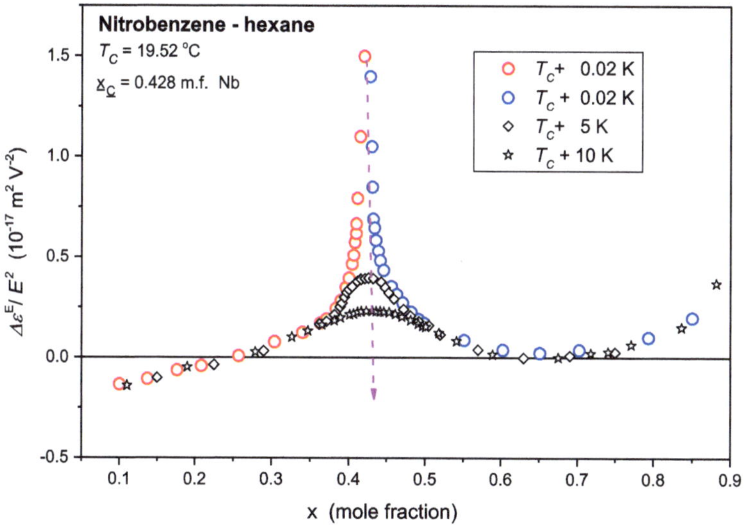

Fig. 9 The isothermal, "concentrational" behavior of nonlinear dielectric effect in the isotropic phase of nitrobenzene–hexane critical mixture ($T_C = 19.5\,°\mathrm{C}$ and $x_C = 0.43$ m. f. of nitrobenzene) for selected isotherms in the homogeneous liquid phase, above T_C. The dashed arrow shows the "propagation" of the maximal impact of the critical point into the homogeneous phase. [11]

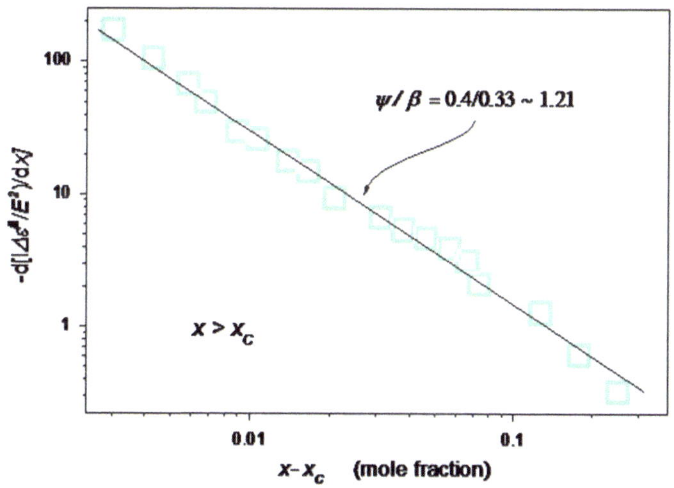

Fig. 10 The critical part of *NDE* for the isotherm $T = T_C + 0.02$ K in the log–log scale to show the validity of parameterization via Eq. (19) [11]

Taking $\psi = 0.39$ and $\beta = 0.33$, i.e., effective values of critical exponent when neglecting the correction-to-scaling term one obtains $\psi/\beta \approx 1.18$—in the fair agreement with results presented in Fig. 9.

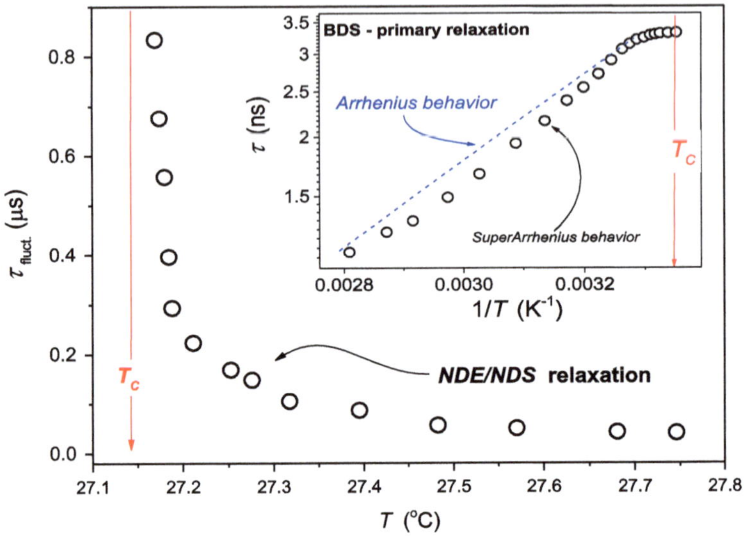

Fig. 11 The inset shows the primary ("structural") relaxation time in the homogeneous phase of a critical mixture of limited miscibility (nitrobenzene–dodecane mixture). The solid (blue line) shows the simple Arrhenius evolution, whereas the distortion from such dependence shows the Super-Arrhenius behavior. The critical effects manifest as a distortion in the immediate vicinity of T_C. The main part of the figure shows the "nonlinear" relaxation time obtained from frequency-related *NDE* studies, well portrayed via Eq. (20). The vertical arrow indicates the critical consolute temperature [11]

The analysis of the *NDE* critical behavior in Fig. 7 reveals two "critical" domains, associated with different values of the critical exponent: (i) in the immediate vicinity of T_C: $\psi \approx 0.39$ and (ii) remote from T_C: $\psi \approx 0.6$. The latter one is in agreement with theoretical predictions from refs. [25, 32, 33], where the "mixed critical" is absent. The temperature of the crossover between domains (i) and (ii) depends on the frequency of the weak measuring field. To explain such phenomenon, one should take into account the presence of two timescales in the dual-field *NDE* studies on critical liquids. The first one is the "sampling" timescale which can be estimated as $\tau_{\text{meas.}} = 1/f_{\text{meas}}$. In practice, the convenient range of frequency is between 20 kHz and 20 MHz, and then the timescale: 50 ns $< \tau_{\text{meas.}} <$ 50 µs. The lifetime of critical fluctuation defines the system timescale [29]:

$$\tau_{\text{fluct}} = \frac{\tau_0}{(T - T_C)^{z\nu}} \propto \frac{1}{(T - T_C)^{1.9}} \tag{20}$$

where $z = 3$ is the dynamic (critical) exponent for the non-conserved order parameter, and the exponent for the correlation length $\nu \approx 0.63$ (the effective value when neglecting correction-to-scaling). Then, for critical mixtures the exponent $z\nu \approx 1.9$ [29, 40].

The system timescale extends between $\tau_{\text{flukt.}} \rightarrow \infty$ for $T \rightarrow T_C$ and sub-nanosecond values for ... For $\tau_{\text{meas.}}/\tau_{\text{fluct.}}/ < 1$, the sampling time is much faster than the lifetime of fluctuations. Consequently, one can detect the "mixed criticality" of fluctuations related to the exponent $\psi \approx 0.37 \div 0.4$. For $\tau_{\text{meas.}}/\tau_{\text{fluct.}} > 1$, several fluctuations appear and disappear during the measurement cycle and one cannot detect their specific features. Consequently, *NDE* measurements detect the average from several fluctuations, what "masks" the mixed criticality and yields the non-classical critical exponent $\psi \approx 0.57 \div 0.6$. Figure 11 shows the evolution of the lifetime of critical fluctuation in nitrobenzene–hexane critical mixture determined from crossover condition $T\ (\tau_{\text{meas.}}/\tau_{\text{fluct}} = 1)$ in frequency-related *NDE* studies. In agreement with Eq. (20), $\tau_{fluct.} \rightarrow \infty$ for $T \rightarrow T_C$.

When testing dynamics via the evolution of the primary (structural, alpha) relaxation time, i.e., estimated from the peak frequency of the primary loss curve $\varepsilon''(f)$ via $\tau = 1/2\pi f_{\text{peak}}$, the largest value of the primary relaxation $\tau(T_C) \propto 4\,\text{ns}$ is reached. Above T_C, the evolution is clearly Super-Arrhenius (SA) [40], with weak distortion in the immediate vicinity T_C. Hence, the qualitative difference between the "linear" ($\tau\ (T \rightarrow T_C) \approx 5\,\text{ns}$) and "nonlinear" ($\tau_{\text{flukt.}}\ (T \rightarrow T_C) \rightarrow \infty$) takes place. It results from the fact that *NDE* is directly coupled to critical fluctuations, i.e., "heterogeneities" which sizes and lifetimes grow up infinitely on cooling toward the critical point. The primary ("linear") relaxation time is linked to the relaxation of a dipole moment of single molecules in the surrounding gradually modified by developing critical fluctuations.

The application of the dual-field *NDE* experimental technique enables addressing one of the most basic problems regarding the dynamics of critical fluctuations: is it homogeneous or heterogeneous?

Figures 12 and 13 present the preliminary evidence supporting the heterogeneous picture. It is worth recalling that in critical mixtures the decay after switching-off the strong electric field is stretched exponential and universal as shown in ref. [34] for EKE:

$$R(t) \propto \exp\left[-\left(\frac{t}{\tau_{\text{fluct.}}}\right)^{x}\right], \tag{21}$$

where the normalized decay after switching-off the strong electric field: $R(t) = \left[\left(\Delta\varepsilon^{E}/E^{2}\right)(t)\right]/\left[\Delta\varepsilon^{2}/E^{2}\right]_{\text{saturation}}$, t is the elapse time after switching-off the electric field, the universal stretched exponent (SE) $x = (2-\eta)/(5-\eta)$ and $\eta \approx 0.04$ is the critical exponent for the correlation length. For the relaxation time $\tau_{\text{flukt.}} \propto \xi^{z} \propto (T - T_C)^{-z\nu} \approx (T - T_C)^{-1.9}$, in agreement with Eq. (20) [34].

The result presented in Fig. 12 clearly confirms the universal, system independent, and stretched-exponential pattern of the *NDE* decay after switch of the DC strong electric field pulse, in agreement with the evidence given in refs. [34, 40]. However, for the selective excitation of critical fluctuations within the homogeneous critical mixture in the immediate vicinity of the critical point via the sine-wave pulse with the given frequency of the strong electric field, one obtains the single exponential decay. It is associated only with one relaxation time. Such behavior is clearly present

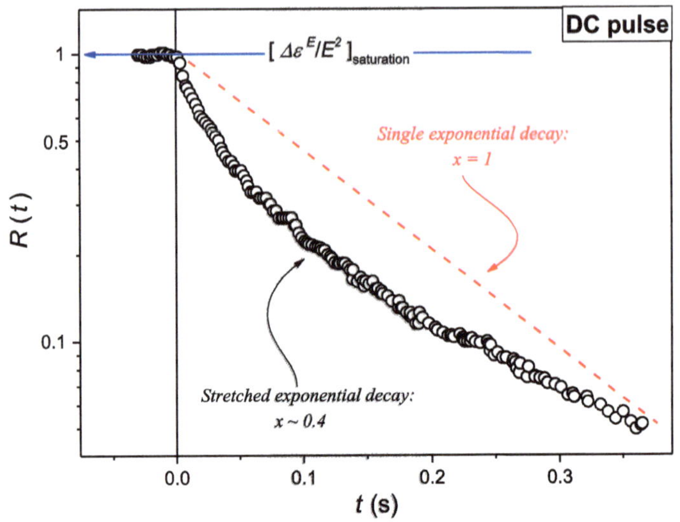

Fig. 12 The decay of *NDE* signal after switching-off the DC pulse of the strong electric field, in nitrobenzene–dodecane critical mixture. Tests for the pressure path of approaching the critical consolute point: the presented result is for the isotherm $T = T_C$ ($P = 0.1$ MPa) and for $P = P_C - 1\ MPa$. The decay is described via the response function Eq. (21) with the universal "stretching exponent" (SE) $x = 0.4 \pm 0.02$. The dashed (red) line shows the "reference" simple Arrhenius behavior [11]

in Fig. 13. In the opinion of the author, Figs. 12 and 13 can be considered as the evidence for the heterogeneous dynamics of critical fluctuations.

Results presented in Figs. 12 and 13 were obtained for the pressure path of approaching the critical consolute point. For the temperature approaching under atmospheric pressure, reaching similar (long) values of the decay time are possible only extremely close to the critical point: $T - T_C < 0.02$ K. Additionally, such experiment requires also extraordinary temperature stabilization. The situation becomes experimentally convenient for the pressure path, due to the fact that for nitrobenzene–decane mixture $\mathrm{d}T_C/\mathrm{d}P = 0.003\,\mathrm{K}/\mathrm{MPa}$. Consequently, for the pressure path, one can much easier reach the very immediate vicinity of (T_C, P_C). It is notable that one can change pressure in 0.1 MPa steps, what is the equivalent of 0.003 K (!) [41].

4 Nonlinear Dielectric Effect in Supercooled Nitrobenzene

Results presented above-recalled topics originating from the first experimental evidences by Arkadiusz Piekara in the 30s [5–10]. Piekara also noted the "inverse" $NDE > 0$ in nitrobenzene and developed the successful molecular explanation of this phenomenon, as indicated in Figs. 1 and 2 [10]. New factors which can influence

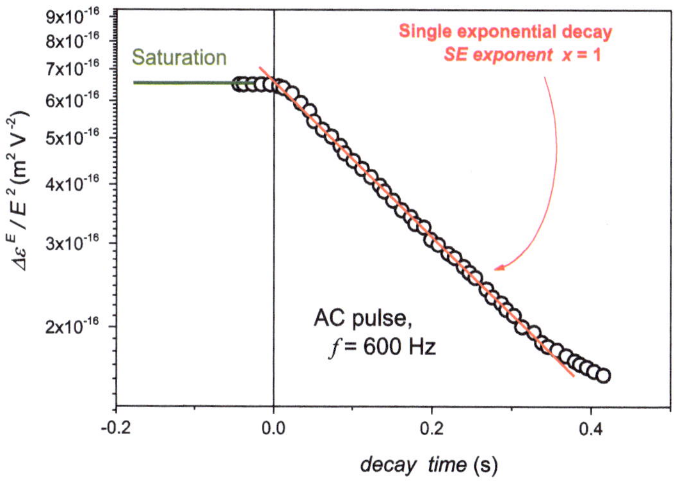

Fig. 13 The decay of *NDE* after switching the AC pulse of the strong electric fields in the immediate vicinity of the critical consolute point in nitrobenzene–decane mixture for the isotherm $T = T_C$ ($P = 0.1$ MPa) and for $\Delta P = 1$ MPa as the distance from the critical point. The strong electric field was applied as the AC pulse, in the form of the sinusoidal wave with the frequency $f = 600$ Hz. In the semi-log scale, the line (in red) indicates the single exponential decay, i.e., $x = 1$ in Eq. (21) [11]

this phenomenon have been noted only recently [38]. This was possible due to *NDE* studies in the broad range temperature, including the supercooled region of nitrobenzene. One should stress that nitrobenzene most easily crystallizes at $T_m \approx 6.5\,°\mathrm{C}$ and its supercooling requires (very) careful cleaning, degassing, and using a specially prepared measurement capacitor. Figure 14 presents results obtained in temperature-related NDE studies.

Five decades ago, Hanus [42] developed the semi-phenomenological mean-field model suggesting that for molecular liquids which do not exhibit a liquid crystalline polymorphism a phase transition to a partially aligned nematic-like mesophase may occur under a strong electric field. The possible appearance of such phenomenon was suggested for molecular liquids with interacting molecules and a relatively high melting temperature, such as nitrobenzene [42]. Although ref. [42] focused on the Kerr effect, the parallel relation for *NDE* can be easily derived, namely:

$$\frac{\Delta n}{E^2}, \frac{\Delta \varepsilon^E}{E^2} \propto F'(\varepsilon)\, C \frac{\Delta\alpha \Delta\alpha'}{T - T^*}, \quad T > T_m^1 \tag{22}$$

where $F'(\varepsilon)$ is the local field factor, for *NDE* $C = 16\pi/45k_B$, $\Delta\alpha$ and $\Delta\alpha'$ are anisotropies dielectric polarizabilities linked to the strong electric field inducing anisotropy (yielding the "prolate", uniaxial structure) and the weak measuring field, respectively. T^* denotes the lowest temperature to which the liquid can be hypothetically supercooled and T_m^1 is the one-dimensional freezing/melting temperature at which a discontinuous transition occurs.

Soon after the report of Hanus [42], de Gennes [43, 44] published results of the Kerr effect and the Cotton–Mouton effect studies in the isotropic phase of a rod-like liquid crystalline material, MBBA, for which he noted:

$$\frac{\Delta n^E}{E^2}, \frac{\Delta n^E}{H^2} \propto \frac{\Delta n \Delta m}{T - T^*} = \frac{A}{T - T^*}, \quad T > T^C = T^* + \Delta T \tag{23}$$

where Δn is the molecular anisotropy of the refractive index and $\Delta m = \Delta\varepsilon$ for the Kerr effect, $\Delta\varepsilon$ is the molecular anisotropy of the dielectric constant, $\Delta m = \Delta\chi_H$ is the molecular anisotropy of the magnetic susceptibility, relevant for the Cotton–Mouton effect. T^* is the temperature of the hypothetical continuous phase transition, i.e., the temperature to which the isotropic liquid can be supercooled. T^C is the clearing temperature, i.e., the temperature of isotropic–nematic (I–N) weakly discontinuous phase transition at which freezing associated with one-dimensional orientation takes place.

The inspiration of above experimental results led to the formulation of the Landau–de Gennes (LdG) model [44], one of the key theoretical tools in the physics of liquid crystals [45] and in the soft matter physics [46]. Stinson and Litster [47] linked to the pattern of Eq. (23) the intensity of the scattered light on approaching the I–N transition: in this case, the amplitude $A \propto \Delta n^2$. In 1992, Rzoska et al.

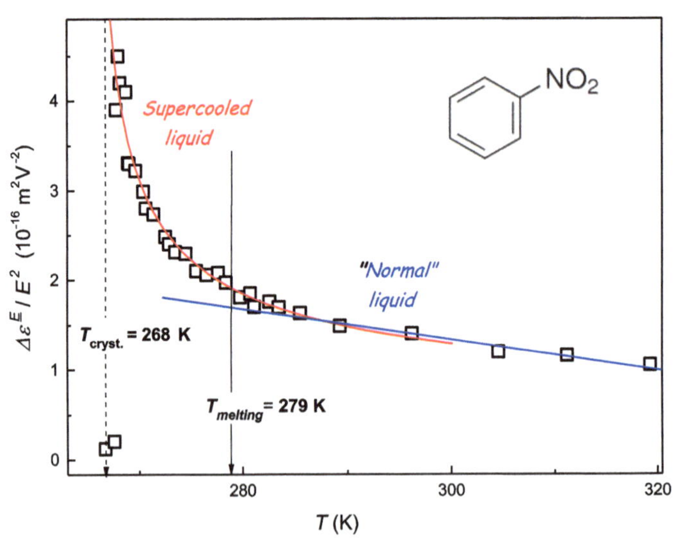

Fig. 14 Temperature behavior of the nonlinear dielectric effect in the liquid, supercooled, and solid states of nitrobenzene. The form of nitrobenzene molecule is shown. The crystallization took place at $T_{L-S} \approx 267.1$ K. The straight line shows the possibility of portrayal via the linear dependence above the melting temperature. Based on results from ref. [38]

[48] applied the Landau–de Gennes model for obtaining the parallel of Eq. (23) also for *NDE*. In this case, the amplitude $A \propto \Delta\varepsilon^f \Delta\varepsilon$, where $\Delta\varepsilon^f$ is the anisotropy of dielectric permittivity for the measurement radio frequency and $\Delta\varepsilon$ is the anisotropy of dielectric constant, related to the strong electric field. Notable is the striking similarity of relations (23) and (22), although the latter was focused on the impact of the strong electric field on non-mesogenic liquids. It is notable that estimations of the intensity of the electric field which can already induce the quasi-nematic structure in nitrobenzene given by Hanus [42] coincide with intensities applied in *NDE* studies based on the dual-field principle: $E \sim 10\,\text{kV/cm}$. For such liquid as CS_2, Hanus [42] suggested the necessity of one decade higher intensities of the electric field to induce similar phenomena. For anisotropic rod-like molecules, for which the orientational freezing in the nematic phase is the inherent feature, the pretransitional behavior predicted by Eqs. (22) and (23) occurs at arbitrary intensity of the strong field (electric, magnetic). The problem which remains is the description of the quasi-critical increasing of *NDE* in the supercooled nitrobenzene. The derivative-based analysis (Eqs. 16–18) of results from Fig. 14 showed that [38]:

$$\frac{\Delta\varepsilon^E}{E^2} = \frac{A}{(T - T^*)^{1/2}} + [a + bT]_{\text{bckg}} \tag{24}$$

Hence, for nitrobenzene $\psi \approx 1/2$, instead $\psi = \gamma = 1$ as in Eqs. (22) and (23). It is notable that such value can be obtained from the dependence defining the *NDE* behavior on approaching the critical consolute point, assuming the dimensionality $d = 3$ and the mean-field, tricritical, values of critical exponents:

$$\psi = d\nu - \gamma = 3 \times 0.5 - 1 = 1/2 \text{ and } \psi = \gamma - 2\beta = 1 - 2 \times 0.5 = 1/2 \tag{25}$$

where the value $\beta = 1/4$ is the order parameter exponent for the tricritical case (TCP), for which $d = 3$ is the border dimensionality between the non-classical and classical descriptions.

5 Nonlinear Dielectric Effect in Liquid Crystals

The first evidence for the pretransitional behavior of *NDE* in a liquid crystalline (LC) material was obtained by Małecki and Zioło [49] for the isotropic phase of MBBA [*N*-(4-methoxybenzylidene)-4-butylaniline]. This is one of the oldest "classical" liquid crystalline materials with the isotropic—(320 K)–nematic—(295 K)–Solid mesomorphism, and the transverse dipole $\mu \approx 2.2D$ [45]. Figure 15 presents the authors' measurements directly recalling ref. [49].

The same plot contains also results of the next classical rod-like compound: 5CB (4- n-pentyl-4'-cyanobiphenyl) also with the isotropic—(308 K)–nematic—(297)–solid mesomorphism, but with the longitudinal dipole moment $\mu \approx 4D$ [45]. The inset in Fig. 15 shows reciprocals of results

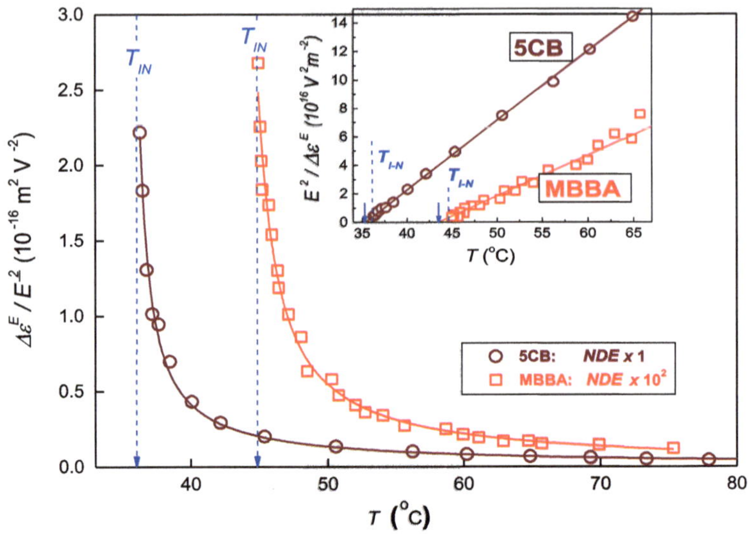

Fig. 15 *NDE* in the isotropic phase of liquid crystalline MBBA and 5CB: results are for the measurement frequency $f = 1$ MHz. The inset shows reciprocals of experimental data from the main part of the plot: dashed arrows show the isotropic–nematic (I–N) discontinuous transition and the solid ones the extrapolation indicating the hypothetical continuous phase transition. Discontinuities of the I–N transitions: $\Delta T = T_{IN} - T^* \approx 1.1$ K for MBBA and $\Delta T = T_{IN} - T^* \approx 1.2$ K for 5CB [11]

from the main part of the plot, confirming the validity of Eq. (23). It is visible that the form of pretransitional effects is the same for both compounds. However, the evolution of dielectric constant in the isotropic phase of MBBA and 5CB is qualitatively different. For the latter, the change from $d\varepsilon/dT < 0 \to dT_C/dT > 0$ for $T \to T^C$ occurs, as shown in Fig. 16. The pretransitional anomaly can be well portrayed by the relation resembling the one known for the homogeneous phase of critical, binary mixtures [50]:

$$\varepsilon(T) = \varepsilon^* + a\left(T - T^*\right) + A\left(T - T^*\right)^{1-\alpha}, \quad \text{for} \quad T > T_{I-N} \tag{26}$$

where the exponent $\alpha = 1/2$ is the exponent for the specific heat, and the internal energy, pretransitional behavior. (ε^*, T^*) denote the extrapolated location of the hypothetical continuous phase transition and $T^* = T_{I-N} - \Delta T$ where ΔT is the measure of the I–N phase transition discontinuity. T_{I-N} is for the I–N transition temperature: it is also known as the clearing temperature T^C which is generally related to the isotropic–mesophase transition.

This behavior can be associated with the growing number of 5CB molecules in prenematic fluctuations, which have to exhibit such basic feature of the nematic phase as the nematic/orientational ordering of rod-like molecules and the equivalence of $\vec{n}$ and $-\vec{n}$ directors [45], leading to the cancellation of dipole moments within fluctua-

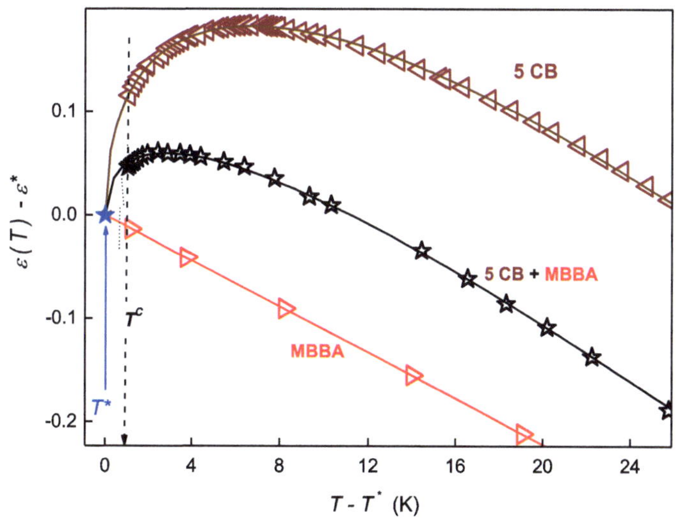

Fig. 16 The behavior of dielectric constant (measurement frequency $f = 100$ kHz) in the isotropic phase of MBBA and 5CB, for $T \to T_{I-N}$. Note the lack of the pretransitional effect for MBBA and the notable pretransitional effect in 5CB, portrayed by Eq. (26). Based on reanalyzed results from ref. [51]

tions. The latter causes that for 5CB the dielectric constant of prenematic fluctuations is notably smaller than for the "isotropic, fluidlike" surrounding: consequently, the total value of dielectric constant decreases on cooling toward T_C. For MBBA, with the transverse dipole moment such mechanism is absent and the dielectric constant of prenematic fluctuations and their fluidlike surrounding are the same. Consequently, for MBBA, there is no pretransitional anomaly for $\varepsilon\,(T \to T_C)$ due to the lack of the contrast factor between fluctuations and the fluidlike surrounding. Regarding *NDE*, it is directly coupled to fluctuations and consequently pretransitional effects for both 5CB and MBBA are strong and have the same form.

Regarding dynamics in the isotropic phase of liquid crystalline materials, there is the qualitative difference between the "linear" (i.e., related to BDS studies: $\varepsilon'(f)$, $\varepsilon''(f)$ and the primary relaxation time) and the "nonlinear" (*NDE, EKE*) case. Figure 17 presents the temperature evolution of the primary relaxation time, determined from the peak frequency of dielectric loss curves from BDS $(\varepsilon'(f), \varepsilon''(f))$ studies. The obtained Super-Arrhenius (SA) behavior can be effectively described by the Vogel–Fulcher–Tammann dependence [40], but the optimal parameterization yields the quasi-critical function [53–55]:

$$\tau\,(T) = \tau_o\,(T - T_X)^{-\phi} \tag{27}$$

with the exponent $\phi \approx 2.3$ and the singular temperature $T_X \approx T_{I-N} - 26$ K.

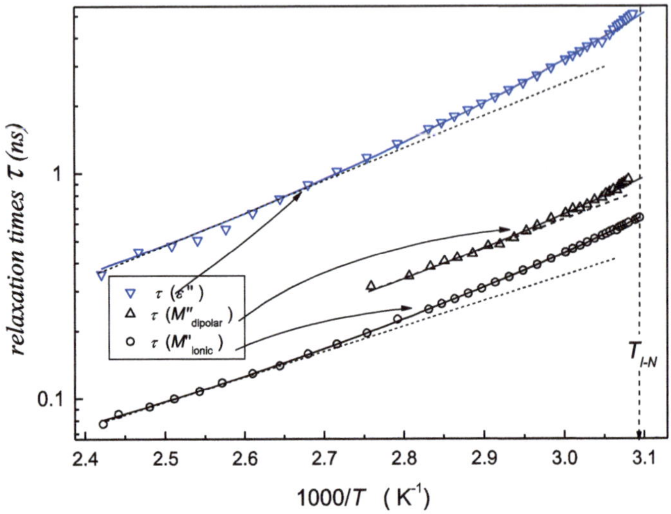

Fig. 17 The temperature evolution of the relaxation time from the BDS measurements in isotropic 9CB: "the linear case". Apart from the primary relaxation time (in blue, the upper plot), the behavior in the modulus (M) representation giving the direct insight into translation-related processes is given. In each case, the SA behavior takes place [11]

The evolution of relaxation time determined from *NDE* measurement is directly coupled to prenematic fluctuations, i.e., internally ordered heterogeneities in the isotropic surrounding, which is given by [40, 45]

$$\tau_{\text{fluct.}} = \tau_0 \left(T - T^*\right)^{-y} = \frac{\tau_0}{T - T^*} \tag{28}$$

where $y = z\nu = 1$; $z = 2$ is the dynamic exponent for the conserved order parameter and the "classical" value of the correlation length exponent $\nu = 1/2$. For this relation, the singular temperature $T^* = T_{I-N} - \Delta T$ and $\Delta T \approx 1.2$ K (for 5CB) and $\Delta T \approx 3$ K for 9CB.

Experimentally, the pretransitional behavior associated with Eq. (28) can be directly detected from *NDE* measurements via the "crossover" analysis comparing the measurement timescale and the system timescale, as in the case of critical, binary mixtures described above. The obtained in this way evidence, in fair agreement with Eq. (28), is presented in Fig. 18.

Figure 19 shows the pretransitional effect in the isotropic phase of 5CB and its chiral isomer 5 * CB (isopentylcyanbiphenyl). Structures of both compounds are given in Fig. 17. Both 5CB and 5 * CB have the same permanent dipole moments and according to the output relation resulted from the basic implementation of the Landau–de Gennes model, their pretransitional effects on approaching the I–N (5CB) and the I–N* (5 * CB) transitions should be the same [40]:

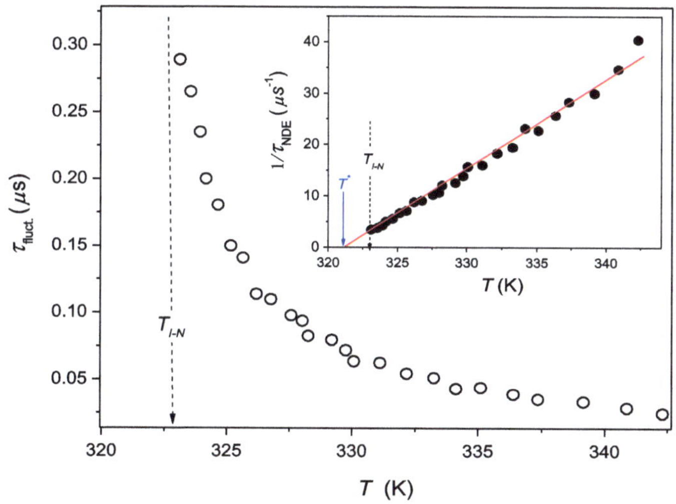

Fig. 18 The "nonlinear" relaxation time resulted from *NDE* measurements, directly linked to prenematic fluctuations. The inset shows the reciprocal of experimental data from the main part of the plot showing the validity of Eq. (28) with the power exponent $y = 1$ [11]. Results for the isotropic phase of nonylcyanobiphenyl (9CB)

$$\frac{\Delta\varepsilon^E}{E^2} = C\chi_0 \frac{\Delta\varepsilon^0 \Delta\varepsilon^f}{T - T^*} = \frac{A_{NDE}}{T - T^*} \tag{29}$$

where C is the model constant, χ_0 is the amplitude of compressibility, $\Delta\varepsilon^0$ is the molecular anisotropy of dielectric constant in the zero-frequency limit (related to the strong electric field), and $\Delta\varepsilon^f$ is the molecular anisotropy of dielectric constant in the zero-frequency limit (related to the weak, measuring electric field).

When discussing the qualitative difference of *NDE* pretransitional effect in 5CB and 5 * CB, one should indicate that the experimentally measured largest value of the relaxation time related to fluctuations in 5CB: $\tau_{flict.}(T_{I-N}) \sim 1\,\mu s$. Similar value can be expected for 5 * CB. Hence, when changing the *NDE* measurement frequency from 30 kHz to 12 MHz, as in Fig. 17, the clear crossover from the domain (A) $t_{meas.}/\tau_{fluct} > 1$ (remote from T^C) to domain (B) $t_{meas.}/\tau_{fluct} < 1$ (close to T^C) occurs. The arrangement of molecules in premesomorphic fluctuation can be expected as shown schematically in Fig. 19. For 5CB, one can expect the prenematic ordering, with weak distortion from the "ideal" nematic arrangement due to the relative high temperatures ($T > T^C$) and large susceptibility to perturbations. Notwithstanding, the "cancellation" of permanent dipole moment within fluctuations, as described above, can be expected. For 5 * CB, such cancellation is not possible. The "steric" restriction causes that for molecules within fluctuations there are two notable components of the dipole moment with respect to the direction of the director $\vec{n}$: $\mu = \mu_{||} + \mu_{\perp}$. Only the parallel can be cancelled within the fluctuations. Consequently, the negative dielectric anisotropy of the premesomorphic

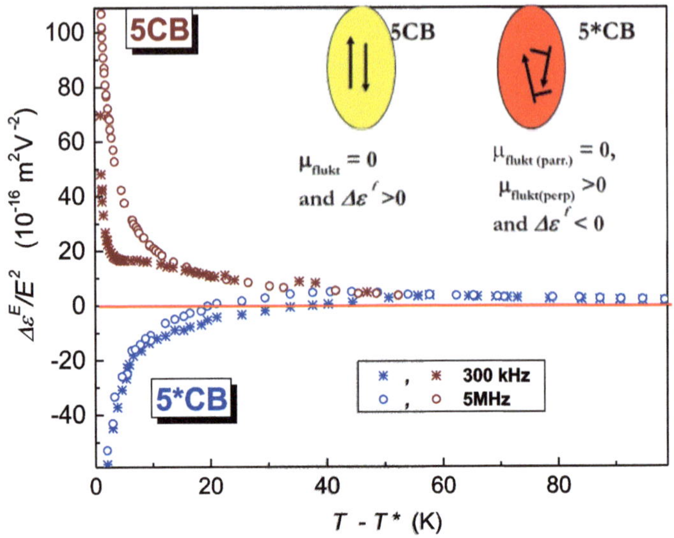

Fig. 19 Nonlinear dielectric effect in the isotropic phase of pentylcyanobiphenyl (5CB) and isopentylcyanobiphenyl (5 * CB), on approaching the clearing point, i.e., $I \rightarrow N$ and $I \rightarrow N^*$(N^*—chiral nematic) phase transitions, respectively. Results are for few frequencies of the weak measuring field given in the figure. The inset shows schematically the structure of 5CB and 5 * CB molecules and its consequence on properties on prenematic fluctuations. Based on reanalyzed and supplemented results from ref. [52]

fluctuations appears. For $T \rightarrow T^*$, the appearance and disappearance of prenematic fluctuations are described by τ_{fluct} as the system timescale takes place. In domain (A), the system timescale is much faster than the measurement process (t_{meas}) and specific features of a single fluctuation cannot be detected. This happens for any applied measurement frequency well above the clearing temperature. For the lowest frequency ($f = 30$ kHz), such condition occurs at any temperature, even $T = T_C$. Under such conditions, *NDE* is positive and described by the mean-field dependence with $A_{NDE} \propto \left(\Delta\varepsilon^0\right)^2$. Heuristically, one can claim that the "ideal" mean-field LdG model behavior is reached due to the applied "detection timescale". However, for higher measurement frequencies (~MHz), the crossover to the domain (B) is possible. The detection process may be faster than the lifetime of fluctuations and their specific features can be directly observed: in the case of 5 * CB this means the detection of the negative dielectric anisotropy of fluctuations. All these can lead to the indication of the alternative way of description of the pretransitional effect in the isotropic phase resulting from Eq. (3):

$$\frac{\Delta\varepsilon^E}{E^2} = C\chi_0 \frac{\langle\Delta M^0\rangle\langle\Delta M^f\rangle}{T - T^*} \tag{30}$$

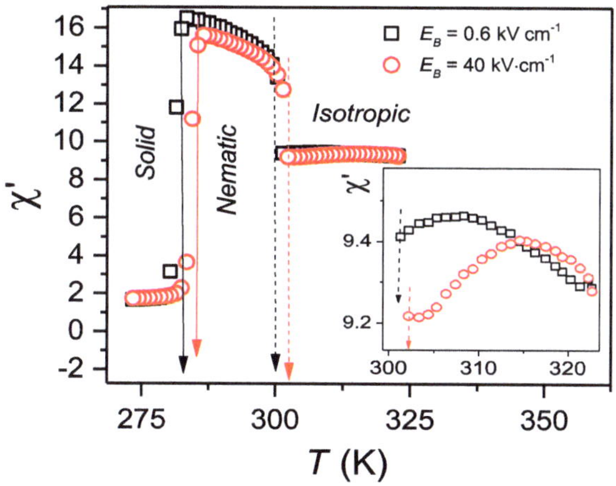

Fig. 20 Results of measurement of dielectric constant ($f = 10$ kHz) for rodlike nematogenes' mixure under two different biasing DC electric fields, as given in the figure. Arrows indicated subsequent phase transitions [11]

Such relation takes as the reference the averaged local fluctuations of the order parameter, which can coincide with the LdG based on Eq. (29) for the ratio of timescale as in the domain (A).

It is notable that for decades the orientation of permanent dipole moment described by the Herweg–Debye model (Eq. 1) was considered as the only source of the negative sign contribution to *NDE* [10]. The above discussion shows the new source of the negative sign contribution to *NDE*—beyond this paradigm.

Finally, we would like to address the question of the impact of the strong electric field on the pretransitional effect and structured mesoscale heterogeneities—fluctuations. Figure 20 shows the behavior of dielectric constant in a rod-like liquid crystalline sample of hexylcyanobiphenyl (6CB), from the same homologous group as 5CB. Measurement was carried out under different biasing fields. We noted the notable change in the form of the pretransitional anomaly when increasing the intensity of the electric field well above 10 kVcm^1. For the highest applied DC electric field, the impact on dielectric constant of the tested sample was dramatic: it can notably change the form of pretransitional effect and then also the form of heterogeneities—fluctuations. Moreover, in the solid phase, a clear pretransitional effect appeared, absent when $E_{\mathrm{Bias}} \to 0$ emerges. Notable is the shift of phase transition temperatures. Properties of 5CB and 6CB liquid crystalline materials can be considered as complex fluids in which properties are strongly influenced by pretransitional fluctuations/heterogeneities resulted from the weakly discontinuous nature of the I–N transitions. It seems that above some threshold value of the electric field, slightly lower than 40 kV/cm in the given case, one should consider rather the new state induced by the electric field than *NDE* associated with a relatively weak distribution for which $\Delta\varepsilon^E \propto E^2$.

6 Conclusions

After the World War II, Arkadiusz Piekara continued studies on the nonlinear dielectric effect, first at the University in Poznań (nowadays Adam Mickiewicz University, Poznań, Poland) and since 1964 at the Warsaw University (Warsaw, Poland) [7]. In the late 70, one of his former key students, August Chełkowski, moved to the University of Silesia (Katowice) which was just organized. He was accompanied by the young assistant Jerzy Zioło, who continued *NDE* studies in Katowice. Since the mid-80s, the *NDE* lab in Katowice was developed by Sylwester J. Rzoska, former student of Zioło, investigating mainly on critical mixtures. In the mid of the 90s to the staff joined Aleksandra Drozd-Rzoska who focused on liquid crystals. In the last decade Szymon Starzonek joined to the team. From several years, the authors of this report have continued *NDE*, *NDS*, and *EKE/TEB* studies in the Institute of High Pressure Physics PAS in Warsaw.

This contribution shows the progress in studies on various types of critical liquids, returning finally to the basic case of nitrobenzene where the "anomalous, positive" value of *NDE* also can be associated with the critical-like pretransitional behavior. It is shown that *NDE*-related phenomena in the homogeneous phase of critical mixtures and in nitrobenzene, as well as in the isotropic phase of rod-like liquid crystal, can be described in a synergic and coherent way when taking into account the impact of pretransitional fluctuations associated with continuous or weakly discontinuous phase transitions. Nonlinear dielectric studies started from the insight into variety of molecular processes in "normal" liquid dielectrics: they are summarized in the monograph by August Chełkowski [10]. From the early 80s, grown up the evidence related to critical mixtures and later to liquid crystals. These studies explored and developed the classical *NDE/NDS* measurement concept which recalls classical studies by Herweg and Piekara [1–10]. It is based on the application of two electric fields: the weak, measuring, with radio frequency and the intensity $E_{\text{meas.}} \sim 1\,\text{kV/m}$ and the strong one $E \sim 1\,\text{MV}\big/\text{m}$. The latter means the voltage $U = 1\,\text{kV}$ for the often used gap of the flat-parallel capacitor $d = 1\,\text{mm}$. The notable feature of the dual-field method is the large sensitivity making it possible to detect effect as small as $\Delta\varepsilon^E/E^2 \sim 10^{-19}\left(\text{m}^2/\text{V}^2\right)$, even for the mentioned above "bulk" gap of the capacitor. It is notable that the dual-field *NDE/NDS* can be considered as the clear counterpart of the electro-optic Kerr effect (transient electric birefringence: *TEB*, *EKE*) but for radio frequencies (kHz–MHz domain). The current state-of-the-art of the dual-field nonlinear dielectric spectroscopy is presented in the appendix.

Nowadays, the key focus of the nonlinear dielectric is located within the domain of the glass transition and in this case mainly single-field *NDS* techniques of measurements are developed. These issues are broadly discussed in other chapters of this book. Results related to the dual-field *NDE/NDS* in glassforming liquids can be found in ref. [56]. The nonlinear dielectric spectroscopy and the nonlinear dielectric effect seem to the natural extension of *BDS*. The latter yields the direct insight to single-molecular processes. *NDE/NDS* can directly detect multimolecular species and processes.

Acknowledgements The preparation of contribution by National Centre for Science (NCN, Poland) via the grant ref. 2016/21/B/ST3/02203 (A. Drozd-Rzoska and S. Starzonek) and grant ref. 2017/25/B/ST3/02458 (Sylwester J. Rzoska).

Appendix

The Dual-Field *NDE* Experimental Technique

There are few principles for measuring changes of dielectric properties under strong electric. This chapter is associated with the dual-field technique which recalls classical experiments by Herweg (1920/1922) [2, 3] and Piekara (1932/1937). Those studies focused on molecular liquids, such as diethyl ether (DEE) or solutions [5–10] nitrobenzene, with the relatively small dielectric constant. Herweg used the resonant circuit to detect very small changes of the electric capacitance $\Delta C\,(E)\,/C\ \sim 10^{-6}$ induced by the strong electric field. Piekara developed the concept toward the application of the superheterodyne-like design [6–10]. The sensitivity of the apparatus made it possible to detect the smallest known *NDE* values in liquids, associated with statistical fluctuations of polarization. Such effects are the only source of *NDE* in nonpolar solvents, such as carbon tetrachloride or benzene [10, 57]:

$$\frac{\Delta\varepsilon^E}{E^2} \approx C\left(\Delta P + \left\langle \Delta M^2 \right\rangle\right)^2 \chi \tag{31}$$

where $\Delta P = P - \langle P \rangle$ is for statistical fluctuations of polarizability and $\Delta\, M^2 = M^2 - \left\langle M^2 \right\rangle$ is related to fluctuations of the (induced) dipole moment, χ_T denotes the isothermal compressibility

Notable is the coherence of Eq. (3) for critical fluctuations and Eq. (30) for statistical fluctuations. The contribution from statistical fluctuations is always positive, what is also the case of critical fluctuations in binary mixtures, although for some liquid crystals the negative sign of the fluctuations-related effect is also possible (for instance in 5 * CB). Statistical fluctuations are related to the smallest known values of *NDE*: ~$10^{-19}\,\text{m}^2/\text{V}^2$. Molecular mechanisms (the orientation of dipole moments, intermolecular couplings, associates, intramolecular rotation, etc.) usually yield *NDE* in the range from 10^{-18} m^2/V^2 to $10^{-16}\,\text{m}^2/\text{V}^2$. Pretransitional fluctuations led to the contribution from $10^{-18}\,\text{m}^2/\text{V}^2$ (isotropic MBBA, critical mixtures where $\varepsilon_1 - \varepsilon_2 < 5$, etc.) to even $10^{-14}\,\text{m}^2/\text{V}^2$ [10]. The superheterodyne-type (with two generators) design of *NDE* apparatus in practice appeared to be (very) sensitive to parasitic distortions such as the electromagnetic "noise" and "imperfections" of the power supply. Moreover, it requires the perfect grounding what is difficult for the "naturally scattered design". The source of a notable systematic biasing error was also the problem with the scaling of the obtained experimentally values of electric capacitance $\Delta C\,(E)$ to estimate the required value of $\Delta\varepsilon^E$. Figure 21 shows the new scheme of the apparatus for the single-generator dual-field *NDE* apparatus. The compact design of the *NDE* front "measuring" module, from which signal is directly

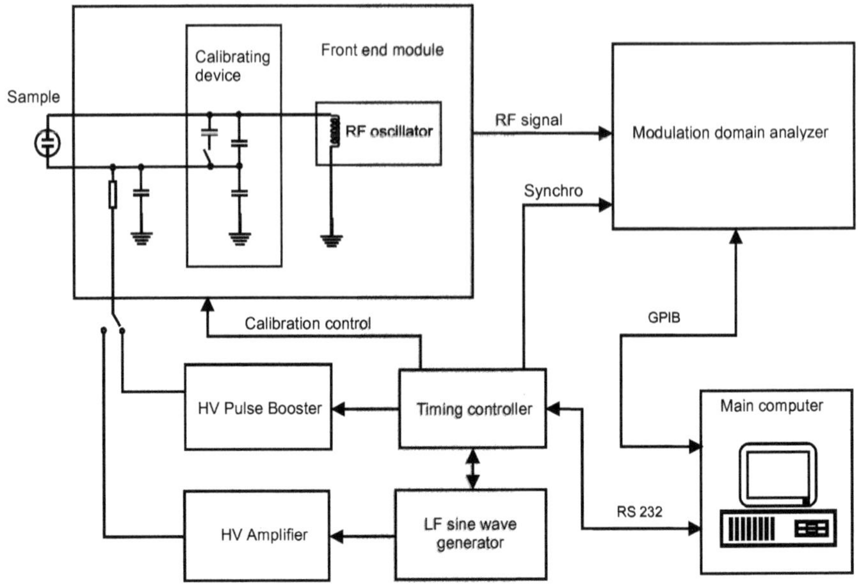

Fig. 21 Single generator dual-field apparatus for *NDE/NDS* measurements applying the modulation domain analyzer enabling the detection of frequency versus time changes of the output signal. In the first version of the apparatus, the MDA (HP) 53310A was used. The strong electric field can be applied in the form of DC pulses, $U < 1200$ V lasting from $\Delta t = 0.5$ ms to several minutes. Additional possibility was the train of sinusoidal $E(t)$ changes: frequencies $f < 20$ kHz, voltage $U_{peak-peak} < 1000$ V: the pulse length depends on the frequency usually it contained ~10 cycles. The apparatus made it possible to detect *NDE* for selected measurement frequencies from ca. 20 kHz to 20 MHz. The *NDE* versus time profile, both in the DC and AC modes of the strong electric field, could be scanned. Such feature shifted the basic *NDE* toward the nonlinear dielectric spectroscopy (*NDS*). The scanned *NDE* versus time outputs are cumulated to increase the signal-to-noise ratio in the final step of the measurement process

directed to them modulation domain analyzer and the "hard" separation of modules, solved the key grounding problem.

The next issue was the new design of the calibration unit, which now can be permanently linked to the resonant circuit, also when applying the high voltage (Fig. 22). The calibration unit is based on the reed delay switch mounted on a copper jacket. One of its contacts sticks out about 1 mm outside the copper coat. The switching on/off results in the change of the capacitance between the stick and the copper surrounding. It can be located directly within the generator due to small dimensions (10 mm diameter and 30 mm length).

The next important issue is the design of measurement capacitors, with the tested liquid dielectric. The high sensitivity of the dual-field *NDE* apparatus requires intensities of the electric field only $E_{Strong} < 10\,\text{kV/cm}$ (often $E_{strong} \sim 2\,\text{kV/cm}$ is satisfactorily *NDE* registration) and for the weak measuring field $E_{Weak} < 10\,\text{V/cm}$. These enable the usage of macro-gaps of the measurement capacitor ranging from 50 μm to 1 mm.

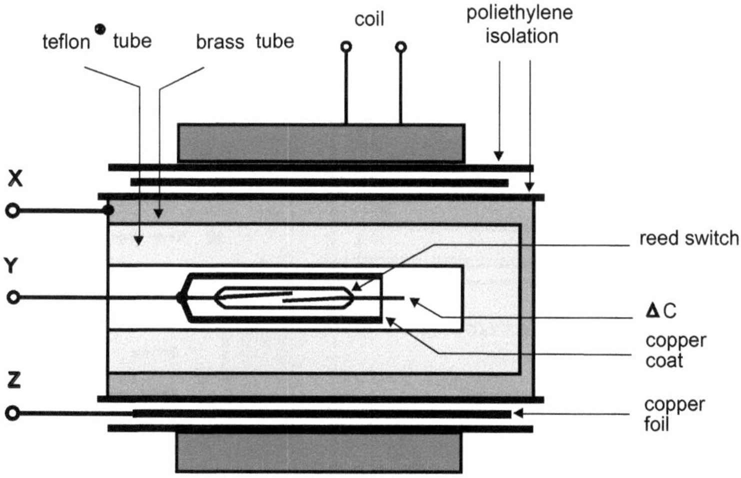

Fig. 22 The facility for scaling direct experimental $\Delta C\,(E)$ values of capacitance to determine related values of $\Delta\varepsilon^{E}$, necessary to calculate *NDE*. The magnetic field from the coil close/open reed switch, introducing well-defined changes of the electric capacitance. For the *NDE* facility of the authors $\Delta C_{\text{calib.}} = 4.7\text{femtoF}$ and the duration $\Delta t = 1$ ms [58]

Such gaps help to avoid gas bubbles which can distort/destroy experimental result. The design of such capacitor, ready also for high-pressure measurements, is shown in Fig. 23.

Apart from the high sensitivity, the nice feature of the dual-field *NDE* technique is the fact that the strong electric field is applied in the form of well-defined pulses. This enables "online" detection of parasitic artifacts: for instance, heating of the sample during the measurement process causes the horizontal shift of the baseline after the application of the strong electric field; gas bubbles cause characteristic "deformation" of the *NDE* output signal. All these enable online detection of experimental problems associated with samples and experimental preparations. The dual-field *NDE*, briefly described above, can be considered as the direct parallel of the "dynamic" electro-optic Kerr effect, but for radio frequencies. Recent years open the possibility of dual-field *NDE/NDS* measurements using only a single generator, and the new generation of modulation domain analyzers has emerged. It offers the possibility of the simultaneous detection of both $\Delta\varepsilon'^{E}$ (real) and $\Delta\varepsilon''^{E}$ (imaginary) components, in the DC and AC modes of the "pulsing" strong electric field with changing frequency. This can be supplemented by the controlled scan of frequencies of the weak measuring field.

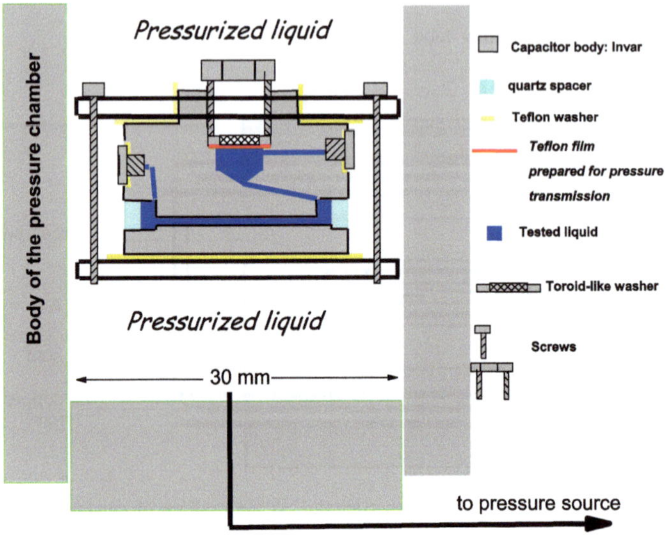

Fig. 23 The measurement capacitor uses for high-pressure studies. The diameter $2r = 16\,\text{mm}$; the distance between plates of the capacitor d: from 50 μm to 1 mm. As the spacer the ring made from quartz is used. Pressure is transmitted to the tested liquid via the deformation of the specially prepared Teflon film. For the given design of the capacitor, there is no possibility of the contact between the pressurized medium and the measured liquid and pressure can be increased and decreased without a risk such that parasitic interaction appears. The undesired gas appearance of gas bubbles can be easily detected

References

1. J.D. Hamblin, *Science in the Early Twentieth Century: An Encyclopedia* (ABC CLIO, Santa Barbara, 2005)
2. J. Herweg, Z. Phys. **3**, 36 (1920)
3. J. Herweg, W. Poetsch, Z. Phys. **29**, 105 (1922)
4. P. Debye, Phys. Z. **36**, 193 (1935)
5. A. Piekara, B. Piekara, C.R. Acad, Sci. Paris **203**, 852 (1936)
6. A. Piekara, C.R. Acad, Sci. Paris **203**, 1058 (1936)
7. A. Piekara, *Nayjaśniejszemu y Naypotężniejszemu Panu—Memoires of Arkadiusz Piekara* (PAX, Warsaw, 1981)
8. A. Piekara, Acta Phys. Polon. **6**, 130 (1937)
9. A. Piekara, S. Kielich, J. Chem. Phys. **28**, 1297 (1956)
10. A. Chełkowski, *Dielectric Physics* (PWN-Elsevier, Warsaw, 1980)
11. New experimental results of the authors of the given contribution
12. A. Piekara, Phys. Rev. **42**, 445 (1932)
13. S.J. Rzoska, Phase Transitions **27**, 1 (1990)
14. J. Chrapeć, S.J. Rzoska, J. Zioło, Chem. Phys. **111**, 155 (1987)
15. V.K. Semenchenko, M. Avimov, Zhur. Fiz. Khim. **XXIX**, 1343–1344 (1955)
16. V.K. Semenchenko, K.V. Arhangelskii, Zhur. Fiz. Khim. **XXXV**, 928 (1961)
17. K.V. Arhangelskii, V.K. Semenchenko, Zhurn. Fiz. Khim. **XLI**, 1303 (1967)
18. M. Givion, I. Pelah, U. Efron, Phys. Lett. A **48**, 1203 (1972)
19. Z. Ziejewska, J. Piotrowska-Szczepaniak, J. Zioło, Acta. Phys. Polon. A **56**, 347 (1979)

20. D.T. Jacobs, S.C. Greer, Phys. Rev. A **24**, 2075 (1981)
21. M. Hollecker, J. Goulon, J.-M. Thiebaut, J.-L. Rivail, Chem. Phys. **11**, 99 (1975)
22. M. Givon, I. Pelah, U. Efron, Phys. Lett. **48**, 12 (1974)
23. R.G. Quinn, C.P. Smyth, J. Chem. Phys. **41**, 2037 (1964)
24. J. Thoen, R. Kindt, W. Van Dael, M. Merabet, T.K. Bose, Phys. A **156**, 92 (1989)
25. J. Goulon, J.-L. Greffe, D.W. Oxtoby, J. Chem. Phys. **70**, 4742 (1979)
26. J.V. Sengers, D. Bedeaux, P. Mazur, S.C. Greer, Phys. A **104**, 573 (1980)
27. S.J. Rzoska, P. Urbanowicz, A. Drozd-Rzoska, M. Paluch, P. Habdas, Europhys. Lett. **45**, 334 (1999)
28. A. Drozd-Rzoska, S.J. Rzoska, A.R. Imre, Phys. Chem. Chem. Phys. **6**, 2291 (2004)
29. M.A. Anisimov, in *Introduction to Phase Transitions in Liquids and Liquid Crystals* (Gordon and Breach, Reading, 1994)
30. S.J. Rzoska, J. Chrapeć, J. Zioło, Phys. A **139**, 569 (1986)
31. J. Chrapeć, S.J. Rzoska, Phys. Lett. A **139**(343), 343 (1989)
32. J.S. Hoye, G. Stell, J. Chem. Phys. **81**, 3200 (1984)
33. A. Onuki, M. Doi, Europhys. Lett. **17**, 63 (1992)
34. R. Piazza, T. Bellini, V. Degiorgio, R.E. Goldstein, S. Leibler, R. Lipowsky, Phys. Rev. B **38**, 7223 (1988)
35. S.J. Rzoska, Phys. Rev. E **48**, 1136 (1993)
36. S.J. Rzoska, V. Degiorgio, M. Giardini, Phys. Rev. E **49**, 5234 (1994)
37. M.E. Fisher, Phys. Rev. Lett. **57**, 1911 (1986)
38. A. Drozd-Rzoska, S.J. Rzoska, Phys. Rev. E **93**, 062131 (2016)
39. A. Drozd-Rzoska, S.J. Rzoska, J.C. Martinez-Garcia, J. Chem. Phys. **141**, 094907 (2014)
40. S.J. Rzoska, V. Mazur (eds.), in *Soft Matter under Exogenic Impacts*, NATO Sci. Series II, vol. 242 (Springer, Berlin, 2007)
41. P. Urbanowicz, S.J. Rzoska, M. Paluch, B. Sawicki, A. Szulc, J. Zioło, Chem. Phys. **201**, 575 (1995)
42. J. Hanus, Phys. Rev. **178**, 420 (1969)
43. P.G. de Gennes, Phys. Lett. A **30**, 454 (1969)
44. P.G. de Gennes, *The Physics of Liquid Crystals* (Claredon. Press, Oxford, New York, 1974)
45. D. Demus, J.W. Goodby, G.W. Gray, H.W. Spiess, V. Vill, in *Handbook for Liquid Crystals*, Fundamentals, vol. 1 (Wiley-VCH, New York, 1998)
46. A.L. Jones, *Soft Matter Physics* (Oxford University Press, Oxford, 2010)
47. T.W. Stinson, J.D. Lister, Phys. Rev. Lett. **30**, 688 (1973)
48. S.J. Rzoska, J. Zioło, Liq. Cryst. **17**(629), 629 (1994)
49. J. Małecki, J. Zioło, Chem. Phys. **35**, 187 (1978)
50. A. Drozd-Rzoska, S.J. Rzoska, J. Zioło, Critical behaviour of dielectric permittivity in the isotropic phase of nematogens. Phys. Rev. E **54**, 6452 (1996)
51. M. Janik, S.J. Rzoska, A. Drozd-Rzoska, J. Zioło, P. Janik, S. Maslanka, K. Czupryński, J. Chem. Phys. **124**, 144907 (2006)
52. A. Drozd-Rzoska, S.J. Rzoska, M. Paluch, S. Pawlus, J. Ziolo, P.G. Santangelo, C.M. Roland, K. Czupryński, R. Dąbrowski, Phys. Rev. E **71**, 011508 (2005)
53. A. Drozd-Rzoska, S.J. Rzoska, M. Paluch, J. Chem. Phys. **129**, 184509 (2009)
54. J.C. Martinez-Garcia, S.J. Rzoska, A. Drozd-Rzoska, J. Martinez-Garcia, A universal description of ultraslow glass dynamics. Nat. Comm. **4**, 1823 (2013)
55. J.C. Martinez-Garcia, S.J. Rzoska, A. Drozd-Rzoska, J. Martinez-Garcia, J.C. Mauro, Sci. Rep. **4**, 5160 (2014)
56. S.J. Rzoska, A. Drozd-Rzoska, J. Phys.: Condens. Matter **24**, 035101 (2011)
57. G. Maroulis, T. Bancewicz, B. Champagne, A.D. Buckingham (eds.), *Atomic and Molecular Nonlinear Optics: Theory, Experiment and Computation: A homage to the Pioneering Work of Stanisław Kielich (1925–1993)* (IOS Press, NY, 2011)
58. M. Górny, J. Zioło, S.J. Rzoska, Rev. Sci. Instrum. **67**, 4290 (1996)

Third and Fifth Harmonic Responses in Viscous Liquids

S. Albert, M. Michl, P. Lunkenheimer, A. Loidl, P. M. Déjardin and F. Ladieu

Abstract We review the works devoted to third and fifth harmonic susceptibilities in glasses, namely $\chi_3^{(3)}$ and $\chi_5^{(5)}$. We explain why these nonlinear responses are especially well adapted to test whether or not some amorphous correlations develop upon cooling. We show that the experimental frequency and temperature dependences of $\chi_3^{(3)}$ and of $\chi_5^{(5)}$ have anomalous features, since their behavior is qualitatively different to that of an ideal gas, which is the high-temperature limit of a fluid. Most of the works have interpreted this anomalous behavior as reflecting the growth, upon cooling, of amorphously ordered domains, as predicted by the general framework of Bouchaud and Biroli (BB). We explain why most—if not all—of the challenging interpretations can be recast in a way which is consistent with that of Bouchaud and Biroli. Finally, the comparison of the anomalous features of $\chi_5^{(5)}$ and of $\chi_3^{(3)}$ shows that the amorphously ordered domains are compact, i.e., the fractal dimension d_f is close to the dimension d of space. This suggests that the glass transition of molecular liquids corresponds to a new universality class of critical phenomena.

S. Albert · F. Ladieu (✉)
SPEC, CEA, CNRS, Université Paris-Saclay, CEA Saclay Bat 772, 91191
Gif-sur-Yvette Cedex, France
e-mail: francois.ladieu@cea.fr

S. Albert
e-mail: samuel.albert@centraliens.net

M. Michl · P. Lunkenheimer · A. Loidl
Experimental Physics V, Center for Electronic Correlations and Magnetism,
University of Augsburg, 86159 Augsburg, Germany
e-mail: marion.michl@physik.uni-augsburg.de

P. Lunkenheimer
e-mail: peter.lunkenheimer@physik.uni-augsburg.de

A. Loidl
e-mail: alois.loidl@physik.uni-augsburg.de

P. M. Déjardin
LAMPS Université de Perpignan, Via Domitia - 52 avenue Paul Alduy,
66860 Perpignan Cedex, France
e-mail: dejardip@univ-perp.fr

R. Richert (ed.), *Nonlinear Dielectric Spectroscopy*, Advances in Dielectrics
https://doi.org/10.1007/978-3-319-77574-6_7

1 Why Measuring Harmonic Susceptibilities? Some Facts and an Oversimplified Argument

Most of our everyday materials are glasses, from window glasses to plastic bottles, and from colloids to pastes and granular materials. Yet the formation of the glassy state is still a conundrum and the most basic questions about the nature of the glassy state remain unsolved, e.g., it is still hotly debated whether glasses are genuine solids or merely hyperviscous liquids.

Over the past three decades, the notion evolved that higher order harmonic susceptibilities are especially well suited to unveil the very peculiar correlations governing the glass formation, yielding information that cannot be accessed by monitoring the linear response. This is illustrated in Fig. 1 displaying the third harmonic cubic susceptibility $\chi_3^{(3)}$—defined in Sect. 2.1—for four very different kinds of glasses [1–6]. In the case of spin glasses [1, 7]—see Fig. 1a— it was discovered in the 80s that $\chi_3^{(3)}$ diverges at the spin glass transition temperature T_{SG}, revealing the long-range nature of the spin glass amorphous order emerging around T_{SG}. Here the expression "amorphous order" corresponds to a minimum of the free energy realized by a configuration which is not spatially periodic. Similar nonlinear susceptibility experiments have been performed by Hemberger et al. [2] on an orientational glass former. In orientational glasses, electric dipolar or quadrupolar degrees of freedom undergo a cooperative freezing process without long-range orientational order [8]. As illustrated in Fig. 1b, the divergence of $|\chi_3^{(3)}|$ is not accompanied by any divergence of the linear susceptibility $|\chi_1|$.

We shall show in Eqs. (1) and (2) that this is intimately related to the very notion of amorphous ordering. For structural glasses, e.g., glycerol, it was discovered [3, 4] less than 10 years ago that $|\chi_3^{(3)}(\omega, T)|$ has a hump close to the α relaxation frequency f_α, and that the height of this hump is increasing anomalously upon cooling. A hump of $|\chi_3^{(3)}|$ has also been recently discovered in a colloidal glass [5, 6], in the vicinity of the β relaxation frequency f_β, revealing that any shear strain connects the system to a nonequilibrium steady state—see [5, 6]. Of course, as detailed balance does not hold in colloids, the comparison of colloidal glasses with spin glasses, orientational glasses, and structural glasses cannot be quantitative. However, the four very different kinds of glasses of Fig. 1 have the common qualitative property that nonlinear cubic responses unveil new information about the glassy state formation.

Let us now give an oversimplified argument explaining why nonlinear responses should unveil the correlations developing in glasses. We shall adopt the dielectric language adapted to this review devoted to supercooled liquids—where detailed balance holds— and consider a static electric field $E_{\rm st}$ applied onto molecules carrying a dipole moment $\mu_{\rm dip}$. At high temperature T, the system behaves as an ideal gas and its polarization P is given by

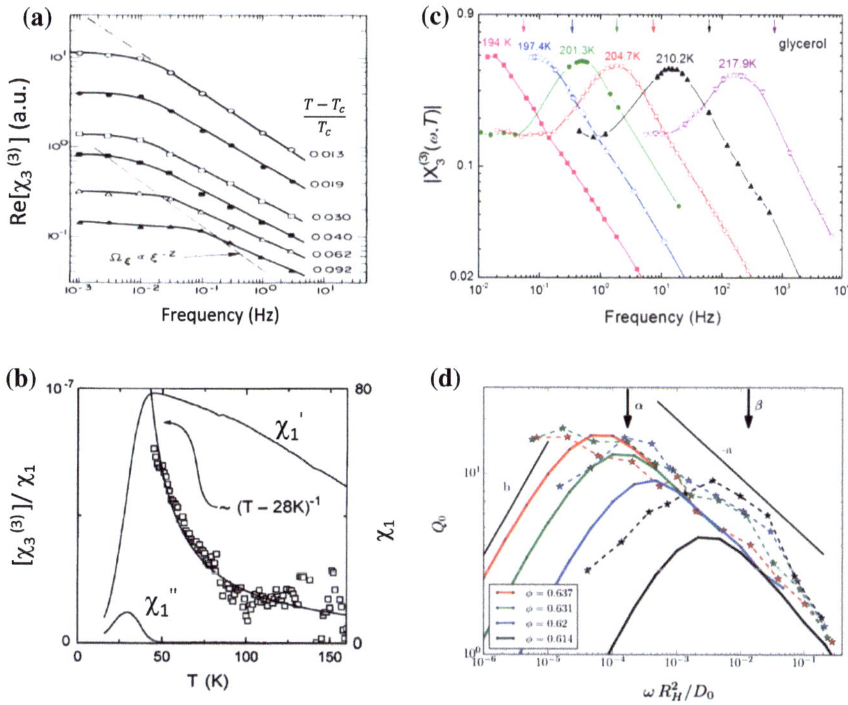

Fig. 1 Third Harmonic susceptibilities of very different types of glasses approaching their glass transition. **a** In the Ag:Mn spin glass [1], the static value of $\chi_3^{(3)}$ diverges when approaching the critical temperature $T_c \simeq 2.94$ K [1]. **b** Similar arguments are used to rationalize the third harmonic dielectric susceptibility of an orientational glass [2]. **c** In glycerol [3, 4], the modulus of the—dimensionless—cubic susceptibility $X_3^{(3)}$ has a peak as function of frequency, which increases anomalously upon cooling. **d** Strain–stress experiment in the colloidal system studied in Refs. [5, 6]. When increasing the volumic density ϕ, the increasing peak of $Q_0 = |\chi_3^{(3)}/\chi_1|$ reveals that any shear strain connects the system to a non-equilibrium steady state—see [5, 6]. In all these four examples, $\chi_3^{(3)}$ unveils informations about the nature of the glassy state that cannot be obtained by studying the linear susceptibility χ_1. From Refs. [1–3, 6]

$$
\begin{aligned}
P &= \frac{\mu_{\rm dip}}{a^d}\mathscr{L}_d\left(\frac{\mu_{\rm dip}E_{\rm st}}{k_{\rm B}T}\right) \\
&\simeq \frac{1}{3}\frac{\mu_{\rm dip}}{a^d}\left(\frac{\mu_{\rm dip}E_{\rm st}}{k_{\rm B}T}\right) - \frac{1}{45}\frac{\mu_{\rm dip}}{a^d}\left(\frac{\mu_{\rm dip}E_{\rm st}}{k_{\rm B}T}\right)^3 + \frac{2}{945}\frac{\mu_{\rm dip}}{a^d}\left(\frac{\mu_{\rm dip}E_{\rm st}}{k_{\rm B}T}\right)^5 + \cdots
\end{aligned} \tag{1}
$$

where a^d is the molecular d-dimensional volume, $\mathscr{L}_d$ is the suitable Langevin function expressing the thermal equilibrium of a single dipole in dimension d, and where the numerical prefactors of the linear-, third-, and fifth-order responses correspond to the case $d = 3$. Assume now that upon cooling some correlations develop over a characteristic lengthscale ℓ, i.e., molecules are correlated within groups containing

$N_{\rm corr} = (\ell/a)^{d_f}$ molecules, with d_f the fractal dimension characterizing the correlated regions. Because these domains are independent from each other, one can use Eq. (1), provided that we change the elementary volume a^d by that of a domain—namely $a^d(\ell/a)^d$—as well as the molecular dipole $\mu_{\rm dip}$ by that of a domain—namely $\mu_{\rm dip}(\ell/a)^{(d_f/2)}$. Here, the exponent $d_f/2$ expresses the amorphous ordering within the correlated regions, i.e., the fact that the orientation of the correlated molecules looks random in space. We obtain

$$\frac{P}{\mu_{\rm dip}/a^d} \simeq \frac{1}{3}\left(\frac{\ell}{a}\right)^{d_f-d}\left(\frac{\mu_{\rm dip}E_{\rm st}}{k_{\rm B}T}\right) - \frac{1}{45}\left(\frac{\ell}{a}\right)^{2d_f-d}\left(\frac{\mu_{\rm dip}E_{\rm st}}{k_{\rm B}T}\right)^3 + \\ + \frac{2}{945}\left(\frac{\ell}{a}\right)^{3d_f-d}\left(\frac{\mu_{\rm dip}E_{\rm st}}{k_{\rm B}T}\right)^5 + \cdots \tag{2}$$

which shows that the larger the order k of the response, the stronger the increase of the response when ℓ increases. As $d_f \le d$, Eq. (2) shows that the linear response *never* diverges with ℓ: it is always, for any ℓ, of the order of $\mu_{\rm dip}^2/(a^d k_{\rm B}T)$. This can be seen directly in Eq. (2) in the case $d_f = d$; while for $d_f < d$ one must add to Eq. (2) the polarization arising from the uncorrelated molecules not belonging to any correlated region. This insensitivity of the linear response to ℓ directly comes from the amorphous nature of orientations that we have assumed when rescaling the net dipole of a domain—by using the power $d_f/2$. By contrast, in a standard para–ferro transition one would use instead a power d_f to rescale the moment of a domain, and we would find that the linear response diverges with ℓ as soon as $d_f > d/2$—which is the standard result close to a second-order phase transition. For amorphous ordering, the cubic response is thus the lowest order response diverging with ℓ, as soon as $d_f > d/2$. This is why cubic responses—as well as higher order responses—are ideally suited to test whether or not amorphous order develops in supercooled liquids upon cooling.

For spin glasses, the above purely thermodynamic argument is enough to relate the divergence of the static value of $\chi_3^{(3)}$—see Fig. 1a—to the divergence of the amorphous order correlation length ℓ. For structural glasses this argument must be complemented by some dynamical argument, since we have seen on Fig. 1c that the anomalous behavior of $\chi_3^{(3)}$ takes place around the relaxation frequency f_α. This has been done, on very general grounds, by the predictions of Bouchaud and Biroli, who anticipated [9] the main features reported in Fig. 1c. BB's predictions will be explained in Sect. 3. Before, we shall review in Sect. 2 the main experimental features of third and fifth harmonic susceptibilities. Because of the generality of Eq. (2) and of BB's framework, we anticipate that χ_3 and χ_5 have common anomalous features that can be interpreted as reflecting the evolution of ℓ—and thus of $N_{\rm corr}$—upon cooling. The end of the chapter, Sect. 4, will be devoted to more specific approaches to the cubic response of glassforming liquids. Beyond their apparent diversity, we shall show that they can be unified by the fact that in all of them, $N_{\rm corr}$ is a key parameter—even though it is sometimes implicit. The Appendix contains some additional material for the readers aiming at deepening their understanding of this field of high harmonic responses.

2 Experimental Behavior of Third and Fifth Harmonic Susceptibilities

2.1 Definitions

When submitted to an electric field $E(t)$ depending on time t, the most general expression of the polarization $P(t)$ of a dielectric medium is given by a series expansion as follows :

$$P(t) = \sum_{m=0}^{\infty} P_{2m+1}(t), \tag{3}$$

where because of the $E \to -E$ symmetry, the sum contains only odd terms, and the $(2m+1)$-order polarization $P_{2m+1}(t)$ is proportional to E^{2m+1}. The most general expression of $P_{2m+1}(t)$ is given by

$$\frac{P_{2m+1}(t)}{\varepsilon_0} = \int_{-\infty}^{\infty} \cdots \int_{-\infty}^{\infty} \chi_{2m+1}(t - t'_1, \ldots, t - t'_{2m+1}) E(t'_1) \ldots E(t'_{2m+1}) \mathrm{d}t'_1 \ldots \mathrm{d}t'_{2m+1}. \tag{4}$$

Because of causality, $\chi_{2m+1} \equiv 0$ whenever one of its arguments is negative. For a field $E(t) = E\cos(\omega t)$ of frequency ω and of amplitude E, it is convenient to replace χ_{2m+1} by its $(2m+1)$-fold Fourier transform and to integrate first over $t'_1, \ldots, t'_{2m+1}$. Defining the onefold Fourier transform $\phi(\omega)$ of any function $\phi(t)$ by $\phi(\omega) = \int \phi(t) \mathrm{e}^{-i\omega t} dt$ (with $i^2 = -1$) and using $\int \mathrm{e}^{-i(\omega_1 - \omega)t} dt = 2\pi\delta(\omega_1 - \omega)$, where δ is the Dirac delta function, one obtains the expression of $P_{2m+1}(t)$. This expression can be simplified by using two properties: (a) the fact that the various frequencies ω_λ play the same role, which implies $\chi_{2m+1}(-\omega, \omega, \ldots, \omega) = \chi_{2m+1}(\omega, -\omega, \ldots, \omega)$; (b) the fact that χ_{2m+1} is real in the time domain implying that $\chi_{2m+1}(-\omega, \ldots, -\omega)$ is the complex conjugate of $\chi_{2m+1}(\omega, \ldots, \omega)$. By using these two properties, we obtain the expression of all the $P_{2m+1}(t)$, and in the case of the third-order polarization this yields the following:

$$\frac{P_3(t)}{\varepsilon_0} = \frac{1}{4} E^3 |\chi_3^{(3)}(\omega)| \cos(3\omega t - \delta_3^{(3)}(\omega)) + \frac{3}{4} E^3 |\chi_3^{(1)}(\omega)| \cos(\omega t - \delta_3^{(1)}(\omega)), \tag{5}$$

where we have set $\chi_3(\omega, \omega, \omega) = |\chi_3^{(3)}(\omega)| \mathrm{e}^{-i\delta_3^{(3)}(\omega)}$, and $\chi_3(\omega, \omega, -\omega) = |\chi_3^{(1)}(\omega)| \mathrm{e}^{-i\delta_3^{(1)}(\omega)}$.

Similarly, for the fifth-order polarization, we obtain

$$\frac{P_5(t)}{\varepsilon_0} = \frac{1}{16}E^5|\chi_5^{(5)}(\omega)|\cos(5\omega t - \delta_5^{(5)}(\omega)) + \frac{5}{16}E^5|\chi_5^{(3)}(\omega)|\cos(3\omega t - \delta_5^{(3)}(\omega)) + \\ + \frac{10}{16}E^5|\chi_5^{(1)}(\omega)|\cos(\omega t - \delta_5^{(1)}(\omega)), \tag{6}$$

where, we have set $\chi_5(\omega, \omega, \omega, \omega, \omega) = |\chi_5^{(5)}(\omega)|e^{-i\delta_5^{(5)}(\omega)}$, and similarly $\chi_5(\omega, \omega, \omega, \omega, -\omega) = |\chi_5^{(3)}(\omega)|e^{-i\delta_5^{(3)}(\omega)}$ as well as $\chi_5(\omega, \omega, \omega, -\omega, -\omega) = |\chi_5^{(1)}(\omega)|e^{-i\delta_5^{(1)}(\omega)}$.

For completeness, we recall that the expression of the linear polarization $P_1(t)$ is $P_1(t)/\varepsilon_0 = E|\chi_1(\omega)|\cos(\omega t - \delta_1(\omega))$ where we have set $\chi_1(\omega) = |\chi_1(\omega)|e^{-i\delta_1(\omega)}$. In the linear case, we often drop the exponent indicating the harmonic, since the linear response $P_1(t)$ is by design at the fundamental angular frequency ω. The only exception to this simplification is in Fig. 11 (see below), where for convenience the linear susceptibility is denoted $\chi_1^{(1)}$.

Up to now, we have only considered nonlinear responses induced by a pure AC field E, allowing to define the third harmonic cubic susceptibility $\chi_3^{(3)}$ and/or the fifth harmonic fifth-order susceptibility $\chi_5^{(5)}$ to which this chapter is devoted. In Sect. 2.3 and Figs. 8 and 9, we shall briefly compare $\chi_3^{(3)}$ with other cubic susceptibilities, namely $\chi_3^{(1)}$ already defined in Eq. (5) as well as $\chi_{2;1}^{(1)}$ that we introduce now.

This supplementary cubic susceptibility is one of the new terms arising when a static field $E_{\rm st}$ is superimposed on top of E. Because of $E_{\rm st}$, new cubic responses arise, both for even and odd harmonics. For brevity, we shall write only the expression of the first harmonic part $P_3^{(1)}$ of the cubic polarization, which now contains the following two terms:

$$\frac{P_3^{(1)}(t)}{\varepsilon_0} = \frac{3}{4}|\chi_3^{(1)}(\omega)|E^3\cos(\omega t - \delta_3^{(1)}(\omega)) + 3|\chi_{2;1}^{(1)}(\omega)|E_{\rm st}^2 E\cos(\omega t - \delta_{2;1}^{(1)}(\omega)), \tag{7}$$

where we have defined $|\chi_{2;1}^{(1)}(\omega)|\exp(-i\delta_{2;1}^{(1)}(\omega)) = \chi_3(0, 0, \omega)$.

For any cubic susceptibility—generically noted χ_3—or for any fifth-order susceptibility—generically noted χ_5—the corresponding dimensionless susceptibility X_3 or X_5 is defined as

$$X_3 \equiv \frac{k_B T}{\varepsilon_0 \Delta\chi_1^2 a^3}\chi_3, \qquad X_5 \equiv \frac{(k_B T)^2}{\varepsilon_0^2 \Delta\chi_1^3 a^6}\chi_5, \tag{8}$$

where $\Delta\chi_1$ is the "dielectric strength", i.e., $\Delta\chi_1 = \chi_{\rm lin}(0) - \chi_{\rm lin}(\infty)$ where $\chi_{\rm lin}(0)$ is the linear susceptibility at zero frequency and $\chi_{\rm lin}(\infty)$ is the linear susceptibility at a-high-frequency, where the orientational mechanism has ceased to operate. Note that X_3 as well as X_5 have the great advantage to be both dimensionless and independent of the field amplitude.

2.2 Frequency and Temperature Dependence of Third Harmonic Susceptibility

In this section, we review the characteristic features of $\chi_3^{(3)}$ both as a function of frequency and temperature. We separate the effects at equilibrium above T_g and those recorded below T_g in the out-of-equilibrium regime.

2.2.1 Above T_g

In the α regime:

Figure 2 shows the modulus $|\chi_3^{(3)}|$ for propylene carbonate [10]. It is an archetypical example of what has been measured in glassforming liquids close to T_g. For a given temperature, one distinguishes two domains:

1. For very low frequencies, $f/f_\alpha \leq 0.05$, a plateau is observed as indicated by the shaded area in Fig. 2, i.e., $|\chi_3^{(3)}|$ does not depend on frequency. This is reminiscent of the behavior of an ideal gas of dipoles where each dipole experiences a Brownian motion without any correlation with other dipoles. In such an ideal gas, $|\chi_3^{(3)}|$ has a plateau below the relaxation frequency and monotonously falls to zero as one increases the frequency. Because the observed plateau in Fig. 2 is reminiscent to the ideal gas case, it has sometimes [3, 4] been called the "trivial" regime. What is meant here is not that the analytical expressions of the various χ_3 are "simple"—see Appendix 2— but that the glassy correlations do not change qualitatively the shape of $\chi_3^{(3)}$ in this range. Physically, an ideal gas of dipoles corresponds to the high-T limit of a fluid. This is why it is a useful benchmark which allows to distinguish the "trivial" features and those involving glassy correlations.
2. When rising the frequency above $0.05 f_\alpha$, one observes for $|\chi_3^{(3)}|$ a hump for a frequency $f_{\text{peak}}/f_\alpha \simeq c$ where the constant c does not depend on T and weakly

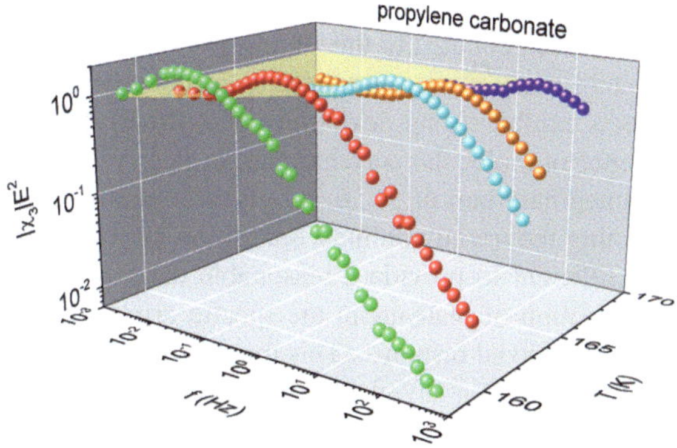

Fig. 2 Third-order harmonic component of the dielectric susceptibility of propylene carbonate [10]. Spectra of $|\chi_3^{(3)}|E^2$ are shown for various temperatures measured at a field of 225 kV/cm. The yellow-shaded plane indicates the plateau arising in the trivial regime

depends on the liquid (e.g., $c \simeq 0.22$ for glycerol and $c \simeq 0.3$ for propylene carbonate). This hump is followed by a power law decrease $|\chi_3^{(3)}| \sim f^{-\beta_3}$ where $\beta_3 < 1$ is close [3] to the exponent governing the decrease of $|\chi_1|$ above f_α. Qualitatively, this hump is important since it exists *neither* in the cubic susceptibility of an ideal gas of dipoles *nor* in the modulus of the linear response $|\chi_1|$ of the supercooled liquids. This is why this hump has been termed the "glassy contribution" to χ_3. On a more quantitative basis, the proportionality of f_{peak} and of f_α has been observed for f_α ranging from 0.01 Hz to 10 kHz—above 10 kHz the measurement of $\chi_3^{(3)}$ is obscured by heating issues, see [11] and Sect. 5.

The consistency of the above considerations can be checked by comparing the third-order susceptibility of canonical glass formers to that of monohydroxy alcohols. The linear dielectric response of the latter is often dominated by a Debye relaxation process, which is commonly ascribed to the fact that part of the molecules are forming chain-like hydrogen-bonded molecule clusters with relatively high dipolar moments [12]. This process represents an idealized Debye relaxation case as it lacks the heterogeneity-related broadening found for other glass formers. Moreover, correlations or cooperativity should not play a significant role for this process, because cluster–cluster interactions can be expected to be rare compared to the intermolecular interactions governing the α relaxation in most canonical glass formers [13]. Thus, this relaxation process arising from rather isolated dipolar clusters distributed in a liquid matrix can be expected to represent a good approximation of the "ideal dipole gas" case mentioned above. The monohydroxy alcohol 1-propanol is especially well suited to check this notion because here transitions between different chain topologies, as found in several other alcohols affecting the nonlinear response [14, 15], do not seem to play a role [15]. Figure 3a shows the frequency-dependent modulus, real, and imaginary part of $\chi_3^{(3)} E^2$ for 1-propanol at 120 K [13, 16]. Indeed, no hump is observed in $|\chi_3^{(3)}|(\nu)$ as predicted for a noncooperative Debye relaxation. The solid lines were calculated according to Ref. [17], accounting for the expected trivial polarization-saturation effect. Indeed, the spectra of all three quantities are reasonably described in this way. In the calculation, for the molecular volume an additional factor of 2.9 had to be applied to match the experimental data, which is well consistent with the notion that the Debye relaxation in the monohydroxy alcohols arises from the dynamics of clusters formed by several molecules.

In marked contrast to this dipole-gas-like behavior of the Debye relaxation of 1-propanol, the $\chi_3^{(3)}$ spectra related to the conventional α relaxation of canonical glass formers exhibit strong deviations from the trivial response, just as expected in the presence of molecular correlations. As an example, Fig. 3b shows the modulus, real, and imaginary part of $\chi_3^{(3)} E^2$ of glycerol at 204 K. Again the lines were calculated assuming the trivial nonlinear saturation effect only [17]. Obviously, this approach is insufficient to provide a reasonable description of the experimental data. Only the detection of plateaus in the spectra arising at low frequencies agrees with the calculated trivial response. This mirrors the fact that, on long time scales, the liquid flow smoothes out any glassy correlations.

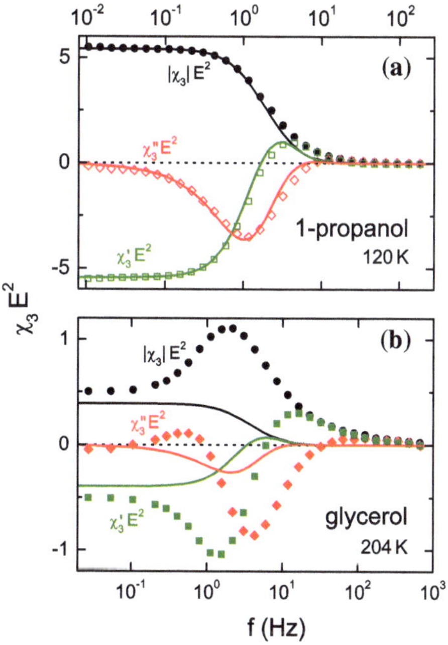

Fig. 3 **a** Modulus, real, and imaginary part of the third-order dielectric susceptibility $\chi_3^{(3)}$ (times E^2) of 1-propanol at 120 K as measured with a field of 468 kV/cm [16]. The solid lines were calculated according to Ref. [17]. **b** Same for glycerol at 204 K and 354 kV/cm [16]

When varying the temperature, two very different behaviors of $\chi_3^{(3)}$ are observed:

1. In the plateau region, the weak temperature dependence of $\chi_3^{(3)}$ is easily captured by converting $\chi_3^{(3)}$ into its dimensionless form $X_3^{(3)}$ by using Eq. (8): one observes [3, 4] that *in the plateau region* $X_3^{(3)}$ *does not depend at all on the temperature.* Qualitatively this is important since in an ideal gas of dipoles $X_3^{(3)}$ does also not depend on temperature, once plotted as a function of f/f_α. This reinforces the "trivial" nature of the plateau region, i.e., the fact that it is not qualitatively affected by glassy correlations.
2. In the hump region, $|X_3^{(3)}(f/f_\alpha)|$ increases upon cooling, again emphasizing the "anomalous"—or "non trivial"—behavior of the glassy contribution to $\chi_3^{(3)}$. This increase of the hump of $|X_3^{(3)}|$ has been related to that of the apparent activation energy $E_{\rm act}(T) \equiv \partial \ln \tau_\alpha / \partial(1/T)$—see Refs. [10, 18]—as well as to $T\chi_T \equiv |\partial \ln \tau_\alpha / \partial \ln T|$ [3, 4, 19, 20]. Note that because the experimental temperature interval is not so large, the temperature behavior of $E_{\rm act}$ and of $T\chi_T$ is extremely similar. Both quantities are physically appealing since they are related to the number $N_{\rm corr}(T)$ of correlated molecules: the line of thought where $E_{\rm act} \sim N_{\rm corr}(T)$ dates back to the work of Adam and Gibbs [21]; while another series of papers [22, 23] proposed a decade ago that $N_{\rm corr} \propto T\chi_T$. Figure 4 illustrates how good is the correlation between the increase of the hump of $|X_3^{(3)}|$—left axis—and $E_{\rm act}(T)$. This correlation holds for five glass formers, of extremely different fragilities, including a plastic crystal, where only the orientational degrees of freedom experience the glass transition [24].

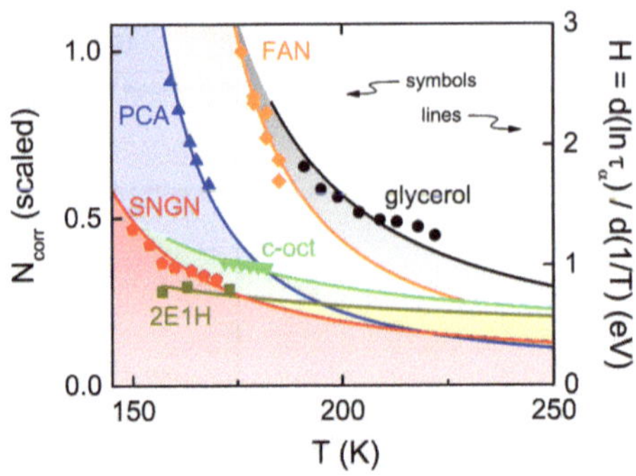

Fig. 4 For several glass formers, $N_{\text{corr}}(T)$ as extracted from the hump of $|X_3^{(3)}|$ (left axis) closely follows $E_{\text{act}}(T)$, deduced from the temperature dependence of the α-relaxation time [10] (right axis). The abbreviations stand for propylene carbonate (PCA), 3-fluoroaniline (FAN), 2-ethyl-1-hexanol (2E1H), cyclo-octanol (c-oct), and a mixture of 60% succinonitrile and 40% glutaronitrile (SNGN). From Ref. [18]

In the excess wing regime:

In the dielectric-loss spectra of various glass formers, at high frequencies the excess wing shows up, corresponding to a second, shallower power law at the right flank of the α peak [25]. Figure 5a shows loss spectra of glycerol, measured at low and high fields up to 671 kV/cm [26, 27], where the excess wing is indicated by the dashed lines (It should be noted that the difference of these loss curves for high and low fields is directly related to the cubic susceptibility $\chi_3^{(1)}$, defined in Eq. (5) [16]). As already reported in the seminal paper by Richert and Weinstein [28], in Fig. 5a at the right flank of the α-relaxation peak a strong field-induced increase of the dielectric loss is found while no significant field dependence is detected at its low-frequency flank. In Ref. [28] it was pointed out that these findings are well consistent with the heterogeneity-based box model (see Sect. 4.3). However, as revealed by Fig. 5a, remarkably in the region of the excess wing no significant nonlinear effect is detected. Time-resolved measurements, later on reported by Samanta and Richert [29], revealed nonlinearity effects in the excess wing region when applying the high field for extended times of up to several 10,000 cycles. Anyhow, the nonlinearity in this region seems to be clearly weaker than for the main relaxation and the nonlinear behavior of the excess wing differs from that of the α relaxation.

To check whether weaker nonlinearity in the excess wing region is also revealed in higher harmonic susceptibility measurements, Fig. 5b directly compares the modulus of the linear dielectric susceptibility of glycerol at 191 K to the third-order susceptibility $|\chi_3^{(3)}|$ (multiplied by E^2) [30] (We show $|\chi_1|$ corrected for $\chi_{1,\infty} = \varepsilon_\infty - 1$ caused by the ionic and electronic polarizability, whose contribution to the modulus strongly superimposes the excess wing). While the linear response exhibits a clear signature of the excess wing above about 100 Hz (dashed line), no trace of this spectral feature is found in $|\chi_3^{(3)}(\nu)|$. Thus, we conclude that possible nonlinearity contributions arising from the excess wing, if present at all, must be significantly

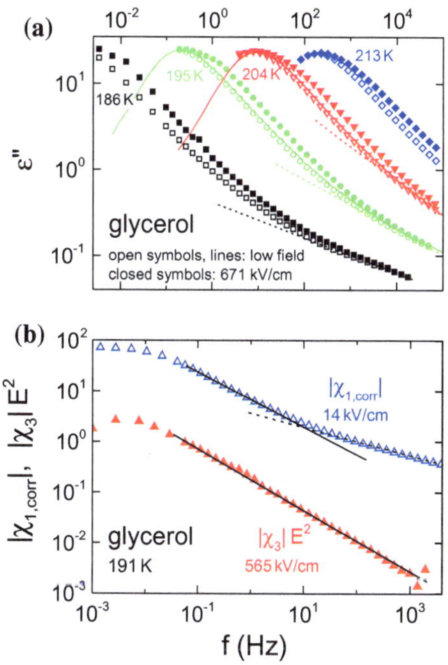

Fig. 5 **a** Dielectric loss of glycerol measured at fields of 14 kV/cm (open symbols) and 671 kV/cm (closed symbols) shown for four temperatures [27]. The solid lines were measured with 0.2 kV/cm [26]. The dashed lines indicate the excess wing. **b** Open triangles: Absolute values of χ_1 (corrected for $\chi_{1,\infty} = \varepsilon_\infty - 1$) at 14 kV/cm for glycerol at 191 K. Closed triangles: $\chi_3^{(3)} E^2$ at 565 kV/cm [30]. The solid lines indicate similar power laws above the peak frequency for both quantities. The dashed line indicates the excess wing in the linear susceptibility at high frequencies, which has no corresponding feature in $\chi_3^{(3)}(\nu)$

weaker than the known power law decay of the third-order susceptibility at high frequencies, ascribed to the nonlinearity of the α relaxation.

The excess wing is often regarded as the manifestation of a secondary relaxation process, partly superimposed by the dominating α-relaxation [31, 32]. Thus, the weaker nonlinearity of the excess wing seems to support long-standing assumptions of the absence of cooperativity in the molecular motions that lead to secondary relaxation processes [33, 34]. Moreover, in a recent work [35] it was pointed out that the small or even absent nonlinear effects in the excess wing region can also be consistently explained within the framework of the coupling model [34], where the excess wing is identified with the so-called "nearly constant loss" caused by caged molecular motions.

2.2.2 Below T_g

Below T_g, the physical properties are *aging*, i.e., they depend on the time t_a elapsed since the material has fallen out of equilibrium, i.e., since the glass transition temperature T_g has been crossed. The mechanism of aging is still a matter of controversy [36–40], owing to the enormous theoretical and numerical difficulties inherent to out-of-equilibrium processes. Experimentally, a few clear-cut results have been obtained in spin glasses [41] where it was shown, by using nonlinear techniques, that the increase of the relaxation time τ_α with the aging time t_a can be rather convincingly attributed to the growth of the number N_{corr} of correlated spins with t_a. Very recently extremely sophisticated numerical simulations have been carried out by the so called Janus international collaboration, yielding, among many other results, a strong microscopic support [42] to the interpretation given previously in the experiments of Ref. [41].

In structural glasses, the aging properties of the linear response have been reported more than one decade ago [44, 45]. More recently, the aging properties of $\chi_3^{(3)}$ were reported in glycerol [43] and its main outputs are summarized in Figs. 6 and 7. A glycerol sample previously well equilibrated at $T_g + 8$ K was quenched to a working temperature $T_w = T_g - 8$ K and its third harmonic cubic susceptibility was continuously monitored as a function of t_a. The dominant effect is the increase of the relaxation time τ_α with t_a. In Ref. [43] τ_α increases by a factor $\simeq$6 between the arrival at T_w—i.e., $t_a = 0$—and the finally equilibrated situation reached for $t_a \gg \tau_{\alpha,\mathrm{eq}}$ where τ_α is equal to its equilibrium value $\tau_{\alpha,\mathrm{eq}}$—and no longer evolves with t_a. This variation of τ_α with the aging time t_a can be very accurately deduced from the shift that it produces on the imaginary part of the linear response $\chi''(f, t_a)$.

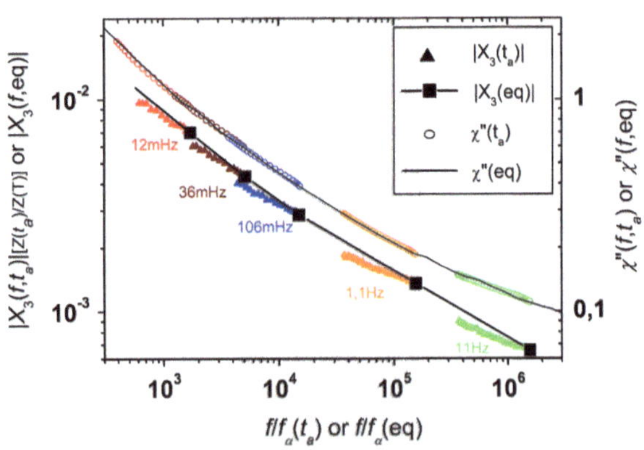

Fig. 6 During the aging of glycerol—at $T_g - 8$K—the increase of τ_α with the aging time t_a is measured by rescaling the aging data—symbols—of χ_1''—right axis onto the equilibrium data—solid black line. The corresponding scaling *fails* for $X_3^{(3)}(f, t_a)$—left axis—revealing the increase of N_{corr} during aging. See [43] for details about the quantity $z(t_a)/z(T)$ which is involved in the left axis but varies by less than 2% during aging. From [43]

This is summarized in Fig. 6 for 5 different frequencies: when plotted as a function of $f/f_\alpha(t_a) \equiv 2\pi f \tau_\alpha(t_a)$, the aging values of $\chi''(f, t_a)$—symbols—are nicely rescaled onto the equilibrium values $\chi''(f, \text{eq})$—continuous line—measured when $t_a \gg \tau_{\alpha,eq}$. The most important experimental result is that this scaling *fails* for $|X_3^{(3)}(f, t_a)|$ as shown by the left axis of Fig. 6: For short aging times, the difference between aging data (symbols) and equilibrium values (continuous line) is largest. This has been interpreted as *an increase of* N_{corr} *with the aging time* t_a. This increase of $N_{\text{corr}}(t_a)$ toward its equilibrated value $N_{\text{corr}}(\text{eq})$ is illustrated in Fig. 7 where the variation of $\delta = N_{\text{corr}}(t_a)/N_{\text{corr}}(\text{eq})$ is plotted as a function of t_a. It turns out to be independent of the measuring frequency, which is a very important self-consistency check.

The increase of N_{corr} during aging can be rather well captured by extrapolating the $N_{\text{corr}}(T)$ variation obtained from the growth of the hump of $|\chi_3^{(3)}|$ measured at equilibrium above T_g and by translating the $\tau_\alpha(t_a)$ in terms of a fictive temperature $T_{\text{fict}}(t_a)$ which decreases during aging, finally reaching T_w when $t_a \gg \tau_{\alpha,\text{eq}}$. This yields the continuous line in Fig. 7, which fairly well captures the data drawn from the aging of $\chi_3^{(3)}$. Because this extrapolation roughly agrees with the aging data, one can estimate that the quench from $T_g + 8$ K to $T_w = T_g - 8$ K corresponds to a doubling of $N_{\text{corr,eq}}$. The approximately 10% increase reported in Fig. 7 is thus the long time *tail* of this increase, while the first 90% increase cannot be measured because it takes place during the quench.

Beyond the qualitative result that N_{corr} increases during aging, these $\chi_3^{(3)}(t_a)$ data can be used to test quantitatively some theories about the emergence of the glassy state. By gathering, in the inset of Fig. 7, the equilibrium data—symbols lying in the [1; 1.3] interval of the horizontal axis—and the aging data translated in terms of $T_{\text{fict}}(t_a)$—symbols lying in the [2; 2.3] interval—one extends considerably the experimental temperature interval, which puts strong constraints onto theories. Summarizing two different predictions by $\ln(\tau_\alpha/\tau_0) = Y N_{\text{corr}}^{\psi/3}/(k_B T)$ with $Y \sim T$; $\psi = 3/2$ for Random First- Order Transition Theory (RFOT) [46] while $Y \sim 1$; $\psi = 1$ for the numerical approach of Ref. [47], Fig. 7 is designed to test these two predictions, see

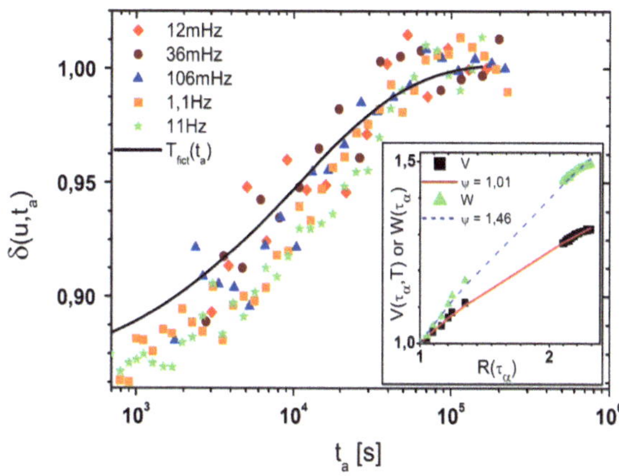

Fig. 7 The values of $\delta = N_{\text{corr}}(t_a)/N_{\text{corr}}(\text{eq})$ extracted from Fig. 6 show the increase of N_{corr} during aging. Inset: different theories are tested gathering equilibrium and aging experiments. From [43]

Ref. [43] for details: it shows that both of them are consistent with experiments—contrary to another prediction relying onto a critical relation $\tau_\alpha \propto N_{\mathrm{corr}}^z$, yielding an unrealistic large value of $z \sim 20$ to account for the experiments.

2.3 Strong Similarities Between Third and First Cubic Susceptibilities

We now come back to equilibrium measurements, i.e., above T_g—and compare the behavior of the third harmonic cubic susceptibility $\chi_3^{(3)}$ as well as the first harmonic cubic susceptibilities $\chi_3^{(1)}$ and $\chi_{2;1}^{(1)}$ introduced in Eq. (7). We remind that $\chi_{2;1}^{(1)}$ corresponds to the case, where a static field E_{st} is superimposed to the AC field $E\cos(\omega t)$.

Figures 8 and 9 show the modulus and the phases of the three cubic susceptibilities for glycerol and for propylene carbonate.

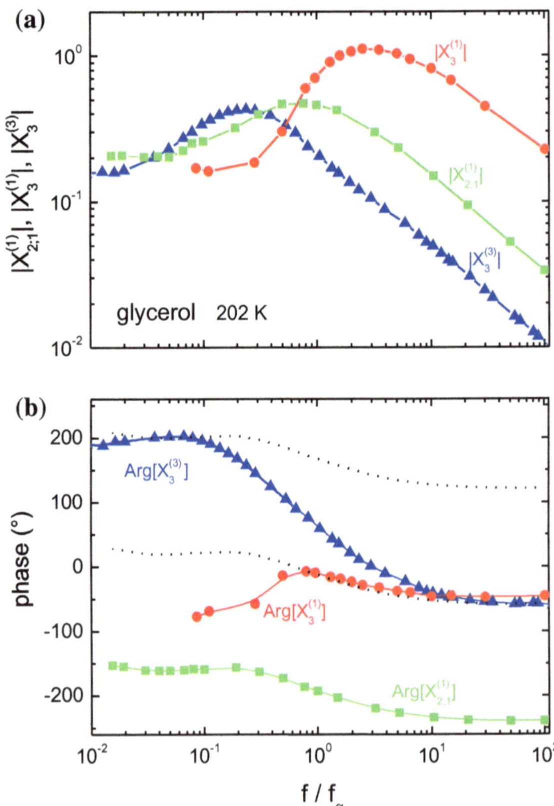

Fig. 8 For glycerol and $f_\alpha \simeq 2$ Hz, modulus—top panel—and phase—bottom panel—of the three cubic susceptibilities defined in Eqs. (5) and (7). The salient features of the three cubic susceptibilities are similar, which strongly suggests a common physical origin—see text. Dotted lines are $\mathrm{Arg}[X_{2;1}^{(1)}] + \pi$ or $+2\pi$ and support Eqs. (9) and (10). From [48]

Fig. 9 Same representation as in Fig. 8 but for propylene carbonate. From [48]

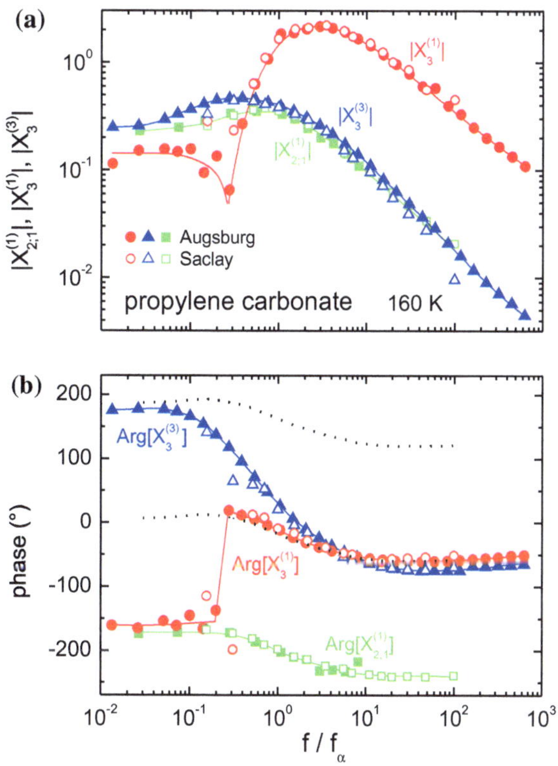

1. For the modulus: At a fixed temperature, the main features of the frequency dependence of $|\chi_3^{(1)}|$ and of $|\chi_{2;1}^{(1)}|$ are the same as those of $|\chi_3^{(3)}|$: when increasing the frequency, one first observes a low-frequency plateau, followed by a hump in the vicinity of f_α and then by a power law decrease $\sim f^{-\beta_3}$. The most important differences between the three cubic susceptibilities are the precise location of the hump and the absolute value of the height of the hump. As for the temperature dependence, one recovers for $|\chi_3^{(1)}|$ and for $|\chi_{2;1}^{(1)}|$ what we have already seen for $|\chi_3^{(3)}|$: once put into their dimensionless forms X_3 the three cubic susceptibilities do not depend on T in the plateau region, at variance with the region of the hump where they increase upon cooling typically as $E_{\rm act}(T) \equiv \partial \ln \tau_\alpha / \partial(1/T)$ which in this T range is very close to $T\chi_T \equiv |\partial \ln \tau_\alpha / \partial \ln T|$ [3, 4, 10, 19, 20, 48].
2. The phases of the three cubic susceptibilities basically do not depend explicitly on temperature, but only on $u = f/f_\alpha$, through a master curve that depends only on the precise cubic susceptibility under consideration. These master curves have the same qualitative shape as a function of u in both glycerol and propylene carbonate. We note that the phases of the three cubic susceptibilities are related to each other. In the plateau region all the phases are equal, which is expected because at low frequency the systems responds adiabatically to the

field. At higher frequencies, we note that for both glycerol and propylene carbonate (expressing the phases in radians):

$$\mathrm{Arg}\left[X_3^{(1)}\right] \approx \mathrm{Arg}\left[X_{2;1}^{(1)}\right] + \pi \quad \text{for } f/f_\alpha \geq 0.5; \tag{9}$$

$$\mathrm{Arg}\left[X_3^{(1)}\right] \approx \mathrm{Arg}\left[X_3^{(3)}\right] \quad \text{for } f/f_\alpha \geq 5 \tag{10}$$

which are quite nontrivial relations.

3. In the phase of $\chi_3^{(1)}$ of propylene carbonate (Fig. 9), a jump of π is observed which is accompanied by the indication of a spikelike minimum in the modulus—see [48] for more details. A similar jump may also be present in glycerol (Fig. 8). This jump in the phase happens at the crossover between the T-independent "plateau" and the strongly T-dependent hump. More precisely in the "plateau" region, one observes a reduction of the real part of the dielectric constant χ_1', while around the hump χ_1' is enhanced. At the frequency of the jump, both effects compensate and this coincides with a very low value of the imaginary part of $X_3^{(1)}$.

2.4 *Frequency and Temperature Dependence of Fifth Harmonic Susceptibility*

In this section, we first explain why measuring $\chi_5^{(5)}$ is interesting for a better understanding of the glass transition. We then see the characteristic features of $\chi_5^{(5)}$ as a function of frequency and temperature.

2.4.1 Interest in the Fifth-Order Susceptibility

In the previous sections, we have seen that the increase of the hump of $|X_3|$ upon cooling has been interpreted as reflecting that of the correlation volume $N_{\rm corr}a^3$. However in practice, this increase of $N_{\rm corr}$ remains modest—typically it is an increase by a factor 1.5—in the range 0.01 Hz$\leq f_\alpha \leq$ 10 kHz, where the experiments are typically performed. Physically, this may be interpreted by the fact that an increase of $N_{\rm corr}$ changes the activation energy, yielding an exponentially large increase of the relaxation time τ_α. Now if one demands, as in standard critical phenomena, to see at least a factor of 10 of increase of $|X_3|$ to be able to conclude on criticality, one is lead to astronomical values of τ_α: extrapolating the above result, e.g., $|X_3| \propto |\partial \ln \tau_\alpha / \partial \ln T|$ and assuming a VFT law for τ_α, one concludes that the experimental characteristic times corresponding to an increase of $|X_3|$ by one order of magnitude is 0.1 ms $\leq \tau_\alpha \leq 10^{18}$ s. This means experiments lasting longer than the age of the universe.

This issue of astronomical timescales can be circumvented by using a less commonly exploited but very general property of phase transitions: close to a critical point all the responses diverge together [49], since the common cause of all these divergences is the growth of the same correlation length. Showing that all the responses of order k behave as a power law of the first diverging susceptibility is another way of establishing criticality. For glasses, we have seen in Eq. (2) that, apart from χ_1 which is blind to glassy correlations, all other responses $\chi_{k\geq 3}$ grow as power laws with the amorphous ordering length ℓ: $\chi_3 \propto (l/a)^{2d_f-d}$ and $\chi_5 \propto (l/a)^{3d_f-d}$. Therefore, assuming that the main cause for the singular responses appearing in the system is the development of correlations, there should be a scaling relation between the third- and fifth-order responses, namely one should observe $\chi_5 \propto \chi_3^{\mu(d_f)}$ where $\mu(d_f) = (3d_f - d)/(2d_f - d)$.

Measuring χ_5 is, of course, extremely difficult, because, for the experimentally available electric fields, one has the hierarchy $|\chi_1|E \gg |\chi_3|E^3 \gg |\chi_5|E^5$. However, this was done in Ref. [50] and we shall now briefly review the corresponding results.

2.4.2 Characteristic Features of the Fifth-Order Susceptibility

The modulus $|\chi_5^{(5)}|$ of glycerol and propylene carbonate [50] can be seen in Fig. 10 as a function of frequency and temperature. Similarly to what has been seen in Sect. 2.2 on $|\chi_3^{(3)}|$, the frequency dependence can be separated in two domains (see also Fig. 11):

1. For very low reduced frequencies ($f/f_\alpha \leq 0.05$), there is a plateau (indicated by the yellow-shaded planes in Fig. 10), where the reduced response $X_5^{(5)}$ depends neither on frequency nor on temperature. In this plateau, the behavior of the supercooled liquid cannot be qualitatively distinguished from the behavior expected from a high-temperature liquid of dipoles, depicted by the "trivial" $X_k^{(k)}$ curves represented as dotted lines in Fig. 11.
2. At higher frequencies, we can observe a hump of $|X_5^{(5)}|$ that remarkably occurs at the same peak frequency f_{peak} as in $|\chi_3^{(3)}|$ in both glycerol and propylene carbonate. Again one finds that, for the five temperatures where the peak is studied, $f_{\text{peak}}/f_\alpha = c$, where the constant c does not depend on T and weakly changes with the liquid. This peak is *much sharper for* $|X_5^{(5)}|$ *than for* $|X_3^{(3)}|$: this is clearly evidenced by Fig. 11 where the linear-, cubic-, and fifth-order susceptibilities are compared, after normalization to their low-frequency value. This shows that the anomalous features in the frequency dependence are stronger in $|X_5^{(5)}|$ than in $|X_3^{(3)}|$: This may be regarded as a sign of criticality since close to a critical point, the larger the order k of the response, the stronger the anomalous features of X_k.

A second, and more quantitative indication of incipient criticality is obtained by studying the temperature dependence of $|X_5^{(5)}|$ and by comparing it with that of $|X_3^{(3)}|$:

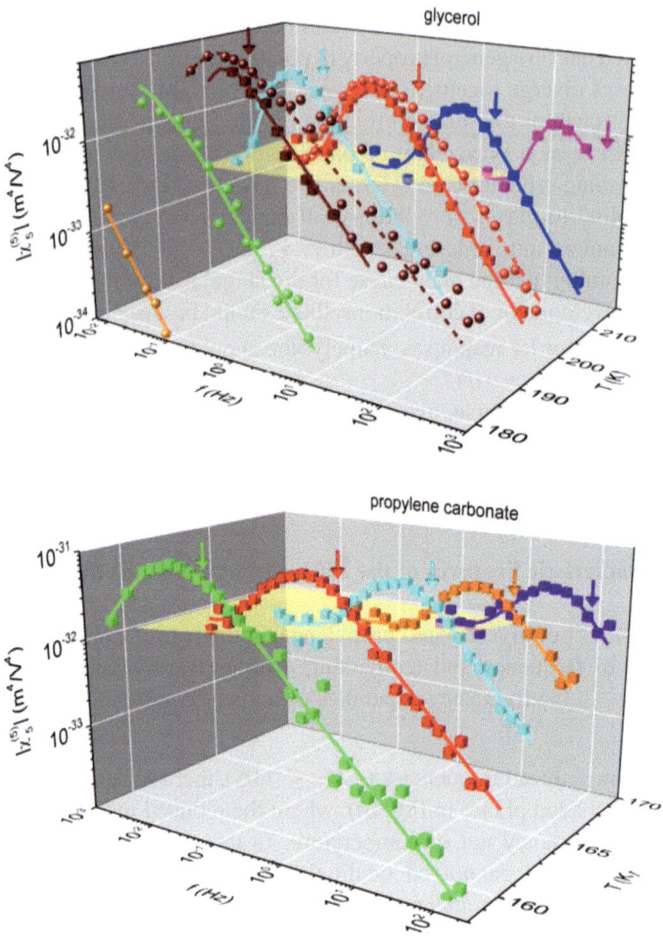

Fig. 10 Measured values of $|\chi_5^{(5)}|$ for glycerol—upper panel—and propylene carbonate—lower panel—(the spheres and cubes in the upper panel indicate results from two different experimental setups). The hump lies at the same frequency as for $|\chi_3^{(3)}|$ and has significantly stronger variations in frequency and in temperature, see Figs. 11 and 12. The arrows indicate the peak positions f_α in the dielectric loss. The yellow-shaded planes indicate the plateau arising in the trivial regime. From Ref. [50]

1. In the plateau region at $f/f_\alpha \leq 0.05$, the value of $|X_5^{(5)}|$ does not depend on the temperature. This shows that the factor involved in the calculation of the dimensionless $X_5^{(5)}$ from $\chi_5^{(5)}$—see Eq. (8)—is extremely efficient to remove all trivial temperature dependences. As the trivial behavior depends on frequency—see the dashed lines of Fig. 11— the "singular" parts of X_3 and of X_5 are obtained as follows:

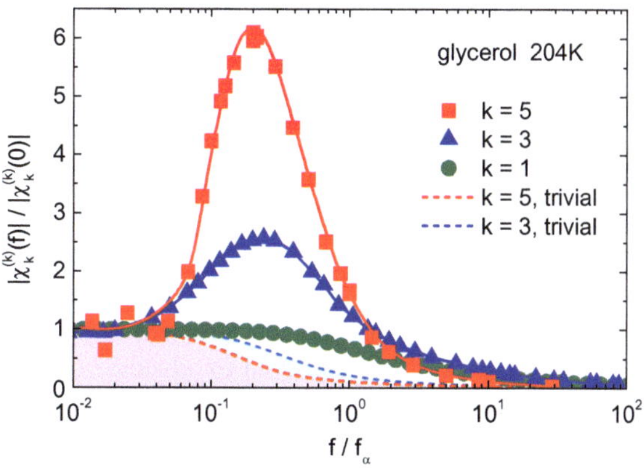

Fig. 11 For glycerol, comparison of the fifth, third and linear susceptibilities—the latter is noted $|\chi_1^{(1)}|$. The hump for $|\chi_5^{(5)}|$ is much stronger than that of $|\chi_3^{(3)}|$. The dashed lines are the trivial contribution—see [50] for details. From Ref. [50]

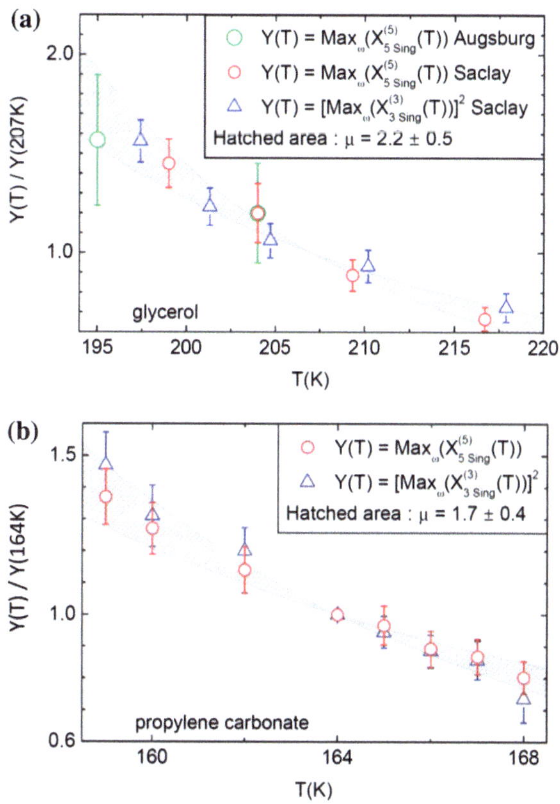

Fig. 12 Temperature evolution of the singular parts of fifth-and third-order responses. All quantities are normalized at a given temperature—namely 207 K for glycerol, upper panel; and 164 K for propylene carbonate, bottom panel. This allows to determine the exponent μ relating $|X_5|$ and $|X_3|^{\mu}$ and to conclude that the amorphously ordering domains are compact —see text. The hatched areas represent the uncertainty on μ. From Ref. [50]

$$X^{(3)}_{3,\text{sing.}} \equiv X^{(3)}_3 - X^{(3)}_{3,\text{trivial}}, \qquad X^{(5)}_{5,\text{sing.}} \equiv X^{(5)}_5 - X^{(5)}_{5,\text{trivial}} \tag{11}$$

which correspond in Fig. 11 to a complex subtraction between the measured data—symbols— and the trivial behavior—dashed lines.

2. Around the hump, the temperature behavior of $|X^{(5)}_{5,\text{sing.}}(f_{\text{peak}})|$ is compared to that of $|X^{(3)}_{3,\text{sing.}}(f_{\text{peak}})|^{\mu}$ where μ is an exponent, that is, determined experimentally by looking for the best overlap of the two series of data in Fig. 12—see [50] for details. This leads us to values of $\mu = 2.2 \pm 0.5$ in glycerol and $\mu = 1.7 \pm 0.4$ in propylene carbonate. Therefore, within experimental uncertainties, results for $|X^{(3)}_3|$ and $|X^{(5)}_5|$ would seem to advocate a value of $\mu \approx 2$. With $\mu = (3d_f - d)/(2d_f - d)$ as seen in Eq. (2)—see also Eq. (13) below—this corresponds to a fractal dimensions of $d_f \approx 3$.

3 Testing Bouchaud–Biroli's Predictions as well as the General Theories of the Glass Transition

Having shown the experimental data for the nonlinear responses, we now move to the interpretation part and start with Bouchaud–Biroli's approach (BB), which is the most general one. The more specific and/or phenomenological approaches of nonlinear responses will be detailed in Sect. 4.

3.1 Bouchaud–Biroli's Predictions

3.1.1 General Considerations About χ_{2k+1}

To illustrate the general relations existing between the susceptibility χ_{2k+1} and the correlation function of order $2k+2$—with $k \geq 0$—in a system at thermal equilibrium, let us consider a sample, submitted to a constant and uniform magnetic field h, containing N spins with an Hamiltonian $\mathcal{H}$ that depends on the spin configuration "c". The elementary relations of statistical physics yield the magnetization $M \equiv \sum_i <S_i>/(Na^3)$, where a^3 is the elementary volume and where the thermal average $<S_i>$ is obtained with the help of the partition function $Z = \sum_c \exp(-\beta\mathcal{H} + \beta h \sum_k S_k)$ by writing $<S_i> = \sum_c S_i \exp(-\beta\mathcal{H} + \beta h \sum_k S_k)/Z$ with $\beta = 1/(k_B T)$. The linear response $\chi_1 \equiv (\partial M/\partial h)_{h=0}$ is readily obtained as follows:

$$Na^3\chi_1 = \frac{1}{\beta Z}\left(\frac{\partial^2 Z}{\partial h^2}\right)_{h=0} - \frac{1}{\beta}\left(\frac{\partial Z}{Z\partial h}\right)^2_{h=0} = \beta\left(\sum_{i1;i2} <S_{i1}S_{i2}> - \left(\sum_{i1} <S_{i1}>\right)^2\right), \tag{12}$$

which shows that the linear response is related to the connected two-point correlation function. Repeating the argument for higher order responses—e.g., $\chi_3 \propto (\partial^3 M/\partial h^3)_{h=0}$—one obtains that χ_{2k+1} is connected to the $(2k+2)$ points correlation function -e.g., χ_3 is connected to a sum combining $<S_{i1}S_{i2}S_{i3}S_{i4}>$, $<S_{i1}S_{i2}S_{i3}>$ $<S_{i4}>$, $<S_{i1}S_{i2}><S_{i3}S_{i4}>$, etc.

3.1.2 The Spin Glass Case

Spin glasses are characterized by the fact that there is frozen disorder, i.e., the set of the interaction constants $\{J_{i;j}\}$ between two given spins S_i and S_j is fixed once and for all, and has a random sign—half of the pairs of spins are coupled ferromagnetically, the other half antiferromagnetically. Despite the fact that the system is neither a ferromagnet, nor an antiferromagnet, upon cooling it freezes, below a critical temperature T_{SG}, into a solid—long-range ordered—state called a spin glass state. This amorphous ordering is not detected by χ_1 which does not diverge at T_{SG}: this is because the various terms of $\sum_{i1;i2} <S_{i1}S_{i2}>$ cancel since half of them are positive and the other half are negative. By contrast, the cubic susceptibility χ_3 contains a term $\sum_{i1;i2} <S_{i1}S_{i2}>^2$, which does diverge since all its components are strictly positive: this comes from the fact that the influence $<S_{i1}S_{i2}>$ of the polarization of spin S_{i1} on spin S_{i2} may be either positive or negative, but it has the same sign as the reverse influence $<S_{i2}S_{i1}>$ of spin S_{i2} on spin S_{i1}. This is why the amorphous ordering is directly elicited by the divergence of the static value of χ_3 when decreasing T toward T_{SG}, as already illustrated in Fig. 1a. By adding a standard scaling assumption close to T_{SG}, one can account for the behavior of χ_3 at finite frequencies, i.e., one easily explains that χ_3 is frequency independent for $\omega\tau_\alpha \leq 1$, and smoothly tends to zero at higher frequencies. Finally, similar scaling arguments about correlation functions easily explain the fact that the stronger $k \geq 1$ the more violent the divergence of χ_{2k+1} in spin glasses, as observed experimentally by Lévy [51].

3.1.3 The Glassforming Liquids Case

The case of glassforming liquids is, of course, different from that of spin glasses for some obvious reasons (e.g., molecules have both translational and rotational degrees of freedom). As it has been well established that rotational and translational degrees of freedom are well coupled in most of liquids, it is tempting to attempt a mapping between spin glasses and glassforming liquids by replacing the spins S_i by the local fluctuations of density $\delta\rho_i$ or by the dielectric polarization p_i. As far as nonlinear responses are concerned, this mapping requires a grain of salt because (a) there is no frozen-in disorder in glassforming liquids, and (b) there is a nonzero value of the molecular configurational entropy S_c around T_g.

The main physical idea of BB's work [9] is that these difficulties have an effect which is important at low frequencies and negligible at high enough frequencies:

1. Provided $f \geq f_\alpha$, i.e., for processes faster than the relaxation time, one cannot distinguish between a truly frozen glass and a still flowing liquid. If some amorphous order is present in the glassforming system, then nontrivial spatial correlations should be present and lead to anomalously high values of nonlinear susceptibilities: this holds for very general reasons—e.g., the Langevin equation for continuous spins which is used in Ref. [9] needs not to specify the detailed Hamiltonian of the system—and comes from an analysis of the most diverging term in the four terms contributing to $\chi_3(\omega)$. If the amorphous correlations extend far enough to be in the scaling regime, one can neglect the subleading terms and one predicts that the nonlinear susceptibilities are dominated by the glassy correlations and given by [9, 50]:

$$X_{2k+1}^{\text{glass}}(f, T) = [N_{\text{corr}}(T)]^{\alpha_k} \times \mathscr{H}_k\left(\frac{f}{f_\alpha}\right) \quad \text{with } \alpha_k = (k+1) - d/d_f, \quad (13)$$

where the scaling functions $\mathscr{H}_k$ do not explicitly depend on temperature, but depend on the kind of susceptibility that is considered, i.e., $X_3^{(1)}$, $X_3^{(3)}$, or $X_{2,1}^{(1)}$ in the third-order case $k = 1$. We emphasize that in Ref. [9] the amorphously ordered domains were assumed to be compact, i.e., $d_f = d$, yielding $\alpha_1 = 1$, i.e., $X_3 \propto N_{\text{corr}}$. The possibility of having a fractal dimension d_f lower than the spatial dimension d was considered in Ref. [50] where the fifth-order response was studied. As already shown in Sect. 2.4.2, the experimental results were consistent with $d_f = d$, i.e., $X_5 \propto N_{\text{corr}}^2$.
2. In the low-frequency regime $f \ll f_\alpha$, relaxation has happened everywhere in the system, destroying amorphous order [52] and the associated anomalous response to the external field and $\mathscr{H}_k(0) = 0$. In other words, in this very-low-frequency regime, every molecule behaves independently of others and X_{2k+1} is dominated by the "trivial" response of effectively independent molecules.

Due to the definition adopted in Eq. (8), the trivial contribution to X_{2k+1} should not depend on temperature (or very weakly). Hence, provided N_{corr} increases upon cooling, there will be a regime where the glassy contribution X_{2k+1}^{glass} should exceed the trivial contribution, leading to hump-shaped nonlinear susceptibilities, peaking at $f_{\text{peak}} \sim f_\alpha$, where the scaling function $\mathscr{H}_k$ reaches its maximum.

3.2 *Experimentally Testing BB's Predictions*

We now briefly recall why all the experimental features reported in Sect. 2 are well accounted for by BB's prediction:

1. The modulus of both the third- order susceptibilities $|\chi_3^{(3)}|$, $|\chi_3^{(1)}|$, $|\chi_{2;1}^{(1)}|$ and of $|\chi_5^{(5)}|$ have a humped shape in frequency, contrary to $|\chi_1|$.

2. Due to the fact that $\mathscr{H}_k$ does not depend explicitly on T, the value of f_{peak}/f_α should not depend on temperature, consistent with the experimental behavior.
3. Because of the dominant role played by the glassy response for $f \geq f_{\rm peak}$, the T-dependence of $|X_{2k+1}|$ will be much stronger above $f_{\rm peak}$ than in the trivial low-frequency region.
4. Finally, because nonlinear susceptibilities are expressed in terms of scaling functions, it is natural that the behavior of their moduli and phases are quantitatively related especially at high frequency where the "trivial" contribution can be neglected, consistent with Eqs. (9) and (10)—see below for a more quantitative argument in the context of the so-called "Toy model" [53].

Having shown that BB's prediction is consistent with experiments, the temperature variation of $N_{\rm corr}$ can be drawn from the increase of the hump of X_3 upon cooling. It has been found [3, 4, 10, 19, 20] that the temperature dependence of $N_{\rm corr}$ inferred from the height of the humps of the three X_3s are compatible with one another, and closely related to the temperature dependence of $T\chi_T$, which was proposed in Refs. [22, 23] as a simplified estimator of $N_{\rm corr}$ in supercooled liquids. The convergence of these different estimates, that rely on general, model-free theoretical arguments, is a strong hint that the underlying physical phenomenon is indeed the growth of collective effects in glassy systems—a conclusion that will be reinforced by analyzing other approaches in Sect. 4.

Let us again emphasize that the BB prediction relies on a scaling argument, where the correlation length ℓ of amorphously ordered domains is (much) larger than the molecular size a. This naturally explains the similarities of the cubic responses in microscopically very different liquids such as glycerol and propylene carbonate, as well as many other liquids [10, 20]. Indeed, the microscopic differences are likely to be wiped out for large $\ell \propto N_{\rm corr}^{1/d_f}$, much like in usual phase transitions.

3.3 *Static Versus Dynamic Length Scale? χ_3 and χ_5 as Tests of the Theories of the Glass Transition*

We now shortly discuss whether $N_{\rm corr}$, as extracted from the hump of $|X_3|$, must be regarded as a purely dynamical correlation volume, or as a static correlation volume. This ambiguity arises because theorems relating (in a strict sense) nonlinear responses to high-order correlation functions only exist in the static case, and that supplementary arguments are needed to interpret the humped shape of X_3 (and of X_5) observed experimentally. In the original BB's work [9], it was clearly stated that $N_{\rm corr}$ was a dynamical correlation volume since it was related to a four-point time-dependent correlation function. This question was revisited in Ref. [50] where it was argued that the experimental results could be accounted for only when assuming that $N_{\rm corr}$ is driven by static correlations. This statement comes from an inspection of the various theories of the glass transition [50]: as we now briefly explain, only the theories where the underlying static correlation volume is driving the dynamical correlation volume are consistent with the observed features of nonlinear responses.

As a first example, the case of the family of kinetically constrained models (KCMs) [54] is especially interesting since dynamical correlations, revealed by, e.g., four-point correlation functions, exist even in the absence of a static correlation length. However in the KCM family, one does not expect any humped shape for nonlinear responses [50]. This is not the case for theories (such as RFOT [46] or Frustration theories [55]) where a nontrivial thermodynamic critical point drives the glass transition: in this case, the incipient amorphous order allows to account [50] for the observed features of X_3 and X_5. This is why it was argued in [48, 50] that, in order for X_3 and X_5 to grow, some incipient amorphous order is needed, and that dynamical correlations in strongly supercooled liquids are driven by static ("point-to-set") correlations [56] —this statement will be reinforced in Sect. 4.2.

4 More Specific Models for Harmonic Susceptibilities

We now review the various other approaches that have been elaborated for the nonlinear responses of glassforming liquids. We shall see that most of them—if not all—are consistent with BB's approach since they involve N_{corr} as a key—implicit or explicit— parameter.

4.1 Toy and Pragmatical Models

The "Toy model" has been proposed in Refs. [19, 57] as a simple incarnation of the BB mechanism, while the "Pragmatical model" is more recent [58, 59]. Both models start with the same assumptions: (i) each amorphously ordered domain is compact and contains N_{corr} molecules, which yields a dipole moment $\propto \sqrt{N_{\text{corr}}}$ and leads to an anomalous contribution to the cubic response $X_3^{\text{glass}} \propto N_{\text{corr}}$; (ii) there is a crossover at low frequencies toward a trivial cubic susceptibility contribution X_3^{triv} which does not depend on N_{corr}. More precisely, in the "Toy model", each amorphously ordered domain is supposed to live in a simplified energy landscape, namely an asymmetric double-well potential with a dimensionless asymmetry δ, favoring one well over the other. The most important difference between the Toy and the Pragmatical model comes from the description of the low-frequency crossover, see Refs. [57, 59] for more details.

On top of N_{corr} and δ, the Toy model uses a third adjustable parameter, namely the frequency f^* below which the trivial contribution becomes dominant. In Ref. [57], *both the modulus and the phase* of $X_3^{(3)}(\omega, T)$ and of $X_3^{(1)}(\omega, T)$ in glycerol were well fitted by using $f^* \simeq f_\alpha/7, \delta = 0.6$ and, for $T = 204$ K, $N_{\text{corr}} = 5$ for $X_3^{(3)}$ and $N_{\text{corr}} = 15$ for $X_3^{(1)}$. Figure 13 gives an example of the Toy model prediction for

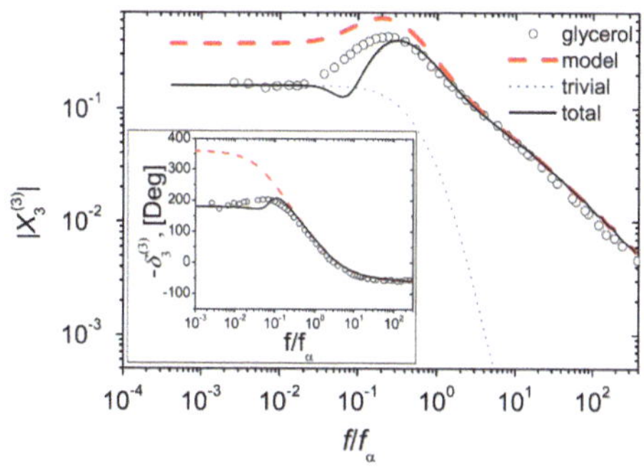

Fig. 13 Fit of the values of $X_3^{(3)}$ measured in glycerol—symbols—at 204 K by using the Toy model with $N_{\mathrm{corr}} = 5$, $\delta = 0.6$ and $f^* \simeq f_\alpha/7$. The prediction of the Toy model is given by the two thick solid lines (main panel for the modulus of $X_3^{(3)}$ and inset for its phase). From Ref. [57]

$X_3^{(3)}$ in glycerol. Besides, in Ref. [19], the behavior of $X_{2,1}^{(1)}(\omega, T)$ in glycerol was fitted with the same values of δ and of f^* but with $N_{\mathrm{corr}} = 10$ (at a slightly different temperature $T = 202$ K). Of course, the fact that a different value of N_{corr} must be used for the three cubic susceptibilities reveals that the Toy model is oversimplified, as expected. However, keeping in mind that the precise value of N_{corr} does not change the behavior of the phases, we note that the fit of the three experimental phases is achieved [19, 57] by using the very same values of f^*/f_α and of δ. This means that Eqs. (9) and (10) are well accounted for by the Toy model by choosing two free parameters. This is a quantitative illustration of how the BB general framework does indeed lead to strong relations between the various nonlinear susceptibilities, such as those contained in Eqs. (9) and (10).

Let us mention briefly the Asymmetric Double-Well Potential (ADWP) model [60], which is also about species living in a double well of asymmetry energy Δ, excepted that two key assumptions of the Toy and Pragmatical models are not made: the value of N_{corr} is not introduced, and the crossover to trivial cubic response is not enforced at low frequencies. As a result, the hump for $|X_3^{(3)}|$ is predicted [60, 61] only when the reduced asymmetry $\delta = \tanh(\Delta/(2k_{\mathrm{B}}T))$ is close to a very specific value, namely $\delta_{\mathrm{c}} = \sqrt{1/3}$, where X_3 vanishes at zero frequency due to the compensation of its several terms. However, at the fifth order [61], this compensation happens for two values of δ very different from δ_{c}: as a result, the model cannot predict a hump happening both for the third and for the fifth order in the same parametric regime, contrarily to the experimental results of Ref. [50]. This very recent calculation of fifth-order susceptibility [61] reinforces the point of view of the Toy and Pragmatical models, which do predict a hump occurring at the same frequency and temperature due to their two key assumptions (N_{corr} and crossover to trivial nonlinear responses at low frequencies). This can be understood qualitatively: because the Toy model predicts [57] an anomalous contribution $X_{2k+1}^{\mathrm{glass}} \sim [N_{\mathrm{corr}}]^k$, provided that N_{corr} is large enough, the magnitude of this contribution is much larger than that of the small trivial contribution $X_{2k+1}^{\mathrm{triv.}} \sim 1$, and the left side of the peak of $|X_{2k+1}|$ arises just because the Toy model enforces a crossover from the large anomalous response to the small trivial response at low frequencies $f \ll f_\alpha$. As for the right side of the peak, it comes from the fact that $|X_{2k+1}| \to 0$ when $f \gg f_\alpha$ for the simple reason that the supercooled liquid does not respond to the field at very large frequencies.

4.2 Entropic Effects

A contribution to nonlinear responses was recently calculated by Johari in Refs. [62, 63] in the case where a static field E_{st} drives the supercooled liquid in the nonlinear regime. Johari's idea was positively tested in the corresponding $\chi_{2;1}^{(1)}$ experiments in Refs. [64–67]—see however Ref. [68] for a case where the agreement is not as good. It was then extended to pure ac experiments—and thus to $\chi_3^{(3)}$—in Refs. [69, 70]. The relation between Johari's idea and N_{corr} was made in Ref. [48].

4.2.1 When a Static Field E_{st} is applied

Let us start with the case of $\chi_{2;1}^{(1)}$ experiments, i.e., with the case where a static field E_{st} is superimposed onto an AC field $E\cos(\omega t)$. In this case, there is a well-defined variation of entropy $[\delta S]_{E_{\text{st}}}$ induced by E_{st}, which, for small E_{st} and a fixed T, is given by

$$[\delta S]_{E_{\text{st}}} \approx \frac{1}{2}\varepsilon_0 \frac{\partial \Delta\chi_1}{\partial T} E_{\text{st}}^2 a^3, \tag{14}$$

where a^3 is the molecular volume. Equation (14) holds generically for any material. However, in the specific case of supercooled liquids close enough to their glass transition temperature T_g, a special relation exists between the molecular relaxation time τ_α and the configurational contribution to the entropy S_c. This relation, first anticipated by Adam and Gibbs [21], can be written as

$$\ln \frac{\tau_\alpha(T)}{\tau_0} = \frac{\Delta_0}{T S_c(T)}, \tag{15}$$

where τ_0 is a microscopic time, and Δ_0 is an effective energy barrier for a molecule. The temperature dependence of $T S_c(T)$ quite well captures the temperature variation of $\ln(\tau_\alpha)$, at least for a large class of supercooled liquids [71].

Following Johari [62, 63] let us now assume that $[\delta S]_{E_{\text{st}}}$ is dominated by the dependence of S_c on field—see the Appendix of Ref. [48] for a further discussion of this important physical assumption. Combining Eqs. (14) and (15), one finds that a static field E_{st} produces a shift of $\ln(\tau_\alpha/\tau_0)$ given by

$$[\delta \ln \tau_\alpha]_{E_{\text{st}}} = -\frac{\Delta_0}{T S_c^2} [\delta S]_{E_{\text{st}}} \,. \tag{16}$$

As shown in Ref. [48] this entropic effect gives a contribution to $X^{(1)}_{2;1}$, which we call $J^{(1)}_{2;1}$ after Johari. Introducing $x = \omega\tau_\alpha$, the most general and model-free expression of $J^{(1)}_{2;1}$ reads

$$J^{(1)}_{2;1} = -\frac{k_B \Delta_0}{6 S_c^2}\left[\frac{\partial \ln(\Delta\chi_1)}{\partial T}\right]\left[\frac{\partial \frac{\chi_{\text{lin}}}{\Delta\chi_1}}{\partial \ln x}\right] \propto \frac{1}{S_c^2}, \tag{17}$$

where χ_{lin} is the complex linear susceptibility.

Equation (17) deserves three comments:

1. $|J^{(1)}_{2;1}|$ has a humped shaped in frequency with a maximum in the region of $\omega\tau_\alpha \simeq 1$, because of the frequency dependence of the factor $\propto \partial\chi_{\text{lin}}/\partial \ln x$ in Eq. (17).
2. The temperature variation of $J^{(1)}_{2;1}$ is overwhelmingly dominated by that of S_c^{-1} because $S_c \propto (T - T_K)$ with T_K the Kauzmann temperature.
3. The smaller S_c, the larger must be the size of the amorphously ordered domains—in the hypothetical limit where S_c would vanish, the whole sample would be trapped in a single amorphously ordered sate and N_{corr} would diverge. In other words, there is a relation between S_c^{-1} and N_{corr}, which yields [48]:

$$J^{(1)}_{2;1} \propto N^q_{\text{corr}}, \tag{18}$$

where it was in shown in Ref. [48] that:

a. the exponent q lies in the [2/3; 2] interval when one combines the Adam–Gibbs original argument with general constraints about boundary conditions [48].
b. the exponent q lies in the [1/3; 3/2] interval [48] when one uses the RFOT and plays with its two critical exponents Ψ and θ. Notably, taking the "recommended RFOT values"—$\Psi = \theta = 3/2$ for $d = 3$—gives $q = 1$, which precisely corresponds to BB's prediction. In this case, entropic effects are a physically motivated picture of BB's mechanism—see [48] for a refined discussion.

4.2.2 When a Pure AC Field $E\cos(\omega t)$ is applied

Motivated by several works [64–67] showing that Johari's reduction of entropy fairly well captures the measured $\chi^{(1)}_{2;1}$ in various liquids, an extension of this idea was proposed in Refs. [69, 70] for pure ac experiments, i.e., for $\chi_3^{(3)}$ and $\chi_3^{(1)}$. This has given rise to the phenomenological model elaborated in Refs. [69, 70] where the entropy reduction depends on time, which is nevertheless acceptable in the region $\omega\tau_\alpha \leq 1$ where the model is used. Figure 14 shows the calculated values for $|\chi_3^{(3)}|$ at three temperatures for glycerol. The calculation fairly well reproduces the hump of the modulus observed experimentally—the phase has not been calculated. As

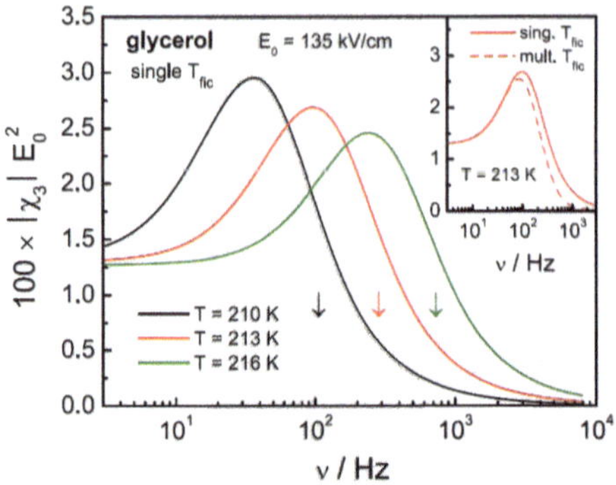

Fig. 14 The model elaborated in Refs. [69, 70] includes three contributions—entropy reduction, Box model, and trivial. It predicts for $|\chi_3^{(3)}|$ the solid lines which account very well for the measured values in glycerol in frequency and in temperature. The peak of $|\chi_3^{(3)}|$ arises because of the entropy reduction effect (noticed "sing. $T_{\rm fic.}$") which completely dominates the two other contributions in the peak region, as shown by the inset. From Ref. [70]

very clearly explained in Ref. [70], the hump displayed in Fig. 14 comes directly from the entropic contribution and not from the two other contributions included in the model (namely the "trivial"—or "saturation"—contribution, and the Box model contribution—see Sect. 4.3 below).

Summarizing this section about entropy effects, we remind the two main assumptions made by Johari: (i) the field-induced entropy variation mainly goes into the configurational part of the entropy; (ii) its effects can be calculated by using the Adam–Gibbs relation. Once combined, these two assumptions give a contribution to $\chi_{2;1}^{(1)}$ reasonably well in agreement with the measured values in several liquids [64–67]. An extension to $\chi_3^{(3)}$ is even possible, at least in the region $\omega\tau_\alpha \leq 1$ and fairly well accounts for the measured hump of $|\chi_3^{(3)}|$ in glycerol [69, 70]—a figure similar to Fig. 11 for $|\chi_5^{(5)}(\omega)/\chi_5^{(5)}(0)|$ is even obtained in Ref. [70]. As shown in Eq. (18), this entropy contribution to cubic responses is related to $N_{\rm corr}$, which is consistent with the general prediction of BB. Additionally, because S_c is a static quantity, Eq. (18) supports the interpretation that the various cubic susceptibilities χ_3 are related to static amorphous correlations, as discussed in Sect. 3.3.

4.3 *Box Model*

4.3.1 Are Nonlinear Effects Related to Energy Absorption?

The "Box model" is historically the first model of nonlinear response in supercooled liquids, designed to account for the Nonresonant Hole Burning (NHB) experiments [72]. When these pioneering experiments were carried out, a central question was whether the dynamics in supercooled liquids is homogeneous or heterogeneous. In the seminal Ref. [72] it was reported that when applying a strong AC field E of angular frequency ω, the changes in the dielectric spectrum are localized close to ω and that they last a time of the order of $1/\omega$. These two findings yield a strong qualitative support to the heterogeneous character of the dynamics, and the Box model was designed to provide a quantitative description of these results. Accordingly, the Box model assumes that the dielectric response comes from "domains"—that will be later called Dynamical Heterogeneities (DH)— each domain being characterized by its dielectric relaxation time τ and obeying the Debye dynamics. The distribution of the various τ's is chosen to recover the measured non Debye spectrum by adding the various linear Debye susceptibilities $\chi_{1,dh} = \Delta\chi_1/(1 - i\omega\tau)$ of the various domains. For the nonlinear response, the Box model assumes that it is given by the Debye linear equation in which $\tau(T)$ is replaced by $\tau(T_f)$ where the fictive temperature $T_f = T + \delta T_f$ is governed by the constitutive equation—see e.g., [28, 73]:

$$c_{dh}\frac{\partial(\delta T_f)}{\partial t} + \kappa\delta T_f = \frac{1}{2}\varepsilon_0\chi''_{1,dh}\omega E^2 \tag{19}$$

with c_{dh} the volumic specific heat of the DH under consideration, κ the thermal conductance (divided by the DH volume v) between the DH and the phonon bath, $\tau_{\text{therm}} = c/\kappa$ the corresponding thermal relaxation time. In Eq. (19), only the constant part of the dissipated power has been written, omitting its component at 2ω which is important only for $\chi_3^{(3)}$—see e.g., [73]. From Eq. (19) one easily finds the stationary value $\delta T_f^\star$ of δT_f which reads as follows:

$$\delta T_f^\star = \frac{\tau_{\text{therm}}}{\tau}\frac{\varepsilon_0\Delta\chi_1 E^2}{2c_{dh}}\frac{\omega^2\tau^2}{1+\omega^2\tau^2}. \tag{20}$$

As very clearly stated in the seminal Ref. [72] because the DH size is smaller than 5 nm, the typical value of τ_{therm} is at most in the nanoseconds range: this yields, close to T_g, a vanishingly small value of τ_{therm}/τ, which, because of Eq. (20), gives fully negligible values for $\delta T_f^\star$. The choice of the Box model is to increase τ_{therm} by orders of magnitude by setting $\tau_{\text{therm}} = \tau$, expanding onto the intuition that this is a way to model the "energy storage" in the domains. The main justification of this choice is *its efficiency*: it allows to account reasonably well for the NHB experiments [72] and thus to bring a strong support to the heterogeneous character of the dynamics in supercooled liquids. Since the seminal Ref. [72], some other works have shown [28, 74–76] that the Box model efficiently accounts for the measured $\chi_3^{(1)}(f > f_\alpha)$ in many glassforming liquids. It was shown also [73] that the Box model is *not* able to fit quantitatively the measured $X_3^{(3)}$ (even though some qualitative features are accounted for), and that the Box model only provides a vanishing contribution to $X_{2,1}^{(1)}$—see [19].

The key choice $\tau_{\text{therm}} = \tau$ made by the Box model has two important consequences for cubic susceptibilities: it implies (a) that $\chi_3^{(1)}$ mainly comes from the energy absorption (since the source term in Eq. (19) is the dissipated power) and

(b) that $\chi_3^{(1)}$ does not explicitly depend on the volume $v = N_{\rm corr} a^3$ of the DH's (see [28, 73]). However, alternative models of nonlinear responses are now available [57, 59] where, instead of choosing $\tau_{\rm therm}$, one directly resolves the microscopic population equations, which is a molecular physics approach, and not a macroscopic law transferred to microscopics. The population equations approach is equivalent to solving the relevant multidimensional Fokker–Planck equation describing the collective tumbling dynamics of the system at times longer than the time between two molecular collisions (called τ_c in Appendix 3). By using this molecular physics approach one obtains that $\chi_3^{(1)}$ is governed by $N_{\rm corr}$ and not by energy absorption. For $\chi_3^{(1)}$, writing loosely $P_3^{(1)} \approx \partial P_1/(\partial \ln \tau)\delta \ln \tau$, one sees that the pivotal quantity is the field-induced shift of the relaxation time $\delta \ln \tau$. Comparing the Box model (BM) and, e.g., the Toy model (TM), one gets respectively:

$$\delta \ln \tau_{BM} \simeq -\frac{1}{2}\chi_T \frac{\varepsilon_0 \Delta\chi_1 E^2}{c_{dh}} \quad ; \quad \delta \ln \tau_{TM} \simeq -\frac{3}{2}\frac{N_{\rm corr}}{T}\frac{\varepsilon_0 \Delta\chi_1 E^2}{k_{\rm B}/a^3} \tag{21}$$

where we remind our definition $\chi_T = |\partial \ln \tau_\alpha / \partial T|$ and where the limit $\omega\tau \gg 1$, relevant for the $\chi_3^{(1)}(f > f_\alpha)$ was taken in the Box model, while the simplest case (symmetric double well with a net dipole parallel to the field) was considered for the Toy model. Equation (21) deserves two comments:

1. One sees that the two values of $\delta \ln \tau$ are similar provided $N_{\rm corr}$ and $T\chi_T$ are proportional— which is a reasonable assumption as explained above and in Refs. [4, 22, 23]. Taking reasonable values of this proportionality factor, it was shown in Ref. [48] that $\chi_3^{(1)}(f > f_\alpha)$ is *the same* in the two models. This sheds a new light on the efficiency of the Box model and on consequence (b).
2. Let us shortly discuss consequence (a). In the Toy model, $\delta \ln \tau$ directly expresses the field-induced modification of the energy of each of the two wells modeling a given DH. It comes from the work produced by E onto the DH and this is why it involves $N_{\rm corr}$: the larger this number, the larger the work produced by the field because the net dipole of a DH is $\propto \sqrt{N_{\rm corr}}$ and thus increases with $N_{\rm corr}$. It is easy to show that the dissipation—i.e., the "energy absorption"—is *not* involved in $\delta \ln \tau$ because dissipation depends only on χ_1'', which in the Toy model does not depend on $N_{\rm corr}$. In the Toy model, as in the Pragmatical model [59] and the Diezemann model [60], the heating is neglected because at the scale of a given DH it is vanishingly small as shown above when discussing $\tau_{\rm therm}$. Of course, at the scale of the whole sample, some global heating arises for thick samples and/or high frequencies because the dissipated power has to travel to the electrodes which are the actual heat sinks in dielectric experiments [11]. This purely exogenous effect can be precisely calculated by solving the heat propagation equation, see e.g., Ref. [11] and Appendix 2, and must not be confused with what was discussed in this section.

4.3.2 Gathering the Three Measured Cubic Susceptibilities

As explained above, in Refs. [69, 70], the three experimental cubic susceptibilities have been argued to result from a superposition of an entropic contribution and of an energy absorption contribution coming from the Box model (plus a trivial contribution playing a minor role around the peaks of the cubic susceptibilities). More precisely, the hump of $|X_{2,1}^{(1)}|$ and of $|X_3^{(3)}|$ would be mainly due to the entropy effect, contrarily to the hump of $|X_3^{(1)}|$ which would be due to the Box model contribution. As noted in Ref. [48], this means that very different physical mechanisms would conspire to give contributions of the same order of magnitude, with phases that have no reason to match as they do empirically, see Eqs. (9) and (10): why should $X_3^{(1)}$ and $X_3^{(3)}$ have the same phase at high frequencies if their physical origin is different?

This is why it was emphasized in Ref. [48] that there is no reason for such a similarity if the growth of $X_3^{(1)}$ and $X_3^{(3)}$ are due to independent mechanisms. Because entropic effects have been related to the increase of $N_{\rm corr}$—see Eqs. (17) and (18)—everything becomes instead very natural if the Box model is recast in a framework where $X_3^{(1)}$ is related to the glassy correlation volume. As evoked above, a first step in this direction was done in Ref. [48] where it was shown that the Box model prediction for $X_3^{(1)}$ at high frequencies is identical to the above Toy model prediction, provided $N_{\rm corr}$ and $T\chi_T$ are proportional. In all, it is argued in Ref. [48] that the only reasonable way to account for the similarity of all three cubic susceptibilities, demonstrated experimentally in Figs. 8 and 9, is to invoke a common physical mechanism. As all the other existing approaches, previously reviewed, relate cubic responses to the growth of the glassy correlation volume, reformulating the Box model along the same line seems to be a necessity.

5 Conclusions

We have reviewed in this chapter the salient features reported for the third and fifth harmonic susceptibilities close to the glass transition. This is a three decades long story, which has started in the mid-80s as a decisive tool to evidence the solid, long range ordered, nature of the spin glass phase. The question of whether this notion of "amorphous order" was just a curiosity restricted to the—somehow exotic—case of spin glasses remained mostly theoretical until the seminal work of BB in 2005. This work took a lot from the spin glass physics, and by taking into account the necessary modifications relevant for glass forming liquids, it has anticipated all the salient features discovered in the last decade for the three cubic susceptibilities X_3. This is why, in most of the works, the increase of the hump of X_3 upon cooling has been interpreted as reflecting that of the glassy correlation volume. Challenging alternative and more specific interpretations have been proposed, but we have seen that most—if not all—of them can be recasted into the framework of BB. The avenue opened by BB's prediction was also used to circumvent the issue of exponentially long timescales—which are the reason why the nature of the glass transition is still debated: this is how the idea of comparing the anomalous features of X_3 and of X_5 has arisen. The experimental findings are finally consistent with the existence of an underlying thermodynamic critical point, which drives the formation of amorphously ordered compact domains, the size of which increases upon cooling. Last we note that this field of nonlinear responses in supercooled liquids has been inspiring both theoretically [5, 77] and experimentally, e.g., for colloidal glasses: the very recent experiments [6] have shed a new light on the colloidal glass transition and shown interesting differences with glassforming liquids.

All these progress open several routes of research. On the purely theoretical side, any prediction of nonlinear responses in one of the models belonging to the Kinetically Constrained Model family will be extremely welcome to go beyond the general arguments given in Ref. [50]. Moreover, it would be very interesting to access χ_3 (and χ_5) in molecular liquids at higher temperatures, closer to the Mode Coupling Transition temperature T_{MCT}, and/or for frequencies close to the fast β process where more complex, fractal structures with $d_f < d$ may be anticipated [78, 79]. This will require a joined effort of experimentalists—to avoid heating issues—and of theorists—to elicit the nature of nonlinear responses close to T_{MCT}. Additionally, one could revisit the vast field of polymers by monitoring their nonlinear responses, which should shed new light onto the temperature evolution of the correlations in these systems. Therefore, there is likely much room to deepen our understanding of the glass transition by carrying out new experiments about nonlinear susceptibilities.

Acknowledgements We thank C. Alba-Simionesco, Th. Bauer, U. Buchenau, A. Coniglio, G. Johari, K. Ngai, R. Richert, G. Tarjus, and M. Tarzia for interesting discussions. The work in Saclay has been supported by the Labex RTRA grant Aricover and by the Institut des Systèmes Complexes ISC-PIF. The work in Augsburg was supported by the Deutsche Forschungsgemeinschaft via Research Unit FOR1394.

Appendix 1: Making Sure that Exogenous Effects Are Negligible

We briefly explain how the nonlinear effects reported here have been shown to be—mainly—free of exogenous effects:

1. The global homogeneous heating of the samples by the dielectric energy dissipated by the application of the strong ac field E was shown to be fully negligible for $X_3^{(3)}$ as long as the inverse of the relaxation time f_α is ≤ 1 kHz, see Ref. [11]. Note that these homogeneous heating effects contribute much more to $X_3^{(1)}$: to minimize them, one can either keep f_α below 10 Hz [4], and/or severely limit the number n of periods during which the electric field is applied—see, e.g., [28, 80]).

2. The contribution of electrostriction was demonstrated to be safely negligible in Refs. [4, 74], both by using theoretical estimates and by showing that changing the geometry of spacers does not affect $X_3^{(3)}$.
3. As for the small ionic impurities present in most of liquids, we briefly explain that they have a negligible role, except at zero frequency where the ion contribution might explain why the three X_3s are not strictly equal, contrarily to what is expected on general grounds—see e.g., Figs. 8 and 9. On the one hand, it was shown that the ion heating contribution is fully negligible in $X_{2,1}^{(1)}$ (see Ref. [19]), on the other hand it is well known that ions affect the linear response χ_1 at very low frequencies (say $f/f_\alpha \leq 0.05$): this yields an upturn on the out-of-phase linear response χ_1'', which diverges as $1/\omega$ instead of vanishing as ω in an ideally pure liquid containing only molecular dipoles. This may be the reason why most of the χ_3 measurements are reported above $0.01 f_\alpha$: at lower frequencies the nonlinear responses are likely to be dominated by the ionic contribution.

Appendix 2: Trivial Third and Fifth Harmonic Susceptibilities

As explained in the main text, in the long time limit—i.e., for $f/f_\alpha \ll 1$—the liquid flow destroys the glassy correlations, making each molecule effectively independent of others. This is why we briefly recall what the nonlinear responses of an ideal gas of dipoles are, where each dipole is independent of others, and undergoes a Brownian rotational motion—of characteristic time τ_D—due to the underlying thermal reservoir at temperature T. The linear susceptibility of such an ideal gas of dipoles is given by the Debye susceptibility $\Delta\chi_1/(1 - i\omega\tau_\mathrm{D})$, hence the subscript "Debye" in the Eq. (22) below. By using Refs. [17], and following the definitions given in the main text, as well as Eqs. (5)–(8) above, one gets for the dimensionless nonlinear responses of such an ideal gas, setting for brevity $x = \omega\tau_\mathrm{D}$:

$$\begin{aligned} X_{3,\mathrm{Debye}}^{(3)} &= \left(\frac{-3}{5}\right) \frac{3 - 17x^2 + ix(14 - 6x^2)}{(1+x^2)(9+4x^2)(1+9x^2)} \\ X_{5,\mathrm{Debye}}^{(5)} &= \frac{432(72 - 2377x^2 - 1979x^4 + 2990x^6)}{1680(1+x^2)(4+x^2)(9+4x^2)(1+9x^2)(9+16x^2)(1+25x^2)} \\ &\quad + i\frac{432x(246 - 737x^2 - 1623x^4 + 200x^6)}{560(1+x^2)(4+x^2)(9+4x^2)(1+9x^2)(9+16x^2)(1+25x^2)}. \end{aligned} \tag{22}$$

In Ref. [50] the trivial response combined the above $X_{k,\mathrm{Debye}}^{(k)}$ with a distribution $\mathcal{G}(\tau)$ of relaxation times τ chosen to account for the linear susceptibility of the supercooled liquid of interest. In Refs. [19, 57] a slightly different modelization was used since $\mathcal{G}(\tau)$ was replaced by the Dirac delta function $\delta(\tau - \tau_\alpha)$, i.e., τ_D was simply replaced by τ_α for the cubic trivial susceptibilities.

Appendix 3: Derivation of the Toy model from Langevin–Fokker–Planck Considerations

In this section, we shall rederive thc phenomenological Toy model of Ladieu et al. [57] starting from the Langevin–Fokker–Planck equation, which is the starting point of BB when they illustrate their general theoretical ideas in the last part of Ref. [9]. We shall idealize the supercooled state of a liquid as follows. At high temperatures, the liquid is made of molecules the interactions between which are completely negligible. On cooling, the molecules arrange themselves in groups, called "dynamical heterogeneities" (DH), between which there are no interactions. Inside a typical group, specific intermolecular interactions manifest themselves dynamically, by which we mean that in a time larger than a characteristic time τ_α, such interactions lose their coherence and the typical behavior of the liquid is that of an ideal gas. Before and around τ_α, these interactions manifest themselves in a frequency range $\omega \approx 1/\tau_\alpha$. Thus, sensu stricto, our modeling of this specific process pertains to the behavior of the various dielectric responses of a DH, linear and nonlinear, near this frequency range. This indeed implies that information regarding the "ideal gas" phase must be added to fit experimental data. It may be shown on fairly general grounds that either for linear and nonlinear responses, such extra information simply superposes onto the specific behavior that has been alluded to above [81]. Now, we consider that (a) a given DH has a given size at temperature T, (b) that a DH is made of certain mobile elements that do interact between themselves, (c) that there are no interactions between DHs, (d) that the dipole moment of a DH is $\mu_d = \mu\sqrt{N_{\text{corr}}}$, and (e) that all constituents of a DH are subjected to Brownian motion.

In order to translate the above assumptions in mathematical language, we assign to each constituent of a DH a generalized coordinate $q_i(t)$, so that each DH is described by a set of generalized coordinates $\mathbf{q}$ at temperature T, viz.,

$$\mathbf{q}(t) = \{q_1(t), \ldots, q_n(t)\}.$$

Inside each DH, each elementary constituent is assumed to interact via a multidimensional interaction potential $V_{\text{int}}(\mathbf{q})$ that possesses a double-well structure with minima at $\mathbf{q}_A$ and $\mathbf{q}_B$, and are sensitive both to external stresses and thermal agitation. The equations of motion may be described by overdamped Langevin equations with additive noise, viz.,

$$\dot{q}_i = -\frac{1}{\zeta}\frac{\partial V_T}{\partial q_i}(\mathbf{q}, t) + \Xi_i(t) \tag{23}$$

where ζ is a generalized friction coefficient, $V_T = V_{int} + V_{\text{ext}}$, V_{ext} is the potential energy of externally applied forces and the generalized forces $\Xi_i(t)$ have Gaussian white noise properties, namely

$$\overline{\Xi_i(t)} = 0, \ \overline{\Xi_i(t)\,\Xi_j(t')} = \frac{2kT}{\zeta}\delta_{ij}\delta\left(t - t'\right). \tag{24}$$

Thus, the dynamics of a DH is represented by the stochastic differential equations (23) and (24), which are in effect the starting point of the Bouchaud–Biroli theory, as stated above. A totally equivalent representation of these stochastic dynamics is obtained by writing down the Fokker–Planck equation [82] for the probability density $W(\mathbf{q}, t)$ to find the system in state $\mathbf{q}$ at time t which corresponds to Eqs. (23) and (24), namely

$$\begin{aligned}\frac{\partial W}{\partial t}(\mathbf{q}, t) &= \frac{1}{2\tau_c}\nabla \cdot \left[\nabla W(\mathbf{q}, t) + \beta W(\mathbf{q}, t)\nabla V_T(\mathbf{q}, t)\right] \\ &= L_{\mathrm{FP}}(\mathbf{q}, t)\, W(\mathbf{q}, t)\end{aligned} \tag{25}$$

where $2\tau_c = \zeta/(kT)$ is the characteristic time of fluctuations, ∇ is the del operator in $\mathbf{q}$ space, and $L_{\mathrm{FP}}(\mathbf{q}, t)$ is the Fokker–Planck operator. We notice that Eq. (25) may also be written as follows:

$$\frac{\partial W}{\partial t}(\mathbf{q}, t) = \frac{1}{2\tau_c}\nabla \cdot \left\{e^{-\beta V_T(\mathbf{q},t)}\nabla\left[W(\mathbf{q}, t)\, e^{\beta V_T(\mathbf{q},t)}\right]\right\}. \tag{26}$$

Now, we use the transformation [83]

$$\phi(\mathbf{q}, t) = W(\mathbf{q}, t)\, e^{\beta V_T(\mathbf{q},t)} \tag{27}$$

so that Eq. (26) becomes

$$\begin{aligned}\frac{\partial \phi}{\partial t}(\mathbf{q}, t) - \beta\frac{\partial V_T}{\partial t}(\mathbf{q}, t)\,\phi(\mathbf{q}, t) &= \frac{1}{2\tau_c}e^{\beta V_T(\mathbf{q},t)}\nabla \cdot \left\{e^{-\beta V_T(\mathbf{q},t)}\nabla\phi(\mathbf{q}, t)\right\} \\ &= L_{\mathrm{FP}}^{\dagger}(\mathbf{q}, t)\,\phi(\mathbf{q}, t)\end{aligned} \tag{28}$$

where $L_{\mathrm{FP}}^{\dagger}(\mathbf{q}, t)$ is the adjoint Fokker–Planck operator [82].

Next, we make the first approximation in our derivation, namely, we assume that the time variation of V_T is small with respect to that of W. If the time dependence of V_T is contained in, say, the application of a time-varying uniform AC field only, this implies immediately that neglecting the second term in the left hand side of Eq. (28) means that W is near its equilibrium value, so restricting further calculations to low frequencies, $\omega\tau_c \ll 1$ (quasi-stationary condition). Hence, Eq. (28) now reads

$$\frac{\partial \phi}{\partial t}(\mathbf{q}, t) \approx L_{\mathrm{FP}}^{\dagger}(\mathbf{q}, t)\,\phi(\mathbf{q}, t) \tag{29}$$

Now, the interpretation of the Fokker–Planck equation (25) [or equally well the Langevin equations (23)] with time-dependent potential in terms of usual population equations with time-dependent rate coefficients has a meaning, since now Eq. (27) means detailed balancing. The polarization of an assembly of noninteracting DH in the direction of the applied field may then be defined as

$$P(t) = \rho_0 \mu_d \int \cos \vartheta (\mathbf{q}) W(\mathbf{q}, t) d\mathbf{q}, \tag{30}$$

where ρ_0 is the number of DH per unit volume, and $\vartheta(\mathbf{q})$ is the angle a DH dipole makes with the externally applied electric field. Because of the double-well structure of the interaction potential, we may equally well write Eq. (30)

$$P(t) = \rho_0 \mu_d \left[\int_{\text{well } A} \cos \vartheta (\mathbf{q}) W(\mathbf{q}, t) d\mathbf{q} + \int_{\text{well } B} \cos \vartheta (\mathbf{q}) W(\mathbf{q}, t) d\mathbf{q} \right]. \tag{31}$$

Now, it is known from the Kramers theory of chemical reaction rates [83] that at sufficiently large energy barriers, most of the contributions of the integrands come from the minima of the wells, therefore, we have

$$P(t) \approx \rho_0 \mu_d \left[\cos \vartheta (\mathbf{q}_A) \int_{\text{well } A} W(\mathbf{q}, t) d\mathbf{q} + \cos \vartheta (\mathbf{q}_B) \int_{\text{well } B} W(\mathbf{q}, t) d\mathbf{q} \right] \tag{32}$$

Now, the integrals represent the relative populations $x_i(t) = n_i(t)/N, \;\; i = A, B$ in each well (we assume that $W(\mathbf{q}, t)$ is normalized to unity), where $n_i(t)$ is the number of DH states in well i, and N the total number of DH. At any time t, we have the conservation law

$$x_A(t) + x_B(t) = 1. \tag{33}$$

Thus, Eq. (32) reads

$$P(t) \approx \rho_0 \mu_d \left[\cos \vartheta (\mathbf{q}_A) x_A(t) + \cos \vartheta (\mathbf{q}_B) x_B(t) \right]. \tag{34}$$

We assume now for simplicity that $\vartheta(\mathbf{q}_B) = \pi - \vartheta(\mathbf{q}_A)$, so that

$$P(t) \approx \rho_0 \mu_d \cos \vartheta (\mathbf{q}_A) \left[x_A(t) - x_B(t) \right] \tag{35}$$

Finally, since $\rho_0 = N/V$ where V is the volume of the polar substance made of DH only, we obtain

$$P(t) \approx \frac{\mu_d \cos \vartheta (\mathbf{q}_A)}{N \upsilon_{\text{DH}}} \left[n_A(t) - n_B(t) \right] \tag{36}$$

where υ_{DH} is the volume of a DH. This is the definition of the polarization in the Toy model.

In order to determine the polarization (36), we need to calculate the dynamics of $n_i(t)$. From the conservation law—Eq. (33)—we have

$$\dot{x}_A(t) = -\dot{x}_B(t) = \int\limits_{\text{well } A} \frac{\partial W}{\partial t}(\mathbf{q}, t)\, d\mathbf{q} \tag{37}$$

By using the Fokker–Planck equation (26) and limiting well A to a closed generalized bounding surface constituting the saddle region ∂A, we have by Gauss's theorem

$$\dot{x}_A(t) = -\dot{x}_B(t) = \frac{1}{2\tau_c} \oint\limits_{\partial A} e^{-\beta V_T(\mathbf{q},t)} \nabla\phi(\mathbf{q}, t) \cdot \nu_{\mathbf{q}} dS_{\mathbf{q}} \tag{38}$$

where $\nu_{\mathbf{q}}$ is the outward normal to the bounding surface and $dS_{\mathbf{q}}$ is a generalized surface element of the bounding surface, and where we have used Eq. (27). Now, we follow closely Coffey et al. [84] and introduce the crossover function $\Delta(\mathbf{q}, t)$ via the equation

$$\phi(\mathbf{q}, t) = \phi_A(t) + [\phi_B(t) - \phi_A(t)]\, \Delta(\mathbf{q}, t) \tag{39}$$

where $\Delta(\mathbf{q}, t) = 0$ if $\mathbf{q} \in$ well A while $\Delta(\mathbf{q}, t) = 1$ if $\mathbf{q} \in$ well B and exhibits strong gradients in the saddle region ∂A allowing the crossing from A to B (and vice versa) by thermally activated escape. By combining Eqs. (38) and (39), we have immediately

$$\dot{x}_A(t) = -\dot{x}_B(t) = \frac{\phi_B(t) - \phi_A(t)}{2\tau_c} \oint\limits_{\partial A} e^{-\beta V_T(\mathbf{q},t)} \nabla\left[\Delta(\mathbf{q}, t)\right] \cdot \nu_{\mathbf{q}} dS_{\mathbf{q}}. \tag{40}$$

Now,

$$x_i(t) = \phi_i(t)\, x_i^s(t), \quad x_i^s(t) = \int\limits_{\text{well } i} W_s(\mathbf{q}, t)\, d\mathbf{q} \tag{41}$$

where $W_s(\mathbf{q}, t)$ is a normalized solution of the Fokker–Planck equation

$$L_{\text{FP}}(\mathbf{q}, t)\, W_s(\mathbf{q}, t) = 2\tau_c \frac{\partial W_s}{\partial t}(\mathbf{q}, t) \approx 0 \tag{42}$$

because the frequencies we are concerned with are very small with respect to the inverse thermal fluctuation time τ_c and because the time-dependent part of the potential V_T is much smaller than other terms in it at any time. We have

$$x_A^s(t) + x_B^s(t) = 1. \tag{43}$$

Using Eqs. (41) and (43), we may easily show that [84]

$$\phi_B(t) - \phi_A(t) = \left(\frac{1}{x_A^s(t)} + \frac{1}{x_B^s(t)}\right)\left[x_B(t)\, x_A^s(t) - x_A(t)\, x_B^s(t)\right]. \tag{44}$$

By combining Eqs. (38) and (44), we readily obtain

$$\dot{x}_A(t) = -\dot{x}_B(t) = \Gamma(t)\left(x_B(t)\,x_A^s(t) - x_A(t)\,x_B^s(t)\right). \tag{45}$$

where the overall time-dependent escape rate $\Gamma(t)$ is given by [84]

$$\Gamma(t) = \frac{1}{2\tau_c}\left(\frac{1}{x_A^s(t)} + \frac{1}{x_B^s(t)}\right)\oint_{\partial A} e^{-\beta V_T(\mathbf{q},t)}\nabla\phi(\mathbf{q},t)\cdot\nu_{\mathbf{q}}dS_{\mathbf{q}}. \tag{46}$$

Finally, by setting

$$\Pi_{AB}(t) = \Gamma(t)\,x_B^s(t), \quad \Pi_{BA}(t) = \Gamma(t)\,x_A^s(t), \tag{47}$$

we arrive at the population equations

$$\dot{n}_A(t) = -\dot{n}_B(t) = -\Pi_{AB}(t)\,n_A(t) + \Pi_{BA}(t)\,n_B(t). \tag{48}$$

The obtaining of a more explicit formula for the various rates involved in Eq. (47) is not possible, due to the impossibility to calculate the surface integral in Eq. (46) explicitly, in turn due to the fact that V_T is not known explicitly. Then, the rates in Eqs. (47) and (48) are estimated using Arrhenius's formula. All subsequent derivations regarding the Toy model of Ladieu et al. [57] follow immediately and will not be repeated here due to lack of room and straightforward but laborious algebra.

References

1. L.-P. Lévy, A.T. Oglieski, Nonlinear dynamic susceptibilities at the spin-glass transition of Ag:Mn. Phys. Rev. Lett. **57**, 3288 (1986)
2. J. Hemberger, H. Ries, A. Loidl, R. Böhmer, Static freezing transition at a finite temperature in a quasi-one-dimensional deuteron glass. Phys. Rev. Lett. **76**, 2330 (1996)
3. C. Crauste-Thibierge, C. Brun, F. Ladieu, D. L'Hôte, G. Biroli, J.-P. Bouchaud, Evidence of growing spatial correlations at the glass transition from nonlinear response experiments. Phys. Rev. Lett. **104**, 165703 (2010)
4. C. Brun, F. Ladieu, D. L'Hôte, M. Tarzia, G. Biroli, J.-P. Bouchaud, Nonlinear dielectric susceptibilities: accurate determination of the growing correlation volume in a supercooled liquid. Phys. Rev. B **84**, 104204 (2011)
5. J.M. Brader, M. Siebenbürger, M. Ballauff, K. Reinheimer, M. Wilhelm, S.J. Frey, F. Weysser, M. Fuchs, Nonlinear response of dense colloidal suspensions under oscillatory shear: mode-coupling theory and Fourier transform rheology experiments. Phys. Rev. E **82**, 061401 (2010)
6. R. Seyboldt, D. Merger, F. Coupette, M. Siebenbürger, M. Ballauff, M. Wilhelm, M. Fuchs, Divergence of the third harmonic stress response to oscillatory strain approaching the glass transition. Soft Matter **12**, 8825 (2016)
7. K. Binder, A.P. Young, Spin glasses: experimental facts, theoretical concepts, an dopen questions. Rev. Mod. Phys. **58**, 801 (1986)
8. U.T. Höchli, K. Knorr, A. Loidl, Orientational glasses. Adv. Phys. **39**, 405 (1990)

9. J.-P. Bouchaud, G. Biroli, Nonlinear susceptibility in glassy systems: a probe for cooperative dynamical length scales. Phys. Rev. B **72**, 064204 (2005)
10. Th. Bauer, P. Lunkenheimer, A. Loidl, Cooperativity and the freezing of molecular motion at the glass transition. Phys. Rev. Lett. **111**, 225702 (2013)
11. C. Brun, C. Crauste-Thibierge, F. Ladieu, D. L'Hôte, Study of the heating effect contribution to the nonlinear dielectric response of a supercooled liquid. J. Chem. Phys. **133**, 234901 (2010)
12. C. Gainaru, R. Meier, S. Schildmann, C. Lederle, W. Hiller, E.A. Rössler, R. Böhmer, Nuclear-magnetic-resonance measurements reveal the origin of the Debye process in monohydroxy alcohols. Phys. Rev. Lett. **105**, 258303 (2010)
13. Th. Bauer, M. Michl, P. Lunkenheimer, A. Loidl, Nonlinear dielectric response of Debye, α, and β relaxation in 1-propanol. J. Non-Cryst. Solids **407**, 66 (2015)
14. L.P. Singh, R. Richert, Watching hydrogen-bonded structures in an alcohol convert from rings to chains. Phys. Rev. Lett. **109**, 167802 (2012)
15. L.P. Singh, C. Alba-Simionesco, R. Richert, Dynamics of glass-forming liquids. XVII. Dielectric relaxation and intermolecular association in a series of isomeric octyl alcohols. J. Chem. Phys. **139**, 144503 (2013)
16. P. Lunkenheimer, M. Michl, Th. Bauer, A. Loidl, Investigation of nonlinear effects in glassy matter using dielectric methods. Eur. Phys. J. Spec. Top. **226**, 3157 (2017)
17. W.T. Coffey, B.V. Paranjape, Proc. Roy. Ir. Acad. **78A**, 17 (1978); J. L. Déjardin, Yu. P. Kalmykov, Nonlinear dielectric relaxation of polar molecules in a strong ac electric field: Steady state response. Phys. Rev. E **61**, 1211 (2000)
18. M. Michl, Th. Bauer, A. Loidl, Nonlinear dielectric spectroscopy in a fragile plastic crystal. J. Chem. Phys. **144**, 114506 (2016)
19. D. L'Hôte, R. Tourbot, F. Ladieu, P. Gadige, Control parameter for the glass transition of glycerol evidenced by the static-field-induced nonlinear response. Phys. Rev. B **90**, 104202 (2014)
20. R. Casalini, D. Fragiadakis, C.M. Roland, Dynamic correlation length scales under isochronal conditions. J. Chem. Phys. **142**, 064504 (2015)
21. G. Adam, J.H. Gibbs, On the temperature dependence of cooperative relaxation properties in glass-forming liquids. J. Chem. Phys. **43**, 139 (1965)
22. L. Berthier, G. Biroli, J.-P. Bouchaud, L. Cipeletti, D. El Masri, D. L'Hôte, F. Ladieu, M. Pierno, Direct experimental evidence of a growing length scale accompanying the glass transition. Science **310**, 1797 (2005)
23. C. Dalle-Ferrier, C. Thibierge, C. Alba-Simionesco, L. Berthier, G. Biroli, J.-P. Bouchaud, F. Ladieu, D. L'Hôte, G. Tarjus, Spatial correlations in the dynamics of glassforming liquids: Experimental determination of their temperature dependence, Phys. Rev. E **76**, 041510 (2007); L. Berthier, G. Biroli, J.-P. Bouchaud, W. Kob, K. Miyazaki, D. R. Reichman, Spontaneous and induced dynamic fluctuations in glass formers. I. General results and dependence on ensemble and dynamics. J. Chem. Phys. **126**, 184503 (2007)
24. R. Brand, P. Lunkenheimer, A. Loidl, Relaxation dynamics in plastic crystals. J. Chem. Phys. **116**, 10386 (2002)
25. P. Lunkenheimer, U. Schneider, R. Brand, A. Loidl, Glassy dynamics. Contemp. Phys. **41**, 15 (2000)
26. U. Schneider, P. Lunkenheimer, R. Brand, A. Loidl, *Dielectric and far-infrared spectroscopy of glycerol*. J. Non-Cryst. Solids **235-237**, 173 (1998)
27. Th. Bauer, P. Lunkenheimer, S. Kastner, A. Loidl, Nonlinear dielectric response at the excess wing of glass-forming liquids. Phys. Rev. Lett. **110**, 107603 (2013)
28. R. Richert, S. Weinstein, Nonlinear dielectric response and thermodynamic heterogeneity in liquids. Phys. Rev. Lett. **97**, 095703 (2006)
29. S. Samanta, R. Richert, Limitations of heterogeneous models of liquid dynamics: very slow rate exchange in the excess wing. J. Chem. Phys. **140**, 054503 (2014)
30. Th. Bauer, *Nichtlineare dielektrische Spektroskopie zum Nachweis von Kooperativitt und Heterogenitt in glasbildenden Flssigkeiten* (Verlag Dr. Hut, Munich, 2015)

31. U. Schneider, R. Brand, P. Lunkenheimer, A. Loidl, Excess wing in the dielectric loss of glass formers: a Johari-Goldstein β relaxation? Phys. Rev. Lett. **84**, 5560 (2000)
32. A. Döß, M. Paluch, H. Sillescu, G. Hinze, From strong to fragile glass formers: secondary relaxation in polyalcohols. Phys. Rev. Lett. **88**, 095701 (2002)
33. M. Beiner, H. Huth, K. Schröter, Crossover region of dynamic glass transition: general trends and individual aspects. J. Non-Cryst. Solids **279**, 126 (2001)
34. K.L. Ngai, An extended coupling model description of the evolution of dynamics with time in supercooled liquids and ionic conductors. J. Phys.: Condens. Matter **15**, S1107 (2003)
35. K.L. Ngai, Interpreting the nonlinear dielectric response of glass-formers in terms of the coupling model. J. Chem. Phys. **142**, 114502 (2015)
36. L. Berthier, G. Biroli, Theoretical perspective on the glass transition and amorphous materials. Rev. Mod. Phys. **83**, 587 (2011)
37. A.Q. Tool, Relation between inelastic deformability and thermal expansion of glass in its annealing range. J. Am. Ceram. Soc. **29**, 240 (1946)
38. O.S. Narayanaswamy, Model of structural relaxation in glass. J. Am. Ceram. Soc. **54**, 491 (1971)
39. C.T. Moynihan, A.J. Easteal, M.A. Bolt, J. Tucker, Dependence of fictive temperature of glass on cooling rate. J. Am. Ceram. Soc. **59**, 12 (1976)
40. V. Lubchenko, P.G. Wolynes, Theory of aging in structural glasses. J. Chem. Phys. **121**, 2852 (2004)
41. F. Bert, V. Dupuis, E. Vincent, J. Hammann, J.-P. Bouchaud, Spin anisotropy and slow dynamics in spin glasses. Phys. Rev. Lett. **92**, 167203 (2004)
42. F. Baity-Jesi et al., Matching microscopic and macroscopic responses in glasses. Phys. Rev. Lett. **118**, 157202 (2017)
43. C. Brun, F. Ladieu, D. L'Hôte, G. Biroli, J.-P. Bouchaud, Evidence of growing spatial correlations during the aging of glassy glycerol. Phys. Rev. Lett. **109**, 175702 (2012)
44. R.L. Leheny, S.R. Nagel, Frequency-domain study of physical aging in a simple liquid. Phys. Rev. B **57**, 5154 (1998)
45. P. Lunkenheimer, R. Wehn, U. Schneider, A. Loidl, Glassy aging dynamics. Phys. Rev. Lett. **95**, 055702 (2005)
46. M. Dzero, J. Schmalian, P.G. Wolynes, in *Structural Glasses and Supercooled Liquids: Theory, Experiment, and Applications*, ed. by P.G. Wolynes, V. Lubchenko (Wiley, 2012), pp. 193–222; G. Biroli, J.-P. Bouchaud, in *Structural Glasses and Supercooled Liquids: Theory, Experiment, and Applications*, ed. by P.G. Wolynes, V. Lubchenko (Wiley, 2012), pp. 31–114; G. Biroli, J.-P. Bouchaud, *The Random First-Order Transition Theory of Glasses: A Critical Assessment*. arXiv:0912.2542v1
47. C. Cammarota, A. Cavagna, G. Gradenigo, T.S. Grigera, P. Verrocchio, Numerical determination of the exponents controlling the relationship between time, length, and temperature in glass-forming liquids. J. Chem. Phys. **131**, 194901 (2009)
48. P. Gadige, S. Albert, M. Michl, Th. Bauer, P. Lunkenheimer, A. Loidl, R. Tourbot, C. Wiertel-Gasquet, G. Biroli, J.-P. Bouchaud, F. Ladieu, Unifying different interpretations of the nonlinear response in glass-forming liquids. Phys. Rev. E **96**, 032611 (2017)
49. Y. Kimura, S. Hara, R. Hayakawa, Nonlinear dielectric relaxation spectroscopy of ferroelectric liquid crystals. Phys. Rev. E **62**, R5907–R5910 (2000)
50. S. Albert, Th. Bauer, M. Michl, G. Biroli, J.-P. Bouchaud, A. Loidl, P. Lunkenheimer, R. Tourbot, C. Wiertel-Gasquet, F. Ladieu, Fifth-order susceptibility unveils growth of thermodynamic amorphous order in glass-formers. Science **352**, 1308 (2016)
51. L.P. Lévy, Critical-dynamics of metallic spin glasses. Phys. Rev. B **38**, 4963–4973 (1988)
52. L.Lévy, Phys. Rev. B **38**, 4963 (1988). This comes from the non zero value of the configurational entropy in supercooled liquids. This configurational entropy does not exist in the case of spin glasses, hence there is a finite probability for an amorphously ordered domain to come back to a given configuration even after having relaxed. As a result, instead of a hump for X_3 around f_α, there is a plateau for $X_3(f \leq f_\alpha) \simeq X_3(0)$, the amplitude of which diverges when aproaching the spin glass transition

53. It would be worth to develop further the MCT analysis of [85] to obtain detailed analytical predictions on the phases
54. D. Chandler, J.P. Garrahan, Dynamics on the way to forming glass: bubbles in space-time. Annu. Rev. Phys. Chem. **61**, 191–217 (2010)
55. G. Tarjus, S.A. Kivelson, Z. Nussinov, P. Viot, The frustration-based approach of supercooled liquids and the glass transition: a review and critical assessment. J. Phys.: Cond. Matt. **17**, R1143–R1182 (2005)
56. The growth of χ_3 and χ_5 can be induced by purely dynamical correlation, as it is the case in MCT theory and in the MCT regime. However, in the activated dynamics regime, the only theory that can explain a growth of χ_3 and χ_5 are the ones in which dynamical correlations are driven by static ("point-to-set") correlations
57. F. Ladieu, C. Brun, D. L'Hôte, Nonlinear dielectric susceptibilities in supercooled liquids: a toy model. Phys. Rev. B **85**, 184207 (2012)
58. U. Buchenau, Retardation and flow at the glass transition. Phys. Rev. E **93**, 032608 (2016)
59. U. Buchenau, Pragmatical access to the viscous flow of undercooled liquids. Phys. Rev. E **95**, 062603 (2017)
60. G. Diezemann, Nonlinear response theory for Markov processes: aimple models for glassy relaxation. Phys. Rev. E **85**, 051502 (2012)
61. G. Diezemann, Nonlinear response theory for Markov processes II: fifth-order response functions. Phys. Rev. E **96**, 022150 (2017)
62. G.P. Johari, Effects of electric field on the entropy, viscosity, relaxation time, and glass-formation. J. Chem. Phys. **138**, 154503 (2013)
63. G.P. Johari, Effects of electric field on thermodynamics and ordering of a dipolar liquid. J. Chem. Phys. **145**, 164502 (2016)
64. S. Samanta, R. Richert, Dynamics of glass-forming liquids. XVIII. Does entropy control structural relaxation times? J. Chem. Phys. **142**, 044504 (2015)
65. A.R. Young-Gonzales, S. Samanta, R. Richert, Dynamics of glass-forming liquids. XIX. Rise and decay of field induced anisotropy in the non-linear regime. J. Chem. Phys. **143**, 104504 (2015)
66. B. Riechers, K. Samwer, R. Richert, Structural recovery in plastic crystals by time-resolved non-linear dielectric spectroscopy. J. Chem. Phys. **142**, 154504 (2015)
67. S. Samanta, R. Richert, Electrorheological source of nonlinear dielectric effects in molecular glass-forming liquids. J. Phys. Chem. B **120**, 7737 (2016)
68. S. Samanta, O. Yamamuro, R. Richert, Connecting thermodynamics and dynamics in a supercooled liquid: Cresolphthalein-dimethylether. Thermochim Acta **636**, 57 (2016)
69. R. Richert, Non-linear dielectric signatures of entropy changes in liquids subject to time dependent electric fields. J. Chem. Phys. **144**, 114501 (2016)
70. P. Kim, A.R. Young-Gonzales, R. Richert, Dynamics of glass-forming liquids. XX. Third harmonic experiments of non-linear dielectric effects versus a phenomenological model. J. Chem. Phys. **145**, 064510 (2016)
71. R. Richert, C.A. Angell, Dynamics of glass-forming liquids. V. On the link between molecular dynamics and configurational entropy. J. Chem. Phys. **108**, 9016 (1998)
72. B. Schiener, R. Böhmer, A. Loidl, R.V. Chamberlin, Nonresonant spectral hole burning in the slow dielectric response of supercooled liquids. Science **274**, 752 (1996); B. Schiener, R.V. Chamberlin, G. Diezemann, R.Böhmer, Nonresonant dielectric hole burning spectroscopy of supercooled liquids. J. Chem. Phys. **107**, 7746 (1997)
73. C. Brun, C. Crauste-Thibierge, F. Ladieu, D. L'Hôte, Third harmonics nonlinear susceptibility in supercooled liquids: a comparison to the box model. J. Chem. Phys. **134**, 194507 (2011)
74. S. Weinstein, R. Richert, Nonlinear features in the dielectric behavior of propylene glycol. Phys. Rev. B **75**, 064302 (2007)
75. L.-M. Wang, R. Richert, Measuring the configurational heat capacity of liquids. Phys. Rev. Lett. **99**, 185701 (2007)
76. A. Khalife, U. Pathak, R. Richert, Heating liquid dielectrics by time dependent fields. Eur. Phys. J. B **83**, 429 (2011)

77. E. Lippiello, F. Corberi, A. Sarracino, M. Zannetti, Nonlinear response and fluctuation-dissipation relations. Phys. Rev. E **78**, 041120 (2008); E. Lippiello, F. Corberi, A. Sarracino, M. Zannetti, Nonlinear susceptibilities and the measurement of a cooperative length. Phys. Rev. E **77**, 212201 (2008)
78. G. Biroli, J.-P. Bouchaud, K. Miyazaki, D.R. Reichman, Inhomogeneous mode-coupling theory and growing dynamic length in supercooled liquids. Phys. Rev. Lett. **97**, 195701 (2006)
79. J.D. Stevenson, J. Schmalian, P.G. Wolynes, The shapes of cooperatively rearranging regions in glass-forming liquids. Nat. Phys. **2**, 268–274 (2006)
80. Th. Bauer, P. Lunkenheimer, S. Kastner, A. Loidl, Phys. Rev. Lett. **110**, 107603 (2013) (see Supplementary Information)
81. P.M. Dejardin, F. Ladieu, Unpublished work
82. H. Risken, *The Fokker-Planck Equation*, 2nd edn. (Springer, Berlin, 1989)
83. H.A. Kramers, Brownian motion in a field of force and the diffusion theory of chemical reactions. Physica **7**, 284 (1940)
84. W.T. Coffey, D.A. Garanin, D.J. McCarthy, Crossover formulas in the Kramers theory of thermally activated escape rates: application to spin systems. Adv. Chem. Phys. **117**, 483 (2001)
85. M. Tarzia, G. Biroli, A. Lefèvre, J.-P. Bouchaud, Anomalous nonlinear response of glassy liquids: general arguments and a mode-coupling approach. J. Chem. Phys. **132**, 054501 (2010)

Dynamic Correlation Under Isochronal Conditions

C. M. Roland and D. Fragiadakis

Abstract Results of various methods of evaluating the dynamic correlation volume in glassforming liquids and polymers are summarized. Most studies indicate that this correlation volume depends only on the α-relaxation time; that is, at state points associated with the same value of τ_α, the extent of the correlation among local motions is equivalent. Nonlinear dielectric spectroscopy was used to measure the third-order susceptibility. Its amplitude, a measure of the dynamic correlation volume, is constant for isochronal state points, which supports the interpretation of the magnitude of the nonlinear susceptibility in terms of dynamic correlation. More broadly, it serves to establish that for non-associated materials, the cooperativity of molecular motions is connected to their timescale.

1 Introduction

Among the many interesting features of glass formation is that it takes place without obvious structural changes on the molecular level. The static structure factor (variance in the density ρ) reflecting short-range correlations is essentially the same above and below the glass transition temperature, T_g [1, 2]. The only apparent effects of vitrification on structure come from changes in volume. For this reason, the glass transition is usually regarded as a dynamic phenomenon. Translational and rotational motions become drastically slower due to their cooperative nature, as molecules cannot move independently. As shown by various experiments [3–7], this has two related consequences: dynamic heterogeneity, referring to the coexistence of fast- and slow-moving molecules, with the mobility variations persisting for times commensurate with (or longer than [8]) the primary (α) relaxation time, τ_α; and dynamic correlation, which refers to the mutual interdependence that extends over some length scale. Molecular dynamics (MD) simulations show that these two properties are correlated for a given material, but not generally (Fig. 1; [9]). Characteristics of the glass transition result from these many-body effects, including the distribution of mobili-

C. M. Roland (✉) · D. Fragiadakis
Naval Research Laboratory, Chemistry Division, Washington, DC 20375-5342, USA
e-mail: mike.roland@nrl.navy.mil

R. Richert (ed.), *Nonlinear Dielectric Spectroscopy*, Advances in Dielectrics
https://doi.org/10.1007/978-3-319-77574-6_8

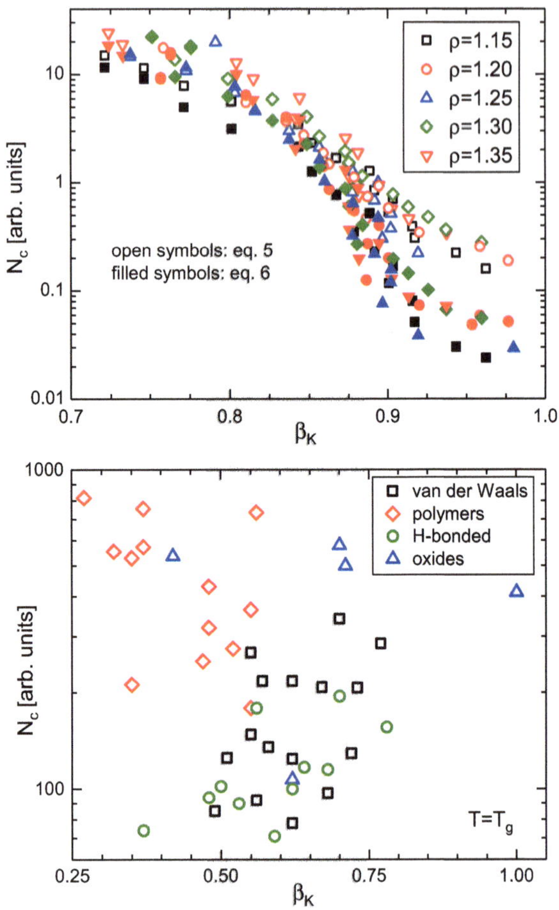

Fig. 1 Comparison of dynamic correlation and the distribution of relaxation times, both reflections of dynamic heterogeneity: (top) MD simulations of the Kob-Andersen binary mixture of Lennard-Jones particles; (bottom) experimental data for different glassforming materials [9]

ties reflected in the breadth of the relaxation dispersion (non-Debye relaxation) and the non-Arrhenius temperature dependence; unsurprisingly, these two properties are connected [10, 11].

Heterogeneous dynamics defines a length scale, ξ, or relatedly the number of dynamically correlated molecules (or polymer segments)

$$N_c \leq \frac{4\pi\rho N_A}{3m}\xi^3 \tag{1}$$

where M is the molecular weight (of the repeat unit for polymers) and N_A is Avogadro's number. Dynamic heterogeneity can be observed directly in colloids [12] and granular systems [13], although for molecular liquids it is difficult to measure because both spatial and temporal correlations are involved. Nevertheless, such information is a prerequisite for "solving" the glass transition problem. If τ_α is coupled to a dynamic correlation length, theoretical models that address τ_α are making predictions, explicitly or otherwise, for ξ. It is for this reason among others that dynamic correlations are an essential component of theoretical efforts [14–18]. Our purpose herein is to review the available data on the connection, if any, between τ_α and ξ or N_c.

2 Determination of Dynamic Correlation Volume

A variety of methods have been used to estimate ξ or N_c. One approach is confinement of the material to spatial dimensions on the order of ξ. Nanoscale confinement can be imposed in one dimension as in supported or freestanding thin films, in two dimensions, e.g., porous glasses, or three dimensions for nano-sized droplets. It is well established that in the absence of adhesion to the walls [19–21], such geometric confinement of supercooled liquids accelerates their dynamics. One interpretation is that the speeding up of local motions occurs when confinement dimensions are on the order of the cooperative length scale. Experiments along these lines yield estimates of ξ in the range of 2–10 nm; that is, several intermolecular distances, increasing with decreasing temperature [22–26].

Using computer simulations, ξ can be estimated in a similar but more rigorous way using a so-called point-to-set construction [27–31]. Typically, in this method a subset of particles from an equilibrium configuration is frozen, forming an amorphous wall, and ξ is defined as the length over which the effect of the wall on liquid dynamics propagates into the liquid. Alternatively, the particles may be frozen to form geometries of a frozen spherical cavity, a thin film of liquid confined between frozen walls, or a set of randomly pinned particles, thus imposing a confining length scale on the system (size of the cavity, film thickness, and distance between pinned particles, respectively), with ξ determined by the dependence of dynamics on the confining length, mirroring experiments on dynamics in confinement. This method has also been experimentally realized on colloidal glasses for the wall and random pinning geometries [32, 33]. The dynamic length scale obtained in this way generally grows on cooling. For some systems in amorphous wall geometries, however, a nonmonotonic temperature dependence of ξ is observed [30, 31]. It is unclear whether this is an intrinsic property of the liquid related perhaps to change in the shape of cooperative rearranging regions with temperature [30], or an effect arising from particularities of dynamics near a wall [34].

Another method relies on structural length scales that are not imposed externally but already exist within the liquid. Polymer chains provide such a length scale: the end-to-end distance (coil size) with a corresponding timescale, the normal mode relaxation. Using the argument that relaxation times and dynamic length scales are correlated, if two processes have the same relaxation time at a temperature T, at this temperature the spatial extent of the molecular motions corresponding to the relaxation will be the same. This line of reasoning leads to the supposition that segmental relaxation times and (extrapolated) relaxation times for the normal mode (end-to-end relaxation) of a polymer are equal at state points for which ξ equals the chain coil size [35]. For polypropylene and polyisoprene near T_g, the method yields $N_c \sim 20$, increasing to $N_c > 200$ for temperatures a few degrees below T_g [6, 36, 37].

The above methods of determining dynamic length scales are indirect; Spiess and coworkers pioneered the use of multidimensional ^{13}C solid-state exchange NMR to directly determine the length scale of dynamic heterogeneities [38–40]. Combining two 2-D spin-echo pulse sequences, the experiment measures the fluctuations within the distribution of relaxation rates. Values of ξ in the range 1–4 nm were obtained at temperatures slightly above T_g [38, 39, 41].

Dynamic heterogeneity reflects spontaneous fluctuations about the average dynamics. By relating fluctuations in the entropy to fluctuations of temperature, Donth and coworkers [42, 43] derived a formula for the number of dynamically correlated molecules in terms of the breadth of the calorimetric glass transition temperature

$$N_c = \frac{k_B N_A}{M} \Delta c_p^{-1} \left(\frac{T}{\delta T} \right)^2 \tag{2}$$

In this equation, k_B is the Boltzmann's constant, Δc_p is the isobaric heat capacity change at T_g, and δT is the half-width of the glass transition in temperature units. A slightly different expression is used by Saiter et al. [44]

$$N_c = \frac{k_B N_A}{M} \Delta \left(c_p^{-1} \right) \left(\frac{T}{\delta T} \right)^2 \tag{3}$$

with a different way of taking into account the heat capacity; there is therefore a certain degree of arbitrariness in the values computed by these methods. Assuming a spherical correlation volume, values of ξ at T_g in the range 1–3 nm have been reported [45].

Linear two-time correlation functions describe only the average dynamics and higher order correlation functions, characterizing fluctuations of dynamics about the mean, are used to probe cooperative motions. The multidimensional NMR experiment described above is a four-time correlation method. The spatial extent of dynamic correlations over the timescale t can be quantified more thoroughly through the use of two-time, two-point functions. An example is for the spatial decay of the temporal correlation of the local density [46, 47]

$$C_4(r, t) = \langle \delta\rho(0, 0)\delta\rho(0, t)\delta\rho(r, 0)\delta\rho(r, t) \rangle - \langle \delta\rho(0, 0)\delta\rho(0, t) \rangle \langle \delta\rho(r, 0)\delta\rho(r, t) \rangle \tag{4}$$

where $\delta\rho(r, t)$ is the deviation from the mean density. $C_4(r, t)$ or its Fourier transform, the dynamic susceptibility, can be measured on colloids [12] and granular systems [13]; however, spontaneous fluctuations of molecular liquids are not directly accessible. Berthier and coworkers [42, 48] have shown that the underlying dynamic heterogeneity can be quantified by analyzing induced fluctuations. If the forcing quantity is temperature [49],

$$N_c = \max\left[\frac{N_\mathrm{A}}{M}\frac{k_\mathrm{B}}{\Delta c_p}T^2\left(\chi_T^\mathrm{NPT}(t)\right)^2 + \chi_4^\mathrm{NPH}(t)\right] \tag{5}$$

The dynamical susceptibility has a maximum as a function of time around $t = \tau_\alpha$, which grows in magnitude on approach to the glass transition. From MD simulations, the contribution of various terms to the dynamic fluctuations can be evaluated, which led to the conclusion that χ_4^NPH, representing fluctuations in the isoenthalpic–isobaric ensemble, is negligible at lower temperatures approaching T_g [44, 50]. This enables the number of correlating molecules to be expressed in terms of experimentally accessible quantities [51]

$$N_c \approx \frac{N_\mathrm{A}}{M}\frac{k_\mathrm{B}}{\Delta c_p}T^2\left(\left.\frac{\partial\phi(t)}{\partial T}\right|_p\right)^2 \tag{6}$$

In Eq. (6), $\phi(t)$ is a linear susceptibility, such as the dielectric relaxation function, $\varepsilon(t)$. Figure 1 shows a comparison of N_c from Eqs. (5) and (6); the agreement is good at longer τ_α. Capaccioli et al. [52] used Eq. (6) to analyze data for a large number of liquids, obtaining values of N_c in the range 100–800 at T_g.

Considering other factors such as the density or enthalpy that contribute to dynamic heterogeneity leads to alternatives for Eq. (5), e.g., [44]

$$N_c = \max\left[\frac{N_\mathrm{A}}{M}\frac{k_\mathrm{B}}{\Delta c_V}T^2\left(\chi_T^\mathrm{NVT}(t)\right)^2 + \frac{N_\mathrm{A}}{M}k_\mathrm{B}T\kappa_T\rho^3\left(\chi_\rho^\mathrm{NPT}(t)\right)^2 + \chi_4^\mathrm{NVE}(t)\right] \tag{7}$$

in which Δc_V is the isochoric heat capacity change at T_g, and κ_T is the isothermal compressibility. Assuming fluctuations in the microcanonical (NVE) ensemble are small (MD simulations provide support for $\chi_4^\mathrm{NVE}(t)$ becoming smaller with decreasing T [53])

$$N_c \approx \frac{N_\mathrm{A}}{M}\frac{k_\mathrm{B}}{\Delta c_V}T^2\left(\left.\frac{\partial\phi(t)}{\partial T}\right|_V\right)^2 + \frac{N_\mathrm{A}}{M}k_\mathrm{B}T\kappa_T\rho^3\left(\left.\frac{\partial\phi(t)}{\partial\rho}\right|_T\right)^2 \tag{8}$$

The first term on the rhs of Eq. (8) represents fluctuations in the NVT ensemble, with the second term the additional contributions from density fluctuations.

The accuracy of the approximate Eqs. (6) and (8) can be tested by comparing to results using the full Eqs. (5) and (7). As shown in Fig. 1, MD simulations support the underlying assumption that $\chi_4^\mathrm{NPH}(t)$ and $\chi_4^\mathrm{NVE}(t)$ are negligible. However, N_c for several liquids computed using both Eqs. (6) and (8) differ by as much as 40% (representative results are displayed in Fig. 2) [54]; that is, the difference between two putatively small contributions is an appreciable amount of the total $\chi_4(t)$. This opens to question both absolute values of the correlation volume extracted from the

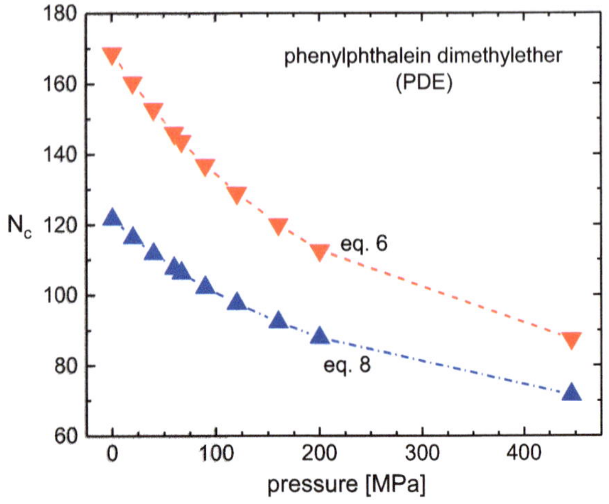

Fig. 2 Number of dynamically correlated molecules calculated assuming contribution from $\chi_4^{NPH}(t)$ (down triangles) or from $\chi_4^{NVE}(t)$ (triangles) is negligible. From Ref. [49]

approximate equations, and more significantly, comparisons of N_c made for different state points.

3 Dynamic Correlation Volume at Constant τ_α

If measurements are carried out as a function of temperature and pressure, a determination is possible of any variation of N_c for state points for which τ_α is constant. This approach is not easily applied to confinement experiments, since it is difficult to achieve hydrostatic conditions for materials in pores or very thin films. However, if the confining geometry is used to vary τ_α, a comparison can be made of isochronal N_c at different temperatures and ambient pressures. Koppensteiner et al. [21] confined salol to porous silica, with T_g differing by 8 K for pore sizes = 2.4 and 4.8 nm. They found that N_c from Eq. (6) varied significantly with temperature; however, at temperatures for which the respective τ_α in the different pores were constant, N_c was essentially constant ($\pm$~5%). An analogous study of multilayered films of polymethylmethacrylate also found that isochronal N_c was the same for film thicknesses varying from 4 nm up to bulk dimensions (Fig. 3) [55].

Compared to experimental confinement studies, using molecular dynamics simulations, it is much more straightforward to use confinement or a point-to-set construction to determine whether N_c is constant under isochronal conditions. For a molecular liquid in a thin film geometry between confining walls [56], the wall induces a slowing down of dynamics which propagates into the liquid a distance ξ. Although the dependence of dynamics on the distance from the wall was not reported, the spatially averaged dynamics of the film was found to be constant at several different state points with the same bulk τ_α, suggesting that ξ defined in this way must also be constant.

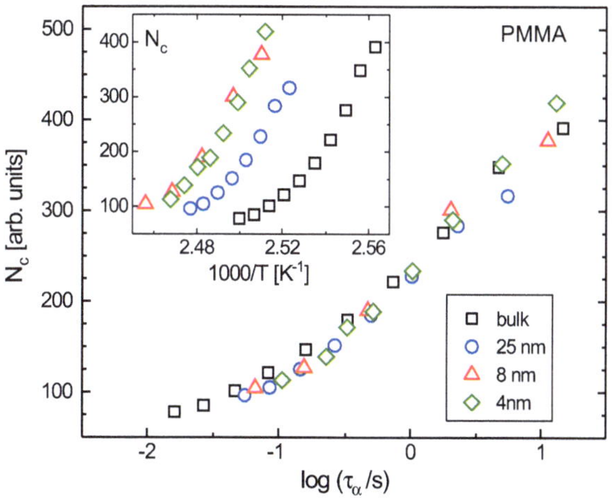

Fig. 3 Number of dynamically correlated segments for polymethylmethacrylate coextruded multilayers. Confinement alters the segmental relaxation time (inset), but N_c from Eq. (6) remains a function of τ_α. From Ref. [50]

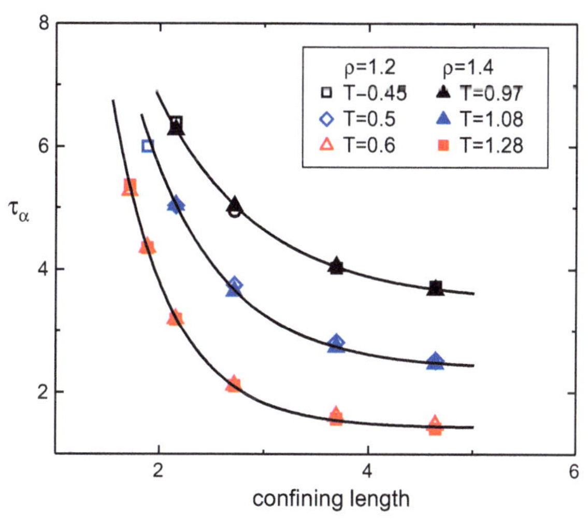

Fig. 4 Relaxation time as a function of confining length $L = c^{1/3}$ of Kob-Andersen Lennard-Jones mixture, with fraction c of pinned particles, simulated at the indicated state points. Each pair of state points is chosen to have the same bulk (unpinned) τ_α

Dynamic correlations can also be investigated using random pinning: pinning a fraction c of atoms or molecules is essentially equivalent to imposing a "confining length scale" equal to the distance between pinned particles $L \sim c^{1/3}$. With increasing pinning fraction, as the confining length scale impinges on ξ, dynamic correlations cause the dynamics of the remaining, mobile particles to increasingly slow down relative to the unconfined bulk liquid. Figure 4 shows new results on random pinning in a Kob-Andersen Lennard-Jones mixture (N = 5000 particles), at three pairs of state points, each pair chosen to share the same bulk τ_α. The α relaxation time increases with increasing pinning fraction (decreasing confining length L), and for state points with higher bulk τ_α, the decrease begins at a larger L reflecting a larger value of ξ. For state points with the same bulk τ_α, the dependence of τ_α on confining length is identical, and therefore ξ is constant at constant τ_α.

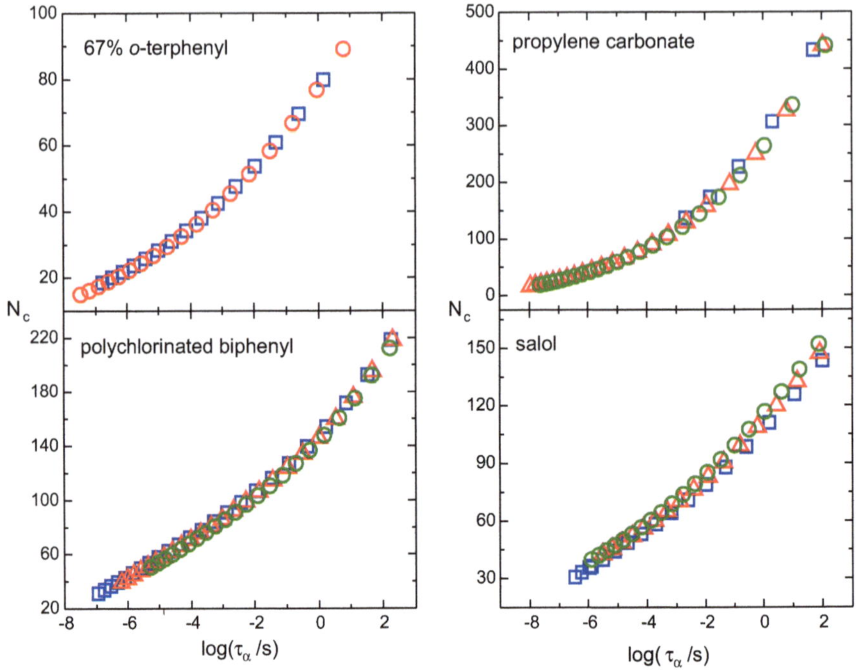

Fig. 5 Constancy of isochronal N_c calculated using Eq. (6) for four liquids. Adapted from Ref. [51]

The merging of the segmental and normal modes in polyisoprene measured at various pressures corresponded to fixed τ_α [32]. If the dynamic correlation volume is equal to the chain coil size at the state point associated with the merging, this result is consistent with constant N_c at fixed τ_α, since the change in the chain radius of gyration with T and P is small (<0.3% based on the measured dielectric strength of the normal mode) [32].

An analysis was carried out on four molecular liquids for which τ_α had been measured as a function of temperature and pressure [57]. As shown in Fig. 5, for a given material, the dynamic correlation volume, evaluated using the approximate Eq. (6), depends on the relaxation time, invariant to T, P, and ρ at fixed τ_α. However, the results in Fig. 5 are at odds with two other studies. Koperwas et al. [54] analyzed dielectric data for three liquids using Eq. (6), and determined that the isochronal N_c for each decreases by as much as 50% for pressures up to a couple hundred MPa. Results for phenylphthalein dimethylether are shown in Fig. 2. Diametrically opposite results were reported by Alba-Simionesco et al. [58], who found that N_c for dibutyl phthalate increased with pressure at constant $\tau_\alpha = 1$ s (Fig. 6). Thus, from measurements on eight liquids, it was concluded that N_c is constant [51], increases [49], or decreases [52] with increasing pressure at constant τ_α. The disparate results of these three studies are not because the behavior of different materials can qualitatively

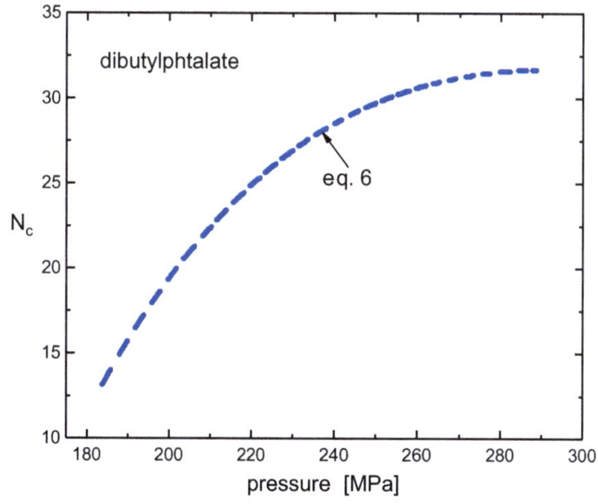

Fig. 6 Number of dynamically correlated molecules calculated using the indicated approximation [52]

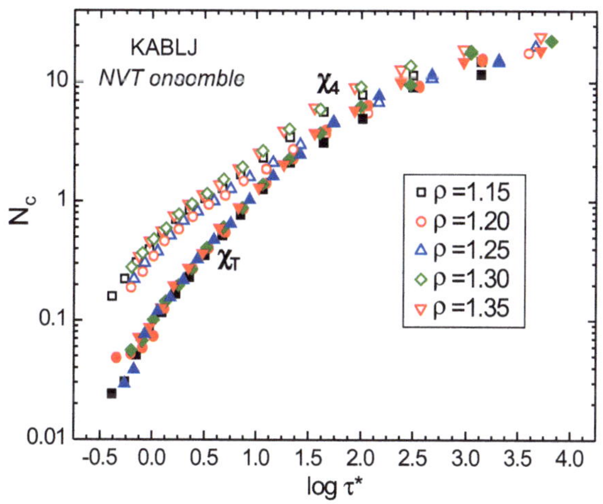

Fig. 7 Number of dynamically correlated particles calculated using Eq. (5) (open symbols) and Eq. (6) (solid symbols) at the indicated densities. From data in [9]

differ, but rather such results cast aspersions on the analysis using the approximate formula for $\chi_4(t)$.

One way to circumvent the ambiguity and inaccuracies in the application of Eqs. (6) and (8) is to obtain the full $\chi_4(t)$, which is possible using computer simulations. It is more convenient to calculate this multipoint dynamic susceptibility as the variance of the self-intermediate scattering function $F_s(k,t)$

$$\chi_4 = N_A\left[\left\langle f_s^2(k,t)\right\rangle - F_s^2(k,t)\right] \tag{9}$$

where $f_s(k,t)$ is the instantaneous value such that $\langle f_s(k,t)\rangle = F_s(k,t)$. Results are shown for a binary LJ liquid calculated in the NVT ensemble [which omits the second term on the rhs of Eq. (7)] in Fig. 7 [9]. The dynamic correlation volume is to good approximation invariant at fixed α-relaxation time. Likewise, MD simulations of diatomic molecules indicated constant isochronal N_c (4% variation in the NVT ensemble; 8% in the NPT) [59].

4 Isochronal N_c from Nonlinear Dielectric Susceptibility

From both MD simulations [9, 51, 53] and the application of approximations to estimate $\chi_4(t)$ from experimental data [19, 32, 50], the conclusion seems to be that N_c is constant for fixed τ_α, independent of T, P, or ρ. However, there are scattered results to the contrary [49, 52], and the reliability of the estimates of N_c might be questioned. Accordingly, an alternative method is desirable.

Bouchaud and Biroli [60] proposed that the nonlinear dielectric susceptibility can be used to measure dynamic correlations in glassforming liquids, specifically that the peak height of the nonlinear susceptibility of glassforming materials increases in proportion to N_c:

$$N_c \propto |\chi_3| \frac{kT}{\varepsilon_0 a^3 (\Delta\chi_1)^2} \tag{10}$$

where ε_0 is the permittivity of free space, a is the molecular volume, $\Delta\chi_1$ is the linear dielectric strength, and $|\chi_3|$ is the modulus of the third-order susceptibility corresponding to polarization cubic in the applied field. The connection between χ_3 and N_c is supported by good agreement of the calculated N_c with the effective activation energy in glassforming liquids [61, 62], and plastic crystals [63]:

$$\left.\frac{\mathrm{d}\ln\tau}{\mathrm{d}T^{-1}}\right|_P = AN_c \tag{11}$$

where A is a constant. This analysis has also been extended to the fifth-order susceptibility χ_5, and results consistent with this analysis are obtained for glycerol and propylene carbonate [64]. The derivation of Eq. (10) is not rigorous, and growing nonlinear susceptibility with similar features to those observed experimentally also appear in mean-field models that lack length scales [65, 66]. An alternative phenomenological model of nonlinear dielectric response that lacks dynamic correlations also produces a growth of the peak in χ_3 proportionally to the apparent activation energy [67, 68]. It is not clear whether these different pictures of nonlinear relaxation are in conflict: based on a careful analysis of the behavior of the three different third-order susceptibilities, it has been suggested that in fact the models of Refs. [67, 68] are consistent with the interpretation leading to Eq. (10), i.e., relating the growth of χ_3 to the growth of cooperatively rearranging regions [69].

Equation (10) provides a method of testing whether N_c is constant under isochronal conditions, by measuring the third-order dielectric susceptibility at ele-

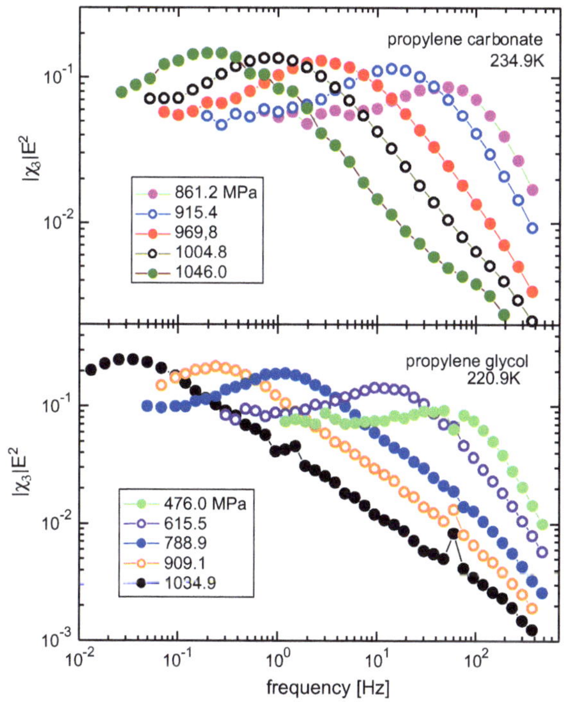

Fig. 8 Representative third-order harmonic spectra of propylene carbonate (top) and propylene glycol (bottom) at the indicated temperature and pressures, the latter increasing from right to left. From Ref. [59]

vated pressures. This was done for two liquids: propylene carbonate and propylene glycol [59].

Propylene carbonate is a non-associated liquid, conforming to isochronal superposition, meaning that its relaxation spectrum is constant for constant relaxation time. For the hydrogen-bonded propylene glycol, the relaxation spectrum becomes broader at for higher pressure and temperature at constant τ. Figure 8 presents $|\chi_3|$ spectra obtained at various pressures. For both liquids, the peak in nonlinear susceptibility grows with decrease in peak frequency, consistent with growth in the correlation volume as the relaxation time becomes longer.

To quantify the dynamic correlations, it is required that the contribution to $|\chi_3|$ from saturation of the dipole orientation be removed from the spectra; this was done following Brun et al. [70] who have shown that $|\chi_3|$ at a frequency 2.5 times the frequency of the maximum in the linear dielectric loss provides a measure of N_c unaffected by dipole saturation. In Fig. 9, N_c for the two liquids are plotted as a function of the linear relaxation frequency. The data for propylene carbonate show the two regimes expected for dynamic correlations—power-law dependences with a steeper slope at higher frequencies [71]. This supports the interpretation of the

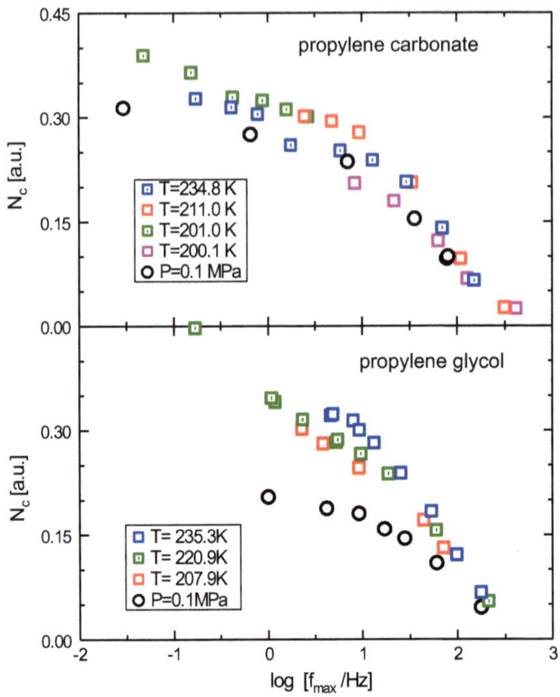

Fig. 9 Number of dynamically correlated molecules (arbitrary units) calculated using Eq. (10) for propylene carbonate (top) and propylene glycol (bottom) as a function of the frequency of the loss peak in the linear spectrum. From Ref. [59]

peak in the nonlinear susceptibility in terms of dynamic correlations. Within the experimental scatter (*ca.* 15%), the number of dynamically correlated molecules for propylene carbonate depends mainly on the relaxation time; there is no systematic variation in N_c with T or P. For propylene glycol, the large concentration of hydrogen bonds changes with thermodynamic conditions, resulting in liquid structure which is not constant at constant τ; indeed, the variety of scaling properties found for simple liquids are absent in associated materials [72]. This is also reflected in substantial variations (exceeding 50%) in N_c for a given τ, specifically a systematic increase in dynamic correlations with increasing temperature or pressure at constant τ.

When high pressures are considered, the values of N_c deviate from the proportionality with effective activation energy which holds at ambient pressure (Fig. 10). Thus, the parameter A in Eq. 10 is pressure dependent. For the case of non-associated liquids such as propylene carbonate, which conform to density scaling, it can be shown that the apparent activation energy is not a constant at constant τ [59]. Since N_c *is* constant (to good approximation), the deviation from strict proportionality of the two quantities can be understood. For associated liquids such as propylene glycol, the decoupling is stronger, additionally reflecting the change in structure at constant τ.

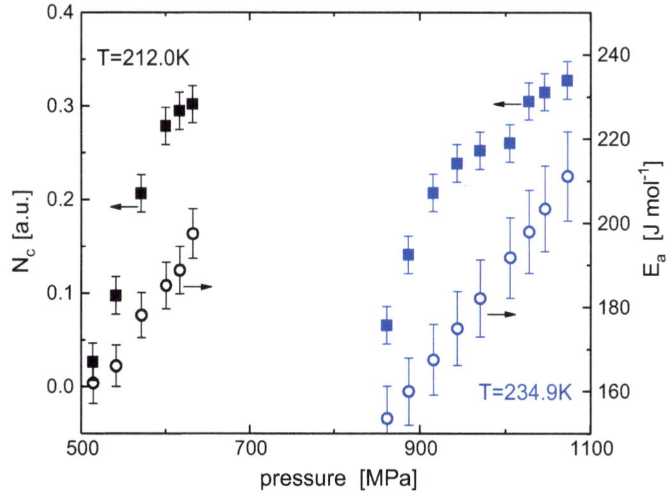

Fig. 10 Comparison of the temperature dependence of the number of dynamically correlated molecules calculated using Eq. (10) (filled squares) and the apparent activation energy (open circles) for propylene carbonate at the indicated temperatures. From Ref. [59]

5 Summary

Approximate analyses of experimental data and molecular dynamic simulations, which of course entail inherent approximations, indicate that ξ and N_c are constant under isochronal conditions. Such results are consistent with nonlinear dielectric measurements interpreting the modulus of the susceptibility as a measure of the dynamic correlation volume. This consistency supports the interpretation of the nonlinear response in terms of dynamic correlation, but more importantly establishes the centrality of dynamic heterogeneity to the glass transition phenomenon.

Acknowledgements This work was supported by the Office of Naval Research.

References

1. B. Frick, C. Alba-Simionesco, K.H. Andersen, L. Willner, Influence of density and temperature on the microscopic structure and the segmental relaxation of polybutadiene. Phys. Rev. E **67**, 051801 (2003)
2. A. Cailliaux, C. Alba-Simionesco, B. Frick, L. Willner, I. Goncharenko, Phys. Rev. E **67**, 010802 (2003)
3. R. Bohmer, Nanoscale heterogeneity of glass-forming liquids: experimental advances. Cur. Opin. Sol. State Mat. Sci. **3**, 378–385 (1998)
4. H. Sillescu, Heterogeneity at the glass transition: a review. J. Non-Cryst. Solids **243**, 81–108 (1999)

5. M.D. Ediger, Spatially heterogeneous dynamics in supercooled liquids. Ann. Rev. Phys. Chem. **51**, 99–128 (2000)
6. H. Sillescu, R. Bohmer, G. Diezemann, G. Hinze, Heterogeneity at the glass transition: what do we know? J. Non-Cryst. Sol. **307–310**, 16–23 (2002)
7. R. Richert, N. Israeloff, C. Alba-Simionesco, F. Ladieu, D. L'Hote, *Experimental approaches to heterogeneous dynamics* in Dynamical Heterogeneities in Glasses, Colloids, and Granular Media, ed. by L. Berthier, G. Biroli, J.-P. Bouchaud, L. Cipelletti, W. van Saarloos (Oxford University Press, Oxford, 2011)
8. K. Kim, S. Saito, Multiple length and time scales of dynamic heterogeneities in model glass-forming liquids: a systematic analysis of multi-point and multi-time correlations. J. Chem. Phys. **138**, 12A506 (2013)
9. C.M. Roland, D. Fragiadakis, D. Coslovich, S. Capaccioli, K.L. Ngai, Correlation of nonexponentiality with dynamic heterogeneity from four-point dynamic susceptibility $\chi_4(t)$ and its approximation $\chi_T(t)$. J. Chem. Phys. **133**, 124507 (2010)
10. R. Böhmer, K.L. Ngai, C.A. Angell, D.J. Plazek, Nonexponential relaxations in strong and fragile glass formers. J. Chem. Phys. **99**, 4201–4209 (1993)
11. K. Niss, C. Dalle-Ferrier, G. Tarjus, C. Alba-Simionesco, On the correlation between fragility and stretching in glass-forming liquids. J. Phys. Cond. Mat. **19**, 076102 (2007)
12. E.R. Weeks, J.C. Crocker, A.C. Levitt, A. Schofield, D.A. Weitz, Three-dimensional direct imaging of structural relaxation near the colloidal glass transition. Science **287**, 627–631 (2000)
13. O. Dauchot, G. Marty, G. Biroli, Dynamical heterogeneity close to the jamming transition in a sheared granular material. Phys. Rev. Lett. **95**, 265701 (2005)
14. G. Adam, J.H. Gibbs, On the temperature dependence of cooperative relaxation properties in glass-forming liquids. J. Chem. Phys. **43**, 139–146 (1965)
15. K.S. Schweizer, E.J. Saltzman, Activated hopping, barrier fluctuations, and heterogeneity in glassy suspensions and liquids. J. Phys. Chem. B **108**, 19729–19741 (2004)
16. J.P. Garrahan, D. Chandler, Dynamics on the way to forming glass: bubbles in space-time. Ann. Rev. Phys. Chem. **61**, 191–217 (2010)
17. V. Lubchenko, P.G. Wolynes, Theory of structural glasses and supercooled liquids. Ann. Rev. Phys. Chem. **58**, 235–266 (2007)
18. A. Heuer, Exploring the potential energy landscape of glass-forming systems: from inherent structures via metabasins to macroscopic transport. J. Phys. Con. Mat. **20**, 373101 (2008)
19. F. Rittig, A. Huwe, G. Fleischer, J. Kärger, F. Kremer, Molecular dynamics of glass-forming liquids in confining geometries. Phys. Chem. Chem. Phys. **1**, 519–523 (1999)
20. G. Dosseh, C. Le Quellec, N. Brodie-Lindner, C. Alba-Simionesco, W. Haeussler, P. Levitz, Fluid–wall interactions effects on the dynamical properties of confined orthoterphenyl. J. Non-Cryst. Sol. **352**, 4964–4968 (2006)
21. J. Koppensteiner, W. Schranz, M.A. Carpenter, Revealing the pure confinement effect in glass-forming liquids by dynamic mechanical analysis. Phys. Rev. B **81**, 024202 (2010)
22. C.L. Jackson, G.B. McKenna, The glass transition of organic liquids confined to small pores. J. Non-Cryst. Sol. **131–133**, 221–224 (1991)
23. Y.B. Melnichenko, J. Schuller, R. Richert, B. Ewen, C.K. Loong, Dynamics of hydrogen bonded liquids confined to mesopores—a dielectric and neutron spectroscopy study. J. Chem. Phys. **103**, 2016–2024 (1995)
24. F. Kremer, A. Huwe, M. Arndt, P. Behrens, W. Schwieger, How many molecules form a liquid? J. Phys. Cond. Mat. **11**, A175–A188 (1999)
25. A. Schonhals, H. Goering, C. Schick, B. Frick, R. Zorn, Glassy dynamics of polymers confined to nanoporous glasses revealed by relaxational and scattering experiments. Eur. Phys. J. E **12**, 173–178 (2003)
26. A. Schonhals, H. Goering, C. Schick, B. Frick, M. Mayorova, R. Zorn, Segmental dynamics of poly(methyl phenyl siloxane) confined to nanoporous glasses. Eur. Phys. J. Spec. Topics. **141**, 255–259 (2007)
27. J.P. Bouchaud, G. Biroli, On the Adam-Gibbs-Kirkpatrick-Thirumalai-Wolynes scenario for the viscosity increase in glasses. J. Chem. Phys. **121**, 7347–7354 (2004)

28. G. Biroli, J.-P. Bouchaud, A. Cavagna, T.S. Grigera, P. Verrocchio, Thermodynamic signature of growing amorphous order in glass-forming liquids. Nat. Phys. **4**, 771–775 (2008)
29. S. Yaida, L. Berthier, P. Charbonneau, G. Tarjus, Point-to-set lengths, local structure, and glassiness. Phys. Rev. E **94**, 032605 (2016)
30. W. Kob, S. Roldán-Vargas, L. Berthier, Non-monotonic temperature evolution of dynamic correlations in glass-forming liquids. Nat. Phys. **8**, 164 (2012)
31. G.M. Hocky, L. Berthier, W. Kob, D.R. Reichman, Static point-to-set correlations in glass-forming liquids. Phys. Rev. E **85**, 011102 (2012)
32. K. Hima Nagamanasa, S. Gokhale, A.K. Sood, R. Ganapathy, Direct measurements of growing amorphous order and non-monotonic dynamic correlations in a colloidal glass-former. Nat. Phys. **11**, 403 (2015)
33. S, Gokhale, K. Hima Nagamanasa, R. Ganapathy, A. K. Sood, Growing dynamic facilitation on approaching the random pinning colloidal glass transition. Nat. Commun. **5**, 4685 (2014)
34. B. Mei, Y. Lu, L. An, H. Li, L. Wang, Nonmonotonic dynamic correlations in quasi-tow-dimensional confined glass-forming liquids. Phys. Rev. E **95**, 050601(R) (2017)
35. A. Schonhals, E. Schlosser, Relationship between segmental and chain dynamics in polymer melts as studied by dielectric spectroscopy. Phys. Scr. **T49**, 233–236 (1993)
36. C. Gainaru, W. Hiller, R. Bohmer, A dielectric study of oligo- and poly(propylene glycol). Macromolecules **43**, 1907–1914 (2010)
37. D. Fragiadakis, R. Casalini, R.B. Bogoslovov, C.G. Robertson, C.M. Roland, Dynamic heterogeneity and density scaling in 1,4-polyisoprene. Macromolecules **44**, 1149–1155 (2011)
38. U. Tracht, M. Wilhelm, A. Heuer, H. Feng, K. Schmidt-Rohr, H.W. Spiess, Length scale of dynamic heterogeneities at the glass transition determined by multidimensional nuclear magnetic resonance. Phys. Rev. Lett. **81**, 2727–2730 (1998)
39. S.A. Reinsberg, X.H. Qiu, M. Wilhelm, H.W. Spiess, M.D. Ediger, Length scale of dynamic heterogeneity in supercooled glycerol near T_g. J. Chem. Phys. **114**, 7299–7302 (2001)
40. S.A. Reinsberg, A. Heuer, B. Doliwa, H. Zimmermann, H.W. Spiess, Comparative study of the NMR length scale of dynamic heterogeneities of three different glass formers. J. Non-Cryst. Solid **307–310**, 208–214 (2002)
41. X.H. Qiu, M.D. Ediger, Length scale of dynamic heterogeneity in supercooled d-sorbitol: comparison to model predictions. J. Phys. Chem. B **107**, 459–464 (2003)
42. E. Donth, The size of cooperatively rearranging regions at the glass transition. J. Non-Cryst. Sol. **53**, 325–330 (1982)
43. K. Schroter, Characteristic length of glass transition heterogeneity from calorimetry. J Non-Cryst. Sol. **352**, 3249–3254 (2006)
44. A. Saiter, L. Delbreilh, H. Couderc, K. Arabeche, A. Schönhals, J.-M. Saiter, Temperature dependence of the characteristic length scale for glassy dynamics: Combination of dielectric and specific heat spectroscopy. Phys. Rev. E **81**, 041805 (2010)
45. E. Hempel, G. Hempel, A. Hensel, C. Schick, E. Donth, Characteristic length of dynamic glass transition near t_g for a wide assortment of glass-forming substances. J. Phys. Chem. B **104**, 2460–2466 (2000)
46. C. Dasgupta, A.V. Indrani, S. Ramaswamy, M.K. Phani, Is there a growing correlation length near the glass transition? Europhys. Lett. **15**, 307–312 (1991)
47. L. Berthier, G. Biroli, Theoretical perspective on the glass transition and amorphous materials. Rev. Mod. Phys. **83**, 587–645 (2011)
48. L. Berthier, G. Biroli, J.-P. Bouchaud, L. Cipelletti, D. El Masri, D. L'Hote, F. Ladieu, M. Pierno, Direct experimental evidence of a growing length scale accompanying the glass transition. Science **310**, 1797–1800 (2005)
49. C. Dalle-Ferrier, C. Thibierge, C. Alba-Simionesco, L. Berthier, G. Biroli, J.P. Bouchaud, F. Ladieu, D. L'Hote, G. Tarjus, Spatial correlations in the dynamics of glassforming liquids: experimental determination of their temperature dependence. Phys. Rev. E **76**, 041510 (2007)
50. L. Berthier, G. Biroli, J.-P. Bouchaud, L. Cipelletti, D. El Masri, D. L'Hôte, F. Ladieu, M. Pierno, Direct experimental evidence of a growing length scale accompanying the glass transition. Science **310**, 1797–1800 (2005)

51. L. Berthier, G. Biroli, J.-P. Bouchaud, W. Kob, K. Miyazaki, D.R. Reichman, Spontaneous and induced dynamic fluctuations in glass formers. 1. General results and dependence on ensemble and dynamics. J. Chem. Phys. **126**, 184503 (2007)
52. S. Capaccioli, G. Ruocco, F. Zamponi, Dynamically correlated regions and configurational entropy in supercooled liquids. J. Phys. Chem. B **112**, 10652–10658 (2008)
53. E. Flenner, G. Szamel, Dynamic heterogeneities above and below the mode-coupling temperature: evidence of a dynamic crossover. J. Chem. Phys. **138**, 12A523 (2013)
54. K. Koperwas, A. Grzybowski, K. Grzybowska, Z. Wojnarowska, A.P. Sokolov, M. Paluch, Effect of temperature and density fluctuations on the spatially heterogeneous dynamics of glass-forming van der Waals liquids under high pressure. Phys. Rev. Lett. **111**, 125701 (2013)
55. R. Casalini, L. Zhu, E. Baer, C.M. Roland, Segmental dynamics and the correlation length in nanoconfined PMMA. Polymer **88**, 133–136 (2016)
56. T.S. Ingebrigtsen, J.R. Errington, T.M. Truskett, J.C. Dyre, Predicting how nanoconfinement changes the relaxation time of a supercooled liquid. Phys. Rev. Lett. **111**, 235901 (2013)
57. D. Fragiadakis, R. Casalini, C.M. Roland, Density scaling and dynamic correlations in viscous liquids. J. Phys. Chem. B **113**, 13134–13147 (2009)
58. C. Alba-Simionesco, C. Dalle-Ferrier, G. Tarjus, Effect of pressure on the number of dynamically correlated molecules when approaching the glass transition. AIP Conf. Proc. **1518**, 527–535 (2013)
59. R. Casalini, D. Fragiadakis, C.M. Roland, Dynamic correlation length scales under isochronal conditions. J. Chem. Phys. **142**, 064504 (2015)
60. J.-P. Bouchaud, G. Biroli, Nonlinear susceptibility in glassy systems: a probe for cooperative dynamical length scales. Phys. Rev. B **72**, 064204 (2005)
61. C. Crauste-Thibierge, C. Brun, F. Ladieu, D. L'Hôte, G. Biroli, J.-P. Bouchaud, Evidence of growing spatial correlations at the glass transition from nonlinear response experiments. Phys. Rev. Lett. **104**, 165703 (2010)
62. Th Bauer, P. Lunkenheimer, A. Loidl, Cooperativity and the freezing of molecular motion at the glass transition. Phys. Rev. Lett. **111**, 225702 (2013)
63. M. Michl, Th Bauer, P. Lunkenheimer, A. Loidl, Nonlinear dielectric spectroscopy in a fragile plastic crystal. J. Chem. Phys. **144**, 114506 (2016)
64. S. Albert, Th Bauer, M. Michl, G. Biroli, J.-P. Bouchaud, A. Loidl, P. Lunkenheimer, R. Tourbot, C. Wiertel-Gasquet, F. Ladieu, Fifth-order susceptibility unveils growth of thermodynamic amorphous order in glass-formers. Science **352**, 1308 (2016)
65. G. Diezemann, Higher-order correlation functions and nonlinear response functions in a Gaussian trap model. J. Chem. Phys. **138**, 12A505 (2013)
66. C. Brun, C. Crauste-Thibierge, F. Ladieu, D. L'Hôte, Third harmonics nonlinear susceptibility in supercooled liquids: a comparison to the box model. J. Chem. Phys. **134**, 194507 (2011)
67. P. Kim, A.R. Young-Gonzales, R. Richert, Dynamics of glass-forming liquids. XX. Third harmonic experiments of non-linear dielectric effects versus a phenomenological model. J. Chem. Phys. **145**, 064510 (2016)
68. R. Richert, Nonlinear dielectric signatures of entropy changes in liquids subject to time dependent electric fields. J. Chem. Phys. **144**, 114501 (2016)
69. P. Gadige, S. Albert, M. Michl, Th Bauer, P. Lunkenheimer, A. Loidl, R. Tourbot, C. Wiertel-Gasquet, G. Biroli, J.-P. Bouchaud, F. Ladieu, Unifying different interpretations of the nonlinear response in glass-forming liquids. Phys. Rev. E **96**, 032611 (2017)
70. C. Brun, F. Ladieu, D. L'Hôte, M. Tarzia, G. Biroli, J.-P. Bouchaud, Nonlinear dielectric susceptibilities: accurate determination of the growing correlation volume in a supercooled liquid. Phys. Rev. B **84**, 104204 (2011)
71. Dynamical Heterogeneities in Glasses, Colloids and Granular Materials, ed. by L. Berthier, G. Biroli, J.-P. Bouchaud, L. Cipelletti, W. van Saarloos (Oxford University Press, Oxford, 2011)
72. C.M. Roland, R. Casalini, R. Bergman, J. Mattsson, Role of hydrogen bonds in the supercooled dynamics of glass-forming liquids at high pressures. Phys. Rev. B **77**, 012201 (2008)

Nonlinear Dielectric Response of Plastic Crystals

P. Lunkenheimer, M. Michl and A. Loidl

Abstract This article summarizes ongoing experimental efforts on nonlinear dielectric spectroscopy on plastic crystals. In plastic crystals, the relevant dipolar orientational degrees of freedom are fixed on a crystalline lattice with perfect translational symmetry. However, while they can reorient freely in the high-temperature plastic phase, they often undergo glassy freezing at low temperatures. Hence, plastic crystals are often considered as model systems for structural glass formers. It is well known that plastic crystals reveal striking similarities with phenomena of conventional supercooled liquids. However, in most cases, they can be characterized as rather strong glass formers. Nonlinear dielectric spectroscopy is an ideal tool to study glass-transition phenomena, providing insight into cooperative phenomena or hidden phase transitions, undetectable by purely linear spectroscopy. In the present article, we discuss dielectric experiments using large electric ac fields probing the nonlinear 1ω and the third-order harmonic 3ω susceptibility. In the 1ω experiments, we find striking differences compared with observations on conventional structural glass formers: at low frequencies plastic crystals do not approach the trivial response, but reveal strong additional nonlinearity. These phenomena document the importance of entropic effects in this class of glassy materials. The harmonic third-order susceptibility reveals a hump-like shape, similar to observations in canonical glass formers, indicating the importance of cooperativity dominating the glass transition. In the frequency regime of the secondary relaxations, only minor nonlinear effects can be detected, supporting arguments in favor of the non-cooperative nature of these faster dynamics processes. Based on a model by Bouchaud and Biroli, from the hump observed in the 3ω susceptibility spectra, the temperature dependence of the number of correlated particles can be determined. We document that the results in plastic crystals perfectly well scale with the results derived from measurements on conventional glass formers, providing evidence that in plastic crystals the non-Arrhenius behavior of the relaxation times also arises from a temperature dependence of the energy barriers due to growing cooperative clusters.

P. Lunkenheimer (✉) · M. Michl · A. Loidl
Experimental Physics V, Center for Electronic Correlations and Magnetism,
University of Augsburg, 86159 Augsburg, Germany
e-mail: peter.lunkenheimer@physik.uni-augsburg.de

R. Richert (ed.), *Nonlinear Dielectric Spectroscopy*, Advances in Dielectrics
https://doi.org/10.1007/978-3-319-77574-6_9

Keywords Plastic crystals · Glassy crystals · Supercooled liquids · Nonlinear dielectric spectroscopy · Harmonic susceptibility · Relaxation dynamics · Glass transition · Non-Arrhenius behavior

1 Introduction

Plastic crystals (PCs) are often considered as model systems for structural glass formers. While in PCs the centers of mass of the molecules are fixed on a crystalline lattice with translational symmetry, their orientational degrees of freedom more or less freely fluctuate at high temperatures and often show glassy freezing at low temperatures (Fig. 1) [1]. The molecules of most PCs have rather globular shape and relatively weak mutual interactions, providing little steric hindrance for reorientational processes. This often leads to high plasticity, thus explaining the term "plastic crystal" first introduced by Timmermans [2] many decades ago. The reorientational relaxation dynamics of PCs, as detected, e.g., by dielectric spectroscopy, in many respects resembles the dynamics of structural glass formers [1, 3, 4]. Especially, often complete orientational ordering at low temperatures can be avoided by supercooling the high-temperature dynamically disordered state. Just as for glassforming liquids, this leads to a continuous slowing down of molecular dynamics over many orders of magnitude, which can be nicely followed by monitoring the reorientational relaxation dynamics by broadband dielectric spectroscopy [1]. For low temperatures, finally a static orientationally disordered state is reached which sometimes is called "glassy crystal" (Fig. 1) [5]. Correspondingly, an "orientational-glass temperature" T_g^o can be defined. It should be mentioned, however, that the term "orientational glass" for this glassy state may be ambiguous because it is often used to exclusively denote the orientationally disordered state of mixed crystals, believed to arise from frustrated interactions due to substitutional disorder [6, 7]. In contrast, the glassy crystal state in PCs is non-ergodic and represents a true analog of the structurally disordered glassy state of conventional glass formers.

The freezing of the molecular dynamics of PCs exhibits many of the puzzling characteristics of canonical (i.e., structural) glass formers. This especially includes the non-exponentiality of the time dependence and the non-Arrhenius behavior of the temperature dependence of this dynamics, both hallmark features of the glassy state of matter [8–12]. Therefore, investigating and understanding the glassy freezing in PCs is an important step on the way to a better understanding of glassforming liquids and the glass transition in general. Indeed, in literature there are various examples for such studies, many of them employing dielectric spectroscopy, which directly senses the reorientational motions [1, 13–17]. Such investigations are usually performed by detecting the *linear* dielectric response of the sample material to an applied ac electrical field of moderate amplitude (typical voltages are of the order of 1 V). However, in recent years it has become clear that the *nonlinear* response of glassforming matter, detected under high fields of up to several 100 kV/cm, can reveal a lot of valuable additional information about glassy freezing [18–27] (see

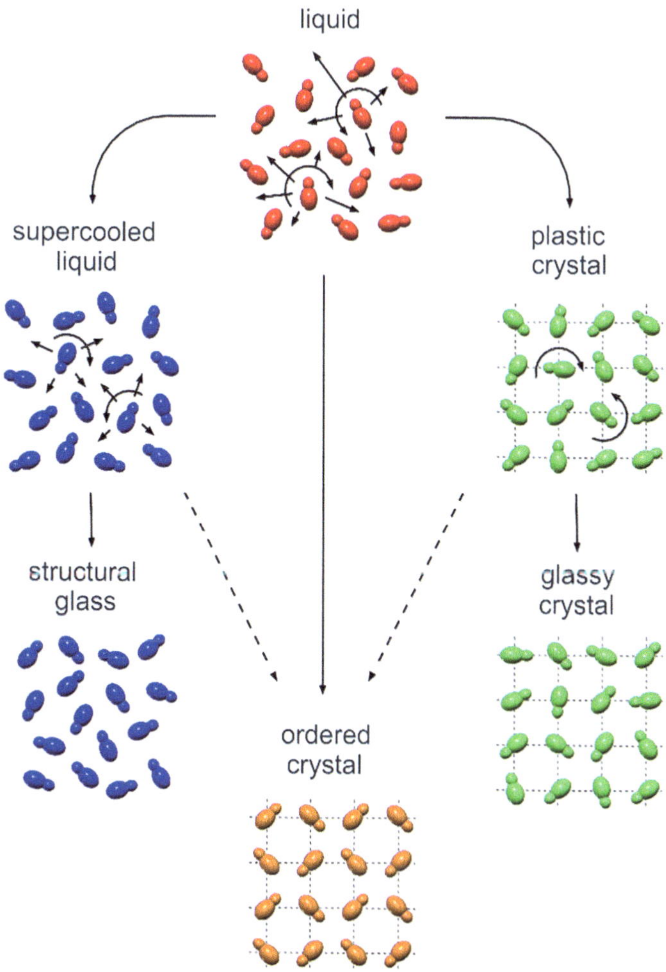

Fig. 1 Schematic representation of the possible transitions of a liquid of dipolar molecules (represented by asymmetric dumbbells) into a structural glass, an ordered crystal, or a glassy crystal [1]

also the other chapters of the present book). Thus, it seemed natural to apply such methods to PCs, too, which indeed was done in several recent works [28–31]. The present chapter will provide an overview of such nonlinear dielectric investigations of PCs.

Among the pioneering nonlinear experiments on *structural* glass formers were dielectric hole-burning experiments, which provided the first experimental verification that the non-exponentiality of the relaxation dynamics in these materials arises from its heterogeneity [32]. Later on, measurements of the field-induced variation of the dielectric permittivity revealed further valuable information on this phenomenon

[20, 33] and on the nonlinear behavior of secondary relaxation effects [25] like the Johari–Goldstein relaxation [34] or the excess wing [11, 35]. Moreover, high-field measurements recently have also provided important hints at the origin of the characteristic non-Arrhenius temperature dependence of glassy dynamics. Particularly, based on a model by Bouchaud, Biroli, and co-workers [36, 37], measurements of the higher order harmonic susceptibilities χ_3 and χ_5 seem to indicate that this phenomenon arises from an increase of molecular cooperativity when approaching T_g at low temperatures [23, 26, 27] (see also the chapter by Albert et al. in the present book). In this way, important clues about the true nature of the glass transition were obtained, which seems to be due to an underlying thermodynamic phase transition [23, 26, 27, 38]. Finally, Johari has recently demonstrated [39, 40] that nonlinear dielectric effects can also arise from the reduction of configurational entropy induced by the external field, leading to an increase of the α relaxation time and, thus, a field-induced rise of the viscosity.

Nonlinear measurements on plastic crystals should be able to reveal analogous information on the role of heterogeneity, cooperativity, and entropy in this class of disordered materials. However, it is not self-evident that the results should be similar to those in structural glass formers: For example, in a material exhibiting translational symmetry, heterogeneity can be expected to be of different nature and, indeed, it was suggested that the dynamics of single molecules in PCs may be intrinsically non-exponential and heterogeneity alone cannot explain the experimental observations [41, 42]. Moreover, the intermolecular interactions that lead to cooperativity also should be influenced by the fact that the molecules are located on fixed lattice positions. Indeed, the deviations of glassy dynamics from thermally activated Arrhenius behavior, often ascribed to cooperativity [8, 43, 44], are generally weaker for PCs than for canonical glass formers [1, 45]. Within Angell's strong-fragile classification scheme [46], this implies that PCs are rather strong glass formers (despite also exceptions are known [47–49]). This is demonstrated in Fig. 2, showing Angell plots of the α-relaxation times τ_α of a number of supercooled liquids [frame (a)] and PCs (b) measured in our group [1, 11, 47–51]. In this type of T_g-scaled Arrhenius plot [52], simple thermally activated behavior, $\tau \propto \exp(E/T)$ (with E the energy barrier in K), leads to a straight line with a slope of about 16 (dashed line) characterized as "strong" dynamics. In contrast, "fragile" glass formers exhibit pronounced curvature in this plot and, consequently, a steep slope m close to T_g. The latter parameter is commonly used to parameterize the deviations from Arrhenius behavior of glass formers [53, 54]; an example for very large fragility with $m = 170$ is shown as dotted line in Fig. 2. Comparing frames (a) and (b) in Fig. 2 nicely reveals that, in general, PCs indeed tend to be stronger in the strong/fragile classification scheme than most canonical supercooled liquids [1, 45]. Note also the interesting case of ethanol, which can be prepared both in a supercooled liquid and a PC state and is clearly stronger in the latter phase [1, 55, 56]. For three of the PC systems included in Fig. 2b (indicated by closed symbols), nonlinear dielectric spectroscopy results will be discussed in the present work.

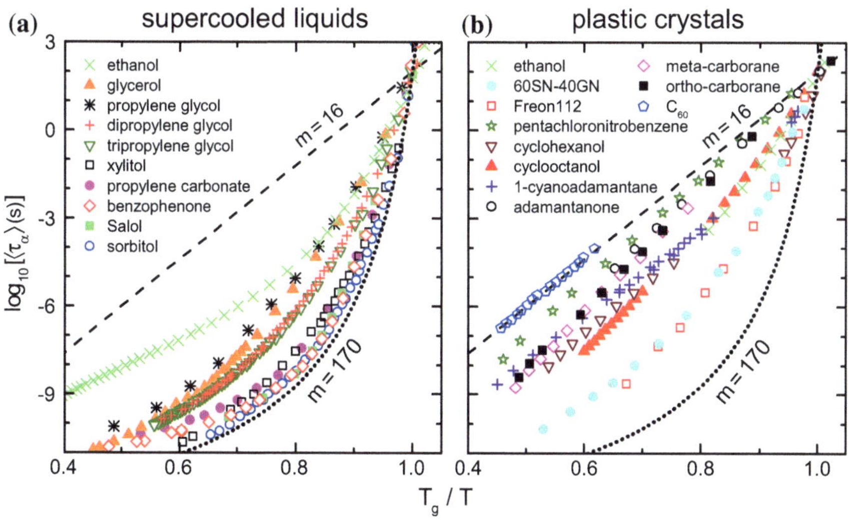

Fig. 2 Angell plot of the α-relaxation times of various supercooled liquids (**a**) [11, 50, 51] and PCs (**b**) [1, 47, 48]. The dashed lines demonstrate maximally strong behavior; the dotted lines indicate extremely high fragility. In **b**, the data for the PCs treated in the present work are shown as closed symbols

In Ref. [1], we suggested that the strong behavior of PCs can be understood when considering the proposed relation of fragility and of the energy landscape in configuration space [57, 58]. Based on the inverse proportionality of effective energy barrier and configurational entropy assumed within the Adam–Gibbs theory [59], the material-dependent variation of fragility of glass formers can be rationalized assuming that higher fragility arises from a higher density of minima in the potential-energy landscape [57, 58]. Within this framework, the overall lower fragility of PCs compared to supercooled liquids may be explained by their additional order due to the existence of a crystalline lattice, which leads to a reduced density of energy minima [1]. As discussed in Refs. [47, 48], the only exceptions are Freon112 and mixtures of succinonitrile and glutaronitrile, where additional conformational and/or substitutional disorder leads to a more complex energy landscape and, thus, more fragile behavior. It is an interesting question how this energy-landscape picture is related to the explanation of non-Arrhenius behavior by increasing molecular cooperativity (causing an increasing effective energy barrier experienced by the molecules; Fig. 3), when a supercooled liquid or PC is cooled towards its T_g. The more complex energy landscape of fragile glass formers as depicted, e.g., in Ref. [57] implies that at high temperatures many different configurational states can be exploited by the system while at lower temperatures only few, low-energy states are accessible. Within the cooperativity framework mentioned above, this may well correspond to the many, relatively small cooperatively rearranging regions (CRRs) [59] assumed to exist on a rather short timescale at high temperatures while close to T_g there are much less,

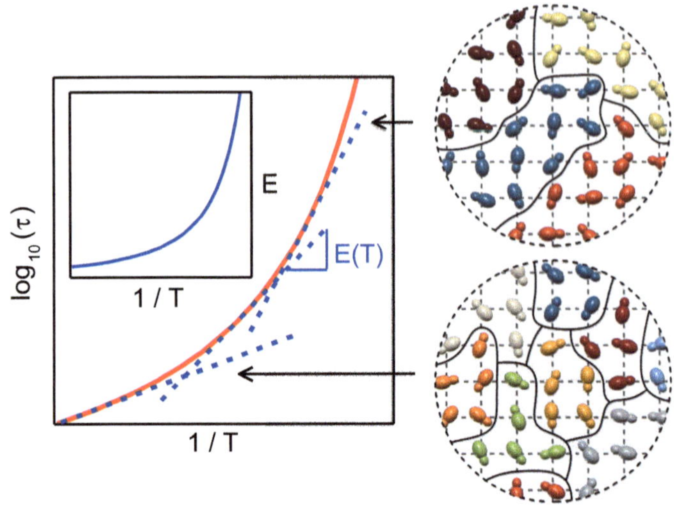

Fig. 3 A possible explanation of the non-Arrhenius behavior of PCs. Left: Schematic Arrhenius representation of the temperature-dependent relaxation time of PCs for Vogel–Fulcher–Tammann behavior (solid line). A temperature-dependent apparent activation energy E (inset) can be estimated from the slope in this plot (indicated for three temperatures by the dashed lines). This increase of E may be ascribed to an increase of the size of CRRs (schematically indicated by molecules of same color at the right)

much larger, and more stable CRRs as schematically indicated in Fig. 3. In contrast, in strong glass formers the variation of accessible states within the less complex energy landscape should be less pronounced, corresponding to a weaker variation of the size and number of the CRRs when approaching T_g. Dielectric measurements of higher harmonic susceptibilities, being able to test possible temperature-dependent variations of the number of dynamically correlated molecules, can give a clue if these variations indeed are rather weak in the strong PCs, which would corroborate the discussed analogy of energy-landscape and cooperativity scenario.

There are various ways to perform nonlinear dielectric measurements of glass-forming materials [18, 19] (see also the other chapters in the present book). In the following, we mainly discuss two different types of experiments applied to PCs: (i) The detection of the complex dielectric permittivity ε^* under high ac fields and its comparison to ε^* measured in the linear regime [28, 29, 31]. (ii) The measurement of higher harmonics of the dielectric susceptibility, especially of the 3ω component χ_3, performed under high ac fields, too [28, 31]. In addition, PCs were also investigated by simultaneously applying a small ac and a high dc bias field [30]. Details on the experimental techniques applied to detect the nonlinear dielectric response can be found, e.g., in [18, 19].

2 Nonlinear Measurements of the Dielectric Permittivity

2.1 Main Relaxation Process

To our knowledge, the first nonlinear dielectric measurements of a plastic-crystalline material reported in literature were performed on cyclo-octanol [28]. It is a typical PC, well characterized by linear dielectric spectroscopy [13, 60]. Its plastic state is most easily investigated by first supercooling the transitions into two different orientationally ordered states, known to occur at 265 and 220 K [61–63], and then performing measurements under heating. T_g^o of cyclo-octanol is about 168 K [60]. Its linear dielectric response in the PC phase is shown by the open symbols in Fig. 4 for four temperatures [28]. The steplike decrease in the dielectric constant $\varepsilon'(\nu)$ and the peak in the loss $\varepsilon''(\nu)$, both shifting to higher frequencies with increasing temperature, reveal the typical signatures of a relaxational process. The latter was identified with the main reorientational motion of the molecules, assigned as α relaxation [13, 60], which exhibits glassy freezing, non-exponentiality, and rather mild non-Arrhenius behavior. In fact, with a fragility parameter $m \approx 33$ [13, 60], cyclo-octanol can be classified as a rather strong glass former (cf. Fig. 2).

The plusses in Fig. 4 represent the spectra obtained for a high electric field of 375 kV/cm. In the region of the α relaxation, a small but significant difference of the high-field and low-field spectra is revealed (see also right insets of Fig. 4). In Fig. 5, the difference of the high- and low-field spectra is plotted. A common way to represent such data is plotting the quantity $\Delta \ln \varepsilon = \ln \varepsilon_{\mathrm{hi}} - \ln \varepsilon_{\mathrm{lo}}$ for $\varepsilon = \varepsilon'$ or $\varepsilon = \varepsilon''$ [20, 25] where $\varepsilon_{\mathrm{hi}}$ and $\varepsilon_{\mathrm{lo}}$ denote the results for high and low fields, respectively. The arrows indicate the α-peak positions ν_α at low fields (cf. Fig. 4b). Obviously, $\Delta \ln \varepsilon''$ exhibits a "V-shaped" behavior with the tip of the "V" in the region of the α-peak frequency. Correspondingly, $\Delta \ln \varepsilon'$ shows a zero-crossing close to ν_α with negative and positive peaks at $\nu < \nu_\alpha$ and $\nu > \nu_\alpha$, respectively. Especially at low frequencies, these difference spectra qualitatively differ from those observed in canonical glass formers, which usually only exhibit a weak nonlinear effect at $\nu < \nu_\alpha$ [20, 25].

This overall behavior seems to be a common motif in nonlinear difference spectra of PCs [28, 29, 31]. An example from Richert and co-workers is provided in Fig. 6. It shows the relative difference of high- and low-field loss spectra of a plastic-crystalline mixture of neopentylglycol and cyclo-hexanol [29] ($T_g^o \approx 155$ K [64]). Here, the quantity $(\varepsilon''_{\mathrm{hi}} - \varepsilon''_{\mathrm{lo}})/\varepsilon''_{\mathrm{lo}}$ is shown, which is comparable to $\Delta \ln \varepsilon''$ if the factor $(\varepsilon''_{\mathrm{hi}}/\varepsilon''_{\mathrm{lo}} - 1)$ is small. Again the V-shape shows up rather close to ν_α, indicated by an arrow in Fig. 6. This is the case for spectra collected at different applied high fields, which can be scaled onto each other as demonstrated in Fig. 6. In Ref. [29], qualitatively similar difference spectra were also reported for plastic-crystalline cyanocyclohexane. Thus, it seems that, especially concerning the pronounced field-induced increase of ε'' at *low* frequencies and the corresponding decrease of ε', PCs seem to exhibit qualitatively different nonlinear behaviors than canonical glass formers [20, 25, 33].

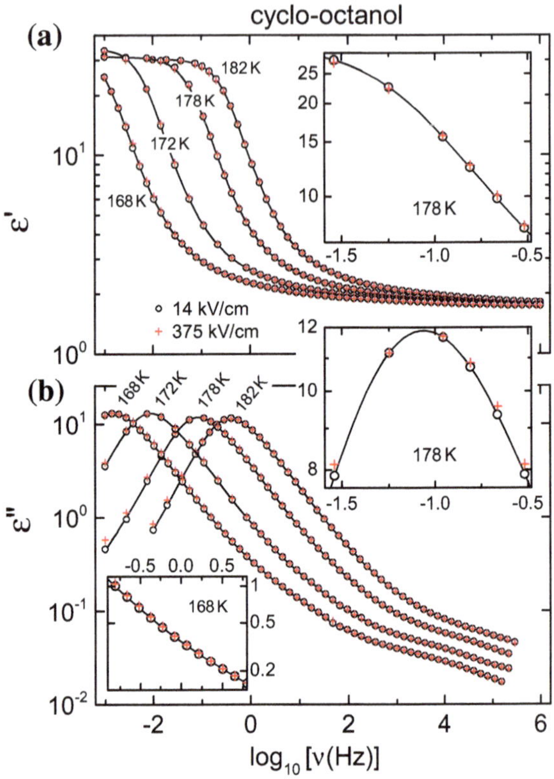

Fig. 4 Dielectric constant (**a**) and loss spectra (**b**) of plastic-crystalline cyclo-octanol measured at various temperatures and at low and high ac fields as indicated in the figure [28]. The insets provide magnified views demonstrating the field dependence in the α-peak region (right insets) and the lack of significant field-induced variation at higher frequencies (lower left inset). The lines are guides to the eyes

As further examples of this quite general nonlinear behavior of PCs, Fig. 7 shows $\Delta \ln \varepsilon''$ for a mixture of 60% succinonitrile and 40% glutaronitrile (60SN-40GN) [31] and for *ortho*-carborane [28, 31], both well-established PCs with orientational-glass temperatures $T_g^o \approx 144$ and 120 K, respectively [48, 65, 66]. While *ortho*-carborane exhibits very strong glassforming characteristics ($m \approx 20$ [65]), 60SN-40GN was shown [48] to be exceptionally fragile ($m \approx 62$) when compared with most other PCs [1] (cf. Fig. 2b). Irrespective of this difference, just as for the other PCs (Figs. 5 and 6) for both materials, a V-shaped behavior of $\Delta \ln \varepsilon''$ is observed, too. In contrast to the other compounds, for 60SN-40GN $\Delta \ln \varepsilon''$ does not approach zero at the minimum. In Ref. [31], this was ascribed to additional contributions from field-induced transitions between different molecular conformations in this material.

The case of *ortho*-carborane illustrates an experimental problem that one may encounter when performing nonlinear measurements of PCs: As the nonlinear dielectric response commonly is much smaller than the linear one, usually very high electrical fields E are necessary to enable its detection. Therefore, accounting for the relation $E \sim 1/d$, the sample thickness d should be as thin as possible. This can be much more easily achieved when the investigated material is liquid during preparation, enabling the use of capacitors with thin spacers (e.g., glass fibers, Teflon foils, or

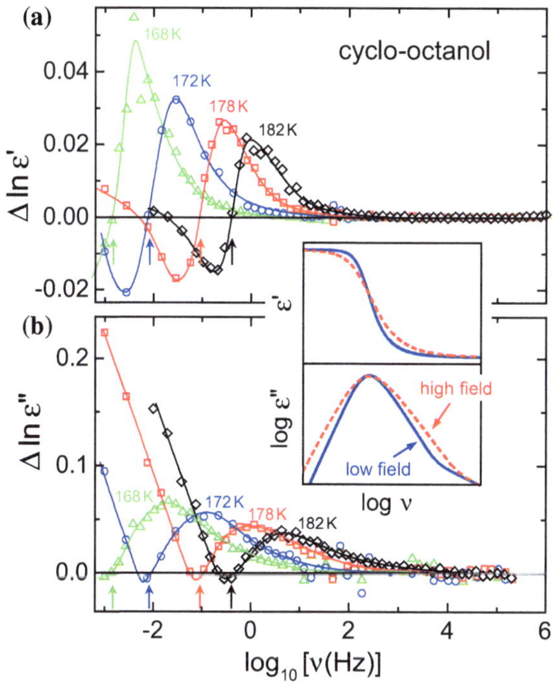

Fig. 5 Difference of the logarithmic high- and low-field dielectric spectra of plastic-crystalline cyclo-octanol as measured for high fields of 375 kV/cm and for various temperatures (cf. Fig. 4) [28]. The arrows indicate the α-peak frequencies. The lines are drawn to guide the eyes. The inset schematically indicates a field-induced broadening, which would lead to qualitatively similar difference spectra as in the main frames

Fig. 6 Relative difference of high- and low-field loss spectra as measured in a mixture of neopentylglycol and cyclo-hexanol [29]. Curves for different high fields are shown which were scaled onto each other. The smooth solid line was calculated according to the box model (see [29] for details) Reprinted from [29] with the permission of AIP Publishing

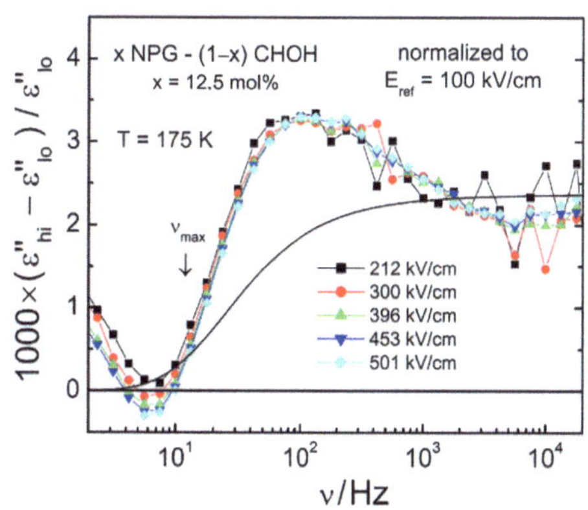

silica microspheres [20, 25, 26, 29–31]). However, in contrast to supercooled liquids, many materials exhibiting a PC phase are not liquid at room temperature. Fortunately, often they can be easily melted at only moderately enhanced temperatures, enabling

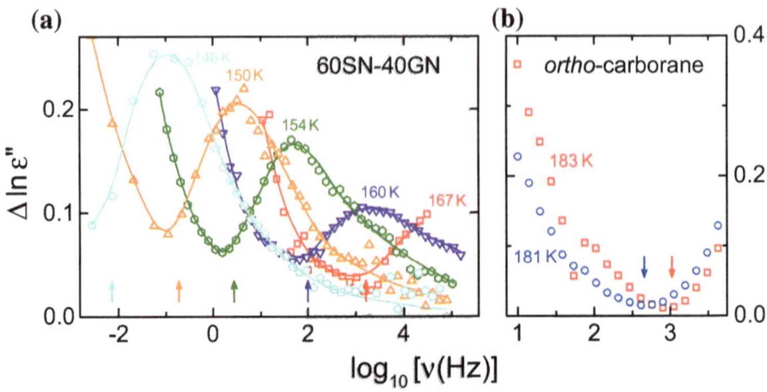

Fig. 7 Difference of the logarithmic dielectric-loss spectra of plastic-crystalline 60SN-40GN and *ortho*-carborane as measured for high fields of 357 and 90 kV/cm, respectively, and for various temperatures [28, 31]. The arrows indicate the α-peak positions. The lines in **a** are drawn to guide the eyes

their filling into capacitors. This is, however, not the case for *ortho*-carborane, which tends to sublimate at elevated temperatures [65]. This material instead has to be prepared from powder, pressing a thin platelet between polished stainless-steel plates that serve as electrodes [28]. The minimum thickness achieved in this way was of the order of 100 μm, much larger than the few μm thickness that can be reached for liquid samples. Consequently, higher fields had to be applied to compensate for this effect, which only was possible in a limited frequency range (Fig. 7b).

It should be noted that the common nonlinear behavior of PCs around the α-peak frequency as documented in Figs. 5, 6, and 7 is consistent with a field-induced broadening of the observed relaxation features. This becomes clear by a comparison with the inset of Fig. 5, which schematically indicates such a scenario with somewhat exaggerated field-induced effects for clarity reasons [28]. The behavior shown there is well consistent with the experimental data. Remarkably, this broadening also occurs at the low-frequency flank of the α peak, causing the mentioned increase of $\Delta \ln \varepsilon''(\nu)$ at low frequencies (Figs. 5b, 6, and 7). Thus, the high-field spectra can no longer be described by the Cole–Davidson (CD) function [67], for which $\varepsilon''(\nu)$ increases linearly (i.e., with a slope one in the log–log plot) at the left flank of the loss peak. The CD function is found to reasonably fit the α relaxation of many glass formers [11] and plastic crystals [1], including cyclo-octanol [60]. Usually, peak broadening observed in the relaxation dynamics of glassy matter is ascribed to heterogeneity arising from the disorder in the material, which induces a distribution of relaxation times [9, 10]. However, as discussed in Ref. [28], it is not clear why a high field should increase heterogeneity and such an explanation of this field-induced broadening seems unlikely.

In contrast to the low-frequency behavior, the field-induced increase of ε'' and ε' at frequencies, $\nu > \nu_\alpha$, found in PCs (Figs. 5, 6 and 7) [28, 29, 31] resembles the behavior in structural glass formers [20, 25, 33]. Within the framework of the box model, considering the dynamic heterogeneity of glassy matter [32, 68], this nonlinear effect was ascribed to a selective transfer of field energy into the heterogeneous regions, accelerating their dynamics and leading to an effective broadening at the right flank of the α relaxation peak (and a corresponding effect in ε') [18, 20, 29, 33]. However, the continuous increase of $\Delta \ln \varepsilon''(\nu)$ found at $\nu < \nu_\alpha$ in PCs cannot be accounted for in this way. This discrepancy is also revealed by the solid line shown in Fig. 6, which was calculated according to the box model [29]. Obviously, within this model the field effects in ε'' are expected to approach zero for low frequencies, in agreement with the behavior in most supercooled liquids but in marked contrast with the findings in the PCs. In contrast, at $\nu > \nu_\alpha$, a positive field effect with the correct order of magnitude is predicted. In Ref. [29], the deviations in this region were ascribed to the suggested intrinsic non-exponentiality of PCs [41, 42]. This notion implies that, in contrast to canonical glass formers, in PCs only part of their non-exponential relaxation behavior arises from heterogeneity.

It should also be noted that the negative values of $\Delta \ln \varepsilon'$ detected at $\nu < \nu_\alpha$ (Fig. 5a) again are at variance with the box model as developed for supercooled liquids. In principle, negative values of $\Delta \ln \varepsilon'$, corresponding to a reduction of the low-frequency dielectric constant, may be explained by saturation effects of the polarization [69–71]. However, instead of the minimum revealed by Fig. 5a, in this case a low-frequency plateau in $\Delta \ln \varepsilon'(\nu)$ is expected as found, e.g., in glassforming 1-propanol [72].

An explanation for the puzzling low-frequency nonlinear properties of PCs discussed above was provided in Ref. [29], based on recent theoretical considerations by Johari [39, 40]. Within this framework, the high external field is assumed to result in a reduction of configurational entropy. Via the relation of entropy and relaxation time that is assumed within the Adam–Gibbs theory of the glass transition [59], this should induce an increase of the relaxation time, resulting in a slight increase of the glass temperature. This entropy effect should, however, only lead to significant nonlinear contributions at low frequencies, $\nu < \nu_\alpha$, because the molecular rearrangements associated with the entropy reduction are too slow to follow the ac field at high frequencies [29]. This is in good accordance with the experimental findings at low frequencies documented in Figs. 5, 6, and 7, which cannot easily be explained by other contributions as discussed above.

Entropy contributions to nonlinear dielectric properties have also been found for various structural glass formers, mostly by performing measurements under a strong bias field [30, 73–75]. However, it seems that entropy-driven nonlinearity effects in PCs are generally stronger than those detected in such supercooled liquids [30, 31]. In Johari's model [39], the impact of a high electrical field on the *reorientational* degrees of freedom of the molecules is considered. It indeed seems reasonable that the field may diminish the reorientational disorder of dipolar molecules and thus influence the entropy of the system. However, while in PCs there is only reorientational disorder, in structural glass formers also *translational* disorder exists, which should be less influenced by the field. In contrast to structural glass formers, for the overall entropy of PCs molecular reorientations are the main source of entropy. Therefore, one may speculate that for PCs the field-induced variation of reorientational disorder is of more importance than for canonical glass formers and that this is the reason for the different low-frequency nonlinear behavior of these two classes of glassy matter [31]. However, in a recent work [40], Johari pointed out that for cyclo-octanol intramolecular degrees of freedom also strongly contribute to its overall entropy.

As noted, e.g., in Ref. [19], from the field-induced variation of the complex dielectric permittivity, the cubic susceptibility $\chi_3^{(1)}$ can be calculated. It is defined via the following relation for the time-dependent polarization $P(t)$, resulting from an applied time-dependent electrical field $E(t) = E_{ac} \cos(\omega t)$ [27, 38]:

$$\begin{aligned}\frac{P(t)}{\varepsilon_0} &= \left|\chi_1^{(1)}\right| E_{ac} \cos(\omega t - \delta_1) + \frac{3}{4}\left|\chi_3^{(1)}\right| E_{ac}^3 \cos\left(\omega t - \delta_3^{(1)}\right) \\ &+ \frac{10}{16}\left|\chi_5^{(1)}\right| E_{ac}^5 \cos\left(\omega t - \delta_5^{(1)}\right) \\ &+ \frac{1}{4}\left|\chi_3^{(3)}\right| E_{ac}^3 \cos\left(3\omega t - \delta_3^{(3)}\right) + \frac{5}{16}\left|\chi_5^{(3)}\right| E_{ac}^5 \cos\left(3\omega t - \delta_5^{(3)}\right) \\ &+ \frac{1}{16}\left|\chi_5^{(5)}\right| E_{ac}^5 \cos\left(5\omega t - \delta_5^{(5)}\right) + \cdots \end{aligned} \quad (1)$$

Here, the lower indices of χ correspond to the exponent of the electrical-field dependence (which defines $\chi_3^{(1)}$ as cubic susceptibility) while the upper ones signal the ω factor. The higher order harmonic susceptibilities $\chi_3^{(3)}$ and $\chi_5^{(5)}$ are often simply denoted as χ_3 and χ_5, respectively. It should be noted that $\chi_3^{(1)}$ contains essentially the same information as $\Delta \ln \varepsilon'(\nu)$ and $\Delta \ln \varepsilon''(\nu)$ plotted in Figs. 5, 6, and 7 and can be directly calculated from the measured low- and high-field permittivities via

$$\left|\chi_3^{(1)}\right| = \frac{4}{3}\frac{1}{E_{ac}^2}\sqrt{(\Delta\varepsilon')^2 + (\Delta\varepsilon'')^2}$$

and

$$\arg\left[\chi_3^{(1)}\right] = \arctan\left(\frac{\Delta\varepsilon''}{\Delta\varepsilon'}\right)$$

with $\Delta\varepsilon' = \varepsilon'_{hi} - \varepsilon'_{lo}$ and $\Delta\varepsilon'' = \varepsilon''_{hi} - \varepsilon''_{lo}$ [19].

Figure 8 shows spectra of the modulus and phase of $\chi_3^{(1)}$ for plastic-crystalline cyclo-octanol. This should be compared to the corresponding results for two supercooled liquids (glycerol and propylene carbonate), recently published in Ref. [76]. The findings in this work were interpreted along similar lines as the higher harmonic susceptibilities reported, e.g., in [23, 26, 27], namely within the theoretical framework by Bouchaud and Biroli [36] assuming molecular cooperativity leading to "amorphous order" that grows when the glass temperature is approached. The humped shape observed in the modulus spectra of various cubic susceptibilities in the region of ν_α can be well understood within this framework. As pointed out in Ref. [76], several common features (e.g., the hump and the similar phase behavior) found in the spectra of different nonlinear susceptibilities point to a common phys-

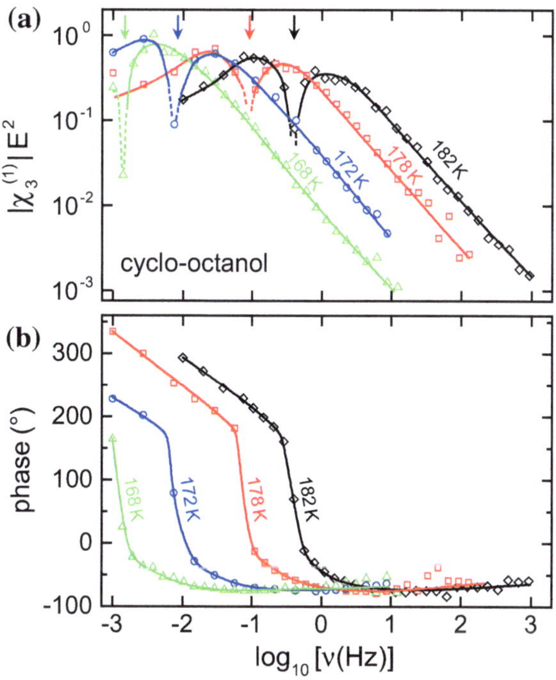

Fig. 8 Modulus (**a**) and phase (**b**) of the cubic susceptibility $\chi_3^{(1)}$ of plastic-crystalline cyclo-octanol, measured at various temperatures for a field of 375 kV/cm. The lines are guides to the eyes

ical origin, dominated by cooperativity effects. Nonlinear entropy contributions as discussed in Johari's model [39] were shown to be consistent with this view.

An inspection of the $\chi_3^{(1)}$ spectra of cyclo-octanol in Fig. 8 reveals some similarities to those in the supercooled liquids [76]: Just as for the latter, a hump shows up in the modulus spectra about half a decade above the α-peak frequency ν_α (Fig. 8a). Moreover, $\left|\chi_3^{(1)}\right|(\nu)$ exhibits a spikelike minimum where it seems to approach zero. Just as for the canonical glass formers, this minimum is accompanied by a strong jump in the phase (Fig. 8b). (In Fig. 2 of Ref. [76], the phase becomes negative at low frequencies but it is a matter of definition if angles above 180° are represented as positive or negative values.) However, Fig. 8 also reveals some differences to the behavior in the supercooled liquids: In plastic-crystalline cyclo-octanol, the minimum in the modulus and the jump in the phase occur at higher frequency, namely just at ν_α, while in Ref. [76] it was found at least a factor of three below ν_α. This effect seems to impede the full formation of the hump in the modulus of the PC. Moreover, the phase at frequencies below the jump is still strongly frequency dependent, while it is nearly constant in the supercooled liquids. In Ref. [76] the anomalies in the $\chi_3^{(1)}$ spectra were ascribed to a transition from the "trivial" saturation response dominating at low frequencies [69–71] to the correlation-dominated regime at higher frequencies. Especially, the saturation effect leads to a reduction of ε', while correlations seem to enhance it. At the frequency of the anomalies, both effects seem to compensate. In the present case of plastic-crystalline cyclo-octanol, similar arguments can be used when assuming a transition from correlation-dominated behavior at $\nu > \nu_\alpha$ to entropy-dominated nonlinearity at $\nu < \nu_\alpha$. Nonlinear saturation and entropy effects are both expected to be most pronounced at low frequencies. In the $\chi_3^{(1)}$ spectra (and thus probably also in the other cubic susceptibilities [76]), their main difference seems to be the essentially frequency-independent behavior of the first compared to the frequency-dependent nonlinear response of the latter. The comparison of the $\chi_3^{(1)}$ results on a PC (Fig. 8) and those on supercooled liquids in Ref. [76] seem to corroborate the notion that entropy effects mainly dominate the low-frequency nonlinear response in PCs, in contrast to saturation effects in supercooled liquids.

2.2 Secondary Relaxations

Just as found for many supercooled liquids, plastic crystals are known to exhibit secondary relaxation processes, termed β relaxations, which are faster than the main reorientational process, usually denoted as α relaxation [1]. As shown by Johari and Goldstein [34], β relaxations seem to be an inherent property of the glassy state of matter. To distinguish such processes from relaxations arising from more trivial effects as, e.g., intramolecular motions, they are nowadays commonly denoted as Johari–Goldstein (JG) relaxations. The microscopic origin of JG relaxations is still controversially discussed. For example, they were ascribed to motions of molecules within "islands of mobility" [34], small-scale motions within a fine structure of the energy landscape experienced by the molecules [77–79], or various other mechanisms (e.g., [80–83]).

The shoulders observed at the high-frequency flanks of the peaks in the loss spectra of cyclo-octanol, shown in Fig. 4b, indicate the presence of two faster secondary processes. Examining data covering a broader temperature and frequency range indeed reveals clear evidence for these processes, which were termed β and γ relaxations [60]. By applying a universal criterion, valid for different classes of glass formers, strong hints were obtained that the β relaxation of cyclo-octanol is a genuine JG relaxation process [83, 84]. As seen in the left inset of Fig. 4, there is no significant difference of the high- and low-field results for the loss in the region of the β relaxation. Obviously, the field-induced variation of ε'' is small in this region and, consequently, the difference spectra shown in Fig. 5 strongly decrease at high frequencies. This is also the case for 60SN-40GN (Fig. 7a) where the nonlinearity also becomes weaker in the regime of its secondary relaxations [31]. For the neopentylglycol/cyclo-hexanol mixture shown in Fig. 6 and for cyanocyclohexane, a decrease of the difference spectra at high frequencies was also observed, which, however, seems to be more gradual than for the other PCs [29].

The weaker nonlinearity of PCs in the regime of their secondary relaxations resembles the corresponding effect found for the excess-wing region of supercooled liquids like glycerol [25, 72]. The excess wing shows up as a second, more shallow power law at the high-frequency flank of the α peak of some glass formers [11, 35, 85]. In various works, it was suggested to be caused by a JG relaxation peak that is partly submerged under the dominating α peak (e.g., [86–88]). Therefore, the reduction or even absence of a nonlinear effect at high frequencies in the PCs may well have the same origin as the negligible nonlinearity in the excess-wing region of the supercooled liquids [25, 72]. As discussed, e.g., in Ref. [25], this finding is consistent with the relation of nonlinearity and molecular cooperativity suggested in Ref. [36] if having in mind that secondary relaxations are often assumed to be of non-cooperative nature [81, 89, 90]. However, it should be noted that in the region of the very strong secondary relaxation of glassforming sorbitol, well-pronounced nonlinear effects were found [19, 91], a fact that could be explained within the framework of the coupling model [92].

As discussed in Ref. [93], based on time-resolved measurements, the number of high-field cycles applied to the sample may play a crucial role in the detection of equilibrium values for the nonlinearity. For example, the degree of nonlinearity in the high-frequency region of supercooled liquids seems to clearly depend on the number of applied cycles [93] and similar behavior was also reported for PCs [29]. However, experiments with different cycle numbers performed for 60SN-40GN did not reveal any differences [31]. While the details of this behavior are not finally clarified, in any case, the nonlinearity in the regime of high-frequency processes seems to be smaller than for the main relaxation in most glasslike materials investigated until now.

Further information on the secondary relaxations in cyclo-octanol can be obtained by transferring the sample into an (at least partly) orientationally ordered state [60–63], which was achieved by heating the sample to 227 K after supercooling and keeping it there for 10 min. As shown in Ref. [60], the secondary relaxations persist in this more ordered state and can be investigated without interference from the dominating α relaxation. Figure 9 shows the results for the loss at 168 K, measured at low and high fields. Above about 1 Hz, in the region of the β and γ relaxations, within experimental resolution no field-induced variation is observed. This finding is in good agreement with those obtained for the plastic-crystalline phase in the secondary relaxation region, discussed above (Figs. 4 and 5). The field dependence observed at the lowest frequencies may be ascribed to the α relaxation arising from residual amounts of the plastic-crystalline phase or to contributions from ionic conductivity, due to small amounts of impurities within the sample. Nonlinearities of ionic conductivity are well-known effects and are discussed in detail by Roling and co-workers (see, e.g., Ref. [94] and the chapter by B. Roling in the present book).

3 Third-Order Harmonic Susceptibility

Another prominent way to detect the nonlinear dielectric response of a material is the measurement of higher harmonics of the dielectric susceptibility: At low fields, the polarization P and field E should be proportional to each other. Therefore, the application of a sinus ac field $E(t)$ results in a sinus polarization $P(t)$ with the same frequency. However, at high fields, P no longer is proportional to E and, thus, an applied sinus field with frequency ω can result in higher harmonics with frequency

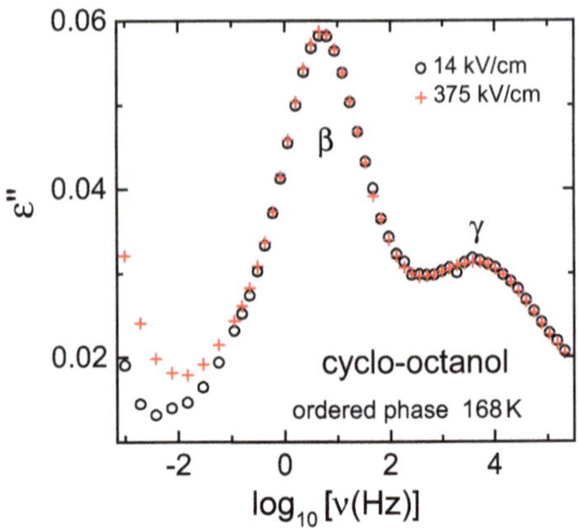

Fig. 9 Dielectric loss spectra of cyclo-octanol in the orientationally ordered state at 168 K, measured at low and high field

3ω, 5ω, etc., which are quantified by the higher order harmonic susceptibilities χ_3, χ_5, etc. as defined in Eq. (1). The even harmonics 2ω, 4ω, etc. should be zero because $P(E)$ should be equal to $-P(-E)$.

Figure 10 shows the modulus of $\chi_3 E^2$ for the PCs cyclo-octanol [28] and 60SN-40GN [31]. In both cases, a hump is observed at a frequency somewhat below ν_α, indicated by the arrows in Fig. 10. As mentioned in the introduction, such a humped shape of the χ_3 spectra is predicted within the model by Biroli and co-workers [36, 37], to arise from molecular cooperativity, which is often assumed to be typical for glassforming systems [8, 43, 44, 59]. Qualitatively, similar spectral shapes of $\chi_3(\nu)$ were also found for various structural glass formers [19, 23, 26, 73]. Within the theoretical framework of Bouchaud and Biroli [36, 37], the detection of a hump in PCs as documented in Fig. 10 indicates that the glassy freezing in PCs is also governed by molecular cooperativity [28]. A possible mechanism for generating molecular correlations in PCs could be lattice strains that reduce the hindering barriers for reorientational motions of neighboring molecules [28]. However, it should be noted that a hump in $|\chi_3|(\nu)$ can also be explained within the framework of other models [73, 95–99].

About one decade below the hump frequencies, the $|\chi_3|E^2$ spectra of 60SN-40GN (Fig. 10b) reveal weak shoulders. In Ref. [31] it was speculated that these spectral features arise from an additional slow relaxation process, for which indications were found in the linear dielectric spectra [48]. The microscopic origin of this process is unknown until now.

In Fig. 11, for cyclo-octanol in addition to the modulus shown in Fig. 10a, the real and imaginary parts of the third-order harmonic susceptibility (times E^2) are presented. This should be compared to the corresponding spectra as found for the supercooled liquids glycerol and 1-propanol [19]. In the latter, the spectra of both quantities could be well described by exclusively considering saturation effects of the polarization at high fields [71], as already treated in very early works on nonlinear dielectric spectroscopy [69, 70]. In 1-propanol, cooperativity seems to be absent for

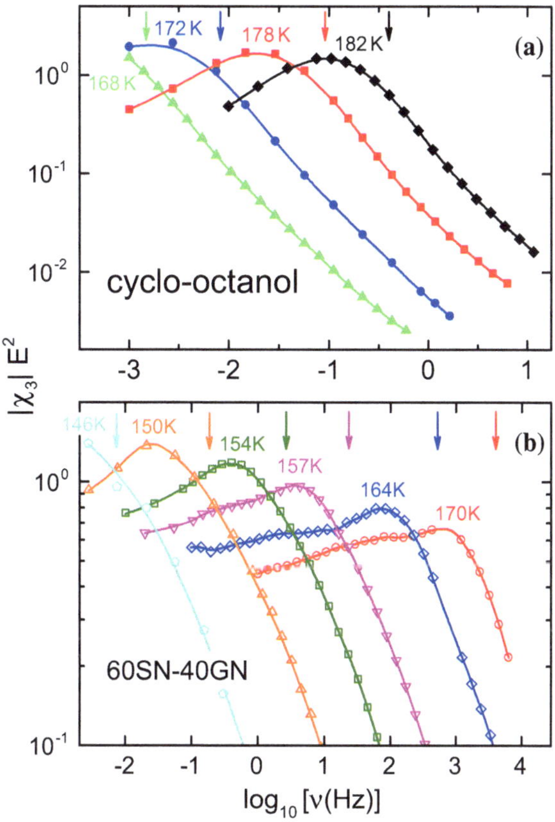

Fig. 10 Modulus of the third-order harmonic component of the dielectric susceptibility (times E^2) of cyclo-octanol (**a**) [28] and 60SN-40GN (**b**) [31], measured at various temperatures. The applied fields were 375 and 357 kV/cm, respectively. The arrows indicate the corresponding α-peak frequencies. The lines are guides for the eyes

the main relaxation process due to the well-known peculiarities of the relaxational dynamics of most monohydroxy alcohols [100, 101]. However, in glycerol, the real and imaginary parts of χ_3 showed clear qualitative deviations from the behavior predicted for entirely saturation-induced nonlinearity, which was ascribed to cooperativity effects [19]. For cyclo-octanol, the spectra of Fig. 11 in many respects resemble those of glycerol and also clearly do not follow the behavior predicted for a purely saturation-dominated system such as 1-propanol. Especially, just as for glycerol, the well-pronounced negative minima, occurring in both the real and the imaginary parts, not far from ν_α, together generate the hump observed in the modulus of χ_3 (Fig. 10a), which is taken as signature for cooperative glassy dynamics [28]. The negative real part of χ_3 found at low frequencies (Fig. 11a) can be assumed to arise from the entropy effects [39] discussed above, in contrast to the saturation effect dominating the low-frequency response in glycerol. At low frequencies, i.e., on long timescales, the liquid flow (directly related to the α relaxation) should destroy glassy correlations [27], leaving room for additional nonlinear contributions becoming dominant, which in the PCs predominantly seem to be entropy effects.

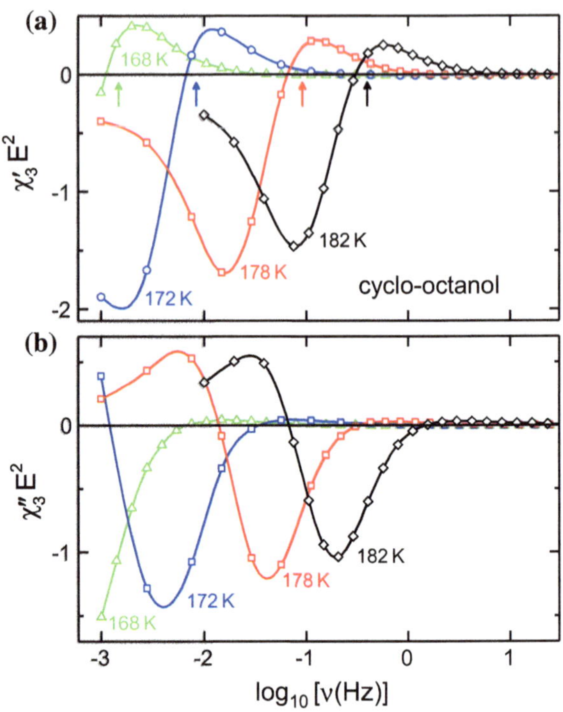

Fig. 11 Real and imaginary parts of $\chi_3 E^2$ of cyclo-octanol at four temperatures measured for a field of 375 kV/cm. The arrows indicate the α-peak frequencies. The lines are guides for the eyes

Within the model by Bouchaud and Biroli [36, 37], a hump observed in the third-order susceptibility χ_3, as documented for the PCs cyclo-octanol and 60SN-40GN in Fig. 10, should be related to the number of correlated molecules N_{corr}. Especially, the dimensionless quantity $X_3^{(3)}$, defined by [23]

$$X_3^{(3)} = \frac{k_B T}{\varepsilon_0 \left(\Delta\chi_1\right)^2 a^3} \chi_3^{(3)},$$

which corrects χ_3 for trivial temperature dependences, should be directly proportional to N_{corr}. (In this equation, $\Delta\chi_1$ is the dielectric strength and a^3 the volume taken up by a single molecule.) Figure 12b presents the temperature dependence of the peak value of this quantity for three supercooled liquids [26] and for the two PCs for which χ_3 data are available (symbols; left scale) [28, 31]. At the peak, $X_3^{(3)}$ should be dominated by the cooperativity contribution to χ_3 and thus correspond to $N_{\mathrm{corr}}(T)$ in arbitrary units. As revealed by Fig. 12b [26, 28, 31], for all these systems, N_{corr} increases with decreasing temperature, implying a growth of correlation length scales. This is in accord with the notion that the glass transition is related to an underlying thermodynamic phase transition [43, 59].

The temperature dependence of N_{corr} for the three PCs, shown in Fig. 12b (closed symbols), fully matches the general scenario found for the supercooled liquids [26]

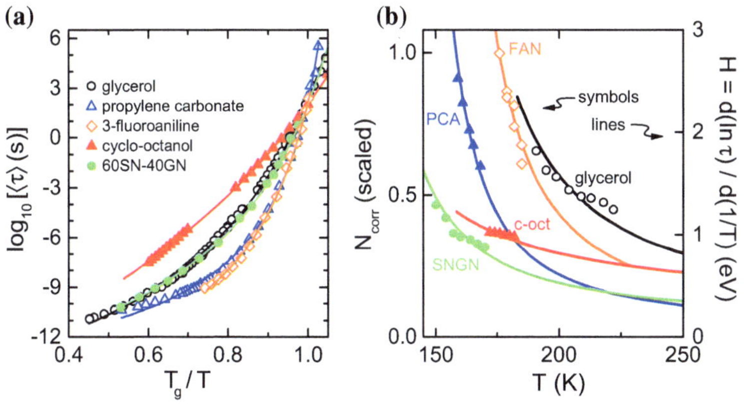

Fig. 12 **a** Temperature-dependent average α-relaxation times of three glassforming liquids and two PCs [48, 50, 60, 89] shown in an Angell plot [52]. The lines are fits with the Vogel–Fulcher–Tammann (VFT) function [50]. **b** Comparison of the effective activation energies H (lines; right scale) with the number of correlated molecules N_{corr} (symbols; left scale) for the same materials as in frame (**a**) [26, 28, 31]. H was determined from the derivatives of the VFT fits of the temperature-dependent relaxation times shown in frame (**a**). N_{corr}, shown in arbitrary numbers, was determined from χ_3 (see text). To match the $H(T)$ curves, N_{corr} was multiplied by separate factors for each material (glycerol: 1.15, propylene carbonate (PCA): 0.72, 3-fluoroaniline (FAN): 1.30, cyclo-octanol: 0.19, 60SN-40GN (SNGN): 1.05). Note that both ordinates start from zero, implying direct proportionality of both quantities

(open symbols): Simply spoken, the more fragile the material, the stronger is the temperature dependence of its N_{corr}. To illustrate their significantly different fragilities, Fig. 12a shows an Angell plot of the relaxation times τ of the same materials as in Fig. 12b. As discussed above, the fragility of a glass former quantifies the degree of deviation of its temperature-dependent relaxation time from Arrhenius behavior. In Fig. 12a, these deviations are revealed to be weakest for glycerol and strongest for the two other glassforming liquids. Indeed, with fragility parameters $m \approx 53$ (glycerol [102]), 90 (3-fluoroaniline), and 104 (propylene carbonate [102]), the supercooled liquids in this plot vary considerably between intermediate and high fragility. Correspondingly, glycerol has significantly weaker temperature dependence of N_{corr} than the other two glass formers (Fig. 12b). As pointed out in Ref. [26], this finding well corroborates the notion that the non-Arrhenius behavior of supercooled liquids arises from increasing molecular cooperativity at low temperatures.

Concerning the PC results included in Fig. 12, cyclo-octanol is known to be a rather strong glass former ($m \approx 33$), which is quite common for this class of glasslike materials [1, 45]. In Fig. 12a, this immediately becomes obvious from the fact that its τ curve is only weakly bended and, for $T > T_g$, lies above the data points of all the other shown materials. Figure 12b reveals that, just as for the supercooled liquids, this strong dynamics of cyclo-octanol is nicely mirrored by the very weak temperature dependence of its N_{corr} as determined from the χ_3 measurements. Thus, it seems that, for this PC, a temperature-dependent variation of cooperativity also is the main factor determining its non-Arrhenius behavior. A crucial test to support this idea is provided by the results on 60SN-40GN. This is one of the very few examples [45, 47, 48], where a relatively *high* fragility ($m \approx 62$) is realized in a PC. Indeed, in Fig. 12a, its $\tau(T)$ curve is similarly bended as for the intermediately fragile glycerol. Therefore, for 60SN-40GN the number of correlated molecules should increase significantly stronger than for cyclo-octanol. In fact, this is observed in Fig. 12b. Within the theoretical framework by Bouchaud and Biroli [36, 37], these results allow to conclude that the origin of the non-Arrhenius behavior in PCs is the same as for structural glass former, namely an increase of cooperativity when approaching the glass transition under cooling.

In Ref. [26], it was demonstrated that, in addition to the qualitative connection of the temperature variations of the α-relaxation time and N_{corr} discussed above, there also seems to be a quantitative relation of both quantities: Let us consider the explanation of the non-Arrhenius behavior of $\tau(T)$ by a temperature-dependent effective energy-barrier H governing molecular motion, as schematically indicated in Fig. 3 [12, 44, 103]. Within this framework, fragile and strong dynamics imply strong or weak temperature dependence of $H(T)$, respectively. Within the time-honored Adam–Gibbs theory [59], it is assumed that the temperature-dependent energy barrier is proportional to the number of molecules within a CRR, i.e., $H \sim N_{\mathrm{corr}}$. As indicated in Fig. 3, $H(T)$ can be estimated by the derivative of the $\log \tau \left(1/T\right)$ curves in the Arrhenius representation, $H = \mathrm{d}(\ln\tau)/\ \mathrm{d}(1/T)$. (To avoid excessive data scatter, usually arising when differentiating experimental curves, derivatives of the fit curves of $\tau(T)$ instead of the experimental data points can be used.) For the materials covered by Fig. 12, the results for $H(T)$ are indicated by the lines shown in frame (b) (right scale). As demonstrated in Ref. [26] for the supercooled liquids, the $N_{\mathrm{corr}}(T)$ data (in arbitrary units) can be reasonably scaled onto the $H(T)$ curves obtained in this way. This obviously is also well fulfilled for the two PCs [28, 31]. It should be noted that both ordinates in Fig. 12 start at zero implying that, indeed, $H(T)$ and $\tau(T)$ are directly proportional to each other. Finally, we want to mention that the scaling factors, applied to match the $N_{\mathrm{corr}}(T)$ to the $H(T)$ curves are of the order of one for the three supercooled liquids and for plastic-crystalline 60SN-40GN (see caption of Fig. 12 for the values). However, for cyclo-octanol, this factor is 0.19 and, thus, significantly smaller. The reason for this difference is not clear at present; seemingly, for the latter compound, the molecular motions are less impeded by a high N_{corr} than in the others. Further nonlinear investigations of canonical PCs are necessary to check if this deviation is a common property of this material class. 60SN-40GN may be suspected to be a special case, due to its strong substitutional disorder.

4 Summary and Conclusions

In the present overview, we have demonstrated a rich variety of nonlinear dielectric phenomena occurring in PCs. We have concentrated on two typical ways of performing nonlinear dielectric experiments, namely the measurement of the 1ω and of the 3ω components of the dielectric susceptibility, both performed under high ac fields. In many respects, PCs reveal similar behavior as found for structural glass formers. Especially, high ac fields lead to an enhancement of the dielectric permittivity at frequencies $\nu > \nu_\alpha$, just as commonly found for supercooled liquids. Therefore, it seems natural to explain this phenomenon in a similar way. Just as for the latter, currently two seemingly different explanations of the nonlinear response at $\nu > \nu_\alpha$ can be invoked, namely a selective transfer of field energy into the heterogeneous regions as considered in the box model [18, 20, 29, 32, 68] or a cooperativity-related origin implying increasing length scales and "amorphous order" when approaching T_g as treated in the model by Bouchaud and Biroli [23, 26, 27, 31, 36, 37, 76]. It should be noted, however, that in a recent work it was proposed that these two approaches even may be compatible [76].

At low frequencies, $\nu < \nu_\alpha$, the nonlinear 1ω response of PCs and supercooled liquids seems to differ markedly. While the latter exhibit only weak nonlinearity in this frequency range, PCs probably are dominated by entropy effects [29] as considered in Johari's theory [39]. To explain this finding, we have speculated about the different relative importance of reorientational degrees of freedom for the entropy in PCs compared to canonical glasses [31] but this issue is still far from clarified.

When approaching high frequencies, in the region of secondary processes as the excess wing or the JG relaxation, for PCs, just as for the supercooled liquids, a gradual reduction of nonlinearity is observed. Within the cooperativity-related framework, this implies less cooperative motions as often assumed for such processes [81, 89, 90].

Of special interest are the results concerning the third-order harmonic susceptibility, characterizing the 3ω dielectric response [28, 31]. For the two PCs for which this quantity was investigated until now, a spectral shape as predicted by the model by Biroli and Bouchaud is found. In this respect, the PCs behave very similar as various supercooled liquids [23, 26, 73]. The results seem to imply that a growth of molecular cooperativity and the approach of amorphous order under cooling is the origin of the non-Arrhenius behavior, not only in supercooled liquids [23, 26, 27] but also in PCs. As discussed in Sect. 1, an energy-landscape scenario [57, 58] was previously invoked to rationalize the commonly less fragile relaxation dynamics of PCs compared to structural glass formers [1, 45, 47, 48]. The found indications for growing length scales when approaching T_g in PCs, based on χ_3 measurements, seem to imply that there must be a relation of this energy-landscape scenario to the cooperativity scenario. A possible rationalization of such a relation was discussed in Sect. 1.

The present work makes clear that quite far-reaching conclusions can be drawn from nonlinear dielectric measurements of PCs, not only concerning this special class of glasslike systems but also concerning the glass transition and glassy state of matter in general. Nevertheless, one should be aware that until now only rather few PC systems have been investigated by nonlinear techniques. Clearly, a broader database is highly desirable to reveal universalities and further help enlightening our understanding of the role of cooperativity and heterogeneity in glassy systems.

Acknowledgements This work was supported by the Deutsche Forschungsgemeinschaft via Research Unit FOR 1394. Stimulating discussions with S. Albert, Th. Bauer, G. Biroli, U. Buchenau, G. Diezemann, G. P. Johari, F. Ladieu, K. L. Ngai, R. Richert, R. M. Pick, and K. Samwer are gratefully acknowledged.

References

1. R. Brand, P. Lunkenheimer, A. Loidl, J. Chem. Phys. **116**, 10386 (2002)
2. J. Timmermans, J. Chim. Phys. **35**, 331 (1938)
3. N.G. Parsonage, L.A.K. Staveley, *Disorder in Crystals* (Oxford University Press, Oxford, 1978)
4. J.N. Sherwood, *The Plastically Crystalline State* (Wiley, New York, 1979)
5. K. Adachi, H. Suga, S. Seki, Bull. Chem. Soc. Jpn **41**, 1073 (1968)
6. U.T. Höchli, K. Knorr, A. Loidl, Adv. Phys. **39**, 405 (1990)
7. A. Loidl, R. Böhmer, in *Disorder Effects on Relaxational Processes*, ed. by R. Richert, A. Blumen (Springer, Berlin, 1994), p. 659
8. M.D. Ediger, C.A. Angell, S.R. Nagel, J. Phys. Chem. **100**, 13200 (1996)
9. H. Sillescu, J. Non-Cryst. Solids **243**, 81 (1999)
10. M.D. Ediger, Annu. Rev. Phys. Chem. **51**, 99 (2000)
11. P. Lunkenheimer, U. Schneider, R. Brand, A. Loidl, Contemp. Phys. **41**, 15 (2000)
12. J.C. Dyre, Rev. Mod. Phys. **78**, 953 (2006)
13. D.L. Leslie-Pelecky, N.O. Birge, Phys. Rev. Lett. **72**, 1232 (1994)
14. M.A. Ramos, S. Vieira, F.J. Bermejo, J. Dawidowski, H.E. Fischer, H. Schober, M.A. González, C.K. Loong, D.L. Price, Phys. Rev. Lett. **78**, 82 (1997)
15. R. Brand, P. Lunkenheimer, U. Schneider, A. Loidl, Phys. Rev. Lett. **82**, 1951 (1999)
16. F. Affouard, M. Descamps, Phys. Rev. Lett. **87**, 035501 (2001)
17. P.-J. Alarco, Y. Abu-Lebdeh, A. Abouimrane, M. Armand, Nature Mater. **3**, 476 (2004)
18. R. Richert, J. Phys.: Condens. Matter **29**, 363001 (2017)
19. P. Lunkenheimer, M. Michl, Th. Bauer, A. Loidl, Eur. Phys. J. Special Topics **226**, 3157 (2017)
20. R. Richert, S. Weinstein, Phys. Rev. Lett. **97**, 095703 (2006)
21. L.-M. Wang, R. Richert, Phys. Rev. Lett. **99**, 185701 (2007)
22. A. Drozd-Rzoska, S.J. Rzoska, J. Zioło, Phys. Rev. E **77**, 041501 (2008)
23. C. Crauste-Thibierge, C. Brun, F. Ladieu, D. L'Hôte, G. Biroli, J-P. Bouchaud, Phys. Rev. Lett. **104**, 165703 (2010)
24. L.P. Singh, R. Richert, Phys. Rev. Lett. **109**, 167802 (2012)
25. Th. Bauer, P. Lunkenheimer, S. Kastner, A. Loidl, Phys. Rev. Lett. **110**, 107603 (2013)
26. Th. Bauer, P. Lunkenheimer, A. Loidl, Phys. Rev. Lett. **111**, 225702 (2013)
27. S. Albert, Th. Bauer, M. Michl, G. Biroli, J.-P. Bouchaud, A. Loidl, P. Lunkenheimer, R. Tourbot, C. Wiertel-Gasquet, F. Ladieu, Science **352**, 1308 (2016)
28. M. Michl, Th. Bauer, P. Lunkenheimer, A. Loidl, Phys. Rev. Lett. **114**, 067601 (2015)
29. B. Riechers, K. Samwer, R. Richert, J. Chem. Phys. **142**, 154504 (2015)
30. S. Samanta, R. Richert, J. Chem. Phys. **142**, 044504 (2015)
31. M. Michl, Th. Bauer, P. Lunkenheimer, A. Loidl, J. Chem. Phys. **144**, 114506 (2016)
32. B. Schiener, R. Böhmer, A. Loidl, R.V. Chamberlin, Science **274**, 752 (1996)
33. S. Weinstein, R. Richert, Phys. Rev. B **75**, 064302 (2007)
34. G.P. Johari, M. Goldstein, J. Chem. Phys. **53**, 2372 (1970)
35. P. Lunkenheimer, A. Loidl, in *Broadband Dielectric Spectroscopy*, ed. by F. Kremer, A. Schönhals (Springer, Berlin, 2002), Chap. 5
36. J.-P. Bouchaud, G. Biroli, Phys. Rev. B **72**, 064204 (2005)
37. M. Tarzia, G. Biroli, A. Lefèvre, J.-P. Bouchaud, J. Chem. Phys. **132**, 054501 (2010)

38. C. Brun, F. Ladieu, D. L'Hôte, M. Tarzia, G. Biroli, J.-P. Bouchaud, Phys. Rev. B **84**, 104204 (2011)
39. G.P. Johari, J. Chem. Phys. **138**, 154503 (2013)
40. G.P. Johari, J. Chem. Phys. **145**, 164502 (2016)
41. M. Winterlich, G. Diezemann, H. Zimmermann, R. Böhmer, Phys. Rev. Lett. **91**, 235504 (2003)
42. G.P. Johari, J. Khouri, J. Chem. Phys. **137**, 104502 (2012)
43. T.R. Kirkpatrick, P.G. Wolynes, Phys. Rev. B **36**, 8552 (1987)
44. P.G. Debenedetti, F.H. Stillinger, Nature **310**, 259 (2001)
45. F. Mizuno, J.-P. Belieres, N. Kuwata, A. Pradel, M. Ribes, C. A. Angell, J. Non-Cryst. Solids **352**, 5147 (2006)
46. C.A. Angell, in *Relaxation in Complex Systems*, ed. by K.L. Ngai, G.B. Wright (Office of Naval Research, Washington DC, 1985), p. 3
47. L.C. Pardo, P. Lunkenheimer, A. Loidl, J. Chem. Phys. **124**, 124911 (2006)
48. Th. Bauer, M. Köhler, P. Lunkenheimer, A. Loidl, C.A. Angell, J. Chem. Phys. **133**, 144509 (2010)
49. M. Götz, Th. Bauer, P. Lunkenheimer, A. Loidl, J. Chem. Phys. **140**, 094504 (2014)
50. P. Lunkenheimer, S. Kastner, M. Köhler, A. Loidl, Phys. Rev. E **81**, 051504 (2010)
51. P. Lunkenheimer, M. Köhler, S. Kastner, A. Loidl, in *Structural Glasses and Supercooled Liquids: Theory, Experiment, and Applications*, ed. by P.G. Wolynes, V. Lubchenko (Wiley, Hoboken, 2012), Chap. 3, p. 115
52. C.A. Angell, W. Sichina, Ann. N.Y. Acad. Sci. **279**, 53 (1976)
53. D.J. Plazek, K.L. Ngai, Macromolecules **24**, 1222 (1991)
54. R. Böhmer, C.A. Angell, Phys. Rev. B **45**, 10091 (1992)
55. A. Srinivasan, F. J. Bermejo, A. de Andrés, J. Dawidowski, J. Zúñiga, A. Criado, Phys. Rev. B **53**, 8172 (1996)
56. R. Brand, P. Lunkenheimer, U. Schneider, A. Loidl, Phys. Rev. B **62**, 8878 (2000)
57. C.A. Angell, J. Phys. Chem. Solids **49**, 863 (1988)
58. R. Böhmer, C.A. Angell, in *Disorder Effects on Relaxational Processes*, ed. by R. Richert, A. Blumen (Springer, Berlin, 1994), p. 11
59. G. Adam, J.H. Gibbs, J. Chem. Phys. **43**, 139 (1965)
60. R. Brand, P. Lunkenheimer, A. Loidl, Phys. Rev. B **56**, R5713 (1997)
61. M. Shablakh, L.A. Dissado, R.M. Hill, J. Chem. Soc. Faraday Trans. II **79**, 369 (1983)
62. O. Anderson, R.G. Ross, Mol. Phys. **71**, 523 (1990)
63. H. Forsmann, O. Anderson, J. Non-Cryst, Solids **131–133**, 1145 (1991)
64. L.P. Singh, S.S.N. Murthy, Phys. Chem. Chem. Phys. **11**, 5110 (2009)
65. P. Lunkenheimer, A. Loidl, J. Chem. Phys. **104**, 4324 (1996)
66. O. Yamamuro, M. Hayashi, T. Matsuo, P. Lunkenheimer, J. Chem. Phys. **119**, 4775 (2003)
67. D.W. Davidson, R.H. Cole, J. Chem. Phys. **18**, 1417 (1950)
68. B. Schiener, R.V. Chamberlin, G. Diezemann, R. Böhmer, J. Chem. Phys. **107**, 7746 (1997)
69. J. Herweg, Z. Phys. **3**, 36 (1920)
70. P. Debye, *Polar Molecules* (Dover Publications, New York, 1929)
71. J.L. Déjardin, YuP Kalmykov, Phys. Rev. E **61**, 1211 (2000)
72. Th. Bauer, M. Michl, P. Lunkenheimer, A. Loidl, J. Non-Cryst. Solids **407**, 66 (2015)
73. P. Kim, A.R. Young-Gonzales, R. Richert, J. Chem. Phys. **145**, 064510 (2016)
74. S. Samanta, R. Richert, J. Phys. Chem. B **120**, 7737 (2016)
75. R. Richert, J. Chem. Phys. **146**, 064501 (2017)
76. P. Gadige, S. Albert, M. Michl, Th. Bauer, P. Lunkenheimer, A. Loidl, R. Tourbot, C. Wiertel-Gasquet, G. Biroli, J.-P. Bouchaud, F. Ladieu, Phys. Rev. E **96**, 032611 (2017)
77. F.H. Stillinger, Science **267**, 1935 (1995)
78. J.S. Harmon, M.D. Demetriou, W.L. Johnson, K. Samwer, Phys. Rev. Lett. **99**, 135502 (2007)
79. C. Gainaru, O. Lips, A. Troshagina, R. Kahlau, A. Brodin, F. Fujara, E.A. Rössler, J. Chem. Phys. **128**, 174505 (2008)
80. R.V. Chamberlin, Phys. Rev. Lett. **82**, 2520 (1999)

81. K.L. Ngai, J. Phys.: Condens. Matter **15**, S1107 (2003)
82. W. Götze, M. Sperl, Phys. Rev. Lett. **92**, 105701 (2004)
83. K.L. Ngai, M. Paluch, J. Chem. Phys. **120**, 857 (2004)
84. K.L. Ngai, Phys. Rev. E **57**, 7346 (1998)
85. R.L. Leheny, S.R. Nagel, Europhys. Lett. **39**, 447 (1997)
86. U. Schneider, R. Brand, P. Lunkenheimer, A. Loidl, Phys. Rev. Lett. **84**, 5560 (2000)
87. K.L. Ngai, P. Lunkenheimer, C. León, U. Schneider, R. Brand, A. Loidl, J. Chem. Phys. **115**, 1405 (2001)
88. A. Döß, M. Paluch, H. Sillescu, G. Hinze, Phys. Rev. Lett. **88**, 095701 (2002)
89. A. Kudlik, S. Benkhof, T. Blochowicz, C. Tschirwitz, E. Rössler, J. Mol. Struct. **479**, 201 (1999)
90. M. Beiner, H. Huth, K. Schröter, J. Non-Cryst. Solids **279**, 126 (2001)
91. S. Samanta, R. Richert, J. Phys. Chem. B **119**, 8909 (2015)
92. K.L. Ngai, J. Chem. Phys. **142**, 114502 (2015)
93. S. Samanta, R. Richert, J. Chem. Phys. **140**, 054503 (2014)
94. B. Roling, L.N. Patro, O. Burghaus, M. Gräf, Eur. Phys. J. Special Topics **226**, 3095 (2017)
95. C. Brun, C. Crauste-Thibierge, F. Ladieu, D. L'Hôte, J. Chem. Phys. **134**, 194507 (2011)
96. G. Diezemann, Phys. Rev. E **85**, 051502 (2012)
97. G. Diezemann, J. Chem. Phys. **138**, 12A505 (2013)
98. R.M. Pick, J. Chem. Phys. **140**, 054508 (2014)
99. U. Buchenau, J. Chem. Phys. **146**, 214503 (2017)
100. C. Gainaru, S. Kastner, F. Mayr, P. Lunkenheimer, S. Schildmann, H. J. Weber, W. Hiller, A. Loidl, R. Böhmer, Phys. Rev. Lett. **107**, 118304 (2011)
101. C. Gainaru, R. Meier, S. Schildmann, C. Lederle, W. Hiller, E.A. Rössler, R. Böhmer, Phys. Rev. Lett. **105**, 258303 (2010)
102. R. Böhmer, K.L. Ngai, C.A. Angell, D.J. Plazek, J. Chem. Phys. **99**, 4201 (1993)
103. S.A. Kivelson, G. Tarjus, Nature Mater. **7**, 831 (2008)

Nonlinear Ionic Conductivity of Solid Electrolytes and Supercooled Ionic Liquids

B. Roling, L. N. Patro and O. Burghaus

Abstract In this chapter, we present experimental and theoretical results for the nonlinear ionic conductivity of solid electrolytes and of supercooled ionic liquids at large electric fields exceeding 100 kV/cm. The nonlinear conductivity was measured by nonlinear ac impedance spectroscopy, i.e., by applying large ac electric fields and analyzing the measured current density spectra, in particular, higher harmonics in the current density spectra. We first review the first and second Wien effect found in classical strong and weak electrolyte solutions as well as the strong nonlinear ion transport effects observed for inorganic ionic glasses and for polymer electrolytes. Then we present models describing the nonlinear ion conductivity of classical electrolyte solutions, ionic glasses, and polymer electrolytes. Finally, recent results are presented for the nonlinear ionic conductivity and permittivity of supercooled ionic liquids. We show that supercooled ionic liquids exhibit anomalous Wien effects, which are clearly distinct from the classical Wien effects. Some ionic liquids exhibit a very strong nonlinearity of the ionic conductivity, manifesting even in seventh-order harmonic currents. We also discuss the frequency dependence of higher-order conductivity and permittivity spectra of these supercooled liquids.

Keywords Ionic conductivity · Solid electrolytes
Ionic liquids · Nonlinear ac impedance spectroscopy · Higher harmonics

1 Introduction

Solid electrolytes are usually single-ion conductors with mobile alkali ions moving in a rigid crystalline or amorphous matrix [1, 2]. Classical liquid electrolytes are composed of a salt dissolved in water or in an organic solvent [3, 4]. For instance, in commercial Li-ion batteries, the electrolyte consists of $LiPF_6$ dissolved in a mixture of organic carbonates. These electrolytes exhibit a high vapor pressure and are,

B. Roling (✉) · L. N. Patro · O. Burghaus
Department of Chemistry, University of Marburg, Hans-Meerwein-Straße 4, 35032 Marburg, Germany
e-mail: roling@staff.uni-marburg.de

R. Richert (ed.), *Nonlinear Dielectric Spectroscopy*, Advances in Dielectrics
https://doi.org/10.1007/978-3-319-77574-6_10

therefore, flammable, leading to safety concerns with Li-ion batteries [5]. Consequently, the replacement of conventional electrolytes by nonvolatile ones, like solid electrolytes or ionic liquid-based electrolytes, would improve battery safety tremendously. Currently, large research efforts are devoted to developing such alternative electrolytes and to characterizing the ion transport mechanisms.

A common method for measuring the ionic conductivity of an electrolyte is ac impedance spectroscopy. In the case of solid electrolytes, not only the total ionic conductivity can be determined, but also different ion transport mechanisms can be distinguished, such as transport across crystalline grains, transport across grain boundaries, as well as Maxwell–Wagner effects due to the existence of different phases with different ionic conductivities [6, 7]. Impedance spectroscopy on liquid and solid electrolytes is usually carried out at low electric field strengths, at which the relation between current density and electric field is linear [8, 9].

Additional information about ion transport mechanism can be obtained from nonlinear ionic conductivity measurements at high electric fields. In the 1920s and 1930s, such measurements were done on classical diluted electrolyte solution and have tremendously contributed to a better understanding of ion transport mechanisms in these solutions [10]. The applied electric fields exceeded 50–100 kV/cm. The results revealed the so-called first Wien effect for strong electrolytes (electrolytes with complete dissociation of the ions in the solvent) and the so-called second Wien effect for weak electrolytes (electrolytes with weak dissociation of salt in the solvent). Theories by Falkenhagen [11], Wilson [12], and Onsager [13] explained these Wien effects and provided new insights into the ion transport mechanisms.

Starting in the 1940s, nonlinear ionic conductivity measurements were also carried out on alkali ion conducting glasses [14–20]. It was found that the increase of the current density with the electric field can be described by a hyperbolic sine function. Different theoretical approaches were suggested to explain this kind of behavior. However, a generally accepted theory does not yet exist.

In recent years, the first nonlinear ionic conductivity measurements on supercooled ionic liquids were carried out [21–24]. These studies will hopefully contribute to a better understanding of ion transport mechanisms in ionic liquids, analogous to the Wien effect studies on diluted electrolyte solutions.

In the first part of this chapter, we start with an overview of the nonlinear ionic conductivity of classical diluted electrolyte solutions, ion conducting glasses, and polymer electrolytes. By using nonlinear ac impedance spectroscopy over broad frequency range, not only the nonlinearity of the long-range ion transport can be studied, but also the frequency-dependent nonlinear effects can be studied related to subdiffusive movements of ions. We also present models for nonlinear ion transport in classical electrolyte solutions, in glasses and in polymer electrolytes. In the second part of the chapter, we summarize the main results obtained for the nonlinear ionic conductivity of supercooled ionic liquids. We discuss anomalous Wien effects observed in the field dependence of the dc ionic conductivity as well as the frequency dependence of higher-order conductivity and permittivity coefficients.

2 Nonlinear Ion Transport in Classical Liquid Electrolytes and in Solid Electrolytes

2.1 *Classical Wien Effects in Diluted Electrolyte Solutions*

Classical liquid electrolytes are solutions of metal salts in water or in organic solvents, like acetonitrile. This class of electrolytes is subdivided into strong and weak electrolytes. Strong electrolytes are characterized by a complete dissociation of the dissolved salt and the formation of solvated cations and anions. Due to the large number density of dissociated ions, the ionic conductivity σ_{ion} of strong electrolytes is relatively high. The molar conductivity $\Lambda = \sigma_{\text{ion}}/c_{\text{salt}}$, with c_{salt} denoting the concentration of the dissolved salt, is a measure for the average mobility of the ions in the solution. For strong electrolytes, the molar conductivity decreases weakly with increasing salt concentration due to the weak Coulomb interactions between the solvated ions, see Fig. 1a. Consequently, the molar conductivity of strong electrolytes is only slightly lower than the molar conductivity at infinite dilution, Λ_∞. The weak Coulomb interactions lead to a weak ion atmosphere effect, i.e., cations are preferentially surrounded by anions and vice versa. In contrast, weak electrolytes are characterized by an incomplete dissociation of the salt. The major part of the ions form pairs or larger aggregates in the solvents, and only a small fraction of dissociated ions are present in the solution, which contribute to the ionic conductivity. Consequently, the ionic conductivity of weak electrolytes is relatively low. The degree of salt dissociation of weak electrolytes depends strongly on the salt concentration, and the molar conductivity decreases with increasing salt concentration according to $\Lambda \propto 1/\sqrt{c_{\text{salt}}}$, see Fig. 1a. Consequently, strong and weak electrolytes can be easily distinguished by analyzing the dependence of the molar conductivity on the salt concentration.

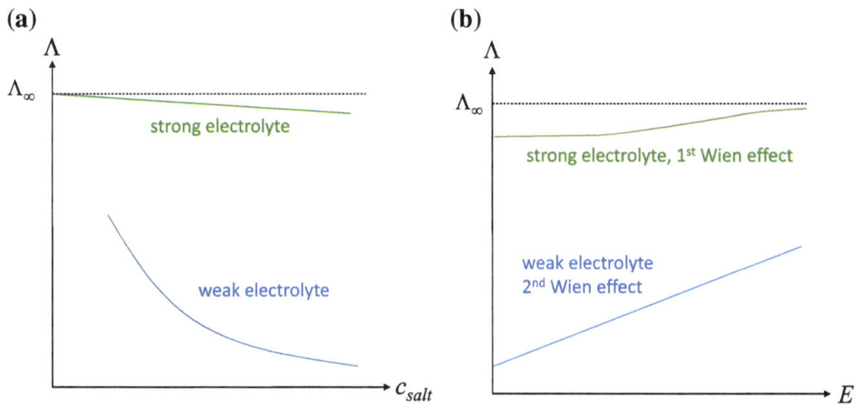

Fig. 1 Dependence of the molar conductivity Λ of classical strong and weak electrolytes on **a** the salt concentration c_{salt} and **b** the electric field E (schematic illustration)

Furthermore, strong and weak electrolytes exhibit a clearly distinct field dependence of the ion conductivity. In the case of strong electrolytes, a strong electric field leads to an enhancement of the molar conductivity, with the molar conductivity approaching the infinite-dilution limit Λ_∞ at very strong fields, see Fig. 1b. This is the so-called first Wien effect. For instance, in the case of a 2.3×10^{-4} M KCl solution in water, the ionic conductivity enhancement at an electric field of 180 kV/cm, with respect to the low-field ion conductivity, is only about 0.45% [25, 26]. At intermediate electric fields, the nonlinear effect increases quadratically with the electric field. The origin of the first Wien effect is the mitigation of the resistance contribution of the ionic atmosphere (ions with opposite charge) around an ion by the strong field. In contrast, weak electrolytes exhibit a much stronger field dependence of the ionic conductivity, see Fig. 1b. This is termed the second Wien effect. For example, the ionic conductivity enhancement of a 7.4×10^{-4} M acetic acid solution in water is 11% at 180 kV/cm [27]. The origin of this effect is a field dependence of the equilibrium constant for the dissociation of ion pairs into free ions. The degree of dissociation and thus the number of free ions increases with increasing field strength. In classical measurements with field strength up to about 200 kV/cm, the conductivity increases with the field in a linear fashion [27].

2.2 Nonlinear DC Ion Transport in Inorganic Glasses and in Polymer Electrolytes

A number of field-dependent ionic conductivity studies were carried out on ion conducting inorganic glasses [14–20]. In these studies, dc electric fields E_{dc} were applied, and the resulting dc currents densities j_{dc} were determined. In general, the obtained $j_{dc}(E_{dc})$ curves could be fitted by a hyperbolic sine function:

$$j_{dc} \propto \sinh\left(\frac{q\ a_{app}\ E_{dc}}{2\ k_B T}\right) \tag{1}$$

Here, q denotes the charge of the mobile ions, while k_B and T are Boltzmann's constant and the temperature, respectively. The quantity a_{app} has the unit of a distance and has been termed as an "apparent jump distance". Equation (1) can be derived theoretically in the framework of a random walk theory with mobile ions carrying out thermally activated hops in a periodic potential landscape, see Sect. 2.4.2. In this framework, a_{app} is identical to the actual jump distance of the mobile ions, i.e., to the distance between neighboring sites. However, the apparent jump distances derived from fits of *experimental data* using Eq. (1) are generally between 15 and 30 Å [14–20]. Thus, a_{app} is much larger than typical distances between neighboring ionic sites in glasses, a. In molecular dynamics simulations, these typical distances were found to be about $a \approx 2.5 - 3$ Å [28, 29]. The physical origin of the large values for a_{app} is most likely related to the amorphous structure of the matrix the ions are moving in. This amorphous structure results in a highly disordered potential

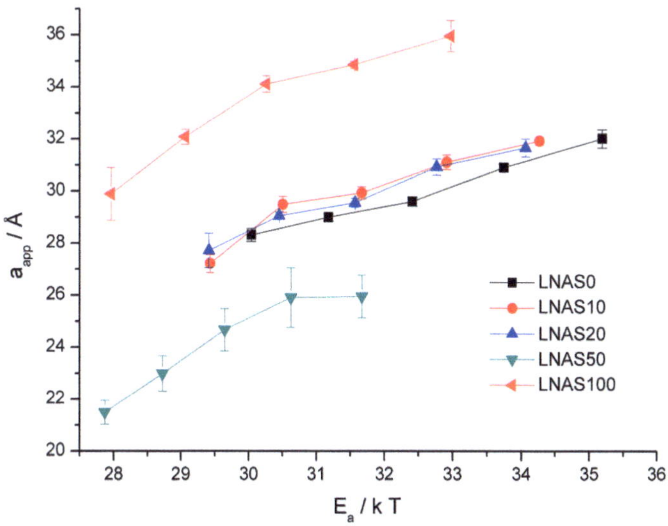

Fig. 2 Apparent jump distance of different Li–Na aluminosilcate glasses with composition $(Li_2O)_{1-x} * (Na_2O)_x * Al_2O_3 * 4SiO_2$ plotted versus inverse temperature. The inverse temperature scale is normalized by the activation energy of the low-field ionic conductivity, E_a. The numbers in the legend are given by $x * 100$. Reprinted with permission from [30]. Copyright 2010 by De Gruyter

landscape with spatially varying site energies and barriers, in contrast to the periodic potential landscape considered in random walk theory. We will come back to this point in Sect. 2.4.2.

The temperature dependence of the apparent jump distance for ionic glasses was analyzed in detail by our group [30]. As a representative example, we show in Fig. 2 the apparent jump distance of Li–Na aluminosilicate glasses plotted versus the inverse temperature. The inverse temperature scale is normalized by the activation energy of the low-field ion conductivity, E_a. Remarkably, the apparent jump distance of all glasses exhibits a similar temperature dependence, i.e., a_{app} increases with decreasing temperature. In a first approximation, the temperature dependence of a_{app} is given by:

$$\frac{a_{\text{app}}}{a} \approx \frac{E_a}{2\,k_{\text{B}}T} \tag{2}$$

A theoretical explanation for this empirical relation does not yet exist.

In the 1990s, Tajitsu studied the nonlinear ion conductivity of a number of polymer electrolytes [31–34]. These polymer electrolytes consisted of alkali salts, mostly lithium salts, dissolved in a polymer matrix, like polyethylene oxide. In homogeneous polymers, the apparent jump distance was found to be in a range 40–50 Å, i.e, similar to the apparent jump distance values of inorganic glasses. However, in a phase-separated polymer electrolyte containing rubber particles with diameter of about 80 nm, Tajitsu obtained very large apparent jump distances of the order of the particle diameter [34].

2.3 Nonlinear AC Ionic Conductivity of Inorganic Glasses

When a *dc* electric field E_{dc} is applied to an ionic conductor, and the stationary current density $j_{dc}(E_{dc})$ is measured in the weak nonlinear regime, $j_{dc}(E_{dc})$ can be expressed by an odd power series of the electric field:

$$j_{dc} = \sigma_{1,dc} \cdot E_{dc} + \sigma_{3,dc} \cdot (E_{dc})^3 + \sigma_{5,dc} \cdot (E_{dc})^5 + \cdots \quad (3)$$

Here, $\sigma_{1,dc}$ is the linear *dc* conductivity, while $\sigma_{n,dc}$ with $n = 3, 5$, etc. denotes the nth-order *dc* conductivity coefficient. A Taylor expansion of Eq. (1) into a power series of the electric field implies that all higher-order dc conductivity coefficients $\sigma_{n,dc}$ with $n \geq 3$ are positive.

When an *ac* electric field $E(t) = E_0 \cdot \cos(\omega t)$ is applied, the current density being in phase with electric field, j_{real}, can be written as follows [22, 23]:

$$\begin{aligned} j_{real} = {} & \sigma^1_{1,real} \cdot E_0 \cdot \cos(\omega t) + \left(\frac{3}{4}\sigma^1_{3,real} \cdot (E_0)^3 \cdot \cos(\omega t) + \frac{1}{4}\sigma^3_{3,real} \cdot (E_0)^3 \cdot \cos(3\omega t)\right) \\ & + \left(\frac{10}{16}\sigma^1_{5,real} \cdot (E_0)^5 \cdot \cos(\omega t) + \frac{5}{16}\sigma^3_{5,real} \cdot (E_0)^5 \cdot \cos(3\omega t) + \frac{1}{16}\sigma^5_{5,real} \cdot (E_0)^5 \cdot \cos(5\omega t)\right) + \cdots \end{aligned} \quad (4)$$

The lower index m of the conductivity coefficient σ^n_m refers to the field dependence of the respective current density term, while the upper index n refers to the number of the harmonic ($n = 1$: base current density; $n \geq 3$: nth-order harmonic current density). The third-order conductivity spectrum $\sigma^3_{3,real}$ can be determined by considering the Fourier components of the current density at 3 ω in Eq. (4). Rearrangement yields:

$$\frac{4 j_{real}(3\omega)}{E_0} = \sigma^3_{3,real} \cdot E_0^2 + \tfrac{5}{4}\sigma^3_{5,real} \cdot E_0^4 + \cdots \quad (5)$$

According to Eq. (5), a plot of $4 j_{real}(3\omega)/E_0$ versus E_0^2 yields a straight line with a slope of $\sigma^3_{3,real}$, if the second term on the right-hand side of Eq. (5) is negligible.

$\sigma^3_{3,real}$ spectra of various ion conducting inorganic glasses were obtained by our group. As an example, Fig. 3a shows a plot of $4 j_{real}(3\omega)/E_0$ versus E_0^2 for a Li–Na aluminosilicate glass with composition $(Li_2O)_{0.8}$ * $(Na_2O)_{0.2}$ * Al_2O_3 * 4 SiO_2 at two different frequencies of the applied ac voltage [35]. At 150 MHz, the data are indeed on a straight line with a slope of $\sigma^3_{3,real}$. In contrast, at a higher frequency of 46 Hz, the data exhibit a curvature and were consequently fitted with a second-order polynomial. In this case, $\sigma^3_{3,real}$ was derived from the linear term.

Typical isotherms of the low-field conductivity spectra $\sigma^1_{1,real}(\nu)$ and of the higher-order conductivity spectrum $\sigma^3_{3,real}(\nu)$ for the glass are shown in Fig. 3b [30, 35] At low frequencies, both the $\sigma^1_{1,real}(\nu)$ spectra and the $\sigma^3_{3,real}(\nu)$

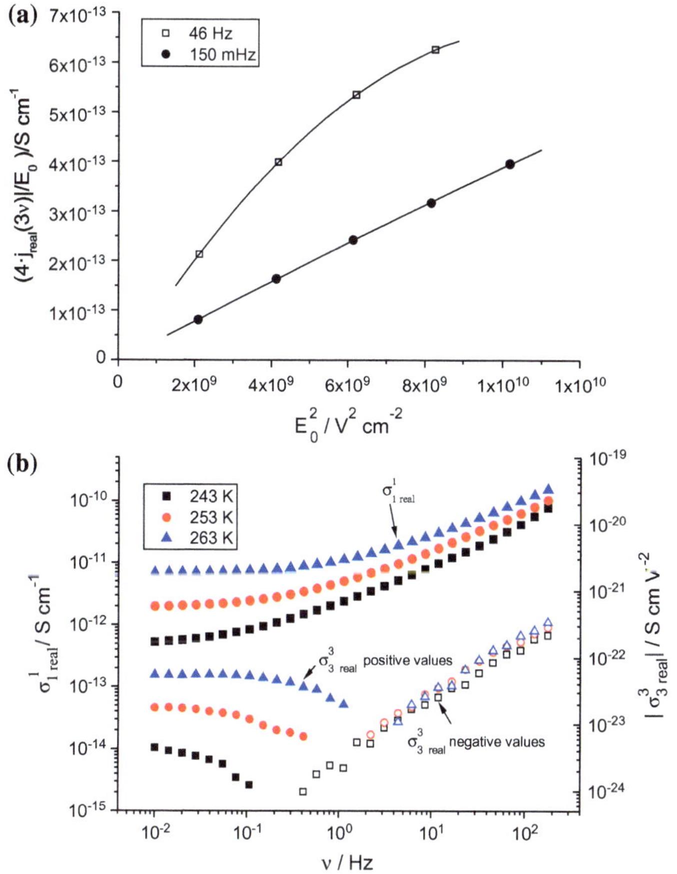

Fig. 3 **a** Plot of $4j_{\text{real}}(3\nu)/E_0$ versus E_0^2 for a $(Li_2O)_{0.8} * (Na_2O)_{0.2} * Al_2O_3 * 4SiO_2$ glass at 273 K and at two different frequencies. Empty symbols denote negative values at high frequencies. The solid lines are second-order polynomial fits. **b** Low-field conductivity spectra $\sigma^1_{1,\text{real}}(\nu)$ and modulus of higher-order conductivity spectra $\left|\sigma^3_{3,\text{real}}(\nu)\right|$ of the glass at different temperatures. Empty symbols denote negative values and filled symbols positive values, respectively. Reprinted with permission from [35]. Copyright 2010 by the American Physical Society

spectra are characterized by dc plateaus originating from long-range ion transport. At higher frequencies, the $\sigma^1_{1,\text{real}}(\nu)$ spectra pass over in the well-known dispersive regime reflecting subdiffusive ion dynamics [9]. In the same frequency range, $\sigma^3_{3,\text{real}}(\nu)$ changes its sign from positive values in the dc regime to negative values in the dispersive regime. Consequently, in the log–log representation shown in Fig. 3b, we have plotted the modulus of $\sigma^3_{3,\text{real}}(\nu)$. In both spectra, $\sigma^1_{1,\text{real}}(\nu)$ and $\sigma^3_{3,\text{real}}(\nu)$, the transition from the dc regime to the dispersive regime shifts to higher frequencies, when the temperature is increased. Remarkably, the slope in the dispersive part of the $\sigma^3_{3,\text{real}}(\nu)$ spectra $p_3 = \mathrm{d}\log\left|\sigma^3_{3,\text{real}}(\nu)\right|/\mathrm{d}\log\nu \approx 0.85 - 0.9$

is significantly larger than the slope in the dispersive part of the $\sigma^{1}_{1,\mathrm{real}}(\nu)$ spectra, $p_1 = \mathrm{d}\log\left|\sigma^{1}_{1,\mathrm{real}}(\nu)\right|/\mathrm{d}\log\nu \approx 0.70$.

2.4 *Models for Nonlinear Ion Transport*

2.4.1 Models for Ion Transport in Diluted Electrolyte Solutions

The field dependence of the molar conductivity of *strong* electrolytes was explained in the well-known theories by Falkenhagen and Wilson [11, 12] At low electric fields, each ion is surrounded by an ion atmosphere with opposite charge exhibiting an average thickness given by the Debye length $L_D = \sqrt{\varepsilon_0\, \varepsilon_r\, k_B T/\left(2N_\mathrm{A}\, c_\mathrm{salt}\cdot e^2\right)}$. Here, ε_0, ε_r, e and N_A denote the vacuum permittivity, the relative permittivity of the electrolyte, the elementary charge, and the Avogadro constant, respectively. At high electric fields, the drift velocity of the ions is so high that the ions travel across many thicknesses of the low-field ion atmosphere within the timescale of the experiment. Consequently, under the influence of a high field, the low-field ion atmosphere, which suppresses ion mobility, is not formed, and the ion mobility is higher than at low fields. Falkenhagen derived the following expression for the field dependence of the molar conductivity:

$$\Lambda = \Lambda_\infty\left(1 - \frac{e\;y}{4\pi\;\varepsilon_0\varepsilon_r(L_D)^2 E}\right) \tag{6}$$

with $y = w/2 - \left(w^2 - 1\right)/4\cdot\ln((w+1)/(w-1))$ and $w^2 = 1 + (2\;k_b T/(e\;E\;L_D))^2$. Since for very high field $E \to \infty$, the quantity y approaches ½, the molar conductivity approaches the molar conductivity at infinite dilution, Λ_∞. Since in diluted solution, the molar conductivity is only slightly lower than Λ_∞, the field dependence of the molar conductivity is weak.

The much stronger field dependence of the molar conductivity of strong electrolytes was explained by Onsager [13]. He considered the field-dependent equilibrium between free ions and ion pairs. Ion pairs are defined as two ions with a separation distance shorter than the Bjerrum length $L_\mathrm{B} = e^2/(4\pi\,\varepsilon_0\varepsilon_r k_\mathrm{B} T)$. Onsager calculated the kinetics of dissociation and recombination of ion pairs under the influence of a strong field in the framework of Brownian dynamics. In this framework, the kinetics of the dissociation of ion pairs is increased by a strong field, while the kinetics of the recombination of ions is not influenced by the field. This leads to a field-dependent equilibrium constant for the reaction: Cation - anion pair $\rightleftarrows$ free cation + free anion:

$$\frac{K(E)}{K(E\to 0)} = \frac{I_1\left(\sqrt{8x}\right)}{\sqrt{2x}} = 1 + x + O\left(x^2\right) \tag{7}$$

with $x = e^3 E/\left(8\pi\varepsilon_0\varepsilon_r(k_B T)^2\right)$ and I_1 denoting a modified Bessel function. In a weak electrolyte with $K \ll 1$, the concentration of free ions and thus the molar conductivity is proportional to $\sqrt{K}$. Thus, in the weak nonlinear regime with $x \ll 1$, the molar conductivity should be given by:

$$\frac{\Lambda}{\Lambda(E \to 0)} = 1 + \frac{1}{2}x + O\left(x^2\right) \tag{8}$$

In experiments, a linear increase of the molar conductivity with increasing field was indeed observed [26]. Recently, the predictions of the Onsager theory were essentially confirmed by Coulomb lattice gas simulations [36].

2.4.2 Models for Nonlinear Ion Transport in Inorganic Glasses and Polymers

The simplest model for nonlinear ion transport in solid materials consists of noninteracting ions carrying out thermally activated hops in a periodic potential landscape under the influence of an electric field E, see Fig. 4. In this model, the current density is given by: $j \propto \sinh(qaE/(2k_B T))$ with a denoting the jump distance of the ion, i.e., the distance between two neighboring sites in the landscape. If real ion conductors could be described by this model, the field-dependent current density would give direct information about the jump distance.

However, in glasses and in polymers, the ions move in a highly disordered potential landscape. In this case, the simplest theoretical approach are single-ion hopping models with a distribution of site energies and/or hopping barriers. For three-dimensional hopping models, analytical expressions for the field dependence of the current density can usually not be derived, but numerical results can be obtained by solving rate equations or by carrying out molecular dynamics or Monte Carlo simulations.

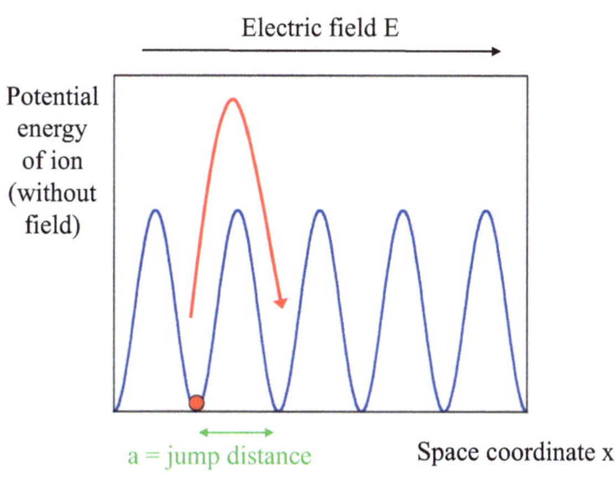

Fig. 4 Hopping of an ion in a periodic potential landscape under the influence of a strong electric field E

It was found that the field dependence of the ionic current density depends strongly on the nature of the disorder. One interesting example are hopping models with random site energy disorder. If the distribution of the site energies is a box function, then third-order dc conductivity $\sigma_{3,dc}$ is negative, whereas $\sigma_{3,dc}$ is positive in the case of a Gaussian distribution of site energies [37]. A negative value for $\sigma_{3,dc}$ has also been obtained in Monte Carlo simulations of a random barrier model with a box distribution of hopping barriers [38]. In [38], the negative sign of $\sigma_{3,dc}$ was explained by the structure of the ion transport pathways in the random barrier landscape. There are only a few percolating pathways and many pathways with dead ends. Particles moving along the percolating pathways migrate preferentially into field direction and give a positive contribution to $\sigma_{3,dc}$. However, ions are often forced by the electric field to move into dead ends and are then trapped in these dead ends for a certain amount of time. This trapping effect gives a negative contribution to $\sigma_{3,dc}$, which overcompensates the positive contribution of the ions on the continuous pathways. Such a trapping effect is also responsible for the negative $\sigma_{3,dc}$ values in the random energy model with box distribution of site energies [37, 39]. On the other hand, in a random energy model with a Gaussian distribution of site energies, a completely different effect is dominant. In this case, a small number of low-energy sites exists, in which the ions are trapped for most of the time. A strong field shuffles ions from these low-energy sites to higher-energy sites, so that these ions become mobile [39]. This effect leads to a positive nonlinear effect $\sigma_{3,dc} > 0$ and is reminiscent of the Poole–Frenkel effect for nonlinear electron transport in disordered solids [40].

The results presented in [37–39] show that the experimental features $\sigma_{3,dc} > 0$ and $a_{app} = 15{-}50$ Å, observed for ionic glasses and for polymer electrolytes are by no means easily reproduced in the framework of ion hopping models. On the contrary, in many disordered potential landscapes, one finds either $a_{app} \approx a$ or even $\sigma_{3,dc} < 0$. Thus, the results of nonlinear conductivity measurements put severe constraints on theoretical models for ion transport in disordered materials.

3 Experimental Setup for Nonlinear AC Impedance Spectroscopy and Derivation of Higher-Order Conductivity and Permittivity Spectra

In the case of solid electrolytes, the nonlinear ac impedance measurements were carried out on thin samples with thicknesses in the range of 50–100 μm [41]. Bulk samples of the solid electrolyte were first cut into cylindrical slices with a thickness in the range of 1 mm using a high-precision cutting machine. Then, the thickness was further reduced by high-precision grinding using a lapping machine. This resulted in a maximum error in the thickness over the faces of a sample of about 2 μm. The thin sample was then attached to a highly resistive quartz glass tube by means of a high-voltage resistant Araldite glue (Vantico). The quartz glass tube was placed inside a quartz glass container, see Fig. 5a [41]. Both the quartz glass tube and the quartz glass container were filled with a 1 M aqueous NaCl solution. Platinum wires connected to the high-voltage measurement system were then dipped into the NaCl solution. Since the ionic resistance of the NaCl solution is many orders of magnitude lower than the ionic resistance of the sample, the NaCl solution acts as an ionode with virtually no voltage drop over the solution. This implies that the voltage applied to the platinum wires drops completely over the sample.

The home-made sample cell for nonlinear ac impedance measurements on supercooled ionic liquids is illustrated in Fig. 5b [22]. In this cell, the ionic liquid is placed between two fine-polished brass electrodes. The distance between these electrodes can be adjusted by the fine rotation of a rotatable top electrode. This rotation is controlled by means of a fine pitch thread (0.35 mm/turn), which is driven by a worn and a bevel gear directly attached inside the sample cell with a reduction factor of 36. A revolution counter, giving a further reduction of 10, on top of the probe head, is connected by a rod to the bevel gear. Overall, one turn of the revolution counter changes the distance between the electrodes by 0.97 μm. The typical distance between the electrodes during the measurements was in the range of 50–60 μm.

Ac electric fields with amplitudes up to about 200 kV/cm were applied to samples, and the resulting current density in phase with the electric fields was analyzed by means of Eqs. (4) and (5). Analogous expressions can be written down for the out-of-phase current density j_{imag}:

$$\begin{aligned} j_{\text{imag}} &= \sigma^1_{1,\text{imag}} \cdot E_0 \cdot \sin(\omega t) + \left(\frac{3}{4}\sigma^1_{3,\text{imag}} \cdot (E_0)^3 \cdot \sin(\omega t) + \frac{1}{4}\sigma^3_{3,\text{imag}} \cdot (E_0)^3 \cdot \sin(3\omega t)\right) \\ &+ \left(\frac{10}{16}\sigma^1_{5,\text{imag}} \cdot (E_0)^5 \cdot \sin(\omega t) + \frac{5}{16}\sigma^3_{5,\text{imag}} \cdot (E_0)^5 \cdot \sin(3\omega t) + \frac{1}{16}\sigma^5_{5,\text{imag}} \cdot (E_0)^5 \cdot \sin(5\omega t)\right) + \cdots \end{aligned} \quad (9)$$

From the real and imaginary parts of the third-order conductivity coefficients in Eqs. (4) and (9), real and imaginary parts of third-order permittivity coefficients were calculated by using the following relations:

$$\varepsilon^3_{3,\text{real}} = \frac{\sigma^3_{3,\text{imag}}}{3\omega\varepsilon_0} \quad \text{and} \quad \varepsilon^3_{3,\text{imag}} = \frac{\sigma^3_{3,\text{real}}}{3\omega\varepsilon_0} \quad (10)$$

4 Nonlinear AC Impedance Measurements on Supercooled Ionic Liquids

Nonlinear ac impedance measurements on supercooled ionic liquids are rare in the literature. To our knowledge, the first nonlinear measurements were carried out in 2009 by Huang and Richert [21]. They found that the dielectric loss of the ionic liquid 1-butyl-3-methylimidazolium hexafluorophosphate in the 1 kHz regime increases when the electric field is changed from 77.4 to 387 kV/cm.

In our group, we carried out nonlinear impedance measurements on the following supercooled ionic liquids consisting of monovalent or divalent cations and monovalent anions: 1-hexyl-3-methylimidazolium bis(trifluoromethanesulfonyl)imide $[C_6mim][NTf_2]$; 1-hexyl-3-methyl-imidazolium

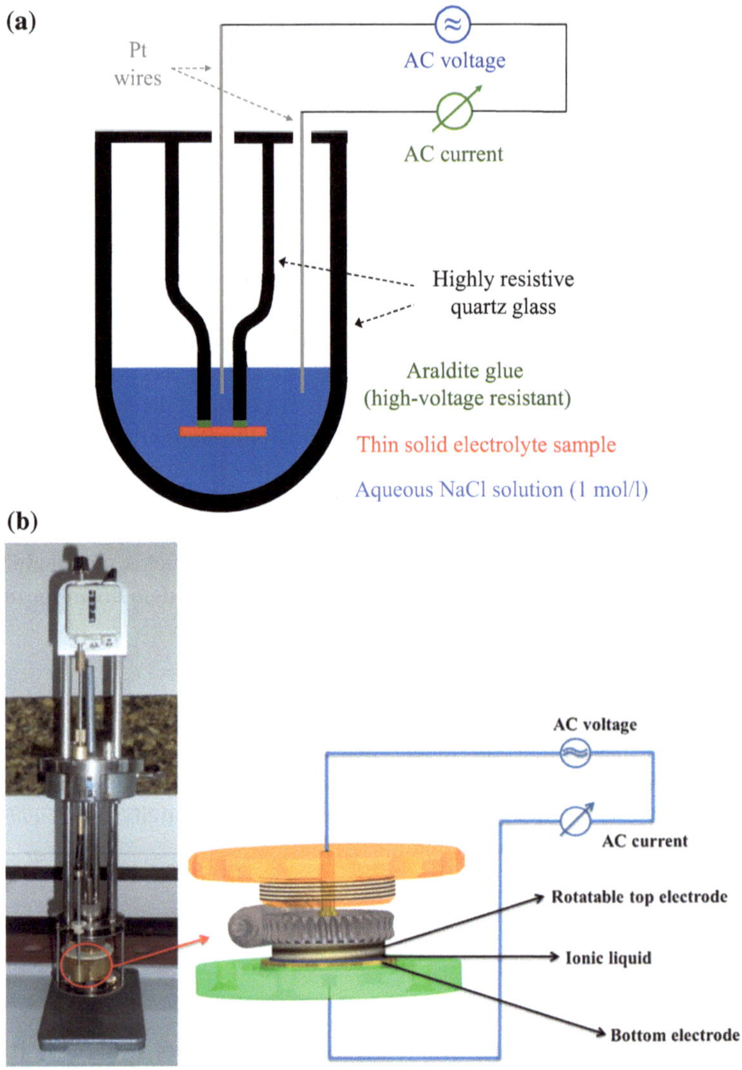

Fig. 5 **a** Schematic illustration of the experimental setup for nonlinear ac impedance measurements on thin solid electrolyte samples. Reprinted from [41]. Copyright (2005), with permission from Elsevier. **b** Left: Photograph of the home-made sample cell for nonlinear ac impedance measurements on supercooled ionic liquids (marked by red circle). The sample cell was integrated in a Novocontrol sample holder BDS 1200. Right: Cross sectional sketch of the sample cell. Reprinted from [22] with the permission of AIP Publishing

chloride [C_6mim][Cl]; Trihexyl(tetradecyl)-phosphonium chloride [$P_{6,6,6,14}$][Cl]; 1,10-bis(2,3-dimethylimidazolium)decane di-bis(trifluoromethanesulfonyl)imide [$(M2I)_2C_{10}$2Im] [NTf_2]$_2$, and 1,10-bis(3-methylimidazolium)decane di-

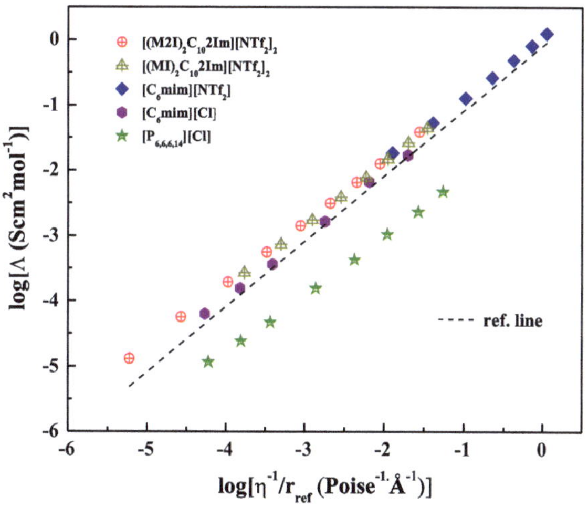

Fig. 6 Walden plot of several monocationic and dicationic ILs. The reference line was calculated by combining the Nernst–Einstein and the Stokes–Einstein relations for strong electrolytes. Reprinted with permission from [23]. Copyright 2016 by the American Physical Society

bis(trifluoromethanesulfonyl)imide $[(MI)_2C_{10}2Im][NTf_2]_2$. These ionic liquids were commercially available and were used after drying under high vacuum conditions (about 10^{-6} mbar) at a temperature of 373 K in order to remove traces of water and of other molecular impurities [23]. In these liquids, the dynamics of cations and anions takes place on the same timescale (no significant decoupling effects). In addition, we carried out viscosity measurements in order to analyze molar conductivity/viscosity relations in a Walden plot [23].

4.1 Nonlinear DC Conductivity of Monocationic and Dicationic Liquids

In Fig. 6, we show a Walden plot for all liquids. The plot contains a reference line, which was calculated by combining the Nernst–Einstein and the Stokes–Einstein relations for strong electrolytes [23]. As seen from the plot, the experimental data of four ILs are close to the reference line. These are the monocationic ILs $[C_6mim][NTf_2]$ and $[C_6mim][Cl]$, as well as the dicationic ILs $[(M2I)_2C_{10}2Im][NTf_2]_2$ and $[(MI)_2C_{10}2Im][NTf_2]_2$. Thus, these "strong" ILs behave like classical strong electrolytes in the sense that ion association effects, which are not taken into account in the Nernst–Einstein and Stokes–Einstein relations, do not seem to play a significant role. In contrast, the experimental molar conductivity values for the ionic liquid $[P_{6,6,6,14}][Cl]$ are about one order of magnitude below the reference line. This point to significant ion association effects in this "weak" IL, like in classical weak electrolytes.

In Fig. 7, we show exemplary results of the nonlinear conductivity spectra of the supercooled dicationic liquid $[(M2I)_2C_{10}2Im]$ $[NTf_2]_2$ at a temperature of 226 K

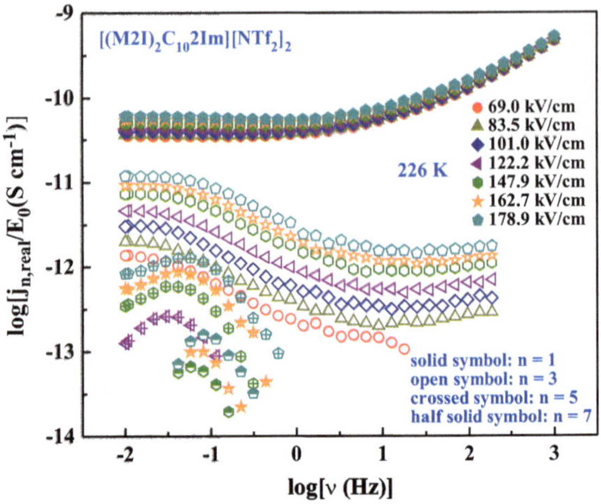

Fig. 7 Real part of frequency-dependent base current density and of frequency-dependent *n*th-order harmonic current densities, all normalized by the field amplitude E_0 at different electric fields for the IL $[(M2I)_2C_{10}2Im][NTf_2]_2$ at 226 K. Reprinted with permission from [23]. Copyright 2016 by the American Physical Society

($T_g = 224$ K) [23]. The nonlinearity of the ionic conductivity at about 180 kV/cm is so strong, that even seventh-order harmonic currents can be detected. The strong nonlinearity manifests also in a strong increase of the base current density measured at ω with increasing field amplitude E_0. By dividing the base current density j_{real} in the low-frequency plateau regime by the field amplitude E_0, we obtain a field-dependent *dc* conductivity $\sigma_{dc}(E_0)$, which is given by the sum of all $\cos(\omega t)$ terms in Eq. (4).

As shown in Fig. 8, the nonlinearity of the ionic conductivity is stronger for the two "strong" divalent ILs $[(M2I)_2C_{10}2Im][NTf_2]_2$ and $[(MI)_2C_{10}2Im][NTf_2]_2$ as compared to the two "strong" monocationic ILs $[C_6mim][NTf_2]$ and $[C_6mim][Cl]$. This result is expected, since also in the case of classical strong electrolytes, the nonlinearity increases with increasing charge number of cations and anions [25, 26]. For instance, a 2.3×10^{-4} M KCl solution in water exhibits a relative conductivity enhancement of about 0.45% at 180 kV/cm, while the conductivity enhancement for a 1.7×10^{-4} M $CdCl_2$ solution in water is about 1% [25, 26]. This finding is understandable, since the force acting on ions at a specific field strength increases with increasing charge number. However, in comparison to these classical strong electrolytes, the "strong" ILs show a much larger nonlinear effect. At 180 kV/cm, the conductivity enhancement is 62% for the IL $[C_6mim][NTf_2]$ and 110% for the dicationic IL $[(M2I)_2C_{10}2Im]$ $[NTf_2]_2$. In contrast, the "weak" IL $[P_{6,6,6,14}][Cl]$ shows a much weaker nonlinearity. At 180 kV/cm, the relative conductivity enhancement is only about 9%. This strength of the nonlinear effect is comparable to that found for the classical weak electrolyte acetic acid. However, in contrast to acetic acid, the nonlinearity increases in a quadratic fashion with the electric field. Thus, the nonlinear effect in $[P_{6,6,6,14}][Cl]$ is at variance with the Onsager theory for classical weak electrolytes.

In summary, we found that both "strong" and "weak" ILs show anomalous Wien effects when compared to classical electrolytes: (i) "Strong" ILs show a much

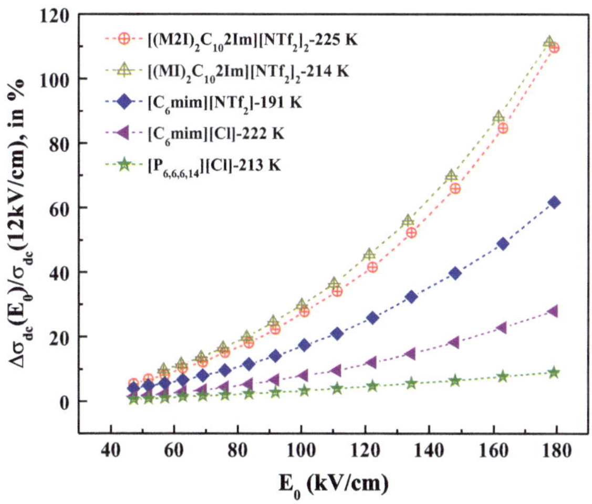

Fig. 8 Electric field amplitude-dependent relative conductivity enhancement of monocationic and dicationic ILs with respect to the conductivity value measured at $E_0 = 12$ kV/cm. Reprinted with permission from [23]. Copyright 2016 by the American Physical Society

stronger nonlinear effect than classical strong electrolytes; In the case of "strong" dicationic ILs, even seventh-order harmonic currents could be detected at *ac* electric field amplitudes of 180 kV/cm. (ii) The "weak" ionic liquid $[P_{6,6,6,14}][Cl]$ shows a nonlinear effect similar to that of classical weak electrolytes. However, the field dependence of the nonlinear effects is clearly distinct from classic weak electrolytes (quadratic vs. linear). (iii) The "strong" ILs show a much stronger nonlinearity than the "weak" IL.

4.2 Frequency Dependence of Nonlinear Permittivity Spectra

The frequency dependence of the third-order permittivity $\varepsilon_3^3 = \varepsilon_{3,\mathrm{real}}^3 + i\varepsilon_{3,\mathrm{imag}}^3$ [see Eq. (10)] was recently studied for supercooled molecular liquids, like glycerol [42–46]. It was shown that the modulus of the third-order permittivity, $|\varepsilon_3^3|$, exhibits a pronounced maximum at frequencies slightly below the α-peak frequency. This maximum was termed as "hump" [42, 43]. The increasing height of the "hump" with decreasing temperature was interpreted as a signature for strongly correlated dynamics of molecules close to the glass transition [42–44, 46]. Richert and coworkers provided an alternative explanation for the "hump" based on the field dependence of the entropy of the supercooled liquids [45]. Diezemann showed that a "hump" is also predicted for the relaxation dynamics of independent molecules in an *asymmetric* double-well potential, but not for the relaxation dynamics in a *symmetric* double-well potential [47].

In order to compare the nonlinear permittivity spectra of supercooled ionic liquids to those of supercooled molecular liquids, we subtracted the third-order dc conductivity contribution, which reflects long-range

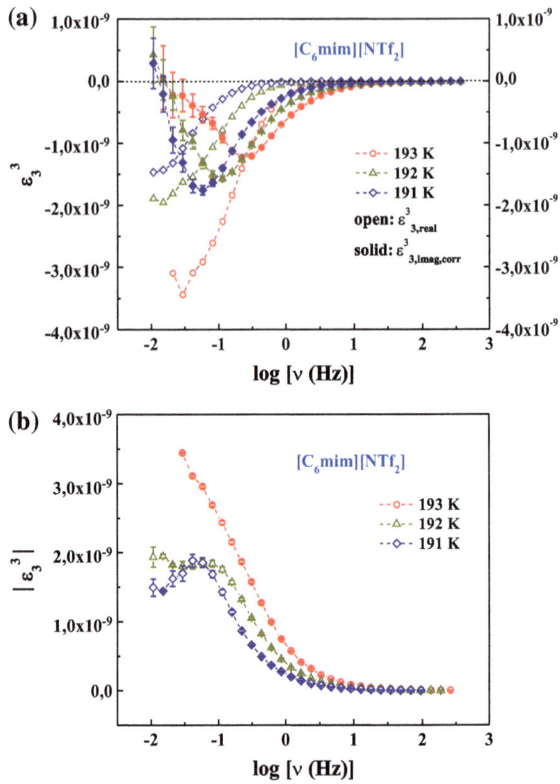

Fig. 9 Third-order permittivity spectra of the monocationic liquid $[C_6mim][NTf_2]$ at temperatures close to the glass transition temperature (189 K): **a** Frequency dependence of the real and the imaginary part of the corrected third-order permittivity coefficient, $\varepsilon^3_{3,\mathrm{real}}$ and $\varepsilon^3_{3,\mathrm{imag,corr}}$; **b** Frequency dependence of the modulus $\left|\varepsilon^3_3\right| = \sqrt{\left(\varepsilon^3_{3,\mathrm{real}}\right)^2 + \left(\varepsilon^3_{3,\mathrm{imag,corr}}\right)^2}$. Reprinted from [24] with the permission of AIP Publishing

ion transport, from the imaginary part of third-order permittivity [24]. Thus, we define a corrected imaginary part of the third-order permittivity by $\varepsilon^3_{3,\mathrm{imag,corr}} = \varepsilon^3_{3,\mathrm{imag}} - \sigma^3_{3,\mathrm{dc}}/(3\omega\varepsilon_0)$ and used this correct imaginary part to calculate the modulus of the third-order permittivity $\left|\varepsilon^3_3\right| = \sqrt{(\varepsilon^3_{3,\mathrm{real}})^2 + \left(\varepsilon^3_{3,\mathrm{imag}} - \sigma_{3,\mathrm{dc}}/(3\omega\varepsilon_0)\right)^2}$. In Fig. 9a, b we show spectra of $\varepsilon^3_{3,\mathrm{real}}$, $\varepsilon^3_{3,\mathrm{imag,corr}}$, and $\left|\varepsilon^3_3\right|$ for the monocationic liquid $[C_6mim][NTf_2]$ at temperatures close to the glass transition temperature $T_g = 189$ K [24]. As seen from Fig. 9b, a "hump" is clearly visible at 191 and 192 K. This hump is caused by a pronounced minimum in the $\varepsilon^3_{3,\mathrm{imag,corr}}$ spectra, which becomes more pronounced with decreasing temperature, see Fig. 9a.

These experimental spectra were compared to model spectra for the relaxation dynamics in asymmetric double-well potentials (ADWP) [24]. As already mentioned, such model spectra do show a "hump" in the modulus of $\left|\varepsilon^3_3\right|$. However, in ADPW, the "hump" is caused by peaks in both the real part and the imaginary part of ε^3_3. The height of both peaks increases with decreasing temperature. This is clearly at variance with the experimental spectra. Based on these results we argue that both the real and the imaginary part of the third-order permittivity should be analyzed when comparing experimental data with theoretical predictions.

In addition, the physical meaning of $\varepsilon^3_{3,\mathrm{imag,corr}} = \varepsilon^3_{3,\mathrm{imag}} - \sigma^3_{3,\mathrm{dc}}/(3\omega\varepsilon_0)$, in the case of ionic conductors should be considered in more detail. Even in the framework of simple hopping models of non-dipolar ions, the quantity $\varepsilon^3_{3,\mathrm{imag,corr}}$ does not simply reflect local hopping movements of the ions, but a nontrivial influence of the long-range transport on this quantity was observed [48]. In addition, the ionic liquids studied here consist of ions with nonzero dipole moments, so that reorientational movements of the ions also contribute to $\varepsilon^3_{3,\mathrm{imag,corr}}$.

5 Summary and Conclusions

In this chapter, we have reviewed the classical Wien effects observed for diluted strong and weak electrolyte solutions as well as the nonlinear ion transport properties of inorganic glasses and polymer electrolytes. The classical Wien effects observed for diluted electrolyte solutions have provided new insights into short-range and long-range interactions and the resulting ion transport mechanisms. In the case of inorganic glasses and polymer electrolytes, strong nonlinear ion transport effects have been found, which are characterized by positive third-order dc conductivities $\sigma_{3,\mathrm{dc}} > 0$ and by apparent jump distances in the range $a_{\mathrm{app}} = 15{-}50$ Å. In the framework of ion hopping models, such strong effects are only observed for disordered potential landscapes with special features, in particular, for landscapes with a small number of low-energy sites, from which trapped ions can be shuffled to high-energy sites by a strong electric field. The mobile ions on the high-energy sites lead to a strong increase of the ionic conductivity. Thus, the results of nonlinear conductivity measurements put severe constraints on ion hopping models.

In the last 10–15 years, a large number of nonlinear ac impedance measurements *over broad frequency ranges* have been carried out. In these measurements, high ac electric fields were applied, and the field dependence of the base-wave current density and of higher harmonics in the current density were analyzed. We have described experimental setups for carrying out such measurements on solid and liquid electrolytes, and we have explained the derivation of higher-order conductivity and permittivity spectra from these measurements.

Finally, we have presented recent results for the nonlinear ionic conductivity and permittivity of supercooled ionic liquids. These are monocationic and dicationic liquids. The nonlinear ionic conductivity of these liquids shows anomalous Wien effects, which are clearly distinct from the classical Wien effects. In particular, "strong" ILs show a much stronger nonlinear effect than classical strong electrolytes. In nonlinear ac impedance measurements, this manifests in harmonic currents up to the seventh order. We have also analyzed the third-order permittivity spectra of supercooled ionic liquids after subtracting the third-order dc conductivity caused by long-range ion transport. Like observed for molecular liquids, the modulus of the corrected third-order permittivity, $|\varepsilon^3_3|$, close to the glass transition temperature shows a "hump", which is caused by a pronounced minimum in the $\varepsilon^3_{3,\mathrm{imag,corr}}$ spectra. We show that although asymmetric double-well potential (ADWP) models predict a hump in $|\varepsilon^3_3|$, the origin of the hump in ADWP spectra is clearly distinct from the origin in the experimental spectra. Consequently, further theoretical work is needed to understand the origin of the "hump" for supercooled ionic liquids.

Acknowledgments We would like to thank the German Science Foundation (DFG) for financial support in the framework of the Research Unit FOR 1394. Valuable discussions with Andreas Heuer and Diddo Diddens are also gratefully acknowledged.

References

1. J.C. Bachman, S. Muy, A. Grimaud, H.-H. Chang, N. Pour, S.F. Lux, O. Paschos, F. Maglia, S. Lupart, P. Lamp, L. Giordano, Y. Shao-Horn, Chem. Rev. **116**, 140 (2016)
2. J.W. Fergus, Solid State Ionics **227**, 102 (2012)
3. A.M. Haregewoin, A.S. Wotango, B.-J. Hwang, Energy Env. Sci. **9**, 1955 (2016)
4. A. Gonzalez, E. Goikolea, J.A. Barrena, R. Mysyk, Renew. Sustain. Energ. Rev. **58**, 1189 (2016)
5. Q. Wang, P. Ping, X. Zhao, G. Chu, J. Sun, C. Chen, J. Power Sources **208**, 210 (2012)
6. N.J. Kidner, N.H. Perry, T.O. Mason, J. Am. Ceram. Soc. **91**, 1733 (2008)
7. P. Bron, S. Dehnen, B. Roling, J. Power Sources **329**, 530 (2016)
8. F. Kohler, *The Liquid State* (Verlag Chemie, Weinheim, 1972)
9. B. Roling, C. Martiny, S. Brückner, Phys. Rev. B **63**, 214203 (2001)
10. M. Wien, Ann. Phys. **73**, 161 (1924)
11. H. Falkenhagen, Phys. Z. **30**, 163 (1929)
12. W.S. Wilson, Dissertation, Yale University, 1936
13. L. Onsager, J. Chem. Phys. **2**, 599 (1934)
14. R.J. Maurer, J. Chem. Phys. **9**, 579 (1941)
15. J. Vermeer, Physica **22**, 1257 (1956)
16. L. Zagar, E. Papanilolau, Glastechn. Ber. **42**, 37 (1969)
17. J.P. Lacharme, J.O. Isard, J. Non-Cryst. Solids **27**, 381 (1978)
18. J.M. Hyde, M. Tomozawa, Phys. Chem. Glasses **27**, 147 (1986)
19. J.L. Barton, J. Non-Cryst. Solids **203**, 280 (1996)
20. J.O. Isard, J. Non-Cryst. Solids **202**, 137 (1996)
21. W. Huang, R. Richert, J. Chem. Phys. **131**, 184501 (2009)
22. L.N. Patro, O. Burghaus, B. Roling J. Chem. Phys. **142**, 064505 (2015)
23. L.N. Patro, O. Burghaus, B. Roling, Phys. Rev. Lett. **116**, 185901 (2016)
24. L.N. Patro, O. Burghaus, B. Roling, J. Chem. Phys. **146**, 154503 (2017)
25. A. Patterson Jr., Proc. Natl. Acad. Sci. **39**, 146 (1953)
26. A. Patterson Jr., H. Freitag, J. Electrochem. Soc. **108**, 529 (1961)
27. H.C. Eckstrom, C. Schmelzer, Chem. Rev. **24**, 367 (1939)
28. S. Balasubramanian, K.J. Rao, J. Non-Cryst. Solids **181**, 157 (1995)
29. A. Heuer, K. Kunow, M. Vogel, R.D. Banhatti, Phys. Chem. Chem. Phys. **4**, 3185 (2002)
30. H. Staesche, B. Roling, Z. Phys. Chem. **224**, 1655 (2010)
31. Y. Tajitsu, J. Mater. Sci. **31**, 2081 (1996)
32. Y. Tajitsu, J. Electrostat. **42**, 203 (1997)
33. Y. Tajitsu, J. Electrostat. **43**, 203 (1998)
34. Y. Tajitsu, J. Mater. Sci. Lett. **18**, 1287 (1999)
35. H. Staesche, B. Roling, Phys. Rev. B **82**, 134202 (2010)
36. V. Kaiser, S.T. Bramwell, P.C.W. Holdsworth, R. Moessner, Nat. Mater. **12**, 1033 (2013)

37. S. Röthel, R. Friedrich, L. Lühning, A. Heuer, Z. Phys. Chem. **224**, 1855 (2010)
38. B. Roling, J. Chem. Phys. **117**, 1320 (2002)
39. A. Heuer, L. Lühning, J. Chem. Phys. **140**, 094508 (2014)
40. J. Frenkel, Phys. Rev. **54**, 647 (1938)
41. S. Murugavel, B. Roling, J. Non-Cryst. Solids **351**, 2819 (2005)
42. C. Crauste-Thibierge, C. Brun, F. Ladieu, D. L'Hote, G. Biroli, J.-P. Bouchaud, Phys. Rev. Lett. **104**, 165703 (2010)
43. C. Brun, F. Ladieu, D. L'HÔte, M. Tarzia, G. Biroli, J.P. Bouchaud, Phys. Rev. B **84**, 104204 (2011)
44. Th Bauer, P. Lunkenheimer, A. Loidl, Phys. Rev. Lett. **111**, 225702 (2013)
45. R. Richert, J. Chem. Phys. **144**, 114501 (2016)
46. S. Albert, Th Bauer, M. Michl, G. Biroli, J.P. Bouchaud, A. Loidl, P. Lunkenheimer, R. Tourbot, C. Wiertel-Gasquet, F. Ladieu, Science **352**, 1308 (2016)
47. G. Diezemann, Phys. Rev. E **85**, 051502 (2012)
48. C. Mattner, B. Roling, A. Heuer, Solid State Ionics **261**, 28 (2014)

Nonlinear Oscillatory Shear Mechanical Responses

Kyu Hyun and Manfred Wilhelm

Abstract Mechanical dynamic oscillatory shear test is generally used to characterize and investigate mechanical properties of complex fluids or soft matters. Especially, small amplitude oscillatory shear (SAOS) tests are the canonical method for probing the linear viscoelastic properties of complex fluids because of the firm theoretical background and the ease of implementing suitable test protocols. Material functions of SAOS tests are analogous with dielectric functions from dielectric spectroscopy. However, recently nonlinear responses under large amplitude oscillatory shear (LAOS) flows are also under the spotlight due to usefulness to characterize complex fluids. In this chapter, LAOS tests are reviewed. The key to successful LAOS test is the analysis and fundamental understanding of the nonlinear mechanical responses. To analyze nonlinear responses, there are several analyzing methods and various nonlinear material functions suggested by several researchers. Among the several methods available, FT (Fourier transform)-rheology is intensively reviewed. Finally, several applications to investigate complex fluids (polymer melt and solution, polymer composite and blend, emulsion and block copolymer, and so on) are introduced.

Keywords SAOS · LAOS · FT-rheology

1 Introduction

Rheology is the study of the relationship between mechanical deformation and stress of materials. In this context, materials refer to "soft mater" or "complex fluids". These complex fluids possess mechanical properties that are intermediate between ordinary

K. Hyun
School of Chemical and Biomolecular Engineering, Pusan National University, Busan 46241, South Korea

M. Wilhelm (✉)
Institute for Chemical Technology and Polymer Chemistry, Karlsruhe Institute of Technology (KIT), Karlsruhe, Germany
e-mail: Manfred.Wilhelm@kit.edu

R. Richert (ed.), *Nonlinear Dielectric Spectroscopy*, Advances in Dielectrics
https://doi.org/10.1007/978-3-319-77574-6_11

liquid and solids, e.g., polymer melts and solutions, block copolymers, biological macromolecules, polyelectrolytes, surfactants, suspensions, emulsions and beyond [1]. From an experimental point of view, deformation can generally be divided into two categories: (1) shear flow and (2) extensional (or elongational) flow. These two characteristic flows may be thought of as the classical flow used in rheological measurements. Shear flow test can be further subdivided into steady and unsteady shear test. From steady shear tests, well-defined material functions, i.e., viscosity, can be measured. In unsteady shear tests, there are several kinds of test methods, e.g., shear stress growth, shear stress decay, shear creep, step shear strain, dynamic oscillatory shear tests, and so on [2]. Among unsteady shear flow tests, dynamic oscillatory shear tests are well-known characterization method to investigate viscoelastic properties of complex fluids [2]. Dynamic oscillatory shear tests are performed by subjecting a material to a sinusoidal shear deformation and measuring the resulting mechanical response as a function of time. Oscillatory shear input ($\gamma(t) = \gamma_0 \sin \omega t$, stress input is also possible, however, the focus in this chapter would only be on the shear deformation input) is very similar to oscillatory electric field input ($\boldsymbol{E}(t) = \boldsymbol{E}_0 \sin \omega t$) for dielectric spectroscopy (see Fig. 1). Oscillatory shear tests can be divided into two regimes. One regime is a linear viscoelastic response (small amplitude oscillatory shear, SAOS), and the second regime is nonlinear material response (large amplitude oscillatory shear, LAOS). As the applied amplitude (of strain or stress) is increased from small to large at a fixed frequency, a transition between the linear and nonlinear regimes can appear [3]. Figure 2 schematically illustrates an oscillatory strain sweep test in which the frequency is fixed and the applied strain amplitude is varied. In Fig. 2, the viscoelastic response is quantified by two material functions, namely, the elastic storage modulus $G'(\omega)$ and the viscous loss modulus $G''(\omega)$. In the linear regime, the strain amplitude is sufficiently small that both viscoelastic moduli are independent of the strain amplitude and the oscillatory stress response is sinusoidal. The strain amplitudes used in linear oscillatory shear tests are generally very small, often on the order of $\gamma_0 \approx 10^{-2} - 10^{-1}$ for homopolymer melts and polymer solutions. For some dispersed systems (emulsions, suspensions, and polymer nanocomposite) or block copolymer solutions, the linear regime is limited to even smaller strain amplitudes, $\gamma_0 < 10^{-2}$. With increasing strain amplitude, the nonlinear regime can appear beyond SAOS. In the nonlinear regime, the storage or loss moduli are a function of strain amplitude [$G'(\gamma_0)$ and $G''(\gamma_0)$] and the resulting periodic stress waveform becomes distorted and deviates from a sinusoidal wave (see Fig. 2). This nonlinear regime becomes apparent at larger strain amplitude; therefore, the nonlinear dynamic test is typically referred to as large amplitude oscillatory shear (LAOS) test [3].

2 Small Amplitude Oscillatory Shear (SAOS)

As SAOS test assumes that the material response is in the linear regime within the accuracy of the rheometer and therefore the material functions, e.g., storage modulus G' and loss modulus G'' as a function of frequency fully describe the material

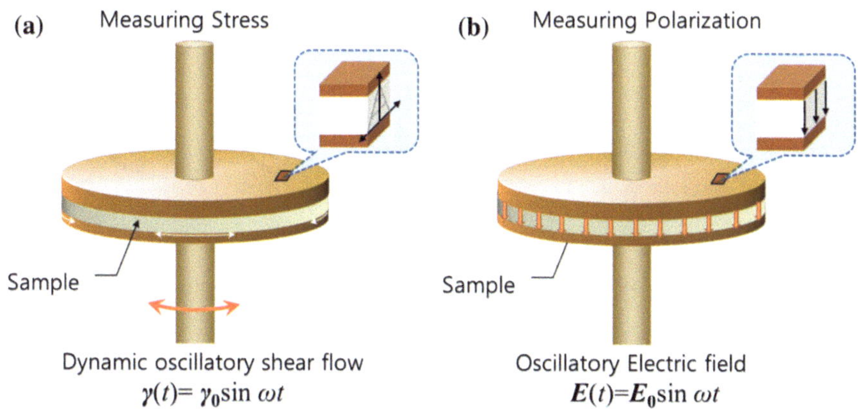

Fig. 1 Schematic illustration of **a** mechanical oscillatory shear measurement and **b** dielectric measurements. Other geometries are also used for both measurements. Just for comparison purposes, parallel disk types are introduced for both cases

response. Since linear viscoelasticity is based on a rigorous theoretical foundation [1, 2, 4–7], SAOS tests provide very useful and convenient the rheological characterization of complex fluids or soft materials.

The sinusoidal shear strain (or shear strain rate) is applied to complex fluid as follows:

$$\gamma(t) = \gamma_0 \sin \omega t \text{ or } \dot{\gamma}(t) = \omega\gamma_0 \cos \omega t. \tag{1}$$

Here, γ_0 is strain amplitude and ω is the angular frequency. The strain amplitude refers to the absolute deformation normalized to the distance between the gaps (see Fig. 1a). When the strain amplitude is small enough (we discuss later about the degree of "small" amplitude in Sect. 4.5), the linear response of complex fluids to this input deformation is between ideal viscous and elastic behavior as follows:

$$\sigma(t) = \sigma_0 \sin(\omega t + \delta), \tag{2}$$

where δ is phase angle, i.e., δ is 0° for pure elastic solid and 90° for viscous liquids, and δ of viscoelastic materials show between 0° and 90°. From simple mathematical calculation,

$$\begin{aligned}\sigma(t) &= \sigma_0 \sin(\omega t + \delta)\\ &= \sigma_0 (\sin \omega t \ \cos\delta + \cos\omega t \ \sin\delta)\\ &= (\sigma_0 \cos\delta) \sin\omega t + (\sigma_0 \sin\delta) \cos\omega t\end{aligned} \tag{3}$$

By splitting up the shear stress in this way, we see that there is a portion of the stress wave that is in phase with the imposed shear strain (i.e., proportional to sinωt;

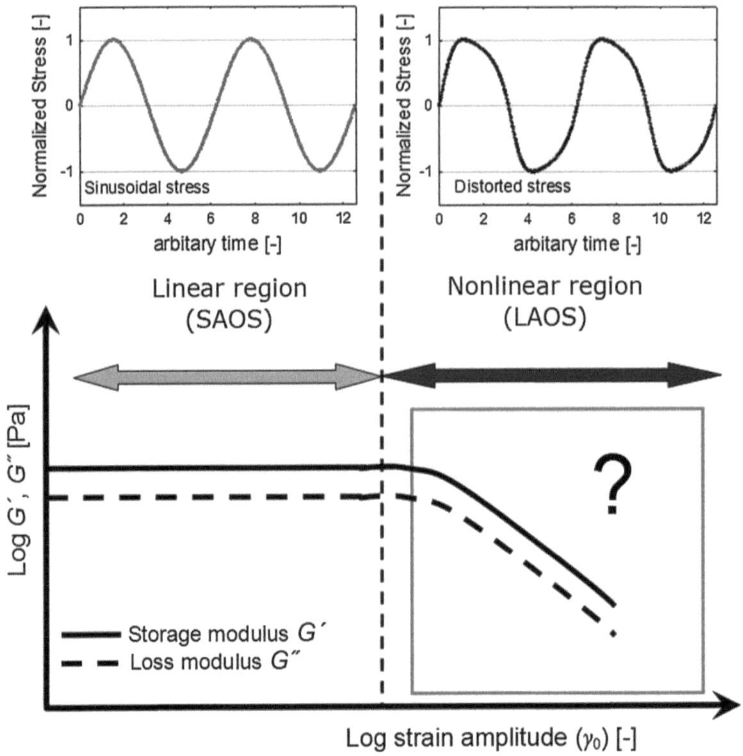

Fig. 2 Schematic illustration of the strain sweep test at a fixed frequency. This sweep test can be used for determining the linear and nonlinear viscoelastic region. In the linear region, the storage (G') and loss (G'') modulus are independent of the applied strain amplitude at a fixed frequency, and the resulting stress is a sinusoidal wave. However, in the nonlinear region, the storage and loss moduli become a function of the strain amplitude [$G'(\gamma_0)$ and $G''(\gamma_0)$] at a fixed frequency, and the resulting stress waveforms are distorted from sinusoidal waves. In the linear region, the oscillatory shear test is called SAOS (small amplitude oscillatory shear), and the application of LAOS (large amplitude oscillatory shear) results in a nonlinear material response. Reproduced by permission of Hyun et al. [3], copyright (2011) of Elsevier

Hooke's law of elasticity) and a portion of the stress wave that is in phase with the imposed shear strain rate (proportional to cosωt; Newton's law for viscosity). Thus, the SAOS tests are ideal for probing viscoelastic materials, defined as materials that show both viscous and elastic properties [2]. The material functions for SAOS are the storage modulus $G'(\omega)$ and the loss modulus $G''(\omega)$, and they are defined as follows:

$$\begin{aligned}\frac{\sigma\,(t)}{\gamma_0} &= \frac{\sigma_0 \cos\delta}{\gamma_0}\sin\omega t + \frac{\sigma_0 \sin\delta}{\gamma_0}\cos\omega t \\ &= G'\,\sin\omega t + G''\cos\omega t\end{aligned} \quad . \tag{4}$$

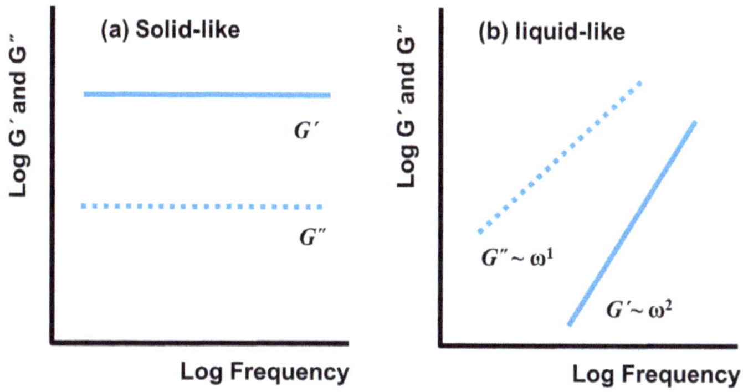

Fig. 3 The storage and loss modus for **a** solid-like materials and **b** liquid-like materials

The storage modulus G' represent "solid-like" behavior and loss modulus G'' represent "liquid-like" behavior. Figure 3 shows the typical frequency-dependent storage and loss modulus. For the "solid-like" fluid, G′ ≫ G″, and G′ is nearly frequency independent (Fig. 3a). For the "liquid-like" fluid, the storage modulus is much smaller than loss modulus, and they scale at low frequency, $G' \propto \omega^2$ and $G'' \propto \omega^1$, respectively (Fig. 3b) [1]. This behavior can be described by a Maxwell constitutive model. It is a mechanical model in which a Hookean spring and a Newtonian dashpot are connected in series.

Figure 4 shows schematically the storage and loss modulus for linear homopolymer melts with entanglements. At low-frequency region, polymer melts behavior like liquid (G′ is smaller than G'', $G' \propto \omega^2$, and $G'' \propto \omega^1$). The low-frequency region called terminal zone. With increasing frequency, G′ become larger, cross over G'' and then larger than G''. The quantity G′ displays a plateau due to entanglements and minimum is observed in G''. The modulus level of the G′ plateau, called the rubbery plateau, is also known as the plateau modulus G_N^0 having a typical value of about 10^5–10^6 Pa. It is inversely proportional to the molecular weight between entanglements (M_e) and the breadth of the rubbery plateau is proportional to the molecular weight. At high frequency, all homopolymers show a glassy modulus (glassy region) of typically 1–3 GPa [2].

2.1 Complex Notation

An alternative way of expressing a periodic function is to use a complex notation. Applied sinusoidal strain can be written as $\gamma(t) = \gamma_0 \exp(i\omega t)$, and the complex modulus is defined as follows:

$$G^*(\omega) = G' + iG'', \tag{5}$$

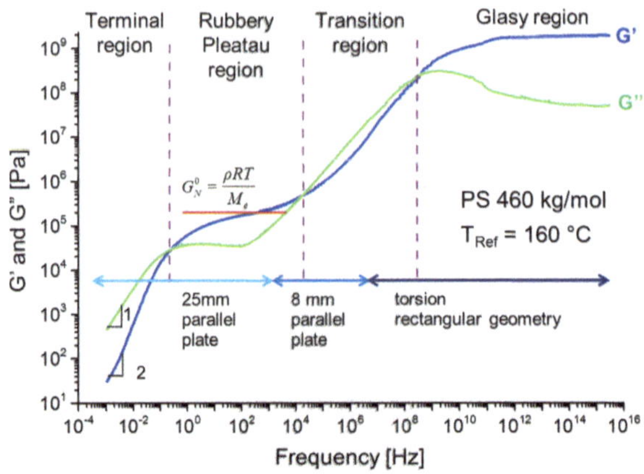

Fig. 4 The storage and loss modus as a function of frequency for linear homopolymer melt with entanglements (Polystyrene $M_w = 460$ kg/mol) at reference temperature 160 °C. The data obtained with several geometries and various temperatures. The data is shifted with time-temperature superposition (TTS) principle

where G′ and G'' have the same definitions before. The complex viscosity $\eta^*(\omega)$ and the complex compliance $J^*(\omega)$ are defined as follows:

$$\eta^*(\omega) = \frac{G^*}{i\omega} = \eta'(\omega) - i\eta''(\omega), \tag{6}$$

$$J^*(\omega) = \frac{1}{G^*} = J'(\omega) - iJ''(\omega). \tag{7}$$

Using the complex notation, we see that the material functions in SAOS are defined analogously to other shear material functions. The complex viscosity is the ratio of shear stress to shear rate, complex modulus is the ratio of shear stress to shear strain, and complex compliance is the ratio of shear strain to shear stress [2]. The magnitudes of complex quantities are found by multiplying a complex number by its complex conjugate and taking the square root

$$\left|G^*\right| = \sqrt{G'^2 + G''^2}. \tag{8}$$

This complex notation is very similar to the dielectric function. In an applied periodic electric field as $\boldsymbol{E}(t) = \boldsymbol{E}_0 \exp(i\omega t)$, the complex dielectric function $\varepsilon^*(\omega)$ and the complex electric modulus $M^*(\omega)$ are defined as follows [8]:

$$\varepsilon^*(\omega) = \varepsilon'(\omega) - i\varepsilon''(\omega), \tag{9}$$

$$M^*(\omega) = \frac{1}{\varepsilon^*} = M'(\omega) + iM''(\omega), \tag{10}$$

where $\varepsilon'(\omega)$ is proportional to the energy stored reversibly in the system per period and the imaginary part $\varepsilon''(\omega)$ is proportional to the energy which is dissipated per period. The definition of linear rheological functions and dielectric functions has similarity.

2.2 Boltzmann Superposition Principle

Strain is presumed to be a linear function of stress, so the total effect of applying several stresses is the sum of the effects of applying each one separately. Ludwig Boltzmann generalized this to give the response to a continuously varying shear deformation,

$$\sigma(\mathrm{t}) = \int_{-\infty}^{t} G\left(t - t'\right) d\gamma\left(t'\right) = \int_{-\infty}^{t} G\left(t - t'\right) \dot{\gamma}\left(t'\right) dt', \tag{11}$$

where $\mathrm{d}\gamma\left(t'\right)$ is the shear strain that occurs between t' and $\mathrm{d}t'$, and $\dot{\gamma}$ is the shear rate during this period, *G(t)* is the relaxation modulus. Equation (11) is the special form of the Boltzmann superposition principle for simple shearing deformations. The Boltzmann superposition principle is valid for very small deformations, but it is also valid for a very slow deformation, even if it is large [9].

In Eq. (11), we can apply sinusoidal shear strain rate (Eq. 1),

$$\begin{aligned} \sigma(t) &= \int_{0}^{\infty} G(s)\,\gamma_0\omega\,\cos(\omega[t - s])\,\mathrm{d}s \quad \text{with } s = t - t' \\ &= \gamma_0\left[\omega\int_{0}^{\infty} G(s)\sin(\omega s)\,\mathrm{d}\right]\sin(\omega t) + \gamma_0\left[\omega\int_{0}^{\infty} G(s)\,cos(\omega s)\,\mathrm{d}s\right] cos(\omega t). \end{aligned} \tag{12}$$

From previous definitions, we can calculate storage and loss modulus from the relaxation modulus as follows:

$$G' = \omega\int_{0}^{\infty} G(s)\sin(\omega s)\,ds \; and \; G'' = \omega\int_{0}^{\infty} G(s)\cos(\omega s)\,ds. \tag{13}$$

A primitive and simple model to describe viscoelastic behavior of materials is the Maxwell model, the relaxation modulus of Maxwell model is $G(t) = G_N^0 \exp(-t/\tau)$ where G_N^0 is the plateau modulus and τ is relaxation time. Applied this modulus in Eq. (13),

$$G' = \frac{G_N^0(\omega\tau)^2}{1 + (\omega\tau)^2} \quad \text{and} \quad G'' = \frac{G_N^0(\omega\tau)}{1 + (\omega\tau)^2}. \tag{14}$$

Table 1 The comparison between dielectric and oscillatory shear measurements

	Dielectric test	Oscillatory shear test
Input	$E(t) = E_0 \sin \omega t$ or $E(t) = E_0 \exp(i\,\omega t)$	$\gamma(t) = \gamma_0 \sin \omega t$ or $\gamma(t) = \gamma_0 \exp(i\,\omega t)$
Response	Polarization, P	Stress, σ
Material functions	$\varepsilon^*(\omega)$, ε', *and* ε'' *or* $M^*(\omega)$, M', *and* M''	$J^*(\omega)$, J', *and* J'' *or* $G^*(\omega)$, G', *and* G''
Simple relaxation model	*Debye model with single relaxation time* (τ_D)	*Maxwell model with single relaxation time* (τ)
	$\varepsilon'(\omega) - \varepsilon_\infty = \frac{\Delta\varepsilon}{1+(\omega\tau_D)^2}$	$G' = \frac{G_N^0(\omega\tau)^2}{1+(\omega\tau)^2}$
	$\varepsilon'' = \frac{\Delta\varepsilon(\omega\tau_D)}{1+(\omega\tau_D)^2}$	$G'' = \frac{G_N^0(\omega\tau)}{1+(\omega\tau)^2}$
	$\Delta\varepsilon = \varepsilon_s - \varepsilon_\infty$ is the dielectric relaxation strength	G_N^0 is plateau modulus
Nonlinear response (see Sect. 3)	$\chi_3, \chi_5 \ldots$ higher order susceptibilities (odd contributions due to $P[-E(t)] = -P[E(t)]$)	$I_3, I_5, \ldots$ higher harmonic intensities (odd contributions due to $\sigma[-\gamma(t)] = -\sigma[\gamma(t)]$)

The two limiting types of behavior are separated by the intermediate crossover region where the system can be regarded typically as viscoelastic. The point at which G' and G'' cross each other determines the place which is related to the relaxation time of the structural units constituting the system ($\tau = 1/\omega$).

In dielectric spectroscopy, the Debye relaxation equation has a similarity (single exponential relaxation with a relaxation time, τ_D) and it can be used to calculate dielectric functions as follows [8]:

$$\varepsilon'(\omega) - \varepsilon_\infty = \frac{\Delta\varepsilon}{1 + (\omega\tau_D)^2} \quad \text{and} \quad \varepsilon'' = \frac{\Delta\varepsilon\,(\omega\tau_D)}{1 + (\omega\tau_D)^2}, \tag{15}$$

where $\Delta\varepsilon = \varepsilon_s - \varepsilon_\infty$ is the dielectric relaxation strength. The Debye relaxation time τ_D is related to the maximum of ε'' similar to the Maxwell relaxation time τ. In Table 1, the linear rheological material functions under dynamics oscillatory shear and dielectric functions are compared and summarized.

As mentioned above, SAOS tests are the canonical method for probing the linear viscoelastic properties of complex fluids because of the firm theoretical background and the ease of implementing suitable test protocols. Linear viscoelastic properties are well related with theoretical background. Linear viscoelastic properties have a resemblance to dielectric properties. However, in most processing operations the deformations can be large and rapid, therefore, it is necessary to investigate nonlinear material properties that control the system response. Nonlinear response under oscillatory shear can give other material characteristic difference with linear viscoelasticity from SAOS test [3].

3 Large Amplitude Oscillatory Shear (LAOS)

It was noted that linear viscoelastic behavior is observed only in deformations that are very small (e.g., $\gamma_0 < 0.01$ for polymer melts) or very slow. The response of a complex fluid to large, rapid deformations is very often nonlinear, which means that viscosity (or modulus) is not a constant any more but the stress response depends on the magnitude, the rate, and the kinematics of the deformation. Thus, the Boltzmann superposition principle is no longer valid, and consequently nonlinear viscoelastic behavior cannot be predicted from linear properties. There exists no general model, i.e., no universal constitutive equation or rheological equation of state that describes all nonlinear behavior [9]. As strain amplitude become large in dynamic oscillatory shear flow, the stress response also transfers from linear to nonlinear regime. It means that stress data should be function of deformation (strain or strain rate). Therefore, nonlinear response can be explained by polynomial or Taylor expansion with respect to the shear strain and strain rate:

$$\sigma(t) = \sum_{i=0}\sum_{j=0} C_{ij}\gamma^{i}(t)\dot{\gamma}^{j}(t). \tag{16}$$

Here, C_{ij} are mathematical constants for expansion of nonlinear stress. But these values are related with material functions at the nonlinear regime. A similar expansion is used to describe nonlinear optics [10]. Nonlinear responses, on the other hand, are well known for the electromagnetic case at much higher, optical frequencies. The stronger electric fields during intense laser irradiation can lead to nonlinear aspects involving optical second and third harmonic generation. The emitted optical radiation of a molecule is directly proportional to the polarization P of a sample. The polarization P can be expressed in a Taylor series with respect to the applied oscillatory E-field ($E(t) = E_0 \sin \omega t$) as follows:

$$P = \chi_1 E + \chi_2 E^2 + \chi_3 E^3 + \cdots. \tag{17}$$

where χ_1, χ_2, χ_3, … are higher order susceptibilities. If electric field strengths are higher than 10^6 V/m, nonlinear effects may take place like Eq. (17) [8]. In Eq. (16), we put applied oscillatory shear strain and strain rate in Eq. (2). The stress response of viscoelastic material is typically independent of the shear direction, i.e., it is assumed that the sign of the shear stress changes as the sign of shearing changes, and therefore, the shear stress must be an odd function of the direction of shearing deformations ($\sigma[-\gamma(t), -\dot{\gamma}(t)] = -\sigma[\gamma(t), \dot{\gamma}(t)]$). Therefore, we may write the shear stress as a function of the *odd* higher order terms in the nonlinear regime as follows:

$$\sigma(t) = \sum_{p,\mathrm{odd}}\sum_{q,\mathrm{odd}}^{p} \gamma_0^{q}\left[a_{pq}\sin q\omega t + b_{pq}\cos q\omega t\right], \tag{18}$$

$$\sigma(t) = \gamma_0 [a_{11} \sin \omega t + b_{11} \cos \omega t] + \gamma_0^3 [a_{31} \sin \omega t + b_{31} \cos \omega t + a_{33} \sin 3\omega t + b_{33} \cos 3\omega t] + O(\gamma_0^5) + \cdots, \tag{19}$$

where $a_{11} = G'(\omega)$ and $b_{11} = G''(\omega)$ in the linear regime. Under these assumptions, therefore, the shear stress waveform contains only *odd* higher harmonic contributions for LAOS (large strain amplitude oscillatory shear) deformations. The polarization P also must be odd function of electric field ($P[-E(t)] = -P[E(t)]$), thus odd higher contributions in nonlinear electric field ($P = \chi_1 E + \chi_3 E^3 + \chi_5 E^5 + \cdots$. The coefficients χ are called higher order susceptibilities) [8]. By contrast, the normal stress differences do not change sign if the shearing direction changes. This means that the normal stress differences must be exclusively *even* functions of shear deformations (i.e., $N_{1,2}[-\gamma(t), -\dot{\gamma}(t)] = N_{1,2}[\gamma(t), \dot{\gamma}(t)]$, where $N_{1,2}$ denotes either the first or second normal stress difference). Thus, the normal stress differences measured under LAOS deformations have only even higher terms of the excitation frequency. Following from Eq. (18), a different notation is also possible for the nonlinear stress, written in terms of amplitude and phase angle. The total nonlinear viscoelastic stress can be expanded as a linear viscoelastic stress characterized by a stress amplitude and phase shift plus the odd higher harmonic contributions (higher stress amplitude and phase shift), consequently, the stress can be represented as

$$\sigma(t) = \sum_{n=1,\mathrm{odd}} \sigma_n \sin(n\omega t + \delta_n), \tag{20}$$

where the harmonic magnitude $\sigma_n(\omega, \gamma_0)$ and the phase angle $\delta_n(\omega, \gamma_0)$ depend on both the strain amplitude γ_0 and the excitation frequency ω. This Eq. (20) clarifies the starting point of "Fourier transform" rheology (FT-rheology) [3]. Note that Eqs. (18) and (20) describe the same nonlinear phenomena using different mathematical descriptions. Giacomin and Dealy [11] referred to Eq. (18) as a power series and Eq. (20) as a Fourier series. One can rewrite each Fourier component from Eq. (20) as components which are in-phase and out-of-phase with the strain input, and factoring out the strain amplitude (γ_0) define a set of nonlinear viscoelastic moduli [11]:

$$\sigma(t) = \gamma_0 \sum_{n,\mathrm{odd}} [G_n'(\omega, \gamma_0) \sin(n\omega t) + G_n''(\omega, \gamma_0) \cos(n\omega t)]. \tag{21}$$

Equation (18) can be rewritten as

$$\sigma(t) = \sum_{n,\mathrm{odd}} \sum_{m,\mathrm{odd}}^{n} \gamma_0^n \left[G_{nm}'(\omega) \sin(m\omega t) + G_{nm}''(\omega) \cos(m\omega t)\right], \tag{22}$$

which nicely separates the strain dependence from the frequency dependence, therefore being distinct from Eq. (21). Giacomin and Dealy [11] reported that the terms of this power series are not simply related to those of the Fourier series. However, either mathematical description can be used to argue for a leading order nonlinear coefficient. The complex mathematics is one of the reasons why there are many ways to interpret the complex nonlinear response under dynamic oscillatory shear.

3.1 G′(γ_0) and G″(γ_0)

As mentioned in the introduction, LAOS tests are typically applied as a strain amplitude sweep at fixed frequency (see Fig. 2). The storage modulus G' and the loss modulus G'' are strictly defined only in the linear viscoelastic regime, and therefore, their values at large strain amplitude may have ambiguous physical meaning. However, the measurements of G' (γ_0) and G'' (γ_0) at a fixed frequency can provide meaningful information. The generic notation G' and G'' will refer to the first harmonic moduli G'_1 and G''_1 [Eq. (21)] which is the most common option for calculating viscoelastic moduli from a non-sinusoidal response and is the typical output of commercial rheometer software. Before quantifying the non-sinusoidal waveforms of LAOS stress responses, we first discuss the interpretation of the amplitude-dependent leading order description of a nonlinear response, i.e., $G'(\gamma_0)$ and $G''(\gamma_0)$. Typical LAOS studies include the results of $G'(\gamma_0)$ and $G''(\gamma_0)$ since this information can be obtained from commercial rheometers even when raw oscillatory data is unavailable. The higher harmonic contributions, e.g., third harmonic contribution $I(3\omega)$ at 3ω, describing the extent of distortion away from a linear sinusoidal stress response are normally not large if compared with the amplitude of the first harmonic (typically $I(3\omega)/I(\omega) < 20\%$). Therefore, the moduli obtained from the first harmonic via Fourier transform analysis are relevant for a leading order description of the viscoelastic properties.

Equating the representation of Fourier series [Eq. (21)] and power series [Eq. (22)], the first harmonic contribution can be calculated as

$$\begin{aligned} \text{1st term} &= [G'_{11}\gamma_0 + G'_{31}\gamma_0^3 + O(\gamma_0^5) + \cdots]\sin\omega t + [G''_{11}\gamma_0 + G''_{31}\gamma_0^3 + O(\gamma_0^5) + \cdots]\cos\omega t \\ &= G'_1(\omega,\gamma_0)\sin\omega t + G''_1(\omega,\gamma_0)\cos\omega t \end{aligned}, \quad (23)$$

which shows that G_1' (ω,γ_0) and G_1'' (ω,γ_0) consist of odd polynomials of the strain amplitude (γ_0) with nonlinear coefficients of frequency (ω). Therefore, we can observe the LAOS behavior of the first term as a function of strain amplitude at a fixed frequency. The nonlinear coefficients from the power series [e.g., $G_{11}'(\omega)$, $G_{31}'(\omega)$, … and $G_{11}''(\omega)$, $G_{31}''(\omega)$, …] in Eq. (23) determine the leading order amplitude dependence of $G_1'(\gamma_0)$ and $G_1''(\gamma_0)$. The relaxation processes which represent the viscoelasticity of the materials are connected with the nonlinear coefficients that are only a function of frequency [$G_{11}'(\omega)$, $G_{31}'(\omega)$, … and $G_{11}''(\omega)$, $G_{31}''(\omega)$, …]. Wyss et al. [12] introduced a technique called strain rate frequency superposition (SRFS) for probing the nonlinear structural relaxation of metastable soft materials. However, the SRFS is a controversial method due to the fact that it significantly overestimates the rate of terminal relaxation and failure of the Kramers–Kronig relations. Additionally, the SRFS master curves only plot the first harmonic of the

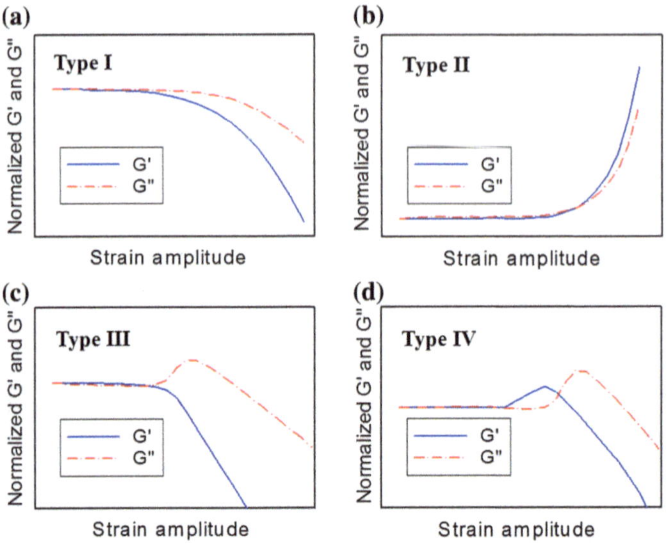

Fig. 5 The four archetypes of LAOS behavior as outlined by Hyun et al. [14]: **a** strain thinning, **b** strain hardening, **c** weak strain overshoot, and **d** strong strain overshoot. Reproduced by permission of Hyun et al. [14], copyright (2002) of Elsevier

storage modulus $G_1{}'$ and the loss modulus $G_1{}''$ from the nonlinear stress data [13]. That is the reason why we carefully investigated nonlinear stress under LAOS flow. Nonetheless, the leading order LAOS behavior is very useful to characterize complex fluids. Hyun et al. [14] observed that the leading order LAOS behavior [$G'(\gamma_0)$ and $G''(\gamma_0)$] of complex fluids could be classified by at least four types of strain amplitude dependence: type I, strain thinning (G' and G'' decreasing); type II, strain hardening (G' and G'' increasing); type III, weak strain overshoot (G' decreasing, G'' increasing followed by decreasing); and type IV, strong strain overshoot (G' and G'' increasing followed by decreasing). The four types of LAOS behavior are schematically shown in Fig. 5 and Hyun et al. [14] documented each class of LAOS behavior from different complex fluids with different microstructures.

3.2 *Nonlinear Stress Curve*

The viscoelastic moduli $G'(\gamma_0)$ and $G''(\gamma_0)$ provide only a leading order characterization of a material (i.e., the first harmonic contribution). Higher order contributions, and nonlinear stress waveforms, can be used to further distinguish and investigate viscoelastic materials. Figure 6 shows an example of how the raw oscillatory stress waveforms can distinguish two commercial polypropylene (PP) melts: one with a linear polymer chain topology and the other consisting of branched polymer chains

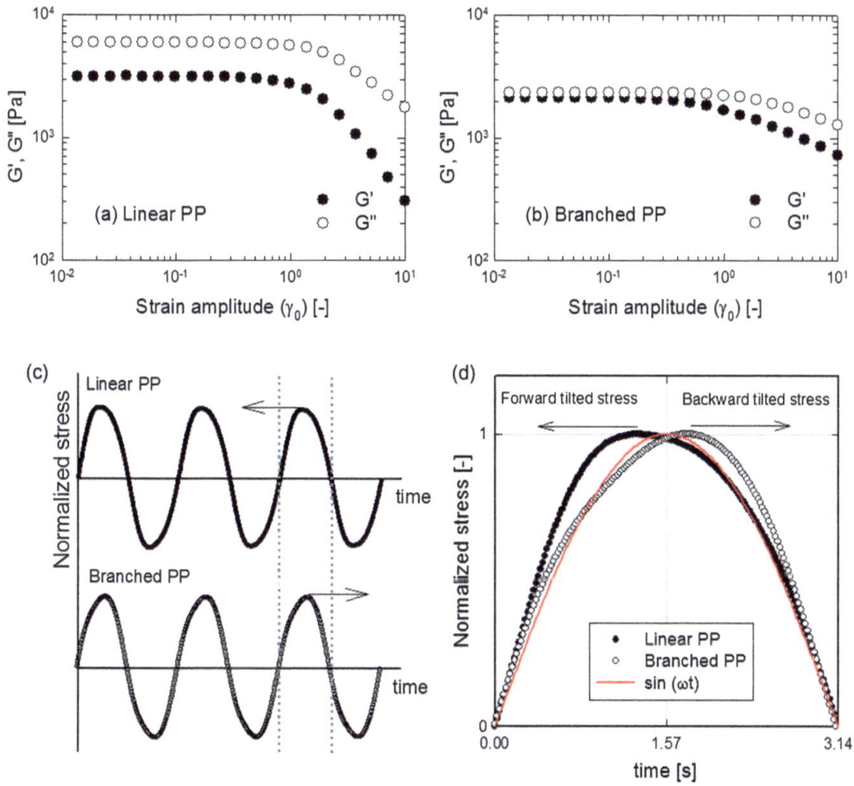

Fig. 6 The viscoelastic moduli $G'(\gamma_0)$ and $G''(\gamma_0)$ at $\omega = 1$ rad/s and $T = 180$ °C for two different polypropylene (PP) samples are shown: **a** linear PP and **b** branched PP. Both linear PP and branched PP display LAOS type I behavior (strain thinning). However, the stress waveform shapes of the linear PP and branched PP samples are different. **c** The oscillatory stress for linear and branched PP at strain amplitude, $\gamma_0 = 7.19$. Both stress shapes are distorted from a single sinusoidal shape. **d** Magnified view of the stress data: linear PP displays a "forward tilted stress" shape whereas the branched PP displays a "backward tilted stress" shape [16]. This difference in the shape of the waveform corresponds to different polymer topology, i.e., linear versus branched chain structure [15]. Reproduced by permission of Hyun et al. [3], copyright (2002) of Elsevier

[15]. Both linear and branched PP display strain thinning behavior when represented simply in terms of $G'(\gamma_0)$ and $G''(\gamma_0)$, i.e., LAOS type *I* (Fig. 6a and b). However, the nonlinear stress waveforms of the molten linear PP and branched PP samples are different (Fig. 6c and d). The linear PP melt displays a "forward tilted stress" shape whereas the branched PP melt displays a "backward tilted stress" shape [15, 16]. The "forward tilted stress" shape was observed in the case of polymer melts and solutions with a linear chain structure whereas the "backward tilted stress" shape was observed for suspensions and polymer melts with branched chains [16]. From this simple example, it is clear that analyzing the shape of the nonlinear stress response provides more structural insight than simply considering the leading order charac-

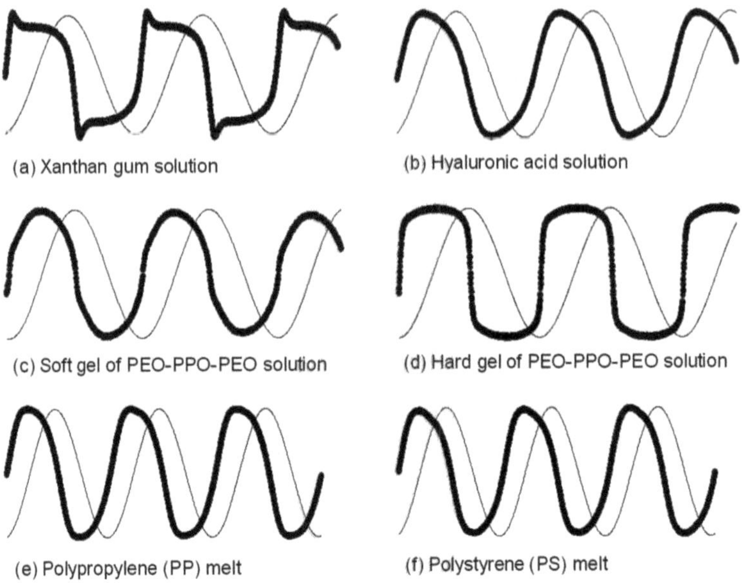

Fig. 7 Stress curves (●) and applied strain curves (—) as a function of time for various complex fluids at large strain amplitude and frequency 1 rad/s. **a** Xanthan gum 4 wt% aqueous solution at strain amplitude $\gamma_0 = 10$. **b** Hyaluronic acid 1% aqueous solution at strain amplitude $\gamma_0 = 7.2$. **c** Soft gel of PEO-PPO-PEO triblock 20 wt% aqueous solution at strain amplitude $\gamma_0 = 10$. **d** Hard gel of PEO-PPO-PEO triblock 20% aqueous solution at strain amplitude $\gamma_0 = 4$. **e** Polypropylene (PP) melt at strain amplitude $\gamma_0 = 10$. **f** Polystyrene (PS) melt at strain amplitude $\gamma_0 = 10$. Reproduced by permission of Hyun et al. [3], copyright (2002) of Elsevier

terization $G'(\gamma_0)$ and $G''(\gamma_0)$. This is because the moduli $G'(\gamma_0)$ and $G''(\gamma_0)$ reflect only the first harmonic contribution from Eqs. (21) and (22). Of course, a wide array of waveform shapes can be observed with LAOS.

Experimental examples of LAOS shear stress responses are now described and shown in Fig. 7. Many classes of complex fluids exhibit nonlinear and distorted stress waveforms under LAOS, for example: polymer melts, polymer blends, polymer solutions, block copolymer solutions, block copolymer melts, suspensions, ER materials, MR (magnetorheological) fluids, biological materials, wormlike micelle solutions, and food products [3]. In Fig. 7, several distorted, non-sinusoidal shear stress waveforms are shown as a function of time for different complex fluids under LAOS, including polymer and block copolymer solutions and polymer melts. A representation of the data that is more amenable to rapid qualitative evaluation is the use of a closed loop plot of stress versus strain (Lissajous curves) or stress versus the rate of strain [7]. The Lissajous curves (stress vs. strain) of various complex fluids subjected to LAOS with a range of strain amplitudes are also displayed in Fig. 8.

From an experimental point of view, the aim of nonlinear oscillatory experiments is to investigate the progressive evolution of the nonlinear response with increasing deformation and to quantify the nonlinear material functions that characterize

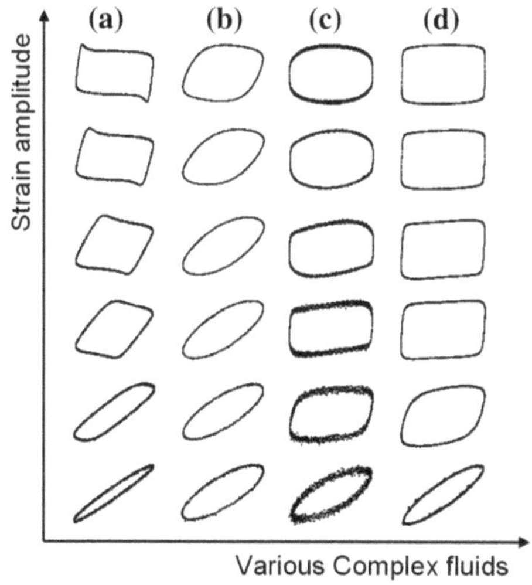

Fig. 8 The Lissajous curves [stress (y axis) versus strain (x axis)] (●) of various complex fluids, arranged from small strain amplitude to large strain amplitude at a fixed frequency, 1 rad/s **a** xanthan gum 4% aqueous solution, **b** hyaluronic acid 1% aqueous solution, **c** soft gel of PEO-PPO-PEO triblock 20% aqueous solution, and **d** hard gel of PEO-PPO-PEO triblock 20% aqueous solution. Reproduced by permission of Hyun et al. [3], copyright (2012) of Elsevier

the material nonlinearity. Furthermore, it is desirable to correlate these nonlinear functions with physical changes in the microstructure or polymer topology. The descriptions of $G_1{'}$ (γ_0) and $G_1{''}$ (γ_0) presented so far have focused only on the evolution in the first harmonic terms and are thus the simplest quantitative method. However, such rankings discard information about the nonlinear stress shape which arises from the higher order odd harmonic terms, for example, the third harmonic contribution. Therefore, several quantitative methods have been proposed for analyzing non-sinusoidal waveforms of shear stresses. For example, (1) Fourier transform [10] (2) decomposition into characteristic waveforms [17] (3) generalized "storage" and "loss" modulus when decomposing the nonlinear stress data [18] (4) Chebyshev polynomials using decomposing stress data [19] and further development of Chebyshev polynomials by Yu et al. [20]. These quantitative methods are well summarized by Hyun et al. [3]. In this chapter, FT-rheology will be reviewed among the several methods.

4 FT-Rheology

4.1 Transform from Time to Frequency Domain

A Fourier transformation (FT) represents the inherent periodic contributions of a time-dependent signal and displays the corresponding amplitudes and phases (or real and imaginary part) as a function of frequency. The experimental setup for a high-performance FT-rheology experiment is based on the modification and extension of a commercial strain-controlled rheometer. In these parallel configurations, the absolute values of the rheometer output signals must be calibrated with respect to the magnitude, phase, and frequency behavior and must also capture the nonlinear contributions introduced by the instrument itself [10]. To avoid such calibration issues, the signal can be normalized to the fundamental frequency, which changes the absolute intensity (an extensive, additive quantity) to a relative intensity (an intensive, nonadditive quantity). The relative intensity is much less vulnerable to non-systematic errors. The reproducibility has been tested and reported to be typically in the range of 0.1% for the intensity of the higher harmonics relative to the intensity of the response at the fundamental frequency I_n/I_1.

A typical FT-rheology spectrum is shown in Fig. 9 (nonlinear stress curve and FT-spectrum). With such a large signal-to-noise ratio, FT-rheology can detect very low levels of nonlinearity in the FT-spectrum. For example, from Fig. 9b, the time-dependent stress curve at an excitation frequency of $\omega_1 = 1$ rad/s and a strain amplitude of $\gamma_0 = 0.37$ appears as a single sinusoid. However, a substantial peak in the Fourier spectra at $3\omega_1$ can be quantified even at this small intensity ($I_3/I_1 < 10^{-2}$). High-performance FT-rheology setups not only have a high sensitivity with respect to the *signal-to-noise ratio*, but can also quantify the system response up to very high multiples of the input signal. Currently, spectra have been recorded spanning up to the 289th harmonic for beer foam (Fig. 10) [21]. Obviously, such a large amount of higher harmonic contributions can lead to a rather complex interpretation. Among the higher harmonics, the relative intensity of the third harmonic [$I_{3/1} \equiv I(3\omega)/I(\omega)$, where ω is the excitation frequency] is generally the most intense.

4.2 Even Harmonics Within the Shear Stress

As discussed in Sect. 3, only odd harmonics of shear stress are expected for typical and idealized material responses to LAOS. However, even harmonics can be observed experimentally, e.g., Figure 9 shows the small peaks in the even higher harmonics of the Fourier transformed shear stress. These even harmonics are relatively small compared with the odd higher harmonics. The occurrence of even higher harmonics within the shear stress in the response signal is often considered as an experimental artifact and the peaks are neglected [22].

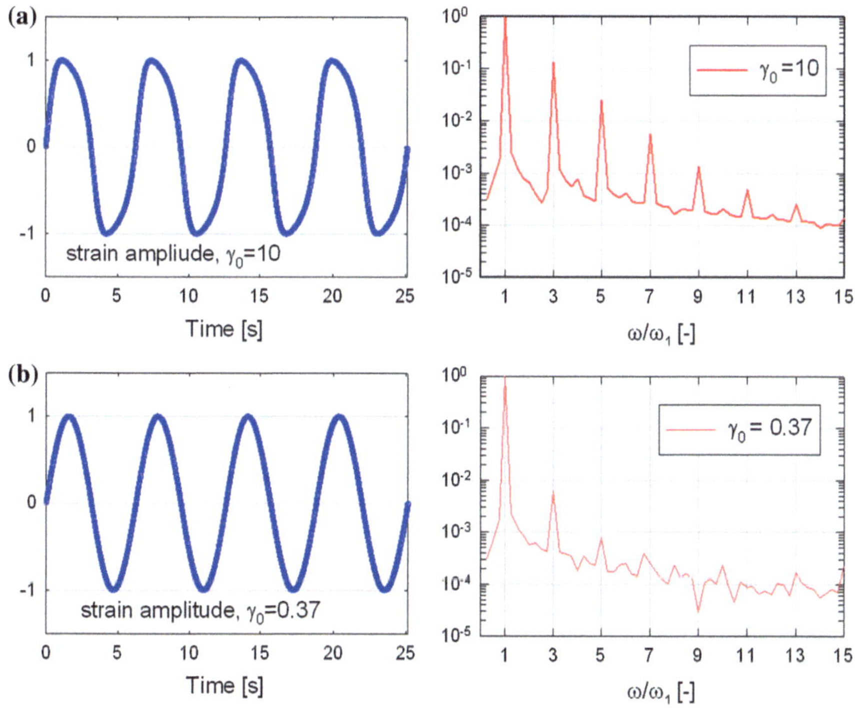

Fig. 9 The stress curve and FT-spectrum of Hyaluronic acid 1% aqueous solution at frequency, $\omega_1 = 1$ rad/s. **a** Normalized stress and FT-spectrum at strain amplitude $\gamma_0 = 10$. A non-sinusoidal shape is observed, and the FT-method quantifies the extent of the nonlinear response via the magnitude (and phase) of the odd higher harmonics. **b** Normalized stress and FT-spectrum at a strain amplitude $\gamma_0 = 0.37$, which corresponds to intermediate strain amplitude. To a naked eye, the time-dependent stress is similar to a perfect sinusoid. However, the presence of a third harmonic is clearly shown by the Fourier spectra. Reproduced by permission of Hyun et al. [3], copyright (2012) of Elsevier

However, even harmonics can be reproducibly generated and quantified using the FT-rheology technique [17]. Wall slip is expected to be one of the main reasons for the occurrence of even harmonic contributions [23–28]. Wilhelm et al. [29] explained the appearance of even harmonics arising from a time-dependent memory effect or a nonlinear elastic contribution in the system. Yosick et al. [30] reported that inertia does not create even harmonics using the upper convected Maxwell (UCM) model supplemented with a kinetic rate equation. Mas and Magnin [31] have argued that a finite yield stress can also be a reason for the occurrence of even harmonics. Yu et al. [20] report that the yield stress is not a sufficient condition to cause even harmonics from the Bingham model even though they observed some even harmonic contributions at the lower limits of their experimental resolution. As we have noted above, it is difficult to accurately measure the relatively small even harmonic contributions compared with the high intensity odd harmonic contributions to the shear stress. For completeness, we note that the Fourier spectrum of the strain only (for a strain-

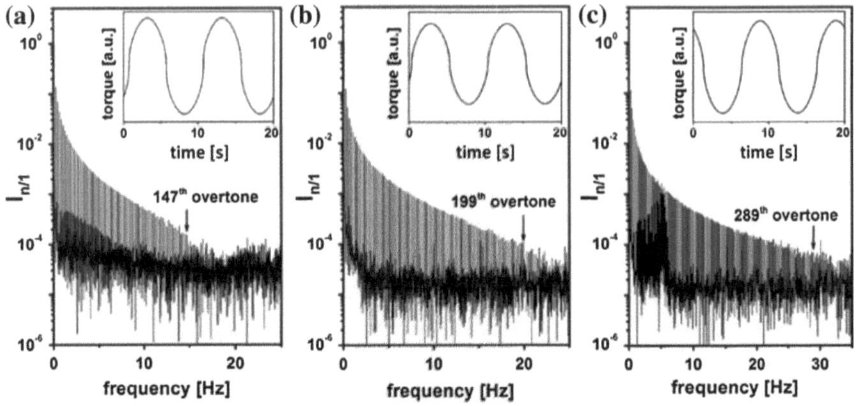

Fig. 10 LAOS measurements on a commercial w/o-emulsion with method A. **a** Measurement of the frequency spectrum with the ARES G1 yielded a maximum of 147 higher harmonics. **b** Comparative measurement to (**a**) with the more sensitive ARES G2 as specified by the manufacturer, method B. The excitation frequency is 0.1 Hz, and the strain amplitude $\gamma_0 = 870$. The cone-plate geometry spanned 50 mm in diameter. The sampling rate was 50 points/s, and the shear stress was recorded over 10 cycles. The number of maximum higher harmonics was increased from 147 to 189 overtones. In **c**, the new record of higher harmonics was measured with the ARES G2. Here, the strain amplitude was maximized to $\gamma_0 = 3000$ in absolute values resulting in 289 overtones. Reproduced by permission of Reinheimer et al. [21], copyright (2012) of Oldenbourg Wissenschaftsverlag, München

controlled test) should not show higher harmonic contributions. However, there are always technical limits to producing perfect sinusoidal signals [32]. These defects which come from non-sinusoidal strain can affect the stress curve itself. Therefore, imperfect excitation can also create even higher harmonics.

We conclude that analysis of even harmonics in the shear stress can give some additional information for microstructured materials (e.g., orientation) or macroscopic changes in materials but care must be taken to avoid systematic experimental artifacts (e.g., fluid inertia, imperfect excitation, or misalignment). For example, defects in polymer solid sample during fatigue test could make anisotropic behavior. As a result, even harmonics can develop. Hirschberg et al. [33] investigated fatigue behavior of polystyrene (PS) and concluded the normalized second harmonic [$I_{2/1} \equiv I(2\omega)/I(\omega)$, where ω is the excitation frequency] is sensitive toward the appearance of a crack. This might open up to use this technique from the original purpose to quantify molecular nonlinearities in polymer melts toward solid polymer fatigue testing as a structural nonlinearity.

4.3 *Nonlinear Quantitative Coefficient,* Q *from FT-Rheology*

If a strain sweep is performed at a fixed frequency, two main regimes can generally be observed. One is the linear regime at small amplitude (SAOS, small amplitude oscillatory shear) in which the rheological properties do not depend on the strain amplitude—at least not in an observable way. This is followed by a nonlinear regime in which viscoelastic properties depend systematically on the strain amplitude (LAOS, large amplitude oscillatory shear) (see also Fig. 2). Hyun et al. [34] subdivided the nonlinear region into two subregions: MAOS (medium amplitude oscillatory shear) and LAOS. MAOS is defined as an intermediate region (between SAOS and LAOS), where a particular scaling behavior is observed for the intensity of third harmonic contribution as a function of strain amplitude. From the series expansions discussed before, the Fourier intensities of the n-th harmonics grow with the corresponding *odd* powers of the strain amplitude ($I_n \propto \gamma_0^n, n = 1, 3, 5 \ldots$) in the small and intermediate strain amplitude range, e.g., Equation (22) (note that each of *decomposed* sine and cosine components, G'_{nm} and G''_{nm}, also scale according to odd powers of the strain amplitude). Therefore, the total intensity of the third harmonic normalized by the first harmonic should be expected to initially scale quadratically with the strain amplitude ($I_{3/1} \equiv I_3/I_1 \propto \gamma_0^3/\gamma_0^1 = \gamma_0^2$). This quadratic relation at the MAOS region was observed from the experiment and simulation results. Thus, Hyun et al. [35] proposed a new nonlinear coefficient Q, defined as

$$Q \equiv I_{3/1}/\gamma_0^2. \tag{24}$$

By convention, the absolute strain amplitude value is used in Eq. (24), not the % strain amplitude. This new nonlinear coefficient provides insight into how a material response develops and transitions from the linear to nonlinear regime. This new nonlinear material coefficient $Q(\omega, \gamma_0)$ characterizes FT-rheology and will be a function of both strain amplitude (γ_0) and frequency (ω). This parameter might be seen in analogy to the nonlinear optical susceptibility. We can define the *nonlinear zero-strain value*, Q_0, as the asymptotic limiting constant value achieved at low strain amplitude,

$$Q_0(\omega) = \lim_{\gamma_0 \to 0} Q(\omega, \gamma_0). \tag{25}$$

Using this coefficient Q_0, we can quantify the intrinsic nonlinearity of complex fluids as a function of frequency.

As already mentioned the underlying mathematics of oscillatory shear deformation is very similar to dielectric spectroscopy in which a sinusoidal electric field is applied and the resulting current is quantified with respect to the dielectric storage ε' and loss ε''. Increasing the magnitude of the electric field leads to detectable dielectric nonlinearities as quantified via higher order susceptibilities. These susceptibilities are material constants independent of the electric field. The rheological nonlinear coefficient Q is analogous to the nonlinear dielectric coefficient χ_3 [10]. As is the case for the nonlinear optical coefficients, the Q coefficient does not vanish but rather approaches a constant value in the limit of zero-strain amplitude. Consequently, this material coefficient reflects the inherent and normalized leading order nonlinear mechanical properties of the material under investigation.

The meaning of the nonlinear coefficient Q can be explored mathematically using the Fourier series [Eq. (20)] and power series [Eq. (22)] with odd higher terms. As a modification to Eq. (20), the shear stress under nonlinear oscillatory shear from FT-rheology can be written as

$$\begin{aligned}\sigma(t) &= \sigma_1 \sin(\omega t + \delta_1) + \sigma_3 \sin(3\omega t + \delta_3) + \cdots \\ &= \sigma_1 \cos\delta_1 \sin\omega t + \sigma_1 \sin\delta_1 \cos\omega t + \sigma_3 \cos\delta_3 \sin 3\omega t + \sigma_3 \sin\delta_3 \cos 3\omega t + \cdots\end{aligned} \tag{26}$$

From the above equation, we can calculate the relative intensity of third harmonic from FT-rheology as

$$I_{3/1} = \frac{I_3}{I_1} = \frac{\sigma_3}{\sigma_1} = \frac{\sqrt{(\sigma_3\cos\delta_3)^2 + (\sigma_3\sin\delta_3)^2}}{\sqrt{(\sigma_1\cos\delta_1)^2 + (\sigma_1\sin\delta_1)^2}}. \tag{27}$$

For the power series [Eq. (22)] of the shear stress waveform,

$$\begin{aligned}\text{1stnonlinearterm} &= [G'_{11}\gamma_0 + G'_{31}\gamma_0^3 + O(\gamma_0^5) + \cdots]\sin\omega t \\ &\quad + [G''_{11}\gamma_0 + G''_{31}\gamma_0^3 + O(\gamma_0^5) + \cdots]\cos\omega t\end{aligned} \tag{28}$$

$$\begin{aligned}\text{3rdnonlinearterm} &= [G'_{33}\gamma_0^3 + G'_{53}\gamma_0^5 + O(\gamma_0^7) + \cdots]\sin 3\omega t \\ &\quad + [G''_{33}\gamma_0^3 + G''_{53}\gamma_0^5 + O(\gamma_0^7) + \cdots]\cos 3\omega t.\end{aligned} \tag{29}$$

These two distinct representations [i.e., Fourier series from Eq. (20) and power series from Eq. (22)] describe the same nonlinear phenomena; therefore, we can define new nonlinear material function Q coefficient by inserting Eqs. (27) and (28) into (26),

$$\begin{aligned}\frac{I_3}{I_1} &= \frac{\sqrt{(G'_{33}\gamma_0^3 + G'_{53}\gamma_0^5 + \cdots)^2 + (G''_{33}\gamma_0^3 + G''_{53}\gamma_0^5 + \cdots)^2}}{\sqrt{(G'_{11}\gamma_0 + G'_{31}\gamma_0^3 + \cdots)^2 + (G''_{11}\gamma_0 + G''_{31}\gamma_0^3 + \cdots)^2}} \\ &= \frac{\sqrt{G'^2_{33}\gamma_0^6 + G''^2_{33}\gamma_0^6 + O(\gamma_0^8)\cdots}}{\sqrt{G'^2_{11}\gamma_0^2 + G''^2_{11}\gamma_0^2 + O(\gamma_0^4)\cdots}} \\ &= \frac{\sqrt{G'^2_{33} + G''^2_{33} + O(\gamma_0^2)\cdots}}{\sqrt{G'^2_{11} + G''^2_{11} + O(\gamma_0^2)\cdots}} \times \frac{\gamma_0^3}{\gamma_0} = Q(\omega, \gamma_0)\cdot\gamma_0^2\end{aligned} \quad . \tag{30}$$

In the limit of the small strain amplitudes, we thus obtain the zero-strain nonlinearity, $Q_0(\omega)$

$$Q_0(\omega) = \lim_{\gamma_0 \to 0} Q(\omega, \gamma_0) = \lim_{\gamma_0 \to 0} \frac{\sqrt{G_{33}'^2 + G_{33}''^2 + O(\gamma_0^2) \cdots}}{\sqrt{G_{11}'^2 + G_{11}''^2 + O(\gamma_0^2) \cdots}} \\ = \frac{\sqrt{G_{33}'^2(\omega) + G_{33}''^2(\omega)}}{\sqrt{G_{11}'^2(\omega) + G_{11}''^2(\omega)}} = \frac{|G_{33}^*(\omega)|}{|G_{11}^*(\omega)|} \quad (31)$$

From Eq. (31), it can be seen that $Q_0(\omega)$ is the normalized third nonlinear complex modulus (third nonlinear term) divided by the linear complex modulus (first nonlinear term). The magnitude of this intrinsic nonlinearity can be evaluated for any complex fluid, as with any other rheological properties. Ironically, $Q_0(\omega)$ is measured by LAOS test; however, this value is not a function of strain amplitude but a function of frequency. Thus, this new nonlinear material functions $Q_0(\omega)$ can be used to investigate relaxation process such as dielectric material functions.

4.4 Q_0 *from Various Constitutive Equations*

Nonlinear stress behavior can be predicted by numerical simulation using various constitutive equations, but it is not easy to obtain analytical solutions by constitutive equations. However, Q_0 can be calculated analytically with various constitutive equations. In Table 2, a variety of Q_0 obtained from various constitutive equations are summarized. From all the results, it is observed $Q_0 \propto \omega^2$ at limiting low frequency. The scaling law ($Q_0 \propto \omega^{k)}$ at limiting high frequency depends on constitutive equations, i.e., $k = -1.0$ for the Pom–Pom model [36] and Giesekus model [37], $k = 0$ for the rigid dumbbell model [38], DE IA [39], MSF [40, 41], corotational Maxwell model [42], and $k = 1.0$ for the White–Metzner model [43]. From experimental observations, Cziep et al. [44] founded $k = -0.35$ for various monodisperse linear homopolymer melts, and Song et al. [45] observed $k = 0$ for diluted monodisperse polystyrene (PS) solutions and $k = -0.23$ for concentrated PS solutions and PS melts. Figure 11 shows Q_0 as a function of De ($\equiv \omega\tau$, τ *is relaxation time, Debora number)* with various constitutive equations. From a variety of Q_0 calculated with the aid of several constitutive equations which represent various non-Newtonian fluids, Q_0 can be a good candidate for nonlinear material functions to characterize various complex fluids.

4.5 *Definition of "Small" Strain Amplitude for a Linear Regime*

The higher harmonic contributions in LAOS emerge according to the quadratic scaling behavior discussed in the previous Sect. (4.3), but eventually deviate from

Table 2 The analytical $Q_0(\omega)$ from various constitutive equations

	Constitutive equation	Reference	Analytical Q_0 ($De = \omega\tau$)
Dilute polymer solution	Rigid dumbbell	Bird et al. (2014) [38]	$\frac{9}{14}\sqrt{\frac{(1+De^2)\left(\begin{array}{r}625+9250De^2+26649De^4\\ +214488De^6+11664De^8\end{array}\right)}{(25+4De^2)(1+4De^2)^2(1+9De^2)^2(25+9De^2)^2}}$
Polymer melt and concentrated polymer solution	DE IA (Doi–Edwards model with independent alignment assumption)	Pearson and Rochefort (1982) [39]	$\frac{15}{42}\frac{De^2}{(1+4De^2)^{1/2}(1+9De^2)^{1/2}}$
	Pom–Pom	Holye et al. (2014) [36]	$\frac{(1-2.5Z^{-1})De^2}{2\pi(1+25De^2Z^{-2})^{0.5}(1+De^2)^{1/2}(1+4De^2)^{1/2}}$
	MSF (molecular stress function)	Wagner et al. (2011) [40] and Abbasi et al. (2013) [41]	$\frac{3}{2}\left(\alpha-\frac{\beta}{10}\right)\frac{De^2}{(1+4De^2)^{1/2}(1+9De^2)^{1/2}}$
	Experimental finding for linear monodisperse homopolymers	Cziep et al. (2016) [44]	$\frac{0.32}{Z^{1/2}}\frac{De^2}{1+33.75Z^{-1}De^{2+0.35}}$
General complex fluids	Corotational Maxwell	Giacomin et al. (2011) [42]	$\frac{1}{4}\frac{De^2}{(1+4De^2)^{1/2}(1+9De^2)^{1/2}}$
	Giesekus	Gurnon and Wagner (2012) [37]	$\frac{\alpha}{4}\frac{De^2(9De^2+4\alpha^2-12\alpha+9)^{1/2}}{(1+De^2)(1+4De^2)^{1/2}(1+9De^2)^{1/2}}$
	White–Metzner with $\eta(\dot{\gamma})=\frac{\eta_0}{(1+(\lambda\dot{\gamma})^2)^{1/2}}$	Merger et al. (2016) [43]	$\frac{1}{8}\frac{De^2}{(1+9De^2)^{1/2}}$

this leading order dependence. As the strain amplitude (γ_o) increases, the variation in the higher harmonics, e.g., $I_{3/1}(\gamma_o)$ is often observed to be a sigmoidal function. Non-sigmoidal behavior has also been observed for $I_{3/1}(\gamma_o)$. For example, with dispersed systems, the sigmoidal behavior of $I_{3/1}$ has shown a "bump" or "overshoot" at intermediate strain amplitude due to the disperse phase [46]. This non-sigmoidal behavior is obtained for systems in which strong interactions occur between a viscoelastic matrix and a dispersed phase, e.g., the major volume phase and carbon black. Nonlinearity, therefore, reflects the superposition of two responses: (1) qualitatively common to all "pure" (unfilled) polymers and (2) related to the "filler" response [47]. Any suitable functional form of $I_{3/1}(\gamma_o)$, whether sigmoidal or non-monotonic, must be continuous and differentiable. These functions describe an asymptotic transition from the linear (SAOS) to the nonlinear regimes (MAOS and LAOS) and the deviation from the limiting value is a measure of the limit of the linear response. This result is striking, since it suggests that any improvement in instrumentation could affect the apparent limit of the linear regime for a specific sample. Within the concept of the Q coefficient and the high sensitivity of FT-rheology a linear regime is only the asymptotic approximation for vanishing nonlinearities. For example, where $Q_0 = 0.01$ with a strain amplitude of $\gamma_0 = 0.01$ the expected nonlinearity of the third harmonic is $I_{3/1} = 10^{-6}$. This value is outside the detection range of any current commercial rheometer, but is evidently nonzero.

This whole argument is recognition that the linear response is only achieved for vanishing deformations, and therefore never precisely achieved in any real experiment. Nevertheless, it is commonly accepted that the linear response can accurately describe the limiting mechanical response. An alternative definition of the linear viscoelastic regime in oscillatory shear might be the regime in which the experimental response obeys the leading order nonlinear scaling and can be extrapolated to the limit of zero-strain amplitude. Additionally, one may define a criteria that the nonlinearity in the signal response must be smaller than a critical threshold, e.g., $I_{3/1} < 0.05\%$ (5×10^{-4}) as determined from FT-rheology, in which case the linear

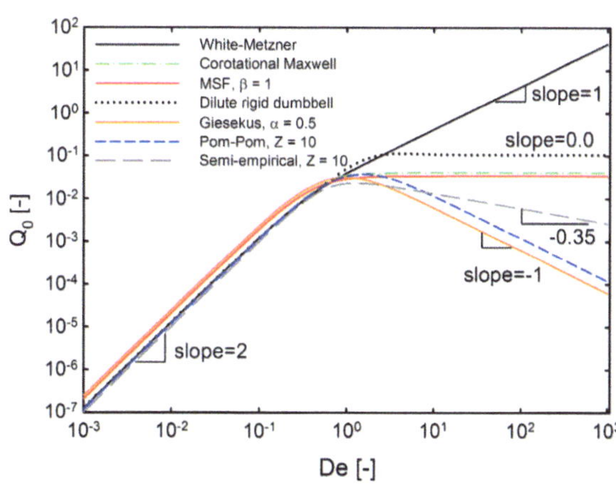

Fig. 11 Analytical Q_0 for the various models in Table 2. In the Pom–Pom model and semiempirical model, Z is the number of entanglements in the backbone

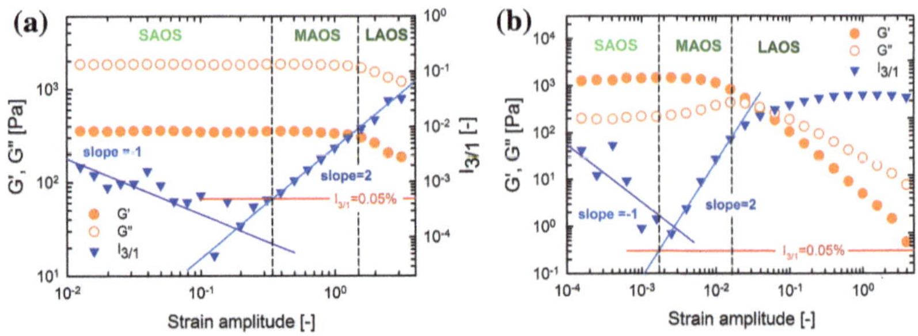

Fig. 12 The G′, G'', and $I_{3/1}$ as a function of strain amplitude. **a** Monodisperse polystyrene (PS) with $M_w = 300$ kg/mol in DOP solution with 40 wt% at a frequency of 5 rad/s at 80 °C. It shows type I (strain thinning) **b** PEO-PPO-PEO triblock copolymer solution at frequency 1 rad/s at 27 °C. It shows type III (weak strain overshoot)

response (i.e., the moduli G' or G'', or the intensity I_1) would describe the overall response by 99.95%. Figure 12 shows the strain sweep results for two different samples: one is monodisperse polystyrene (PS) solution [PS in dioctyl phthalate (DOP)], and the other is PEO-PPO-PEO triblock copolymer aqueous solution. The PS solution shows strain thinning (type I) and the block copolymer solution weak strain overshoot (type III). Depending on the materials (monodisperse polymer solution and triblock copolymer solution), the linear regime begins at different strain amplitudes. From new definition with $I_{3/1} < 0.05\%$ (which is very close to current instrument limitations), monodisperse polymer solution shows linear regime by strain amplitude 0.25 (25%), and triblock copolymer solution shows linear regime by 0.0017 (0.17%). At SAOS regime, I_3 value is lower than noise level; thus, I_3 shows a constant noise value. From the $I_{3/1}$ plot, the SAOS region with $I_{3/1} \equiv I_3/I_1 \propto \gamma_0^0/\gamma_0^1 = \gamma_0^{-1}$ is observed (however there are a lot of noise). At the MAOS regime, quadratic behavior of $I_{3/1}$ ($I_{3/1} \equiv I_3/I_1 \propto \gamma_0^3/\gamma_0^1 = \gamma_0^2$) is observed and then deviation from quadratic behavior of $I_{3/1}$ is observed. From this strain amplitude, the LAOS regime is defined. This new definition of the linear regime under oscillatory conditions may be helpful in the unambiguous determination of the limit of a linear response in a clear and reproducible way that is independent of the instrumentation [3].

5 Applications

We have reviewed FT-rheology to analyze the nonlinear response of materials undergoing large amplitude oscillatory shear flow. FT-rheology can be applied to a wide range of different material systems. Because the nonlinear response of each subclass of complex fluids (e.g., polymer solution or melt, polymer composite and blend, dispersed system, and block copolymer) can be very different, each approach has its own merits and disadvantages. Hyun et al. [3] introduced many applications to investigate complex fluids with nonlinear material functions under LAOS flow. In this chapter, several recently published LAOS investigations with FT-rheology of different complex fluids are surveyed.

5.1 Polymer Melt and Solution

5.1.1 Entangled Polymer Melt

FT-rheology under LAOS is now recognized as a very sensitive characterization method for detecting long chain branching (LCB)—or more generally to distinguish different polymer topologies. Experiment results have shown that the ratio $I_{3/1}$ and the phase angle of the third harmonic (Φ_3) are sensitive to macromolecular architecture, specifically the molecular weight distribution (MWD), number of branches, and their length [48–51]. However, these early publications did not systematically explore LAOS using a well-defined, entangled homopolymer to provide a baseline for our rheological understanding of the influence of excitation frequency, temperature, or molecular weight. Therefore, Cziep et al. [44] systemically investigated the effect of molecular weight, polydispersity, and monomer of linear homopolymer melts on intrinsic nonlinearity Q_0. They used linear polystyrene (PS), poly(p-methylstyrene) (PpMS), polyisoprene (PI), poly(methyl methacrylate) (PMMA), poly(2-vinylpyridine) (P2VP), poly(ethylene oxide) (PEO), and HDPE (high density polyethylene) samples. The general procedure to obtain nonlinear master curves ($Q_0(\omega)$) is illustrated schematically in Fig. 13. The raw stress time data of an oscillating shear experiment are recorded and transformed into a frequency spectrum via Fourier transformation (Fig. 13, schemes 1 and 2). The $I_{3/1}$ is calculated and plotted against the strain amplitude γ_0 (Fig. 13, scheme 3). This procedure is repeated for different frequencies and/or different temperatures to cover a maximum experimental range in the $Q_0(\omega)$ frequency space. From each $I_{3/1}$ plot as a function of strain amplitude, and the parameter $Q(\omega, \gamma_0) = I_{3/1}/\gamma_0^2$ is calculated. In a $Q(\gamma_0)$ plot, a plateau can be identified, where the average value is extrapolated to infinitely small strain amplitudes, and eventually yields $\lim_{\gamma_0 \to 0} Q \equiv Q_0$ (Fig. 13, scheme 4). Each $Q_0(\omega)$ value is plotted against the applied frequency, and a nonlinear master curve is obtained via the TTS (time-temperature superposition) principle (Fig. 13, scheme 5).

From these processes, they plot nonlinear mater curve as a function of De (Fig. 14). With nonlinear mater curve and constitutive equation research (Pom–Pom and MSF model in Table 2), Cziep et al. [44] suggested semiempirical equations for entangled monodisperse linear polymer melt as follows:

$$Q_0\left(De\right) = \frac{0.32}{Z^{1/2}} \frac{De^2}{1 + 33.75 Z^{-1} De^{2+0.35}}, \tag{32}$$

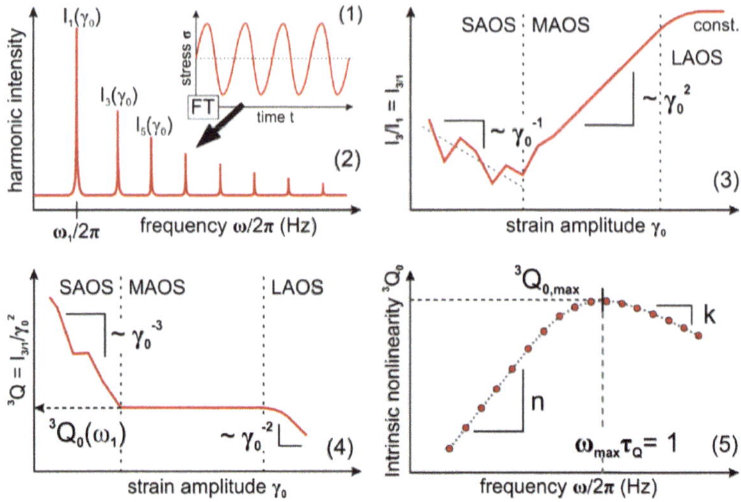

Fig. 13 Scheme of a five-step procedure from raw data (1) to nonlinear master curve (5). (1) Nonlinear stress time data of an oscillatory shear experiment. (2) After Fourier transformation of the time data, a magnitude frequency spectrum with odd higher harmonics can be obtained. (3) The ratio $I_{3/1}(\gamma_0, \omega)$ of the first and third harmonic is proportional to γ_0^2 in the MAOS region. (4) Extrapolation of $Q(\gamma_0, \omega)$ to small amplitudes gives the intrinsic nonlinearity $Q_0(\omega)$. (5) A nonlinear master curve can be created by plotting several values of different excitation frequencies, which are shifted to a reference temperature, utilizing the TTS principle. Reproduced by permission of Cziep et al. [44], copyright (2016) of American Chemical Society

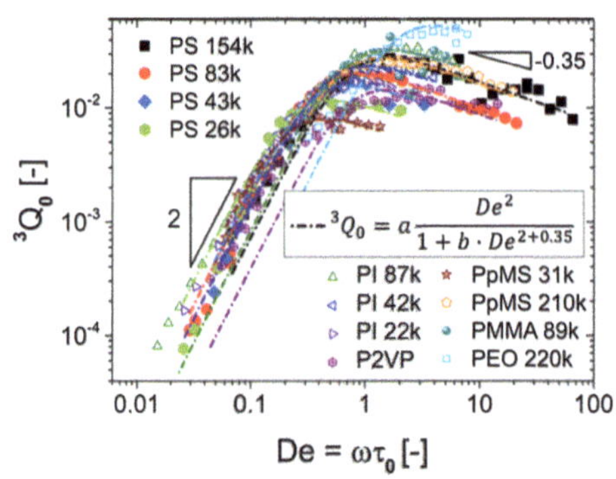

Fig. 14 Nonlinear master curves of monodisperse (PDI $\leq$ 1.07) linear melts and related fits via equation in figure. Reproduced by permission of Cziep et al. [44], copyright (2016) of American Chemical Society

where $Z = M/M_e$ is the number of entanglements. At low frequency, $Q_0(\omega)$ scales quadratically with frequency [$Q_0(\omega) \propto \omega^2$]. It is confirmed by several constitutive equations (Sect. 4.4). For high frequencies, it was experimentally found that $Q_0(\omega)$ v scales with $Q_0(\omega) \propto \omega^k$, with $k = -0.35$, which is in between the values predicted by the two constitutive models that forecast either a behavior with a scaling of -1

(Pom–Pom) or 0 (MSF). Experimentally, the maximum Q_0 ($\equiv Q_{0,\max}$) is observed and corresponds to the longest relaxation time. This semiempirical equation is very helpful for nonlinear behavior of entangled polymer melt and can lead to new developments in constitutive modeling and computer simulations, especially molecular dynamic simulations of polymer melts.

Additionally, Hyun and Wilhelm [34] investigated the effect of polymer topology on the $Q_0(\omega)$ values. They used anionically synthesized monodisperse linear and comb polystyrenes (PS). The correlation of the rheological properties with the comb topology is of special interest for the determination of the degree of branching. The PS comb series consists of a linear backbone with weight-average molecular weight of the backbone $M_b = 275$ kg/mol, and approximately $q = 25$–30 linear branches of varying molecular weight arms with M_a varying from 11.7 to 47 kg/mol (Table 3). In Fig. 15, the values of Q_0 for the linear and comb PS samples are plotted at a reference temperature of $T_{ref} = 190$ °C. The data for the monodisperse linear PS as a function of frequency displays a single local peak value and terminal quadratic behavior ($Q_0 \propto \omega^2$) at low frequencies (Fig. 15b). As the molecular weight increases, the transition to terminal behavior shifts to lower frequencies (analogous to the familiar frequency shift observed in conventional linear viscoelastic properties), and the peak becomes increasingly broad. In the case of the comb PS sample with unentangled branch chains (C622 in Table 3), $Q_0(\omega)$ displays a similar shape as a function of frequency (with one maximum value and a terminal regime ($Q_0 \propto \omega^2$). In contrast to the linear samples, however, the maximum value of Q_0 is lower than for the linear monodisperse PS melts. The authors conjectured that this might result from the dynamic tube dilution (DTD) induced by the side branches. For the comb PS with entangled side branches (C632 and C642 in Table 3), $Q_0(\omega)$ has two peak values: one corresponding to the branches' disentanglement at higher frequencies and the second arising from backbone relaxation at lower frequencies (see Fig. 15a). As a consequence of having entangled branches, the maximum value of Q_0 can be associated with the backbone relaxation ($Q_{0,b}$) and is much lower than that of the comb PS with unentangled branches. As the entangled branch chain length becomes longer, the value of $Q_{0,b}$ drops progressively and the frequency dependence becomes narrower and sharper (see Fig. 15b). In the case of the comb PS series, the volume fraction of the backbone chain decreases as the side branch length increases. From the viewpoint of dynamic tube dilution, the fully relaxed side branches act as an effective solvent for the unrelaxed backbone chain. The increasing length of the side branches has a similar effect to decreasing the concentration of the main backbone chain in a viscous solvent. Quantitative measurement of $Q_0(\omega)$ can thus effectively probe frequency dependence in the relaxation processes associated with disentanglement for a range of polymer melts.

Kempf et al. [52] further investigated comb PS and PpMS. They founded that various relaxation times (reptation time, Rouse time of the backbone, and the branch relaxation time) were directly extracted from the corresponding maxima and minimum in Q_0 curve (Fig. 16). It was also found that the reptation time extracted from the nonlinear master curve did not correspond to the crossover point of G′ and G″ in the linear master curve in the case of branched polymers. The correspondence

Table 3 Molecular characteristics of the samples used

Sample	M_b (kg/mol) backbone	M_a (kg/mol) branch	Q (branches/ backbone)	M_{total} (kg/mol)	$\langle \tau_G \rangle_w$ (s)[a] at 190 °C	Molecular structure
PS 76 k	75.9	–	–	75.9	0.05	Linear
PS 100 k	100	–	–	100	0.14	Linear
PS 220 k	214	–	–	214	2.66	Linear
PS 330 k	330	–	–	330	12.44	Linear
C622-PS	275	11.7	30	624	11.63	Comb
C632-PS	275	25.7	25	913	28.59	Comb
C642-PS	275	47	29	1630	102.06	Comb

[a]The terminal relaxation time was evaluated from linear moduli data at 190 °C.

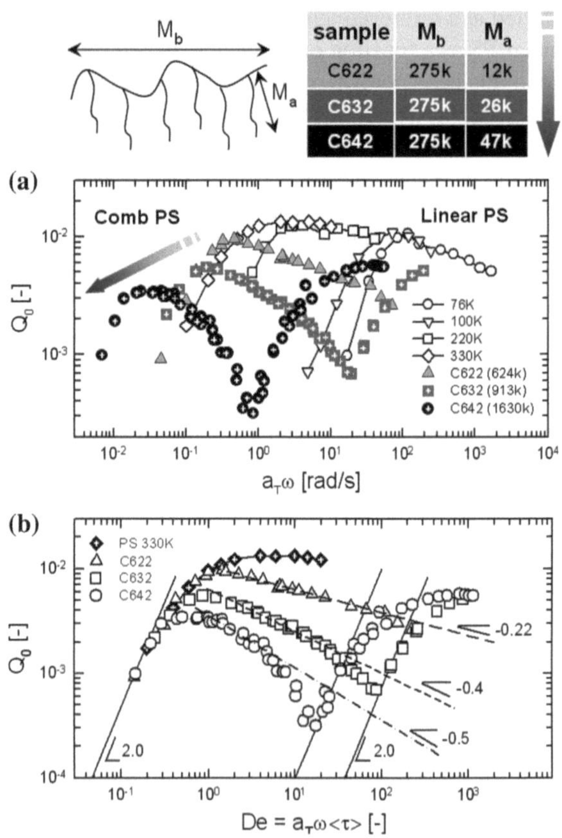

Fig. 15 **a** Frequency dependence of the zero-strain nonlinearity Q_0 for linear PS chains (PS 76, 100, 220, 330 K) and PS combs (C622, C632, and C642) at $T_{ref} = 190$ °C. With increasing molecular weight of the branched chain (M_a), the maximum corresponding backbone chain ($Q_{0,b}$) is decreasing. **b** For clear comparison, the coefficient Q_0 is plotted against Deborah number ($De = a_T \omega <\tau>$) of linear PS 330 K and comb PS (C622, C632 and C642) at T_{ref} = 190 °C. Reproduced by permission of Hyun et al. [34], copyright (2009) of American Chemical Society

of the reptation time with the maximum $Q_{max,bb}(\omega)$ was confirmed via Pom–Pom model simulations for branched polymers. It can be concluded that the reptation time can be extracted from the nonlinear master curve in contrast to the values obtained

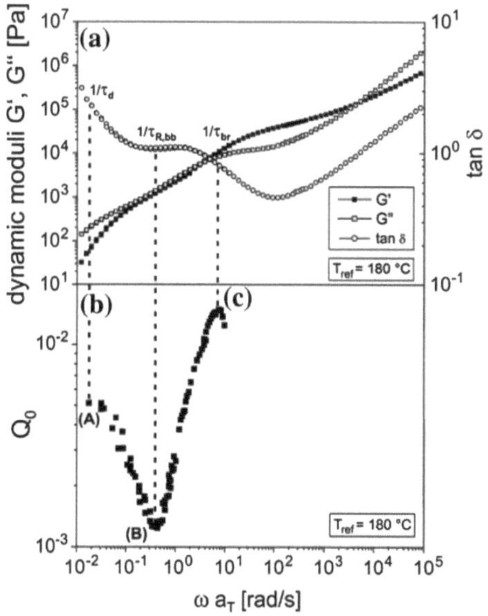

Fig. 16 Comparison between **a** the linear and **b** the intrinsic nonlinear LAOS master curve (Tref = 180 °C) for PpMS197 k-14-42 k (backbone molecular weight of 197 kg/mol with a total of 14 arms with a molecular weight of 42 kg/mol). The maxima and minima present in the intrinsic nonlinear master curve corresponded to relaxation times observed in the linear master curve and, thus, experimentally determined relaxation times. The following points are of interest: **a** reptation time τ_d, **b** backbone Rouse time $\tau_{R,bb}$, and plateau modulus $G_{N,bb}$, and **c** branch relaxation time τ_{br}. Reproduced by permission of Kempf et al. [52], copyright (2013) of American Chemical Society

from the linear measurement data. Therefore, reptation times and relaxation times (Rouse time, branch relaxation time) can be obtained using the nonlinear master curve, even if those times are not accessible from the SAOS data. The experimental accessibility of relaxation times clears the way for a better physical understanding of the underlying relaxation processes and can also be used to improve linear and nonlinear rheological modeling. Using the maximum $Q_{max,bb}(\omega)$ in the nonlinear master curve, even branched polymers with a small number of branches (here in the case of two branches) could be distinguished from the linear polymer topology. A linear dependency of the $Q_{max,bb}(\omega)$ value with the number of branches was found for comb polymers with similar molecular weight of the branches and, respectively, for the molecular weight of the branches for combs with similar number of branches. Comparing the different rheological measurement techniques, it can be concluded that this technique is highly sensitive to determine even low degrees of branching and qualitative correlations can be established.

5.1.2 Polymer Solution

Song et al. [45] investigated intrinsic nonlinearity Q_0 for monodisperse polystyrene (PS) solutions at various concentrations, which were classified as unentangled or entangled solution in a semi-dilute regime. These two types of PS solutions displayed different shapes when Q_0 was plotted as a function of frequency (ω). Unentangled solutions showed increases of Q_0 with frequency at low-frequency regions and plateau behavior at high-frequency region. On the contrary, entangled solutions showed an increase of Q_0 before the MAOS terminal relaxation time and a subsequent decrease, which is similar to that observed for entangled linear polymer melts (Fig. 17). The $Q_0(\omega)$ curves of each group were superposed in a dimensionless coordinate ($Q_0/Q_{0,\max}$ vs. *De*), so that transition from the plateau of Q_0 to decreasing Q_0 at high-frequency region might indicate the onset of entanglement in polymer solution. In particular, all unentangled solutions had the same $Q_{0,\max}$ value (0.006) regardless of polymer concentration and molecular weight, because Q_0 responds to Rouse-like relaxation process only, which is featured as no interchain interaction and chain stretching. However, the $Q_{0,\max}$ values of entangled solutions were dependent on the number of entanglements (Z). The master curve of $Q_{0,\max}$ as a function of Z showed that $Q_{0,\max}$ was constant at low entanglement numbers (few or virtually no entanglements), and then increased with the beginning of entanglement to approach a limiting value at high entanglement numbers, where reptation is the dominant linear relaxation process. From the master curve, the experimental line fitting can be described as follows:

$$Q_{0,\max} = \begin{cases} 0.006 & \text{at } Z^{\mathrm{sol}} < 1 \\ \frac{0.026}{1+3.59\left(Z^{\mathrm{sol}}\right)^{-1.01}} & \text{at } Z^{\mathrm{sol}} \geq 1 \end{cases} \tag{33}$$

Figure 18 shows that the data sets obtained for melts and solutions of linear homopolymers superposed on the master curve within experimental error. The $Q_{0,\max}$ predicted by DE IA (Doi–Edwards with independent alignment assumption) was indicated as 0.040. This deviation is removed when stretching effect of chains is introduced in the model as MSF (molecular stress function) model. The $Q_{0,\max}$ predicted by MSF model is 0.023 which is the same as the prediction of Eq. (31) at infinite Z^{sol}. The MSF model introduces two parameters for predicting intrinsic nonlinear behavior in the MAOS region. The strain measure of DE IA model for linear polymers resulted in a constant value $\alpha = 5/21$, which explains the affine orientation of network strands. The MSF model considers an additional contribution of isotropic strand extension using stretching parameter β. By definition, the β value is fixed as 1 for linear chains. Thus, two parameters remain constant during MAOS measurements for linear chains.

In addition, the master curve of $Q_{0,\max}$ as a function of Z was used to quantify the degree of tube dilation based on the dynamic tube dilution (DTD) theory. Direct comparison of the $Q_{0,\max}$ values of semi-dilute solutions and melts showed that they

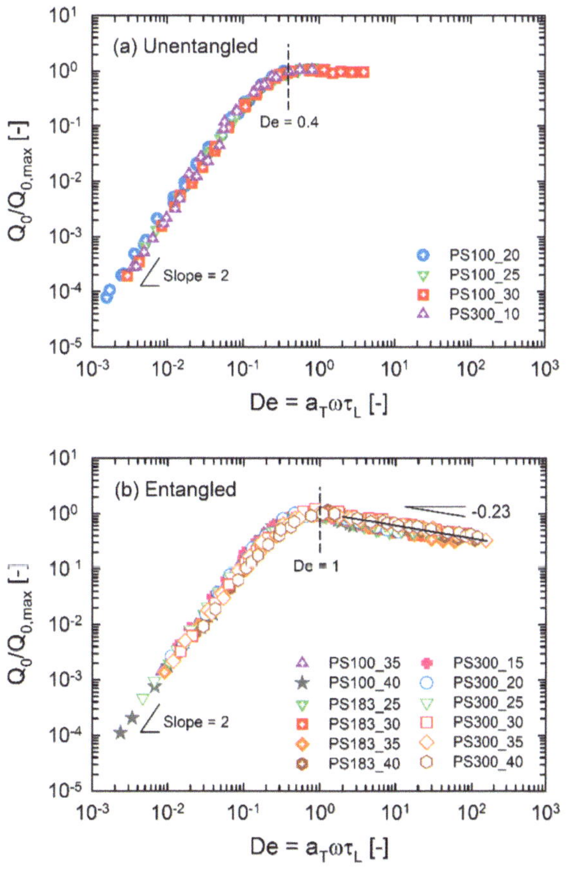

Fig. 17 Dimensionless plots of normalized intrinsic nonlinearity as a function of De (= $a_T\omega\tau_L$). **a** Unentangled solutions show plateau behavior at $De > 0.4$, whereas **b** entangled solutions show a decrease of Q_0 with a constant average slope of -0.23 at $De > 1$. Reproduced by permission of Song et al. [45], copyright (2017) of American Chemical Society

followed the same molecular dynamics in MAOS flow like SAOS (small amplitude oscillatory shear) flow. Comparison between static and dynamic dilutions using the $Q_{0,\mathrm{max}}$ master curve suggested that this curve could characterize the effective number of entanglements per backbone chain for branched polymers. Because it was confirmed again that $Q_0(\omega)$ is highly sensitive to various relaxation processes, MAOS tests may provide a new means of investigating molecular dynamics.

The intrinsic nonlinearity $Q_0(\omega)$ seems to be highly sensitive to the characteristic relaxation processes of polymers and can be regarded a parameter that maximizes delicate changes observed in linear viscoelastic moduli. Thus, the nonlinear response under MAOS shear flow can cast light on unsolved problems about relaxation processes, such as reptation, contour length fluctuation (CLF), and constraint release (CR). To this end, the nonlinear behaviors of various well-defined polymers with different topologies (star, H-shaped, comb-shaped, and so on) need to be further investigated from a molecular dynamics perspective.

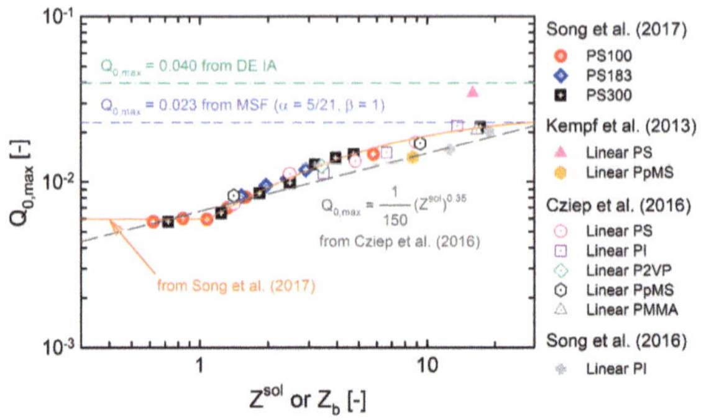

Fig. 18 The master curve of $Q_{0,\max}$ as a function of Z^{sol} and Z_b from recent polymer solution and melt data. Z_b represents the entanglement numbers of backbone chains in melts and Z^{sol} the solution entanglement number. For unentangled solutions, $Q_{0,\max}$ is constant at ~0.006. It increases gradually from $Z^{sol} = 1$ and seems to reach a limiting value. The resultant logistic-type fitting equation for the $Q_{0,\max}$ of entangled solutions is shown in the plot (solid line). For comparison, several predictions obtained by Doi–Edwards (DE IA), molecular stress function (MSF), and semiempirical equation (from Cziep et al. [29]) are plotted together. All data sets on a master curve follow Eq. (34) within acceptable deviations. Reproduced by permission of Song et al. [45], copyright (2017) of American Chemical Society

5.2 *Polymer Composites*

Lim et al. [53] investigated the nonlinear viscoelastic responses of polymer composite systems containing different shaped nanoparticles, e.g., polycaprolactone (PCL)/MWNT (multiwall carbon nanotube, 1-D thread shape), PCL/OMMT (organo-modified montmorillonite, 2-D plate shape with high aspect ratio), and PCL/PCC (precipitated calcium carbonate, 3-D cubic shape). They evaluated the PCL/MWNT composites using several analyzing methods, including Lissajous analysis, stress decomposition, and FT-rheology. The microstructure of PCL/MWNT was estimated using TEM images, the conductivity, and linear viscoelastic properties. An electrical percolation threshold was observed in DC conductivity and the storage modulus at low frequency rapidly changed near the electrical percolation threshold. The stress signals were distorted with increasing MWNT concentration and strain amplitudes. The shape of the elastic stress in the stress decomposition changed from sinusoidal to triangular under LAOS flow as the MWNT concentration increased. In FT-rheology, $I_{3/1}$ increased with strain amplitude and showed a maximum. It was related with the change of microstructure as evidence by the measurement of DC conductivity. The maximum peak of $I_{3/1}$ was also observed in PCL/OMMT, but was not observed in PCL/PCC. As to the particle shape, $I_{3/1}$ in the polymer composites containing the particles of high aspect ratio (MWNT, OMMT) dramatically increased with particle concentration. They calculated the nonlinear parameter Q_0 from the fitting results of Q ($\equiv (I_{3/1})/\gamma_0^2$) by a mathematical model that was similar to the "Carreau–Yasuda" viscosity equation as follows:

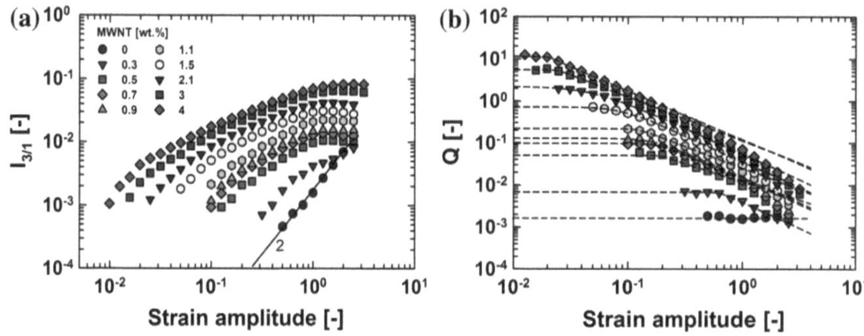

Fig. 19 Nonlinearity of PCL/MWNT composites as a function of strain at 1 rad/s and 130 °C: **a** relative intensity of the third harmonic ($I_{3/1}$) and **b** coefficient Q evaluated from the $I_{3/1}$. The dotted lines are the fitted results (—) with Eq. (34). Reproduced by permission of Lim et al. [53], copyright (2013) of AIP

$$Q = Q_0\left(1 + (C_1\gamma_0)^{C_2}\right)^{(C_3-1)/C_2}, \tag{34}$$

where Q_0 is the *zero-strain Q parameter* (Q_0), C_1 is the inverse of critical strain amplitude (γ_{0c}), and C_3 is degree of strain thinning (C_3). Figure 19a shows that the $I_{3/1}$ of the composites increased with increasing MWNT concentration at the same strain amplitude, and Fig. 19b shows Q as a function of strain amplitude fitted with Eq. (34).

In the cases of MWNT and OMMT, which had a high aspect ratio, Q_0 increased with particle concentration, whereas Q_0 of PCC increased slightly with an increase of particle concentration. They compared linear and nonlinear viscoelastic properties of polymer composites and a PVA/Borax system to understand the effect of internal structure on the amplification of viscoelastic properties. For better comparison between linear and nonlinear viscoelastic properties, they defined a new parameter, the nonlinear–linear viscoelastic ratio (**NLR**) as follows:

$$\text{NLR} = \frac{Q_0(\varphi)/Q_0(0)}{G^*(\varphi)/G^*(0)}. \tag{35}$$

In the case of NLR = 1, the effect of nonlinear viscoelasticity is the same as the effect of linear viscoelasticity with the increase in concentration. In the case of NLR > 1, the nonlinear parameter is amplified more than the linear parameter due to the internal structure. In the case of NLR < 1, the nonlinear parameter is amplified less than the linear parameter. As the concentration of particles and that of borax increased, the NLR deviated from one. The NLR might depend on the internal structure of the polymeric systems. They calculated NLR, and the results are shown in Fig. 20. As the concentration increased, the NLR increased and reached a plateau (see guidelines in Fig. 20), which was roughly NLR (PCL/OMMT) ≈ 2205 > NLR (PCL/MWNT) ≈ 934 > NLR (PCL/PCC) ≈ 1.46 > NLR = 1 > NLR (PVA/Borax) ≈ 0.18. In the case of poly-

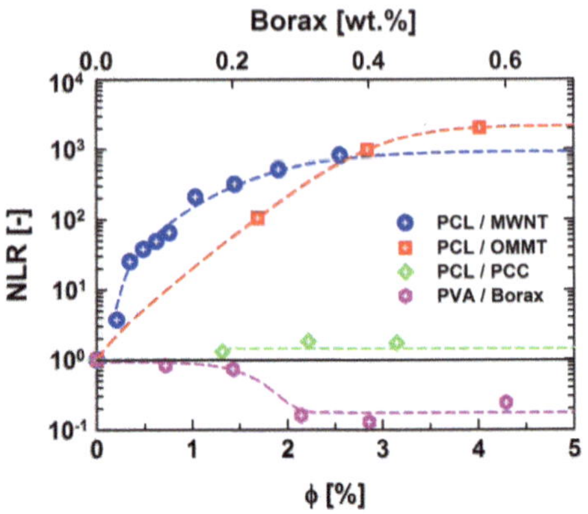

Fig. 20 NLR (nonlinear–linear viscoelastic ratio) of PCL/MWNT (●), PCL/OMMT (■), PCL/OMMT (◆) composites, and PVA/Borax (●) as function of concentration (φ). The lines are added for visual simplification. Reproduced by permission of Lim et al. [53], copyright (2013) of AIP

mer composites, the NLR was larger than one. In contrast, the NLR of the polymer network (PVA/Borax) was smaller than one. This might be due to the difference in the internal structure of polymer composite (heterogeneous phase) and polymer network system (homogeneous phase). In the case of PVA/Borax system, a strong network structure enhanced the first stress contribution among the higher harmonics; in contrast, the network structure suppressed the distortion of the stress. Therefore, the NLR value of PVA/Borax was less than one. In the case of polymer composites, the well-dispersed PCL/OMMT had a larger NLR than the other nanocomposites. It might be inferred that the surface area of PCL/OMMT was larger than the other composites. The larger surface area might increase the interaction between the particles and polymer chains. Based on these results, it could be suggested that the NLR be used as a quantitative parameter to explain the effect of nanoparticles on the polymer composites, including the assessment of dispersion quality or internal structure.

Schwab et al. [54] investigated styrene butadiene rubber (SBR) filled with carbon black (CB) under large amplitude oscillatory shear (LAOS), in which they analyzed it in terms of the nonlinear parameter $I_{3/1}$. Rubber materials filled with reinforcing fillers display nonlinear rheological behavior at small strain amplitudes below $\gamma_0 < 0.1$. Nevertheless, rheological data are analyzed mostly in terms of linear parameters, such as shear moduli (G′, G″), which lose their physical meaning in the nonlinear regime. They used three different CB grades and the filler load was varied between 0 and 70 phr (parts per hundred rubber; relative mass contribution of CB normalized to the rubber content as generally used concentration in rubber industry). The influence of the CB volume fraction φ on the rheological behavior of unvulcanized rubber compounds in terms of nonlinear rheological parameters was investigated. Figure 21 shows the relative third higher harmonic contribution $I_{3/1}(\gamma_0)$. Here, four different regions can be identified. Region I is below the instrument's sensitivity as illustrated in Fig. 13 (3). Region II is dominated by a broad peak in $I_{3/1}(\gamma_0)$. Its origin has not yet been clarified, but measurements with linear homopolymer melts indicate that the peak is due to instrumental problems, because they also show this peak only when measured at the SIS V50 rheometer but it does not occur with open

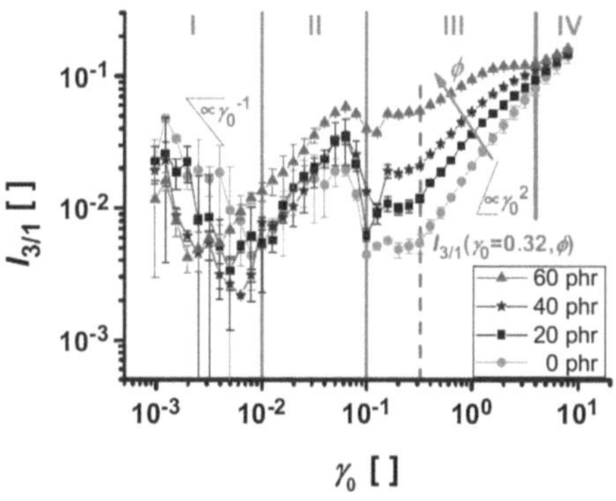

Fig. 21 Relative third higher harmonic contribution $I_{3/1}(\gamma_0)$ of samples filled with different amounts of N339 carbon black. Four different zones can be identified. In region I (below $\gamma_0 = 10^{-2}$), the higher harmonic contribution I (3ω) is below the sensitivity limit of the rheometer and is therefore dominated by random noise. As a consequence $I_{3/1}(\gamma_0)$ is decreasing. Region II ($10^{-2} < \gamma_0 < 10^{-1}$) is dominated by a broad peak of $I_{3/1}(\gamma_0)$, which is most probably due to instrumental limitations. The nonlinear contribution is changing with increasing filler content φ in region III, most pronounced at $\gamma_0 = 0.32$. At very high strain amplitudes (region IV, $\gamma_0 > 4$), the curves merge ($T = 80$ °C, $\omega/2\pi = 0.2$ Hz). Reproduced by permission of Schwab et al. [54], copyright (2013) of WILEY-VCH

cavity rheometers (ARES G2, TA Instruments). Since both Regions I and II are dominated by instrumental limitations, all samples showed similar behavior. In region III, $I_{3/1}(\gamma_0)$ increased with increasing strain amplitude. For the sample without CB, the increase of $I_{3/1}(\gamma_0)$ was nearly linear on the log–log scale. For the N339 samples at 80 °C, the scaling exponent α of $I_{3/1}$ as function of strain amplitude, $I_{3/1}(\gamma_0) \propto \gamma_0^{\alpha}$, is remarkably lower (α = 0.5–1.2) depending on the CB content. The smaller slope than one of polymer melt and solution (α = 2.0) as mentioned before is due to the additives. These organic and inorganic additives are added in quantities as high as 13.35 phr and act partly as plasticizers. They are increasing polymer mobility and thereby may affect the nonlinear behavior. Additionally, not all additives might be soluble in the rubber matrix and different phases can be present in the compound. Therefore, the compounds morphology could also be changed by the mechanical force applied during the LAOS experiment. The presence of CB results in a higher value for $I_{3/1}$ in region III similar to polymer composites and scales with CB loading. This filler effect is most pronounced at amplitudes around $\gamma_0 = 0.32$. In region IV, the filler N339 (CB) is almost inactive in the sense that the nonlinear parameter $I_{3/1}(\gamma_0)$ of the filled systems approaches that of the sample without CB. This indicates a severe destruction of the physical network structure in the compound.

With previous results, $I_{3/1}$ at $\gamma_0 = 0.32$ is used to investigating CB grade effect. Thus, the nonlinear parameter $I_{3/1}(\gamma_0 = 0.32)$ of all samples tested is plotted in Fig. 22 as a function of the internal surface accessible for polymer–filler interactions. The internal surface (in milliliter adsorbed oil per 100 g of rubber) was calculated by the OAN (oil adsorption number) times the weight fraction (Φ_w) of CB in the

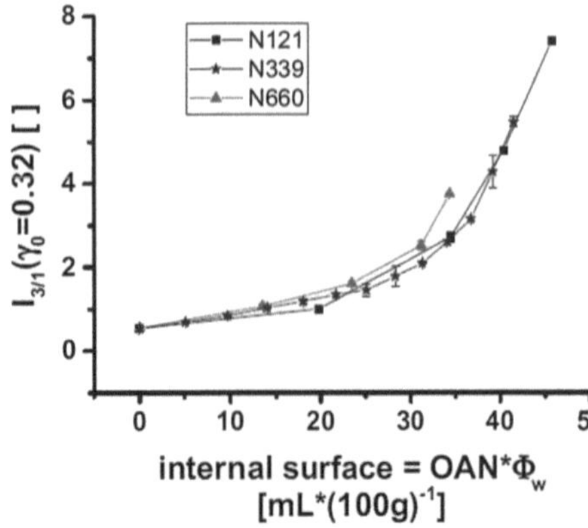

Fig. 22 Third higher harmonic contribution $I_{3/1}$ at a strain amplitude of $\gamma_0 = 0.32$ as a function of the internal surface between CB and the polymer. This internal surface was calculated as the product of the oil adsorption number (OAN) and the weight fraction of CB in the respective compound. The results of all three filler grades fall on one line. This demonstrates the importance of the interface area on the nonlinear contribution ($\gamma_0 = 0.32$, $T = 80$ °C, $\omega/2\pi = 0.2$ Hz). Reproduced by permission of Schwab et al. [54], copyright (2013) of WILEY-VCH

respective compound. All compounds fall on one line independent of the filler grade, so the assumption that the nonlinear contribution is correlated to the amount of polymer–filler interactions seems to be true. With this relation, it is also possible to get information about the size of the internal surface in a CB filled rubber compound by measuring the nonlinear parameter $I_{3/1}$ at a strain amplitude of $\gamma_0 = 0.32$.

Nonlinear contributions to the rheological behavior of filled rubber systems are significant even at low strain amplitudes and understanding the nonlinear behavior can lead to more insights into these compounds.

5.3 Emulsion and Polymer Blends

5.3.1 Emulsion

Small amplitude oscillatory shear tests are a reliable way of extracting a characteristic droplet size for emulsions [55]. Carotenuto et al. [56] proposed using LAOS to determine not only the characteristic dimension of an immiscible polymer blend but also to infer the size distribution of the drops. The principal idea is that even an emulsion formed from two immiscible Newtonian fluids will exhibit a viscoelastic response due to the interfacial tension, and this response will become nonlinear when sufficiently large amplitude shear is applied to the emulsion droplets. Consequently, the droplet size and the size distribution will drastically affect the intensity and phase of the different higher harmonics in the FT-rheology spectra. Reinheimer et al. [57, 58] investigated emulsion system with $I_{5/3}$ instead of $I_{3/1}$. They suggested that as the fundamental peak I_1 is mainly determined by the Newtonian behavior of the neat Newtonian matrix and the dispersed phase, i.e., the viscosity of the two single phases. It was not useful in characterizing the interfacial tension or size and distribution of the included phase and was, therefore, excluded in their analysis. They used intrinsic nonlinearity ${}^{5/3}Q_0$ calculated using I_3 and I_5 as follows:

$$ {}^{5/3}Q_0 = \frac{{}^5Q_0}{{}^3Q_0} = \lim_{\gamma_0 \to 0} \frac{\frac{I_5/I_1}{\gamma_0^4}}{\frac{I_3/I_1}{\gamma_0^2}} = \lim_{\gamma_0 \to 0} \frac{I_5/I_3}{\gamma_0^2}. \tag{36} $$

Furthermore, they also defined ${}^{7/5}Q_0$ with same way. Because concentrated emulsions can produce a very large number of overtones which means a large number of higher harmonic intensities $I_{n/1}$ with $\underline{n} > 7$ are present in the frequency spectrum. They found a relationship between intrinsic nonlinear ratio ${}^{5/3}Q_0$ and ${}^{7/5}Q_0$ as well as the emulsion properties, i.e., η_m is matrix viscosity and η_d dispersed phase viscosity, $\lambda = \eta_d/\eta_m$, Γ is the interfacial tension, and R is the droplet size, which is expressed as follows:

$$ \frac{{}^{5/3}Q_0}{\omega^2} = 0.64\lambda^{1.63} \frac{\eta_m^2 \langle R \rangle_{4,3}^2}{\Gamma^2} \quad \text{and} \quad \frac{{}^{7/5}Q_0}{\omega^2} = 0.64\lambda^{1.63} \frac{\eta_m^2 \langle R \rangle_{5,4}^2}{\Gamma^2}. \tag{37} $$

It is determined simulation and experimental results from LAOS test. They concluded that nonlinear oscillatory shear experiments combined with numerical simulations represent a new approach for characterizing the volume average droplet size and the width of the droplet size distribution.

5.3.2 Polymer Blends

From previous results, the droplet size in an immiscible blend relates to the mechanical nonlinearity. Reinheimer et al. [57, 58] excluded I_1 value in their analysis due to investigating the neat Newtonian matrix and the dispersed phase (for example, ${}^{5/3}Q_0$ calculated using I_3 and I_5 see previous section). In immiscible polymer blend, however, normalized third harmonic $I_{3/1}$ still has a meaning because the polymer already shows non-Newtonian behavior (e.g., shear thinning). Reza et al. [59–61] investigated PP (polypropylene)/PS (polystyrene) blend with inorganic compatibilizer (silica and clay) using $I_{3/1}$. They used NLR value [Eq. (35)] similar to polymer composite. Reza et al. [59] investigated the relationship between NLR value and PS droplet size in the PP matrix domain. From the TEM images, clay was located mostly at the interface or partially inside the PS drops (see TEM picture in Fig. 23), thereby reinforcing the compatibilization effect. Therefore, clay increased the dispersion morphologies of the PP/PS blends. In contrast, fumed silica was located mostly inside the PS droplets, which means the morphologies of PP/PS blends were not improved (TEM picture in Fig. 23). Linear viscoelasticity of both PP/PS/Clay and PP/PS/Silica increased with increasing particle concentrations. NLR values for

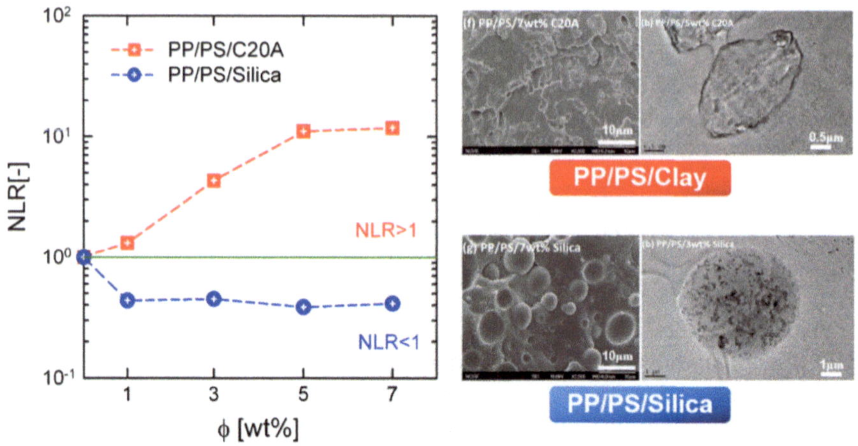

Fig. 23 Comparison of NLR values of PP/PS/C20A and PP/PS/Silica blends as a function of weight fraction of the particles. TEM images of the PP/PS/clay and PP/PS/Silica. Clay is located at the interface between PP and PS phase, in contrast silica particles is located in the PS phase. Reproduced by permission of Reza et al. [59], copyright (2014) of American Chemical Society

the PP/PS/Clay blends were larger than 1 (NLR > 1), whereas NLR values for the PP/PS/Silica blends were less than 1 (NLR < 1). Therefore, NLR could be classified into two categories depending on the morphology. Based on these results, NLR can be used to distinguish between the effects of two different types of nanoparticles on the morphologies of PP/PS blends. Furthermore, Reza et al. [60, 61] investigated the effect various silica (hydrophobic and hydrophilic) and clays (clay hydrophobicity) on the NLR value.

They show how the inverse of droplet size ($1/R_n$) varies as a function of NLR for PP/PS (80/20) blends filled with different clays and silicas at different concentrations in Fig. 24. For fitting the experimental results, they suggest the following empirical equation:

$$\frac{1}{R} = \frac{1}{R_0} + a\left[1 - \exp\left(-b \times NLR\right)\right], \tag{38}$$

where R_0 is the minimal droplet size and a and b are fitting parameters. As can be seen in Fig. 24, $1/R_n$ increases when nanoparticles (silica and clay) were added and the NLR values approached their maxima when the droplet size approached R_0. Interestingly, with the exception of D17 (hydrophobic silica nanoparticle)-filled blends, all blends exhibited an exponential relationship between the inverse of the droplet size and NLR, indicating that droplet sizes at particular concentrations can be predicted from NLR values, which are determined by extrapolation. However, in case of PP/PS/D17, different trends were observed, indicating a rapid reduction in droplet size and the importance of droplet size for determining NLR values in this blend, in which the interface is completely covered with D17 particles. This is in

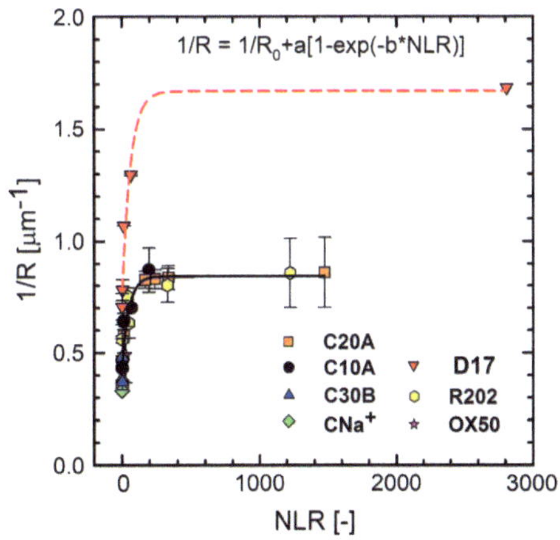

Fig. 24 Inverse droplet radii ($1/R_n$) of PP/PS (80/20) blends containing different clays and silicas at different concentrations as a function of NLR values. Data were fitted using the three parameter exponential equation shown in Figure.$1/R_0 = 0.43$, $a = 0.4$ and $b = 0.02$ for the universal curve (except for D17), respectively. For D17, the obtained parameters were: $\frac{1}{R_0} = 0.75$, $a = 0.91$ and $b = 0.01$. Reproduced by permission of Reza et al. [61], copyright (2016) of American Chemical Society

accordance with the fact that the best compatibilization is obtained when particles are absorbed onto phase interfaces. Therefore, according to our results, D17 is the best compatibilizer, among those examined, for (80/20) PP/PS blends because it inhibits droplet coalescence. Ock et al. [62] investigated NLR values of poly(lactic acid) (PLA) and natural rubber (NR) blends compatibilized with organoclay according to clay contents, mixing conditions, and types of clay. They also found that the NLR value displayed a similar trend as the drop size reduction and was related to the inverse of the drop size for all variables (clay concentration, mixing condition, and types of clay).

5.4 *Block Copolymers and Liquid Crystals*

Certain types of soft matter, such as liquid crystals, amphiphiles, and block copolymers, self-assemble into nanostructured morphologies below their order–disorder transition temperature (T_{ODT}) as a way to minimize highly unfavorable enthalpic contributions to the free energy. Self-assembly usually leads to a polydomain structure with locally anisotropic ordered domains (grains) that are randomly orientated throughout the whole sample, resulting in a macroscopically isotropic material. However, many practical applications including functional membranes, anisotropically charged materials, and photon conductors require that the final material is macroscopically anisotropic. Application of an external stimulus, such as an electric, magnetic, or mechanical field, can then be used to obtain the preferred macroscopic alignment [63]. Among the mechanical fields, LAOS flow has been used to study the orientation/reorientation processes in microphase-separated lamellar structure. The alignment kinetics can be studied in detail via online monitoring of the degree of mechanical nonlinearity exhibited during the orientation process—as determined via $I_{3/1}$ in FT-rheology as a function of time. Oelschlaeger et al. [64] investigated PS-*b*-PI diblock copolymers as well as in diblock and triblock copolymers of styrene and butadiene (PS-PB; PS-PB-PS) under LAOS flow. The evolution of the microstructure during the flow alignment process can be easily quantified using the FT-rheology. For di- and triblocks, parallel alignment is achieved at low frequency and temperatures below the order/disorder transition temperature T_{ODT}. The kinetics of orientation can be quantified by the intensity of the third harmonic $I_{3/1}$(t). $I_{3/1}$(t) can be described by a stretched exponential function with a characteristic relaxation time τ.

$$I_{3/1} = y_0 + A \exp\left[-(t/\tau)^{\beta}\right], \tag{39}$$

where y_0 is the $I_{3/1}$ value at infinite times, A is the decay amplitude, τ is the representative alignment time for lamellar layers, and β is related to the width of the distribution. For di- and triblock copolymers, the kinetics of orientation strongly depend on the strain amplitude, and the time constant varies with a scaling exponent: $\tau \propto \gamma_0^{-4}$ for the diblock and $\tau \propto \gamma_0^{-2.85}$ for the triblock. This scaling exponent greatly exceeds the expected scaling of $\tau \propto \gamma_0^{-1}$ which corresponds to a physical process in which the total applied deformation is responsible for the observed orientation. The larger scaling exponent might be explained by the cooperative nature of the underlying processes. Analysis of the time-dependent variation in the phase difference related to the third harmonic ($\delta_3(t)$) enables further differentiation between the diblock and triblock for the PS-PB and PS-PB-PS model systems. However, Meins et al. [63] found a different scaling exponent $\tau \propto \gamma_0^{-2}$ for a PS-PI diblock copolymer. While Oelschlaeger et al. [64] investigated parallel orientation near T_{ODT}, Meins et al. [63] investigated a perpendicular orientation of a PS-PI system. The different orientation evolution is reflected by the different scaling of alignment time.

Lee et al. [65] investigated liquid crystal (8CB, 4-4'-n-octyl-cyanobiphenyl) in lamellar smectic A phase under LAOS flow. The storage modulus G′(t) and loss modulus G''(t) from the conventional rheometer program under various LAOS flow conditions (different strain amplitude and frequency) decreased and reached equilibrium as a function of time [Fig. 25, normalized modulus can be fitted with Eq. (39)]. This could be attributed to shear alignment of the lamellar smectic A structure. On the contrary, with G′(t) and $G''(t)$, the nonlinearity $I_{3/1}(t)$ showed three different behaviors depending on the magnitude of strain amplitude (Fig. 25): (1) region I (γ_0 < 0.6): increased (increased and reached equilibrium), (2) region II (0.6 < γ_0 < 2.0): increased and decreased (showed maximum value; decreased and reached equilibrium), and (3) region III (γ_0 > 2.0): decreased (decreased and reached equilibrium) as a function of time. These three different time-dependent behaviors of nonlinearity [$I_{3/1}(t)$] were shown to be related with the alignment behavior of the lamellar structure. The reduction of $I_{3/1}(t)$ was observed during the 10 h of macroscopic orientation in both SB (PS-PB; polystyrene–polybutadiene) block copolymers and SBS (PS-PB-PS) triblock copolymers under LAOS flow. The alignment time differed according to molecular size (alignment time of 8CB ~1 h faster than that of block copolymers, ~10 h). Both the SB and SBS samples aligned from a disoriented to parallel alignment. However, in case of Lee et al.'s experimental results, this behavior for the nonlinearity $I_{3/1}(t)$ was observed only at large strain amplitude (region III). Struth et al. [66] reported lamellar domain formation with three distinct order directions at a frequency of $\omega/2\pi = 1$ Hz and a strain amplitude of $\gamma_0 = 1.0$ from six reflections in the X-ray scattering pattern, which was in contrast with the two reflections under steady shear flow. Compared to steady shear flow, oscillatory shear flow periodically increased and decreased the strain (and the shear rate) as well as it changed direction. Therefore, smaller strain amplitude shear flow did not result in a well-oriented lamellar structure compared with steady shear flow. Thus, three different lamellar clusters showed various microscopic stress levels, especially at the grain boundary (boundary at domains), under dynamic oscillatory shear flow. Thus, macroscopic stress curves were more distorted than at large strain amplitude. Thus, the nonlinearity may become larger with time. In the case of block copolymers, the nonlinearity ($I_{3/1}$) increases upon reorientation from parallel to perpendicular alignment [64]. By changing the parallel layer microstructure to a perpendicular one, several oriented domains can be made, resulting in distortion of the macroscopic stress curves. In their results [65], the nonlinearity ($I_{3/1}$) at smaller strain amplitude, e.g., from strain amplitudes of $\gamma_0 = 0.3$ to 0.6 and frequency of $\omega/2\pi = 1$ Hz (see Fig. 26a). With an equilibrium value of 3600 s, the G′, G'', and nonlinearity ($I_{3/1}$) were plotted as a function of strain amplitude, γ_0. Interestingly, $I_{3/1}$ (γ_0) increased and then decreased (maximum) even though G′ (γ_0) and G'' (γ_0) only decreased with increasing strain amplitude (Fig. 26d). From these results, it can be concluded that LAOS analysis of nonlinear stress, especially $I_{3/1}$ from FT-rheology, is more sensitive to microstructure and the related change than storage modulus G′ and loss modulus G''.

5.5 Solid Polymers (Fatigue Test)

Typically, FT-rheology is used to quantify the nonlinear response of polymer melts resulting from molecular dynamics. To expand FT-rheology in the solid regime, first one has to realize that a solid polymer is dominated by elastic response. The elastic response is much better described by a simple Hookean spring model, and consequently, the amount and intensity of higher harmonics are low. Nevertheless, this method opens up to quantify the evolution of structural nonlinearities. Hirschberg et al. [33] investigated the fatigue behavior of polystyrene (PS) in a strain-controlled torsion rectangular oscillatory tests via FT-rheology. Most data in the literature on mechanical fatigue are based on stress—number of cycle (*S-N*) or so-called Wöhler curves—only the conditions for complete part failure under a certain load can be found with high experimental uncertainties. Therefore, the prediction of failure onset is highly important. The tests were performed under large amplitude oscillatory

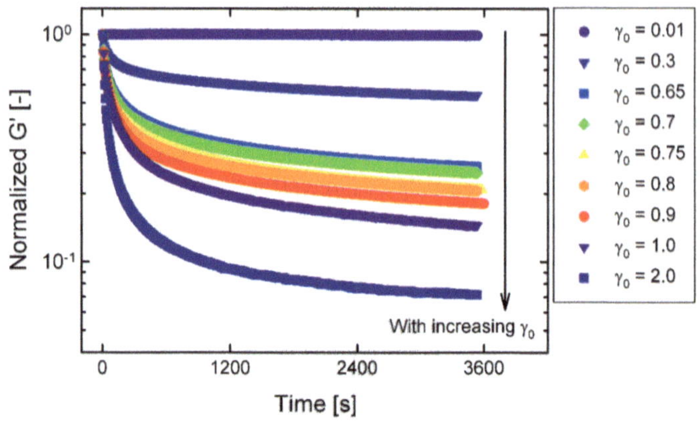

Fig. 25 The normalized G′ data [≡ G′/G′(time = 0)] from time sweep test of 8CB at a fixed frequency of ω = 1 Hz and various strain amplitudes for 3600 s with 50-mm parallel plates. With increasing strain amplitude, the normalized G′ data decreased very quickly and then reached a plateau value. Reproduced by permission of Lee et al. [65], copyright (2015) of AIP

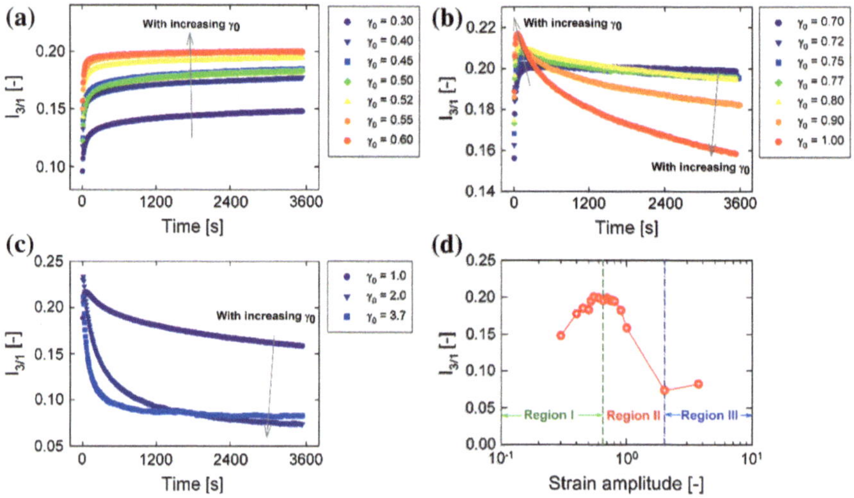

Fig. 26 Nonlinearity $I_{3/1}$ from FT-rheology at various strain amplitudes for 8CB at a fixed frequency of $\omega = 1$ Hz and 25 °C. **a** Increase in $I_{3/1}$ at a strain amplitude from 0.3 to 0.6. **b** Increasing and then decreasing (transition region) of $I_{3/1}$ at strain amplitude from 0.7 to 1.0. **c** Decreasing of $I_{3/1}$ at strain amplitude 2.0 and 3.7. The $I_{3/1}$ at strain amplitude 1.0 is added for comparison. **d** The nonlinearity ($I_{3/1}$) at 60 min (3600 s) as a function of strain amplitude. Reproduced by permission of Lee et al. [65], copyright (2015) of AIP

shear (LAOS), so the stress response was no longer perfectly sinusoidal, and higher harmonics could be detected and quantified in the FT spectra as function of time or number of cycles N, respectively. In Fig. 27, linear parameter such as the storage modulus (G′) was analyzed, as well as nonlinear parameters like the normalized second ($I_{2/1}$) and third ($I_{3/1}$) harmonics as a function of number of cycles N. Most of nonlinear response, odd higher harmonics are important. Even though even harmonics could be detected, the value is usually substantially smaller than odd harmonics. Even harmonics means nonhomogeneous deformation or anisotropic response of materials, usually, they are considered experimental errors. Interestingly Hirschberg et al. investigated even harmonics on purpose. In Fig. 27a, the $G'(N)$ decreases relatively rapidly for about the first 100 cycles, then decreases more slowly for about 600 cycles (from the 100th to the 700th cycle) with a constant slope, $dG'/dN = \text{constant}$. In contrast, the nonlinear parameter $I_{3/1}(N)$ increases for about the first 100 cycles and still increases with a smaller constant slope until about the 700th cycle. Then, a crack occurred (Fig. 27c picture 2) producing a substantial change in the slope of both curves. This is followed by a period of crack propagation up to a point (around 1300 cycles) where the sample is totally broken (complete failure). In this case, the growth of a sidewise crack, as seen in Fig. 27c picture 4 and 5, was observed after about 900 cycles. When the first macroscopic crack occurs in the sample (around 700 cycles), the intensity of $I_{2/1}$ rises abruptly and rises further when the crack propagates. The nonlinear parameter $I_{2/1}(N)$ is very low for undamaged samples, but its intensity was found to increase when defects are created in the structure to a point where cracks became visible in the sample. Figure 28 shows other results. In this figure, $I_{2/1}(N)$ is very sensitive than other rheological properties (G' and G'', and $I_{3/1}$). However, when a crack appeared (see video picture in the figure), $I_{2/1}(N)$ began to increase. As mentioned before (Sect. 4.2), even harmonics are the result of an asymmetry in the deformation flow. This can be attributed to sample anisotropy due to the presence of cracks. Consequently, the even harmonics increase, and especially $I_{2/1}$, can be explained by crack initiation and propagation in the sample. Both parameters $I_{2/1}(N)$ and $I_{3/1}(N)$ are proposed as new criteria to detect the onset of a part failure under the conditions tested and can be used as safety limits for partial damage.

6 Conclusions

In this chapter, the mechanical nonlinear responses of complex fluids under large amplitude oscillatory shear (LAOS) flow are presented and reviewed. Linear response under small amplitude oscillatory shear (SAOS) flow is well known and has its analogy is behavior of dielectric response. It is very useful to characterize complex fluids with firm theoretical background. LAOS tests are substantially more complex rheological probes than SAOS tests because of the nonlinear responses. The complexity of the material response to LAOS is both the strength and weakness of the technique. The additional information obtained can help characterize and quantify the response of complex fluids to nonlinear deformation, but it also makes the results more difficult to interpret. To analyze the nonlinear response, several quantitative methods have been suggested. Among the methods mentioned, in this chapter, FT-rheology is intensively reviewed. We also introduced applications for investigating various complex fluids (polymer melts and solutions, polymer composite and blends, emulsions,

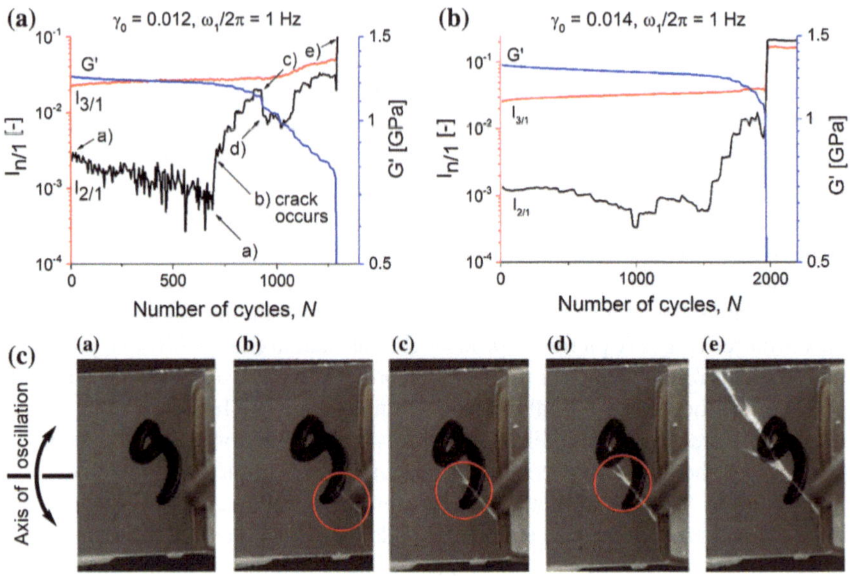

Fig. 27 Storage modulus (G′) and the intensities of the higher harmonics ($I_{2/1}$ and $I_{3/1}$) during a fatigue test at $\omega/2\pi = 1$ Hz and $\gamma_0 = 0.012$ (left) or 0.014 (right), and room temperature (RT). The pictures below are taken from a video of the fatigue test and failure of the sample as labeled in the plot a) above. Reproduced by permission of Hirschberg et al. [33], copyright (2017) of Elsevier

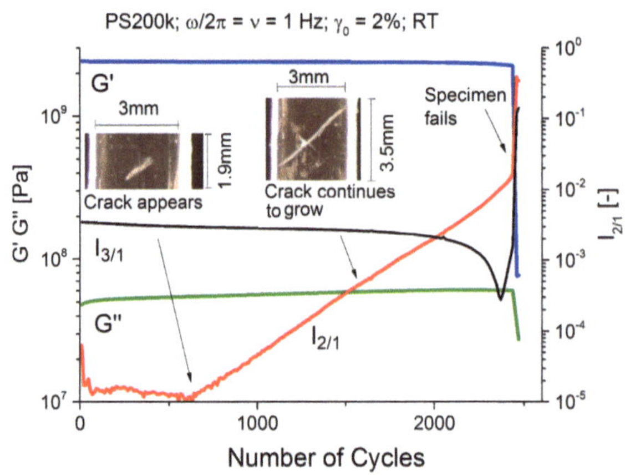

Fig. 28 Storage modulus (G′) and loss modulus (G'') and the intensities of the higher harmonics ($I_{2/1}$ and $I_{3/1}$) during a fatigue test at $\omega/2\pi = 1$ Hz and $\gamma_0 = 0.02$ and room temperature (RT). Clearly, at the point where the first crack starts to appear, the second harmonic $I_{2/1}$ increases its value from 10^{-5} relative contribution

and block copolymers and liquid crystals). From these results, it confirmed that nonlinear material functions under LAOS flow are a very powerful tool to characterize complex fluids. However, the theoretical underpinnings of the nonlinear responses observed under LAOS flow are still poorly understood. Thus, the physical interpretation of nonlinear responses is still developing. We hope to encourage the exchange of idea between nonlinear dielectric spectroscopy and the mechanical measurement.

Acknowledgements The KH acknowledge the financial support of the Alexander von Humboldt Foundation. The authors thank Valerian Hirschberg, Miriam Cziep, and Hyeong Yong Song for supplying figures and Carlo Botha for English proofreading.

Notes

Substantial parts (especially Sect. 3 and 4) of this chapter are taken from a rheological review [3] where rheological nonlinearities are explained in more detail but might not be read by scientists with a background in dielectric spectroscopy. Consequently, this chapter will be very helpful for the reader with a dielectric background to envision the similar concepts of both methodologies.

References

1. R.G. Larson, *The structure and rheology of complex fluids* (Oxford University Press, New York, 1999)
2. F.A. Morrison, *Understanding Rheology* (Oxford University Press, New York, 2001)
3. K. Hyun, M. Wilhelm, C.O. Klein, K.S. Cho, J.G. Nam, K.H. Ahn, S.J. Lee, R.H. Ewoldt, G.H. McKinley, A review of nonlinear oscillatory shear tests: Analysis and application of large amplitude oscillatory shear (LAOS). Prog. Polym. Sci. **36**, 1697–1753 (2011)
4. J.D. Ferry, *Viscoelastic Properties of Polymers* (Wiley, NY, 1980)
5. R.B. Bird, R.C. Armstrong, O. Hassager, *Dynamics of polymeric Liquids*, vol. 1 (Wiley, NY, 1987)
6. N.W. Tschoegl, *The phenomenological theory of linear viscoelastic behavior: an introduction* (Springer-Verlag, NY, 1989)
7. J.M. Dealy, K.F. Wissbrun, *Melt rheology and its role in plastics processing: theory and applications* (VNR, NY, 1990)
8. F. Kremer, A. Schönhals, *Broadband Dielectric Spectroscopy* (Springer, Berlin, 2003)
9. Dealy J.M., Larson R.G. Structure and rheology of molten polymers (2006)
10. M. Wilhelm, Fourier-transform rheology. Macromol. Mater. Eng. **287**, 83–105 (2002)
11. A.J. Giacomin, J.M. Dealy, Large-amplitude oscillatory shear, in *Techniques in Rheological Measurements, Chapter 4*, ed. by A.A. Collyer (Chapman and Hall, London, 1993)
12. H.M. Wyss, K. Miyazaki, J. Mattsson, Z. Hu, D.R. Reichmann, D.A. Weitz, Strain-Rate Frequency Superposition: A Rheological Probe of Structural Relaxation in Soft Materials. Phys. Rev. Lett. **98**, 238303 (2007)
13. Y.H. Wen, J.L. Schaefer, L.A. Archer, Dynamics and Rheology of Soft Colloidal Glasses. ACS Macro Lett. **4**(1), 119–123 (2015)
14. K. Hyun, S.H. Kim, K.H. Ahn, S.J. Lee, Large amplitude oscillatory shear as a way to classify the complex. J. Non-newtonian Fluid Mech. **107**, 51–65 (2002)
15. M. Sugimoto, Y. Suzuki, K. Hyun, K.H. Ahn, T. Ushioda, A. Nishioka, T. Taniguchi, K. Koyama, Melt rheology of long-chain-branched polypropylenes. Rheol. Acta **46**, 33–44 (2006)
16. K. Hyun, J.G. Nam, M. Wilhelm, K.H. Ahn, S.J. Lee, Nonlinear response of complex fluids under LAOS (large amplitude oscillatory shear) flow. Korea-Australia Rheology J **15**, 97–105 (2003)
17. O.C. Klein, H.W. Spiess, A. Calin, C. Balan, M. Wilhelm, Separation of the nonlinear oscillatory response into a superposition of linear, strain hardening, strain softening, and wall slip response. Macromolecules **40**, 4250–4259 (2007)
18. K.S. Cho, K. Hyun, K.H. Ahn, S.J. Lee, A geometrical interpretation of large amplitude oscillatory shear response. J. Rheol. **49**, 747–758 (2005)

19. R.H. Ewoldt, A.E. Hosoi, G.H. McKinley, New measures for characterizing nonlinear viscoelasticity in large amplitude oscillatory shear. J. Rheol. **52**, 1427–1458 (2008)
20. W. Yu, P. Wang, C. Zhou, General stress decomposition in nonlinear oscillatory shear flow. J. Rheol. **53**, 215–238 (2009)
21. K. Reinheimer, J. Kübel, M. Wilhelm, Optimizing the sensitivity of FT-Rheology to quantify and differentiate for the first time the nonlinear mechanical response of dispersed beer foams of light and dark beer. Z. Phys. Chem. **226**, 547–567 (2012)
22. S. Onogi, T. Masuda, T. Matsumoto, Nonlinear behavior of viscoelastic materials. I. Disperse systems of polystyrene solution and carbon black. Trans. Soc. Rheol. **14**, 275–294 (1970)
23. S.G. Hatzikiriakos, J.M. Dealy, Wall slip of molten high density polyethylene. I. Sliding plate rheometer studies. J. Rheol. **35**, 497–523 (1991)
24. S.G. Hatzikiriakos, J.M. Dealy, Role of slip and fracture in the oscillating flow of HDPE in a capillary. J. Rheol. **36**, 845–884 (1992)
25. M.D. Graham, Wall slip and the nonlinear dynamics of large amplitude oscillatory shear flows. J. Rheol. **39**, 697–712 (1995)
26. D.W. Adrian, A.J. Giacomin, The quasi-periodic nature of a polyurethane melt in oscillatory shear. J. Rheol. **36**, 1227–1243 (1992)
27. A.S. Yoshimura, R.K. Prud'homme, Wall slip effects on dynamic oscillatory measurements. J. Rheol. **32**, 575–584 (1988)
28. K. Atalık, R. Keunings, On the occurrence of even harmonics in the shear stress response of viscoelastic fluids in large amplitude oscillatory shear. J. Non-newtonian Fluid. Mech. **122**, 107–116 (2004)
29. M. Wilhelm, D. Maring, H.W. Spiess, Fourier-transform rheology. Rheol. Acta. **37**, 399–405 (1998)
30. J.A. Yosick, A.J. Giacomin, W.E. Stewart, F. Ding, Fluid inertia in large amplitude oscillatory shear. Rheol. Acta **37**, 365–373 (1998)
31. R. Mas, A. Magnin, Experimental validation of steady shear and dynamic viscosity relation for yield stress fluids. Rheol. Acta **36**, 49–55 (1997)
32. J.L. Leblanc, Investigating the nonlinear viscoelastic behavior of rubber materials through Fourier transform rheometry. J. Appl. Polym. Sci. **95**, 90–106 (2005)
33. V. Hirschberg, M. Wilhelm, D. Rodrigue, Fatigue Behavior of Polystyrene (PS) analyzed from the Fourier Transform (FT) of its Stress Response: First evidence of $I_{2/1}$(N) and $I_{3/1}$(N) as new fingerprints. Polym. Test. **60**, 343–350 (2017)
34. K. Hyun, E.S. Baik, K.H. Ahn, S.J. Lee, M. Sugimoto, K. Koyama, Fourier-transform rheology under medium amplitude oscillatory shear for linear and branched polymer melts. J. Rheol. **51**, 1319–1342 (2007)
35. K. Hyun, M. Wilhelm, Establishing a New Mechanical Nonlinear coefficient Q from FT-Rheology: first investigation on entangled linear and comb polymer model systems. Macromolecules **42**, 411–422 (2009)
36. D.M. Holye, D. Auhl, O.G. Harlen, V.C. Barroso, M. Wilhelm, T.C.B. McLeish, Large amplitude oscillatory shear and Fourier transform rheology analysis of branched polymer melts. J. Rheol. **58**, 969–997 (2014)
37. A.K. Gurnon, N.J. Wagner, Large amplitude oscillatory shear (LAOS) measurements to obtain constitutive equation model parameter: giesekus model of banding and nonbanding wormlike micelles. J. Rheol. **56**, 333–351 (2012)
38. R.B. Bird, A.J. Giacomin, A.M. Schmalzer, C. Aumnate, Dilute rigid dumbbell suspensions in large-amplitude oscillatory shear flow: shear stress response. J. Chem. Phys. **140**, 074904 (2014)
39. D.S. Pearson, W.E. Rochefort, Behavior of concentrated polystyrene solutions in large-amplitude oscillating shear fields. J. Polym. Sci. Polym. Phys. Ed. **20**, 83–98 (1982)
40. M.H. Wagner, V.H. Rolón-Garrido, K. Hyun, M. Wilhelm, Analysis of medium amplitude oscillatory shear data of entangled linear and model comb polymers. J. Rheol. **55**, 495–516 (2011)

41. M. Abbasi, N.G. Ebrahimi, M. Wilhelm, Investigation of the rheological behavior of industrial tubular and autoclave LDPEs under SAOS, LAOS, and transient shear, and elongational flows compared with predictions from the MSF theory. J. Rheol. **57**, 1693–1714 (2013)
42. A.J. Giacomin, R.B. Bird, L.M. Johnson, A.W. Mix, Large-amplitude oscillatory shear flow from the corotational Maxwell model. J. Non-Newt. Fluid Mech. **166**, 1081–1099 (2011)
43. D. Merger, M. Abbasi, J. Merger, A.J. Giacomin, Ch. Saengow, M. Wilhelm, Simple scalar model for large amplitude oscillatory shear. Appl. Rheol. **26**, 53809 (2016)
44. M.A. Cziep, M. Abbasi, M. Heck, L. Arens, M. Wilhelm, Effect of molecular weight, polydispersity and monomer of linear homopolymer melts on the intrinsic mechanical nonlinearity 3Q0(w) in MAOS. Macromolecules **49**, 3566–3579 (2016)
45. H.Y. Song, S.J. Park, K. Hyun, Characterization of Dilution Effect of Semi-dilute Polymer Solution on Intrinsic Nonlinearity Q_0 via FT-rheology. Macromolecules **50**, 6238–6254 (2017)
46. J.L. Leblanc, Large amplitude oscillatory shear experiments to investigate the nonlinear viscoelastic properties of highly loaded carbon black rubber compounds without curatives. J. Appl. Poly. Sci. **109**, 1271–1293 (2008)
47. J.L. Leblanc, Non-linear viscoelastic characterization of natural rubber gum through large amplitude harmonic experiments. J. Rubber. Res. **10**, 63–88 (2007)
48. G. Fleury, G. Schlatter, R. Muller, Nonlinear rheology for long chain branching characterization, comparison of two methodologies: fourier Transform rheology and relaxation. Rheol. Acta **44**, 174–187 (2004)
49. G. Schlatter, G. Fleury, R. Muller, Fourier transform rheology of branched polyethylene: experiments and models for assessing the macromolecular architecture. Macromolecules **38**, 6492–6503 (2005)
50. I. Vittorias, M. Parkinson, K. Klimke, B. Debbaut, M. Wilhelm, Detection and quantification of industrial polyethylene branching topologies via Fourier-transform rheology. NMR and simulation using the Pom-pom model Rheol. Acta **46**, 321–340 (2007)
51. T. Neidhöfer, S. Sioula, N. Hadjichristidis, M. Wilhelm, Distinguishing linear from star-branched polystyrene solutions with Fourier-Transform rheology. Macromol. Rapid Commun. **25**, 1921–1926 (2004)
52. M. Kempf, D. Ahirwal, M. Cziep, M. Wilhelm, Synthesis and linear and nonlinear melt rheology of well-defined comb architectures of PS and PpMS with a low and controlled degree of long-chain branching. Macromolecules **46**, 4978–4994 (2013)
53. H.T. Lim, K.H. Ahn, J.S. Hong, K. Hyun, Nonlinear viscoelasticity of polymer nanocomposites under large amplitude oscillatory shear flow. J. Rheol. **57**, 767–789 (2013)
54. L. Schwab, N. Hojdis, J. Lacayo-Pineda, M. Wilhelm, Fourier-Transform rheology of unvulcanized, carbon black filled styrene butadiene rubber. Macromol. Mat. Eng. **301**, 457–468 (2016)
55. W. Yu, M. Bousmina, C. Zhou, Note on morphology determination in emulsions via rheology. J. Non-newtonian Fluid. Mech. **133**, 57–62 (2006)
56. C. Carotenuto, M. Gross, P.L. Maffetone, Fourier transform rheology of dilute immiscible polymer blends: a novel procedure to probe blend morphology. Macromolecules **41**, 4492–4500 (2008)
57. K. Reinheimer, M. Grosso, F. Hetzel, J. Kübel, M. Wilhelm, Fourier Transform Rheology as a universal non-linear mechanical characterization of droplet size and interfacial tension of dilute monodisperse emulsions. J. Colloid Interface Sci. **360**, 818–825 (2011)
58. K. Reinheimer, M. Grosso, F. Hetzel, J. Kübel, M. Wilhelm, Fourier Transform Rheology as an innovative morphological characterization technique for the emulsion volume average radius and its distribution. J. Colloid Interface Sci. **380**, 201–212 (2012)
59. R. Salehiyan, Y. Yoo, W.J. Choi, K. Hyun, Characterization of morphologies of compatibilized polypropylene/polystyrene blends with nanoparticles via nonlinear rheological properties from FT-rheology. Macromolecules **47**, 4066–4076 (2014)
60. R. Salehiyan, H.Y. Song, W.J. Choi, K. Hyun, Characterization of effects of silica nanoparticles on (80/20) PP/PS blends via nonlinear rheological properties from fourier transform rheology. Macromolecules **48**, 4669–4679 (2015)

61. R. Salehiyan, H.Y. Song, M. Kim, W.J. Choi, K. Hyun, Morphological evaluation of pp/ps blends filled with different types of clays by nonlinear rheological analysis. Macromolecules **49**, 3148–3160 (2016)
62. H.G. Ock, K.H. Ahn, S.J. Lee, K. Hyun, Characterization of compatibilizing effect of organoclay in poly(lactic acid) and natural rubber blends by FT-rheology. Macromolecules **49**, 2832–2842 (2016)
63. T. Meins, N. Dingenouts, J. Kübel, M. Wilhelm, In-situ Rheo-Dielectric, ex-situ 2D-SAXS and FT-Rheology investigations of the shear induced alignment of Poly(styrene-b-1,4-isoprene) diblock copolymer melts. Macromolecules **45**, 7206–7219 (2012)
64. C. Oelschlaeger, J.S. Gutmann, M. Wolkenauer, H.W. Spiess, K. Knoll, M. Wilhelm, Kinetics of shear microphase orientation and reorientation in lamellar diblock and triblock copolymer melts as detected via FT-Rheology and 2D-SAXS. Macromol. Chem. Phys. **208**, 1719–1729 (2007)
65. S.H. Lee, H.Y. Song, K. Hyun, Lee JH. Nonlinearity from FT-rheology for liquid crystal 8CB under large amplitude oscillatory shear (LAOS) flow. J. Rheol. **59**, 1–19 (2015)
66. B. Struth, K. Hyun, E. Kats, T. Meins, M. Walther, M. Wilhelm, G. Grübel, Observation of New States of Liquid Crystal 8CB under Nonlinear Shear Conditions as Observed via a Novel and Unique Rheology/Small-Angle X-ray Scattering Combination. Langmuir **27**, 2880–2887 (2011)

Index

R. Richert (ed.), *Nonlinear Dielectric Spectroscopy*, Advances in Dielectrics,
https://doi.org/10.1007/978-3-319-77574-6

MIX
Papier aus verantwortungsvollen Quellen
Paper from responsible sources
FSC® C105338

If you have any concerns about our products,
you can contact us on
ProductSafety@springernature.com

In case Publisher is established outside the EU,
the EU authorized representative is:
Springer Nature Customer Service Center GmbH
Europaplatz 3, 69115 Heidelberg, Germany

Printed by Libri Plureos GmbH
in Hamburg, Germany